W9-DAF-501

Organic Reactions

Organic Reactions

VOLUME 16

JOHN WILEY & SONS, INC.
NEW YORK · LONDON · SYDNEY

Library of Congress Catalog Card Number: 42-20265

PRINTED IN THE UNITED STATES OF AMERICA

PREFACE TO THE SERIES

In the course of nearly every program of research in organic chemistry the investigator finds it necessary to use several of the better-known synthetic reactions. To discover the optimum conditions for the application of even the most familiar one to a compound not previously subjected to the reaction often requires an extensive search of the literature; even then a series of experiments may be necessary. When the results of the investigation are published, the synthesis, which may have required months of work, is usually described without comment. The background of knowledge and experience gained in the literature search and experimentation is thus lost to those who subsequently have occasion to apply the general method. The student of preparative organic chemistry faces similar difficulties. The textbooks and laboratory manuals furnish numerous examples of the application of various syntheses, but only rarely do they convey an accurate conception of the scope and usefulness of the processes.

For many years American organic chemists have discussed these problems. The plan of compiling critical discussions of the more important reactions thus was evolved. The volumes of *Organic Reactions* are collections of chapters each devoted to a single reaction, or a definite phase of a reaction, of wide applicability. The authors have had experience with the processes surveyed. The subjects are presented from the preparative viewpoint, and particular attention is given to limitations, interfering influences, effects of structure, and the selection of experimental techniques. Each chapter includes several detailed procedures illustrating the significant modifications of the method. Most of these procedures have been found satisfactory by the author or one of the editors, but unlike those in *Organic Syntheses* they have not been subjected to careful testing in two or more laboratories. When all known examples of the reaction are not mentioned in the text, tables are given to list compounds which have been prepared by or subjected to the reaction. Every effort has been made to include in the tables all such compounds and references; however, because of the very nature of the reactions discussed and their frequent use as one of the several steps of syntheses in which not all of the intermediates have been isolated, some instances may well have been missed. Nevertheless, the investigator will be able to use the tables and their accompanying bibliographies in place of most or all of the literature search so often required.

Because of the systematic arrangement of the material in the chapters and the entries in the tables, users of the books will be able to find information desired by reference to the table of contents of the appropriate chapter. In the interest of economy the entries in the indices have been kept to a minimum, and, in particular, the compounds listed in the tables are not repeated in the indices.

The success of this publication, which will appear periodically, depends upon the cooperation of organic chemists and their willingness to devote time and effort to the preparation of the chapters. They have manifested their interest already by the almost unanimous acceptance of invitations to contribute to the work. The editors will welcome their continued interest and their suggestions for improvements in *Organic Reactions.*

PREFACE TO VOLUME 16

Volume 16 is a departure in pattern from its predecessors for it deals with a single reaction. The original plan for a chapter on a limited aspect of the Aldol Condensation was abandoned when it became clear that a single long chapter would be more useful and less repetitious than several short chapters. It is not the intention of the editors to follow the pattern of Volume 16 regularly.

CONTENTS

Organic Reactions

THE ALDOL CONDENSATION

ARNOLD T. NIELSEN
Michelson Laboratory, Naval Weapons Center, China Lake, California
AND
WILLIAM J. HOULIHAN
Sandoz Pharmaceuticals, Hanover, New Jersey

CONTENTS

INTRODUCTION

The aldol condensation takes its name from aldol (3-hydroxybutanal), a name introduced by Wurtz who first prepared this β-hydroxy aldehyde from acetaldehyde in 1872.[1] The aldol condensation includes reactions producing β-hydroxy aldehydes (β-aldols) or β-hydroxy ketones (β-ketols)

$$2\ CH_3CHO \xrightarrow{\text{Aq. HCl}} CH_3CHOHCH_2CHO$$

[1] A. Wurtz, *Bull, Soc. Chim. France,* [2] **17,** 436 (1872); *Ber.,* **5,** 326 (1872); *Compt. Rend.,* **74,** 1361 (1872).

by self-condensations or mixed condensations of aldehydes and ketones, as well as reactions leading to α,β-unsaturated aldehydes or α,β-unsaturated ketones, formed by dehydration of intermediate β-aldols or β-ketols. Formation of mesityl oxide by self-condensation of acetone, a reaction discovered by Kane in 1838, is the first known example of ketone self-condensation.[2]

$$2\,CH_3COCH_3 \xrightarrow{H_2SO_4} (CH_3)_2C{=}CHCOCH_3 + H_2O$$

The Claisen-Schmidt condensation is an aldol condensation discovered by Schmidt[3] in 1880 (condensation of furfural with acetaldehyde or acetone) and developed by Claisen (1881–1899).[4–6] It is most often taken to be the condensation of an aromatic aldehyde with an aliphatic aldehyde or ketone to yield an α,β-unsaturated aldehyde or ketone, usually in the presence of a basic catalyst.

$$\text{(2-furyl)}CHO + CH_3CHO \xrightarrow{NaOH} \text{(2-furyl)}CH{=}CHCHO + H_2O$$

$$C_6H_5CHO + CH_3COCH_3 \xrightarrow{NaOH} C_6H_5CH{=}CHCOCH_3 + H_2O$$

However, the term has been extended to include many types of aldehyde-ketone condensations (e.g., chalcone formation[5]) employing either acidic

$$C_6H_5CHO + CH_3COC_6H_5 \xrightarrow{H^{\oplus} \text{ or } OH^{\ominus}} \underset{\text{Chalcone}}{C_6H_5CH{=}CHCOC_6H_5} + H_2O$$

or basic catalysts. Schmidt was first to employ a basic catalyst for the aldol condensation.[3]

The term aldol condensation has sometimes been applied to many so-called "aldol-type" condensations involving reaction of an aldehyde or ketone with a substance R_2CHX or $RCH(X)Y$ (X or Y = an activating group such as CO_2R, CONHR, CN, NO_2, SO_2CH_3; R = alkyl, aryl or hydrogen). Mechanistically, of course, these reactions are like aldol condensations. They produce a hydroxyl compound or its dehydration product and include the Claisen,[7, 8] Knoevenagel,[9, 9a] Doebner,[9, 9a]

[2] R. Kane, *Ann. Physik Chem.*, [2] **44**, 475 (1838); *J. Prakt. Chem.*, **15**, 129 (1838).

[3] J. G. Schmidt, *Ber.*, **13**, 2342 (1880); **14**, 1459 (1881).

[4] L. Claisen and A. Claparède, *Ber.*, **14**, 349 (1881).

[5] L. Claisen, *Ber.*, **20**, 655 (1887).

[6] L. Claisen, *Ann.*, **306**, 322 (1899).

[7] C. R. Hauser and B. E. Hudson, Jr., *Org. Reactions*, **1**, 266–302 (1942).

[8] C. R. Hauser, F. W. Swamer, and J. T. Adams, *Org. Reactions*, **8**, 59–196 (1954).

[9] J. R. Johnson, *Org. Reactions*, **1**, 210–265 (1942).

[9a] G. Jones, *Org. Reactions*, **15**, 204–599 (1967).

Perkin,[9] Stobbe,[10] and Reformatsky[11] reactions.[12] However, since the products are not aldols or ketols or dehydration products thereof, it is believed that the term aldol condensation should not be applied to reactions forming them. The present review includes examples where X and Y = CHO or COR only, i.e., solely condensation reactions of aldehydes and ketones. Included are 1,3-diketones, β-keto aldehydes, and α- and γ-ω-keto compounds (e.g., RCH_2COCO_2H and $RCH_2CO(CH_2)_nX$; $n > 1$, X = any group); some of these reactions might be described as both Knoevenagel and aldol condensations.

Excluded arbitrarily from this review, with few exceptions, are certain Knoevenagel-aldol condensations involving β-keto compounds, $RCOCH_2X$, where the α-methylene group of a ketone is activated *additionally* by some activating group other than the carbonyl group (e.g., β-keto esters). With few exceptions, reaction sequences starting with aldehydes or ketones which involve transient aldol intermediates that are not isolated are excluded; an example would be the formation of certain heterocyclic compounds as in the Robinson-Schöpf reaction.[13] Most reactions of aldehydes or ketones leading indirectly through other reactions to an aldol condensation product are also arbitrarily excluded. One exception is the Michael reaction[14] which may generate 1,5-diketones or keto aldehydes which, *in situ*, undergo an aldol condensation to a cyclic product. Except for a few important and closely related methods, syntheses of β-aldols and β-ketols and their dehydration products by alternative methods of preparation are not discussed. One such method is the Wittig reaction[15] which may be applied to syntheses of α,β-unsaturated aldehydes and ketones.

In France, the term *β-hydroxycarbonylation*[16] has been applied to condensations leading to β-aldols and β-ketols. However, this terminology has not been used extensively elsewhere.

MECHANISM

Simple Base Catalysis. The aldol condensation is catalyzed by acids and bases, the latter being more frequently employed. The

[10] W. S. Johnson and G. H. Daub, *Org. Reactions*, **6**, 1–73 (1951).

[11] R. L. Shriner, *Org. Reactions*, **1**, 1–37 (1942).

[12] For recent summaries of aldol condensation and related reactions, cf. H. O. House, *Modern Synthetic Reactions*, pp. 216–256, Benjamin, New York, 1965; R. L. Reeves, in *The Chemistry of the Carbonyl Group*, ed. by S. Patai, pp. 580–600, Interscience, New York, 1966.

[13] L. A. Paquette and J. W. Heimaster, *J. Am. Chem. Soc.*, **88**, 763 (1966).

[14] E. D. Bergmann, D. Ginsburg, and R. Pappo, *Org. Reactions*, **10**, 179–555 (1959).

[15] A. Maercker, *Org. Reactions*, **14**, 332–334 (1965).

[16] H. Gault, *Bull. Soc. Chim. France*, 302 (1955); "*L'Hydroxycarbonylation*," *Colloques Internationaux du Centre National de la Recherche Scientifique*, Paris, May 31–June 5, 1954. A group of 26 papers published in *Bull. Soc. Chim. France*, 250–302 (1955).

base-catalyzed mechanism, which has received much study,[17, 18] may be pictured by the following steps (R and R″ = alkyl or aryl; R′ and R‴ = alkyl, aryl, or hydrogen).

$$\underset{\mathbf{1}}{\mathrm{RCH_2COR'}} + \mathrm{B} \underset{k_{-1}}{\overset{k_1}{\rightleftharpoons}} \underset{\mathbf{2}}{\mathrm{RCH{=}C(O^{\ominus})R'}} + \mathrm{BH^{\oplus}} \quad \text{(Eq. 1)}$$

$$\underset{\mathbf{3}}{\mathrm{R''COR'''}} + \underset{\mathbf{2}}{\mathrm{RCH{=}C(O^{\ominus})R'}} \underset{k_{-2}}{\overset{k_2}{\rightleftharpoons}} \underset{\mathbf{4}}{\mathrm{R''{-}C(O^{\ominus})(R''')}{-}\mathrm{CH(R)COR'}} \quad \text{(Eq. 2)}$$

$$\mathbf{4} + \mathrm{BH^{\oplus}} \underset{k_{-3}}{\overset{k_3}{\rightleftharpoons}} \underset{\mathbf{5}}{\mathrm{R''{-}C(OH)(R''')}{-}\mathrm{CH(R)COR'}} + \mathrm{B} \quad \text{(Eq. 3)}$$

$$\mathbf{5} + \mathrm{B} \underset{k_{-4}}{\overset{k_4}{\rightleftharpoons}} \underset{\mathbf{6}}{\mathrm{R''{-}C(OH)(R''')}{-}\mathrm{C(R){=}C(O^{\ominus}){-}R'}} + \mathrm{BH^{\oplus}} \quad \text{(Eq. 4)}$$

$$\underset{\mathbf{6}}{\mathrm{R''{-}C(OH)(R''')}{-}\mathrm{C(R){=}C(O^{\ominus}){-}R'}} + \mathrm{BH^{\oplus}} \underset{k_{-5}}{\overset{k_5}{\rightleftharpoons}} \underset{\mathbf{7}}{\mathrm{R''{-}C(R'''){=}C(R){-}C({=}O){-}R'}} + \mathrm{H_2O} + \mathrm{B} \quad \text{(Eq. 5)}$$

The question of the rate-limiting step for aldol and ketol formation has been examined.[17–20] For formation of an aldol in concentrated solutions (from acetaldehyde,[19, 21] propanal,[22] and butanal[23]) the proton removal step (Eq. 1) is rate-limiting ($k_1 < k_2$). However, at low aldehyde concentrations the rate of ionization is not much greater than the rate of condensation, and the reaction becomes second order in aldehyde.[19–27]

17 C. K. Ingold, *Structure and Mechanism in Organic Chemistry*, pp. 676–699, Cornell University Press, Ithaca, New York, 1953.

18 A. A. Frost and R. G. Pearson, *Kinetics and Mechanism*, 2nd ed., pp. 335–350, Wiley, New York, 1961.

19 R. P. Bell, *The Proton in Chemistry*, pp. 137–138, Cornell University Press, Ithaca, New York, 1959.

20 J. Hine, J. G. Houston, J. H. Jensen, and J. Mulders, *J. Am. Chem. Soc.*, **87**, 5050 (1965).

21 A. Broche and R. Gibert, *Bull. Soc. Chim. France*, 131 (1955).

22 J. Jiminez and A. Broche, *Bull. Soc. Chim. France*, 1499 (1961).

23 S. E. Rudakova and V. V. Zharkov, *Zh. Obshch. Khim.*, **33**, 3617 (1963) [*C.A.*, **60**, 7883 (1964)].

24 H. Matsuyama, *Proc. Japan Acad.*, **27**, 552 (1951) [*C.A.*, **46**, 3838 (1952)].

25 J. Hine, J. G. Houston, and J. H. Jensen, *J. Org. Chem.*, **30**, 1184, 1328 (1965).

26 R. P. Bell and M. J. Smith, *J. Chem. Soc.*, 1691 (1958).

27 R. P. Bell and P. T. McTigue, *J. Chem. Soc.*, 2983 (1960).

For condensations of certain aldehydes with ketones the reaction is first order in aldehyde at low concentrations, but zero order in aldehyde at high concentrations: examples are the reaction of *p*-nitrobenzaldehyde with 2-butanone, cyclohexanone, and desoxybenzoin.[28,29] However, for many condensations involving ketones, at a wide range of concentrations, the slow step is the condensation step in which the new carbon-carbon bond is formed (Eq. 2); $k_1 > k_2$. Kinetic evidence supporting this view has been presented for the formation of diacetone alcohol from acetone,[26,30] and for the condensations of aromatic aldehydes with acetophenone,[31] of formaldehyde[32,33] and acetaldehyde[33] with 1,3-diketones, and of formaldehyde[34] with acetone. When the first step (Eq. 1) is rate-limiting, general base catalysis may be important.[35] When the second step (Eq. 2) is relatively slow, specific solvent anion (hydroxide, ethoxide) catalysis is observed.

The α,β-unsaturated aldehydes and ketones (**7**) often found as reaction products can arise from the aldol or ketol **5** by a carbanion elimination (E1cb) mechanism which requires an initial proton removal, followed by elimination of hydroxide ion (Eqs. 4 and 5). The equilibrium of Eq. 4 may lie far to the left ($k_{-4} \gg k_4$; $K_{eq}^{(4)} \ll 1$) and deuterium exchange into **5** would be expected to occur readily during formation of **7** in cases where $k_{-4} \gg k_5$.[36,37] In certain examples, rates k_4 and k_5 appear to be of the same order of magnitude. Factors met in certain aldol condensations (R = alkyl, R″ = aryl; protic solvents) and favoring facile loss of hydroxide ion[37] from **6** could also disfavor proton removal[38,39] from **5**. A comparison of the overall rate of base-catalyzed dehydration (**5** → **7**; rate = $K_{eq}^{(4)}k_5$) for 1-hydroxy-1-phenyl-3-pentanone (**5a**)[40] with rates of proton removal of certain aliphatic ketones under similar reaction

[28] S. F. Nelsen and M. Stiles, unpublished results. We are indebted to Professor Stiles for making some of his results available to us prior to publication.

[29] This situation is illustrated by the lack of deuterium incorporation into *threo*- and *erythro*-4-$O_2NC_6H_4CHOHCH(C_6H_5)COC_6H_5$ in D_2O—NaOD during their interconversion; L. Traynor, Ph.D. thesis, University of Michigan, 1964; *Dissertation Abstr.*, **25**, 6972 (1965).

[30] D. S. Noyce and W. L. Reed, *J. Am. Chem. Soc.*, **81**, 624 (1959).

[31] E. Coombs and D. P. Evans, *J. Chem. Soc.*, 1295 (1940).

[32] P. Rumpf and M. Diard, *Compt. Rend.*, **248**, 823 (1959).

[33] M. Laloi, *J. Rech. Centre Natl. Rech. Sci. Lab. Bellevue (Paris)*, **55**, 141 (1961) [*C.A.*, **58**, 4393 (1963)].

[34] S. Nagase, *Kogyo Kagaku Zasshi*, **64**, 1008 (1961) [*C.A.*, **57**, 2064 (1963)].

[35] D. C. Gutsche, R. S. Buriks, K. Nowotny, and H. Grassner, *J. Am. Chem. Soc.*, **84**, 3775 (1962).

[36] R. G. Pearson and R. L. Dillon, *J. Am. Chem. Soc.*, **75**, 2439 (1953).

[37] J. F. Bunnett, *Angew. Chem., Intl. Ed. Engl.*, **1**, 225 (1962).

[38] A. K. Mills and A. E. Wilder Smith, *Helv. Chim. Acta*, **43**, 1915 (1960).

[39] H. Shechter, M. J. Collis, R. Dessy, Y. Okuzumi, and A. Chen, *J. Am. Chem. Soc.*, **84**, 2905 (1962).

[40] M. Stiles, D. Wolf, and G. V. Hudson, *J. Am. Chem. Soc.*, **81**, 628 (1959).

conditions[38] suggests that k_4 and k_5 could be comparable. The observation that 3-methyl-4-hydroxy-4-phenyl-2-butanone (**5b**) undergoes retrogression to reactants rather than dehydration in basic medium[40] suggests

$$\underset{\mathbf{5a}}{C_6H_5CHOHCH_2COCH_2CH_3} \qquad \underset{\mathbf{5b}}{C_6H_5CHOH\overset{\overset{\displaystyle CH_3}{|}}{C}HCOCH_3}$$

that k_4 is relatively slow ($k_4 \ll k_{-3}k_{-2}/k_3$) in this case; k_5 is not, however, necessarily slow relative to k_4.[40a] The rate of retrogression of **5b** to reactants is about four times faster than that of **5a**,[40] under conditions where the rate of proton removal (k_4) from **5b** would be expected to be 10^2 to 10^3 times slower than from **5a**.[29, 36, 39] Although α,β-unsaturated ketones may be hydrated in basic media, their formation is effectively irreversible in most instances ($k_5 \gg k_{-5}$). This circumstance may offset previous unfavorable equilibria and allow the condensation to proceed to completion (formation of **7**). A detailed understanding of the mechanism of the base-catalyzed dehydration of aldols and ketols awaits further study.

In protic solvents (water, ethanol) the aldol condensation is reversible and the various equilibria are often quite sensitive to the nature of substituents.[40–42] Since in practice the preparations are generally so conducted that thermodynamic control determines the products, the effects of substituents on the several equilibria are of great importance. In aprotic solvents (ethers, hydrocarbons) equilibration occurs exceedingly slowly, particularly with lithium and magnesium enolates.[43–45] Thus kinetically controlled aldol condensations should be more favorable under these conditions.

Amine Catalysis. Primary and secondary amines (especially in the presence of added acid) exert a pronounced catalytic effect on condensation of aldehydes and ketones with active methylene compounds (Knoevenagel condensation).[9a, 46] Pyrrolidine and piperidine are very effective. Tertiary amines are without effect or are very poor aldol catalysts except when general base catalysis is observed.[35, 47, 48] Because of their weak base strength most amines provide only low concentrations

[40a] P. Carsky, P. Zuman, and V. Horak, *Collection Czech. Chem. Commun.*, **30**, 4316 (1965).

[41] J. E. Dubois and H. Viellard, *Tetrahedron Letters*, 1809 (1964).

[42] D. S. Noyce and L. R. Snyder, *J. Am. Chem. Soc.*, **81**, 620 (1959).

[43] H. O. House and B. M. Trost, *J. Org. Chem.*, **30**, 1341 (1965).

[44] H. O. House and B. M. Trost, *J. Org. Chem.*, **30**, 2502 (1965).

[45] H. O. House and V. Kramar, *J. Org. Chem.*, **28**, 3362 (1963).

[46] E. Knoevenagel, *Ann.*, **281**, 25 (1894).

[47] F. H. Westheimer and H. Cohen, *J. Am. Chem. Soc.*, **60**, 90 (1938).

[48] T. A. Spencer, H. S. Neel, T. W. Flechtner, and R. A. Zayle, *Tetrahedron Letters*, 3889 (1965).

of solvent anion base (hydroxide, ethoxide), and the marked catalytic effect of primary and secondary amines must be explained on another basis.

At least two mechanisms appear to be operative for most amine-catalyzed aldol condensations. Many of these reactions, especially those involving secondary amines and reactive aldehydes or ketones capable of forming enamines, require an enamine intermediate in the condensation step leading to an aldol, as well as in the step leading to an α,β-unsaturated carbonyl compound.[49–56] Evidence supporting an enamine mechanism has been presented for the self-condensation of propanal to 2-methyl-2-pentenal[50, 54–56] and for certain Robinson annelation reactions.[48] Enamines have been condensed with aldehydes and ketones to give aldol condensation products.[53a–56b] An acid catalyst (preferably acetic acid), usually employed in an amount equivalent to the amine, is required to effect the condensation. It is reported that enamines fail to undergo aldol condensation with aldehydes in the absence of an added acid catalyst.[56]

$$R''COR''' + RCH{=}C(\mathrm{NR_2})R' \rightleftharpoons R''C(O^{\ominus})(R''')CH(R)C(=\overset{\oplus}{N}R_2)R' \xrightarrow{CH_3CO_2H} R''C(OH)(R''')CH(R)C(=\overset{\oplus}{N}R_2)R' \quad CH_3CO_2^{\ominus}$$

$$\underset{\mathbf{5}}{R''C(OH)(R''')\text{—}CH(R)COR'} \xleftarrow{H_2O} R''C(OH)(R''')CH(R)C(=\overset{\oplus}{N}R_2)R' \quad CH_3CO_2^{\ominus} \xrightarrow{-H_2O}$$

$$R''(R''')C{=}C(R)C(=\overset{\oplus}{N}R_2)R' \quad CH_3CO_2^{\ominus} \xrightarrow{H_2O} \underset{\mathbf{7}}{R''\text{—}C(R''')\text{=}C(R)COR'}$$

[49] T. A. Spencer and K. K. Schmiegel, *Chem. Ind.* (*London*), 1765 (1963).

[50] I. V. Mel'nichenko and A. A. Yasnikov, *Ukr. Khim. Zh.*, **30,** 838 (1964) [*C.A.*, **61,** 14490 (1964)].

[51] D. J. Goldsmith and J. A. Hartman, *J. Org. Chem.*, **29,** 3520, 3524 (1964).

[52] J. Szmuszkovicz, *Advan. Org. Chem.*, **4,** 1–114 (1963).

[53] G. Stork, A. Brizzolara, H. Landesman, J. Szmuszkovicz, and R. Terrell, *J. Am. Chem. Soc.*, **85,** 207 (1963).

[53a] L. Birkofer, S. M. Kim, and H. D. Engels, *Chem. Ber.*, **95,** 1495 (1962).

[54] A. A. Yasnikov and K. I. Matkovskii, *Ukr. Khim. Zh.*, **28,** 210 (1962). [*C.A.*, **58,** 3306 (1963)].

[55] T. S. Boiko, N. V. Volkova, and A. A. Yasnikov, *Ukr. Khim. Zh.*, **29,** 1179 (1963) [*C.A.*, **60,** 3964 (1964)].

[56] N. V. Volkova and A. A. Yasnikov, *Dokl. Akad. Nauk SSSR*, **149,** 94 (1963) [*C.A.*, **59,** 5011 (1963)].

[56a] L. A. Paquette and H. Stucki, *J. Org. Chem.*, **31,** 1232 (1966).

[56b] F. T. Bond and W. E. Musa, *Chem. Ind.* (*London*), 1380 (1966).

A second amine-catalyzed mechanism, involving attack of an imine or immonium ion on the methylene group (enol) of an aldehyde or ketone, is important in certain aldol condensations (Knoevenagel-aldol);[9a] see p. 43.[12, 13, 56c–e] This process is favored by use of primary amines with aldehydes or ketones which readily form imines and in condensations with carbonyl compounds of high enol content such as 1,3-diketones.[56, 57] Aldimines[58] and ketimines[59] condense readily with certain active methylene compounds, including ketones[58] ("alkylidenation reaction"[59]); α,β-unsaturated ketones may be prepared by this reaction which is catalyzed by acids.[58]

$$\mathrm{ArCH{=}\overset{\oplus}{N}{<}} + \mathrm{RCH{=}\underset{\displaystyle OH}{\underset{|}{C}}R'} \rightarrow \mathrm{ArCH\underset{\displaystyle HN^{\oplus}{<}}{\underset{|}{C}}H(R)COR'} \rightarrow \mathrm{ArCH{=}C(R)COR'} + \mathrm{H_2\overset{\oplus}{N}{<}}$$

A third mechanism is possible, that of condensation of an immonium ion with an enamine. An immonium ion-dienamine reaction step is postulated in the pyrrolidinium perchlorate-catalyzed self-condensation of 2-cyclohexen-1-one.[59a] The formation of α,β-unsaturated ketones by self-condensation of enamines derived from methyl alkyl ketones may involve such a reaction (see p. 20);[60] a dienamine intermediate can be isolated. An immonium ion-enamine reaction may occur in the condensation of acetone with 1-propenylpiperidine (added acetic acid) to yield 2,4,5-trimethyl-2,4-hexadienal (see p. 27).[54]

Acid Catalysis. The mechanism of the acid-catalyzed reaction[17, 18] is similar to that of the base-catalyzed reaction in that an initial proton removal step (Eq. 7) is involved, though from a protonated intermediate **8** (R and R″ = alkyl or aryl; R′ and R‴ = alkyl, aryl, or hydrogen). The intermediate enol **9** reacts in a rate-limiting step (Eq. 8)[42] with a second (protonated) carbonyl molecule to produce an aldol or ketol **5**. However, α,β-unsaturated carbonyl compounds **7** (rather than aldols

[56c] J. Hine, B. C. Merron, J. H. Jensen, and J. Mulders, *J. Am. Chem. Soc.*, **88**, 3367 (1966).

[56d] M. L. Bender and A. Williams, *J. Am. Chem. Soc.*, **88**, 2502 (1966).

[56e] J. Hine, F. C. Kokesh, K. G. Hampton, and J. Mulders, *J. Am. Chem. Soc.*, **89**, 1205 (1967).

[57] A. A. Yasnikov, K. I. Matkovskii, and E. M. Gaivoronskaya, *Ukr. Khim. Zh.*, **28**, 88 (1962) [*C.A.*, **58**, 1326 (1963)].

[58] A. H. Blatt and N. Gross, *J. Org. Chem.*, **29**, 3306 (1964).

[59] G. Charles, *Bull. Soc. Chim. France*, 1559 (1963).

[59a] N. J. Leonard and W. J. Musliner, *J. Org. Chem.*, **31**, 639 (1966).

[60] G. Bianchetti, P. Dalla Croce, and D. Pocar, *Tetrahedron Letters*, 2039 (1965).

$$\underset{\underset{\text{O}}{\|}}{\text{RCH}_2\text{CR}'} + \text{H}^{\oplus} \rightleftharpoons \underset{\underset{\text{OH}^{\oplus}}{\|}}{\text{RCH}_2\text{CR}'} \leftrightarrow \underset{\underset{\textbf{8}}{\underset{\text{OH}}{|}}}{\text{RCH}_2\overset{\oplus}{\text{C}}\text{R}'} \qquad \text{(Eq. 6) Fast}$$

$$\underset{\underset{\textbf{8}}{\underset{\text{OH}}{|}}}{\text{RCH}_2\overset{\oplus}{\text{C}}\text{R}'} + \text{B} \rightleftharpoons \underset{\underset{\textbf{9}}{\underset{\text{OH}}{|}}}{\text{RCH=CR}'} + \text{BH}^{\oplus} \qquad \text{(Eq. 7) Fast}$$

$$\underset{\underset{\textbf{10}}{\underset{\text{OH}}{|}}}{\text{R}''\overset{\oplus}{\text{C}}\text{R}'''} + \underset{\underset{\textbf{9}}{\underset{\text{OH}}{|}}}{\text{RCH=CR}'} \rightleftharpoons \underset{\textbf{5}}{\text{R}''\text{—}\overset{\text{OH}}{\underset{\text{R}'''}{\text{C}}}\text{—CH(R)COR}'} + \text{H}^{\oplus} \qquad \text{(Eq. 8) Slow}$$

or ketols) are the most frequently encountered products, because acid-catalyzed dehydration is very rapid and in most instances not easily reversed. The dehydration step (Eq. 9) involves protonation of the hydroxyl group, followed by loss of water and then loss of a proton. The dehydration may proceed, in kinetically distinguishable paths, via a carbonium ion or enol intermediate (Eqs. 9a and 9b).[61–62a] Dehydration

$$\underset{\textbf{5}}{\text{R}''\overset{\text{OH}}{\underset{\text{R}'''}{\text{C}}}\text{—CH(R)COR}'} \underset{}{\overset{\text{H}^{\oplus}}{\rightleftharpoons}} \text{R}''\overset{\overset{\oplus}{\text{O}}\text{H}_2}{\underset{\text{R}'''}{\text{C}}}\text{—CH(R)COR}' \xrightarrow{-\text{H}_2\text{O}} \text{R}''\overset{\oplus}{\underset{\text{R}'''}{\text{C}}}\text{—CH(R)COR}' \xrightarrow{-\text{H}^{\oplus}} \underset{\textbf{7}}{\text{R}''\text{C}=\overset{\text{R}}{\text{C}}\text{COR}'} \quad \text{(Eq. 9a) Fast}$$

$$\underset{\textbf{5}}{\text{R}''\overset{\text{OH}}{\underset{\text{R}'''}{\text{C}}}\text{—CH(R)COR}'} \overset{\text{H}^{\oplus}}{\underset{\text{3 steps}}{\rightleftharpoons}} \text{R}''\overset{\overset{\oplus}{\text{O}}\text{H}_2}{\underset{\text{R}'''}{\text{C}}}\text{—}\underset{\text{R}}{\text{C}}=\overset{\text{OH}}{\text{C}}\text{R}' \xrightarrow{-\text{H}_2\text{O}} \text{R}''\text{C}=\underset{\text{R}'''\ \text{R}}{\text{C}}\text{—}\overset{\text{OH}^{\oplus}}{\underset{}{\text{C}}}\text{R}' \overset{-\text{H}^{\oplus}}{\rightleftharpoons} \underset{\textbf{7}}{\text{R}''\text{C}=\overset{\text{R}}{\text{C}}\text{COR}'} \quad \text{(Eq. 9b) Fast}$$

is generally more rapid than reversal of the rate-determining condensation step (Eq. 8).[42] The course of the acid-catalyzed reaction, unlike the base-catalyzed process, is not characterized by the influence of reversibility in all steps, but principally by reactivity and stability of intermediate enols.[42]

An important and useful acid-catalyzed procedure employs hydrogen chloride in the absence of solvent, or in an aprotic solvent such as benzene,

[61] D. S. Noyce and W. L. Reed, *J. Am. Chem. Soc.*, **80**, 5539 (1958).

[62] M. Stiles and A. Longroy, *Tetrahedron Letters*, 337 (1961).

[62a] S. Cabani and N. Ceccanti, *J. Chem. Soc., Phys. Org.*, 77 (1966).

to produce a β-chloroketone which may be dehydrohalogenated stereoselectively to the more stable *trans* **7** (*trans* disposition of bulkier β-group and α-carbonyl group).[63]

$$\underset{\substack{|\\ R'''}}{\overset{\substack{Cl\\ |}}{R''C}}-\underset{\substack{|\\ H}}{\overset{\substack{R\\ |}}{C}}COR' \xrightarrow{B^{\ominus}} \underset{\substack{|\\ R'''}}{R''C}=\overset{\substack{R\\ |}}{C}COR' + BH + Cl^{\ominus}$$

7

Stereochemistry. Present knowledge of the stereochemistry of the aldol condensation is limited.[48, 64–75c] Aldols and ketols are susceptible to retrogression to form the reactants or to dehydration to yield the corresponding α,β-unsaturated carbonyl compounds. The facile equilibration of aldols through their enolate anions, as well as by retrogression, contributes to the paucity of examples of kinetically controlled aldol formation.[64–72, 76] Available data suggest a general lack of stereoelectronic control or stereospecificity in the C—C bond-forming process. Mixtures of epimers often result under kinetically controlled conditions in condensations leading to acyclic[64] or alicyclic[66, 69–71] products. In the two ketols and the aldol shown,[66, 69, 71] mild conditions and short reaction times lead to mixtures of epimers involving substituents about the bond shown by a broken line. Slightly more vigorous conditions or longer reaction times result in complete conversion to the most stable epimer in each example.[66, 69,70] In the formation of acyclic products, equilibration may lead to mixtures of epimers different from the mixtures formed

[63] H. O. House and R. S. Ro, *J. Am. Chem. Soc.*, **80**, 2428 (1958).

[64] M. Stiles, R. R. Winkler, U. Chang, and L. Traynor, *J. Am. Chem. Soc.*, **86**, 3337 (1964).

[65] A. C. Huitric and W. D. Kumler, *J. Am. Chem. Soc.*, **78**, 1147 (1956).

[66] A. T. Nielsen, *J. Org. Chem.*, **30**, 3650 (1965).

[67] J. A. Marshall and W. I. Fanta, *J. Org. Chem.*, **29**, 2501 (1964).

[68] T. A. Spencer, K. K. Schmiegel, and K. L. Williamson, *J. Am. Chem. Soc.*, **85**, 3785 (1963).

[69] I. Vogel, *J. Chem. Soc.*, 594 (1927).

[70] M. B. Rubin, *J. Org. Chem.*, **29**, 3333 (1964).

[71] W. S. Johnson, J. J. Korst, R. A. Clement, and J. Dutta, *J. Am. Chem. Soc.*, **82**, 614 (1960).

[72] W. S. Johnson, J. Ackerman, J. F. Eastham, and H. A. DeWalt, *J. Am. Chem. Soc.*, **78**, 6302 (1956).

[73] W. F. Johns, *J. Org. Chem.*, **26**, 4583 (1961).

[74] K. Tanabe and Y. Morisawa, *Chem. Pharm. Bull.* (*Tokyo*), **11**, 536 (1963) [*C.A.*, **59**, 7600 (1963)].

[75] K. Tanabe, R. Hayashi, and R. Takasaki, *Chem. Pharm. Bull.* (*Tokyo*), **9**, 1 (1961) [*C.A.*, **60**, 9331 (1964)].

[75a] J. J. Basselier, C. Gueremy, and S. Julia, *Bull. Soc. Chim. France*, 2988 (1965).

[75b] G. L. Buchanan and G. W. McLay, *Tetrahedron*, **22**, 1521 (1966).

[75c] F. Johnson, N. A. Starkovsky, and A. A. Carlson, *J. Am. Chem. Soc.*, **87**, 4612 (1965).

[76] T. A. Spencer, H. S. Neel, D. C. Ward, and K. L. Williamson, *J. Org. Chem.*, **31**, 434 (1966).

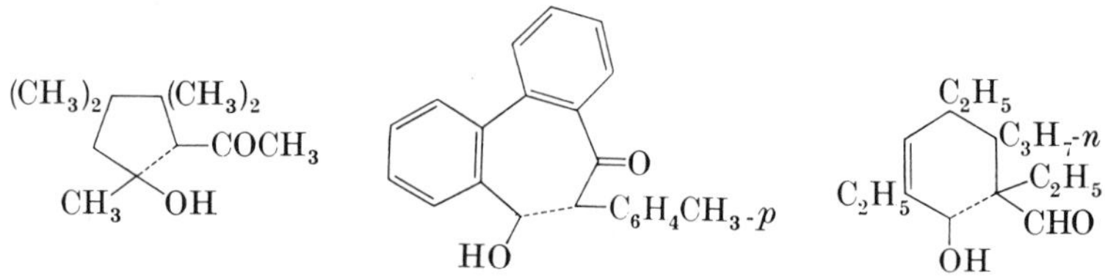

under kinetic control.[64, 65] These observations suggest that the transition state for the condensation has a relatively long developing C—C bond and does not resemble the products.

Other asymmetric centers in the reactant(s) may influence the stereochemistry of the aldol condensation leading to very stereoselective processes as in certain examples of the Robinson annelation reaction;[67, 68, 76, 77] it is not known to what extent these results represent kinetic control.[68]

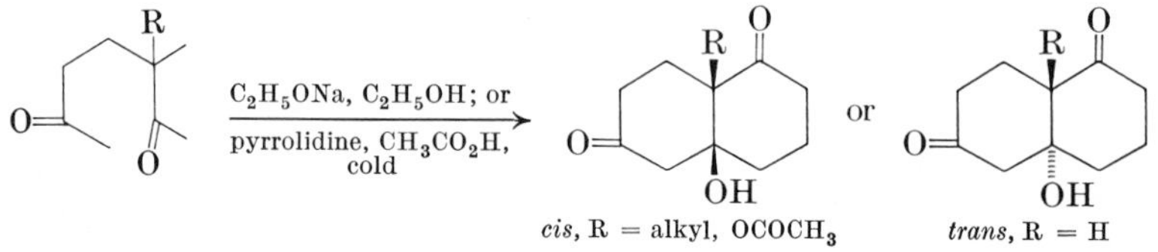

C_2H_5ONa, C_2H_5OH; or pyrrolidine, CH_3CO_2H, cold

cis, R = alkyl, $OCOCH_3$ or *trans*, R = H

In many cases the aldol condensation appears to be more stereoselective under conditions of thermodynamic control (except in certain acyclic systems)[64, 65] than under conditions of kinetic control.

Another important stereochemical question concerns *cis-trans* isomerism of α,β-unsaturated carbonyl compounds derived from ketols and aldols. It is clear from the known examples[63, 78–90a] that the most highly favored (and often the most stable) product is the *trans* isomer (*trans* disposition of the bulkier β-group and the α-carbonyl group); the formation of

[77] T. A. Spencer and T. W. Flechtner, unpublished results; we wish to thank Professor Spencer for making some of his results available to us prior to publication.

[78] C. Y. Chen and R. J. W. Le Fèvre, *J. Chem. Soc.*, 5528 (1965).

[79] S. V. Tsukerman, A. I. Artemenko, and V. F. Lavrushin, *Zh. Obshch. Khim.*, **34**, 3591 (1964) [*C.A.*, **62**, 4736 (1965)].

[80] A. Hassner and T. C. Mead, *Tetrahedron Letters*, 1223 (1962).

[81] R. Heilmann, *Bull. Soc. Chim. France*, [5] **4**, 1064 (1937).

[82] J. E. Dubois and M. Dubois, *Compt. Rend.*, **256**, 715 (1963).

[83] H. Kwart and L. G. Kirk, *J. Org. Chem.*, **22**, 116 (1957).

[84] R. A. Abramovitch and A. Obach, *Can. J. Chem.*, **37**, 502 (1959).

[85] R. Mecke and K. Noack, *Chem. Ber.*, **93**, 210 (1960).

[86] L. Crombie, *Quart. Rev. (London)*, **6**, 101 (1962).

[87] R. E. Buckles, G. V. Mock, and L. Locatell, Jr., *Chem. Rev.*, **55**, 659 (1955).

[88] G. B. Payne and P. H. Williams, *J. Org. Chem.*, **26**, 651 (1961).

[89] G. Gamboni, V. Theus, and H. Schinz, *Helv. Chim. Acta*, **38**, 255 (1955).

[90] V. Theus, W. Surber, L. Colombi, and H. Schinz, *Helv. Chim. Acta*, **38**, 239 (1955).

[90a] J. Sotiropoulos and P. Bédos, *Compt. Rend.*, **263**, Ser. C, 1392 (1966).

3-methyl-*trans*-3-penten-2-one is illustrative.[63] *cis* Isomers may sometimes be isomerized to *trans* isomers with acid or basic catalysts[63, 82, 91];

$$CH_3CHO + CH_3CH_2COCH_3 \xrightarrow{H^{\oplus} \text{ or } OH^{\ominus}}$$

$$\underset{trans}{(CH_3)HC{=}C(CH_3)COCH_3} \xleftarrow{H^{\oplus} \text{ or } OH^{\ominus}} \underset{cis}{(CH_3)HC{=}C(COCH_3)CH_3}$$

the reverse transformation is frequently effected by irradiation with ultraviolet light.[92, 93] The mechanism of the heterolytic isomerization can involve hydration of the olefinic double bond,[63, 91] or formation of a dienolate anion if a γ hydrogen atom is present, followed by equilibration of the α,β- and the β,γ-unsaturated ketones.[94]

SCOPE AND LIMITATIONS

Self-Condensation of Aldehydes

The aldol condensation produces β-hydroxy aldehydes from saturated aliphatic aldehydes having at least one α hydrogen atom. Formaldehyde

$$2\ RR'CHCHO \rightarrow RR'CHCHOHC(R)(R')CHO$$

undergoes self-condensation to polyhydroxyaldehydes (Butlerow reaction);[95–99a] although the initial step leading to glycolaldehyde is not an aldol condensation,[97] subsequent steps do constitute an aldol condensation.

$$2\ HCHO \xrightarrow{OH^{\ominus}} HOCH_2CHO \xrightarrow{OH^{\ominus}} \text{sugars}$$

Aldols produced from aldehydes having only one α hydrogen atom (e.g., isobutyraldol) cannot be dehydrated to α,β-unsaturated aldehydes.

$$2\ (CH_3)_2CHCHO \xrightarrow{KOH} (CH_3)_2CHCHOHC(CH_3)_2CHO$$

[91] D. S. Noyce and M. J. Jorgenson, *J. Am. Chem. Soc.*, **83**, 2525 (1961).
[92] L. v. Hahn and J. F. Miguel, *Compt. Rend.*, **257**, 1948 (1963).
[93] R. E. Lutz and R. H. Jordan, *J. Am. Chem. Soc.*, **72**, 4090 (1950).
[94] H. C. Volger and W. Brackman, *Rec. Trav. Chim.*, **84**, 1017 (1965).
[95] A. M. Butlerow, *Ann.*, **120**, 295 (1861).
[96] E. Pfeil and H. Ruckert, *Ann.*, **641**, 121 (1961).
[97] H. W. Wanzlick, *Angew. Chem., Intl. Ed. Engl.*, **1**, 79 (1962).
[98] J. F. Walker, *Formaldehyde*, 3rd ed., pp. 215–217, Reinhold, New York, 1964.
[99] H. Ruckert, E. Pfeil, and G. Scharf, *Chem. Ber.*, **98**, 2558 (1965).
[99a] K. Runge and R. Mayer, Ger. (East) pat., 44,094 [*C.A.*, **64**, 17426 (1966)].

Unbranched homologs of acetaldehyde (n-alkanals) up to hexanal produce aldols in a normal manner with basic catalysts if reaction conditions are mild; for this particular reaction basic ion-exchange resins appear to be the most effective catalysts.[100] Propanal and butanal have been condensed to aldols in good yields, but the more vigorous conditions (higher temperature, stronger base) needed to effect condensation of higher homologs often cause extensive dehydration of the aldols, leading to α,β-unsaturated aldehydes. For example, heptanal with a mild basic ion-exchange catalyst reacts only at elevated temperature to produce 2-pentyl-2-nonenal, not the aldol.[101] Yields of aldols decrease as the

$$2\ n\text{-}C_3H_7CHO \xrightarrow[25°]{\text{Basic ion-exchange resin}} \underset{66\%}{n\text{-}C_3H_7CHOHCH(C_2H_5)CHO}$$

$$2\ n\text{-}C_6H_{13}CHO \xrightarrow[120\text{–}150°]{\text{Basic ion-exchange resin}} \underset{57\%}{n\text{-}C_6H_{13}CH{=}C(C_5H_{11}\text{-}n)CHO} + H_2O$$

carbon chain increases in length. Phosphorus oxychloride and thionyl chloride have been reported to condense heptanal to the aldol in 41–46% yield at 0–10°.[102, 103]

Numerous side reactions (discussed on pp. 61–62) frequently accompany attempts to apply vigorous conditions to the self-condensation of alkanals. Aldols dimerize to 1,3-dioxane derivatives on standing and should be redistilled immediately before use (see p. 59).

Although general procedures are not available to form aldols in good yields from n-alkanals of more than six carbon atoms, the derived α,β-unsaturated aldehydes have been obtained in excellent yields in all cases reported.[103a] The condensation of octanal to 2-hexyl-2-decenal is an example.[104]

$$2\ n\text{-}C_7H_{15}CHO \xrightarrow{C_2H_5ONa} \underset{79\%}{n\text{-}C_7H_{15}CH{=}C(C_6H_{13}\text{-}n)CHO} + H_2O$$

Aldols having a hydrogen atom α to the carbonyl group may readily be dehydrated, by heating with an acid catalyst such as iodine or oxalic acid, to α,β-unsaturated aldehydes in good yield.[100] Heating strongly in

$$RCH_2CHOHCH(R)CHO \xrightarrow[\text{Heat}]{H^{\oplus}} RCH_2CH{=}C(R)CHO + H_2O$$

[100] M. J. Astle and J. A. Zaslowsky, *Ind. Eng. Chem.*, **44**, 2869 (1952).

[101] G. Durr, *Ann. Chim.* (*Paris*), [13] **1**, 84 (1956).

[102] M. Backès, *Compt. Rend.*, **196**, 277 (1933).

[103] M. Backès, *Bull. Soc. Chim. France*, [5] **9**, 60 (1942).

[103a] W. J. Porter, Jr., J. A. Wingate, and J. A. Hanan, U.S. pat., 3,248,428 [*C.A.*, **65**, 2128 (1966)].

[104] F. J. Villani and F. F. Nord, *J. Am. Chem. Soc.*, **69**, 2605 (1947).

the absence of a catalyst may dissociate aldols to form the parent aldehydes.

The self-condensation of α,β-unsaturated aliphatic aldehydes is an important synthetic route to conjugated polyenals, although yields are often low. The condensation products isolated have been formed by attack on the γ rather than the α carbon atom;[105–113] mild basic catalysts, such as piperidinium acetate, are preferred.[114] With amine catalysts dienamine intermediates are probably involved. These are attacked preferentially on the α position in many reactions, including Michael addition and alkylation.[52] The attack on the γ position observed in aldol condensation suggests a thermodynamically controlled product composition favoring a linear, rather than a cross-conjugated, polyene. 3-Methyl-2-butenal forms dehydrocitral in 14.5% yield.[115] Polyenals

$$2\,CH_3\underset{\displaystyle CH_3}{\underset{|}{C}}{=}CHCHO \xrightarrow{C_5H_{12}N^{\oplus}\ {}^{\ominus}OCOCH_3} CH_3\underset{\displaystyle CH_3}{\underset{|}{C}}{=}CHCH{=}CH\underset{\displaystyle CH_3}{\underset{|}{C}}{=}CHCHO$$

have been prepared in this manner from benzaldehyde and 2-butenal.[116] The Wittig reaction is a useful alternative route to polyenals.[15]

Mixed Condensation of Aldehydes

The condensation of formaldehyde with aliphatic aldehydes (Tollens condensation[117, 118]) can readily produce methylol aldehydes in which all α hydrogen atoms have been replaced by methylol groups. It is difficult to prevent complete methylolation; the second methylol group is evidently introduced more rapidly than the first,[119] and the equilibrium also

[105] F. G. Fischer and K. Löwenberg, *Ann.*, **494**, 263 (1932).

[106] F. G. Fischer and K. Hultzsch, *Ber.*, **68**, 1726 (1935).

[107] T. Reichstein, C. Ammann, and G. Trivelli, *Helv. Chim. Acta*, **15**, 261 (1932).

[108] K. Bernhauer, K. Irrgang, K. Adler, M. Mattauch, P. Miller, and F. Neiser, *Ann.*, **525**, 43 (1936).

[109] F. G. Fischer, K. Hultzsch, and W. Flaig, *Ber.*, **70**, 370 (1937).

[110] R. Kuhn and C. Grundmann, *Ber.*, **71**, 2274 (1938).

[111] C. Grundmann, *Chem. Ber.*, **81**, 510 (1948).

[112] D. N. Kursanov and Z. N. Parnes, *Dokl. Akad. Nauk SSSR*, **103**, 847 (1955) [*C.A.*, **50**, 9326 (1956)].

[113] D. N. Kursanov and Z. N. Parnes, *Dokl. Akad. Nauk SSSR*, **91**, 1125 (1953) [*C.A.*, **48**, 10549 (1954)].

[114] H. Shingu and T. Okazaki, *Bull. Inst. Chem. Res. Kyoto Univ.*, **27**, 69 (1951) [*C.A.*, **47**, 2124 (1953)].

[115] I. N. Nazarov and Z. A. Krasnaya, *Izv. Akad. Nauk SSSR, Otd. Khim. Nauk*, 238 (1958) [*C.A.*, **52**, 12792 (1958)].

[116] J. Schmitt, *Ann.*, **547**, 270 (1941).

[117] B. Tollens and P. Wigand, *Ann.*, **265**, 316 (1891).

[118] P. Rave and B. Tollens, *Ann.*, **276**, 58 (1893).

[119] Y. Ishikawa and T. Minami, *Kogyo Kagaku Zasshi*, **63**, 277 (1960) [*C.A.*, **56**, 2322 (1962)].

favors the polymethylol product. Monomethylolalkanals can be obtained from branched (α- and/or β-substituted) and higher alkanals ($>C_5$) under mild conditions (e.g., potassium carbonate, aqueous ether, low temperature).[120] *n*-Alkanals (butanal through nonanal) have been condensed with formaldehyde to yield 2,2-*bis*-methylol alkanals.[121, 122]

$$RCH_2CHO \xrightarrow{HCHO} RCH(CH_2OH)CHO \xrightarrow{HCHO} RC(CH_2OH)_2CHO \xrightarrow{HCHO} RC(CH_2OH)_3$$

Unless reaction conditions are sufficiently mild, however, reduction of the aldehyde group by formaldehyde often leads, irreversibly, to diols or triols (crossed Cannizzaro reaction);[123, 123a] lower-molecular-weight aliphatic aldehydes are apparently most susceptible to this reduction which is also favored by excess formaldehyde. Although basic condensing agents (hydroxides) are most frequently employed, the reaction is also effected by sulfuric acid.[124]

α-Alkylacroleins are obtained in 46–62% yield by vapor-phase condensation of formaldehyde with *n*-alkanals;[125] the subject has been reviewed.[126]

$$HCHO + RCH_2CHO \xrightarrow[275\text{–}300^\circ]{Na_2O,\ SiO_2} CH_2{=}C(R)CHO + H_2O$$

The condensation of pentanal and hexanal with formaldehyde in water at 100° (18–20 hr.) in the absence of added catalyst has been reported to yield α-*n*-propyl- and α-*n*-butyl-acrolein in 64–66% yield.[127] Dehydration of monomethylolalkanals by means of an acid catalyst usually leads to polymers and low yields of α-alkylacroleins.[120] α-Alkylacroleins are more conveniently prepared from Mannich intermediates, $R'CH(CH_2NR_2)CHO$.[120, 128]

[120] C. S. Marvel, R. L. Myers, and J. H. Saunders, *J. Am. Chem. Soc.*, **70**, 1694 (1948).

[121] T. Shono, H. Ono, and R. Oda, *Kogyo Kagaku Zasshi*, **59**, 960 (1951) [*C.A.*, **52**, 9952 (1958)].

[122] O. Neunhoeffer and H. Neunhoeffer. *Ber.*, **95**, 102 (1962).

[123] T. A. Geissman, *Org. Reactions*, **2**, 94–113 (1944).

[123a] K. Ko, S. Kunimoto, Y. Shimono, and T. Yamaguchi, Japan pat., 8769 [*C.A.*, **65**, 12109 (1966)].

[124] D. Bertin, H. Fritel, and L. Nedelec, *Bull. Soc. Chim. France*, 1068 (1962).

[125] S. Malinowski, H. Jablezynska-Jedrzejewska, S. Basinski, and S. Benbenek, *Chim. Ind. (Paris)*, **85**, 885 (1961) [*C.A.*, **56**, 2321 (1962)].

[126] G. S. Mironov and M. I. Farberov, *Usp. Khim.*, **33**, 649 (1964) [*C.A.*, **61**, 6912 (1964)].

[127] L. M. Korobova and I. A. Livshits, *Zh. Obshch. Khim.*, **34**, 3419 (1964) [*C.A.*, **62**, 2699 (1965)].

[128] F. F. Blicke, *Org. Reactions*, **1**, 303–341 (1942).

Mixed condensations of hydroxy aldehydes have been used for sugar syntheses.[35, 95–98, 129–132]

$$HOCH_2CHO + OHCCHOHCHO \xrightarrow{Ca(OH)_2} \text{aldopentoses}$$

In mixed condensations of aliphatic aldehydes four different aldol products are possible, but one often predominates. Concurrent formation of α,β-unsaturated aldehydes often leads to more complex mixtures. From the limited number of examples available it appears that, at least for most simple aliphatic aldehydes reacting at ambient temperature, the aldol results from attack by the aldehyde with the lesser number of α substituents on the α carbon atom of the aldehyde having the greater

$$CH_3CHO + CH_3CH_2CHO \xrightarrow{KOH} CH_3CHOHCH(CH_3)CHO \xrightarrow{-H_2O} CH_3CH{=}C(CH_3)CHO$$

number of α substituents[133–139] (Lieben's rule).[140, 141] At higher temperatures different behavior may be expected. Isobutyraldehyde, for example, forms aldols (which cannot dehydrate) according to Lieben's rule under mild conditions (25°),[133] but more vigorous conditions (80°) produce α,β-unsaturated aldehydes formed by condensation at the isobutyraldehyde carbonyl group (e.g., 2-ethyl-4-methyl-2-pentenal from butanal).[136] Since the formation of α,β-unsaturated carbonyl compounds

$$CH_3CH_2CHO + (CH_3)_2CHCHO \xrightarrow{NaOH,\ 25°} CH_3CH_2CHOHC(CH_3)_2CHO$$

$$CH_3CH_2CH_2CHO + (CH_3)_2CHCHO \xrightarrow{NaOH,\ 80°} (CH_3)_2CHCH{=}C(C_2H_5)CHO$$

is usually irreversible, this orientation is to be expected from mixed condensations effected under vigorous conditions (thermodynamic control). Some β,γ-unsaturated isomer may be expected in carbonyl products having both α and γ substituents.[142]

[129] L. Hough and J. K. N. Jones, *J. Chem. Soc.*, 3191 (1951).
[130] R. Schaffer and A. Cohen, *J. Org. Chem.*, **28,** 1929 (1963).
[131] W. Pigman, *The Carbohydrates*, pp. 113–114, Academic, New York, 1957.
[132] M. L. Wolfrom, *Advan. Carbohydrate Chem.*, **11,** 193–196 (1956).
[133] M. Kohn, *Monatsh. Chem.*, **22,** 21 (1901).
[134] A. Lilienfeld and S. Tauss, *Monatsh. Chem.*, **19,** 61 (1898).
[135] E. Swoboda and W. Fossek, *Monatsh. Chem.*, **11,** 383 (1890).
[136] Badische Anilin- and Soda-Fabrik, A.–G., Brit. pat. 734,000 [*C.A.*, **50,** 7845 (1956)].
[137] V. Grignard and P. Abelmann, *Bull. Soc. Chim. France*, [4] **7,** 638 (1910).
[138] M. B. Green and W. J. Hickinbottom, *J. Chem. Soc.*, 3262 (1957).
[139] L. K. Evans and A. E. Gillam, *J. Chem. Soc.*, 565 (1943).
[140] A. Lieben, *Monatsh. Chem*, **22,** 289 (1901).
[141] J. E. Dubois, *Bull. Soc. Chim. France*, [5] **20,** C17 (1953).
[142] H. Gilman, *Organic Chemistry*, Vol. I, 2nd ed., pp. 1041–1043, Wiley, New York, 1943.

It appears that increased substitution of bulky groups in the β position of one reactant favors attack at the less substituted aldehyde carbonyl group even at low temperatures. For example, condensation of acetaldehyde with undecanal (sodium amide in ether) leads to nearly equal amounts of the two possible mixed aldols;[143] this result may be due in part to use of the aprotic solvent ether which prevents rapid enolate equilibration.[43–45] 3-(4-Methyl-3-cyclohexenyl)butanal has been reported to condense with acetaldehyde at the methyl group (potassium hydroxide in aqueous methanol).[144] Careful quantitative analyses of product mixtures obtained from mixed aliphatic aldehyde condensations are evidently limited in number, and the subject deserves further study.

$$n\text{-}C_9H_{19}CH_2CHO + CH_3CHO \xrightarrow[(C_2H_5)_2O,\ -4^\circ \text{ to } 18^\circ]{NaNH_2} \underset{14\%}{n\text{-}C_9H_{19}CH_2CHOHCH_2CHO} + \underset{19\%}{CH_3CHOHCH(C_9H_{19}\text{-}n)CHO}$$

$$CH_3\text{-}C_6H_8\text{-}CH(CH_3)CH_2CHO + CH_3CHO \xrightarrow[5\text{–}25^\circ]{KOH,\ aq.\ CH_3OH} CH_3\text{-}C_6H_8\text{-}CH(CH_3)CH_2CH{=}CHCHO$$

Roles of solvent, catalyst, and temperature as well as structure are important in determining product composition. Equilibration of substrate protons (in protic solvents) relative to non-equilibration (in aprotic solvents), which affects concentrations of intermediate enolates and alkoxides, is a factor in these experiments.

Aromatic aldehydes condense with aliphatic ones to produce cinnamaldehydes in fair to good yields. Aldols seldom are isolated. Basic catalysts are employed most frequently. Electron-withdrawing groups in the aromatic ring generally favor the condensation; electron-releasing groups

$$ArCHO + RCH_2CHO \xrightarrow{OH^{\ominus}} \underset{H}{\overset{Ar}{>}}C{=}C\underset{CHO}{\overset{R}{<}} + H_2O$$

disfavor it. As the alkyl group R of the alkanal becomes bulkier, yields decrease. The product has been shown to have the β-aryl group *trans* to the carboxaldehyde group for R = H or C_6H_5.[88, 89]

[143] M. Stoll, *Helv. Chim. Acta*, **30**, 991 (1947).

[144] R. B. Wearn and C. Bordenca, U.S. pat. 2,519,327 [*C.A.*, **45**, 649 (1951)].

An aldehyde enamine has been condensed with propanal in the presence of acetic acid to give a 98% yield of 2-methyl-2-pentenal.[56] This reaction

$$C_2H_5CHO + CH_3CH{=}CHN\langle\text{piperidine ring}\rangle \xrightarrow{CH_3CO_2H} \underset{98\%}{C_2H_5CH{=}C(CH_3)CHO}$$

is of potential value in aldol condensations because it could be applicable to mixed aldehyde condensations if equilibration of the enamine and the aldehyde could be prevented. The Wittig reaction is applicable to the preparation of α,β-unsaturated aldehydes, RCH = C(R′)CHO.[15]

Self-Condensation of Ketones

The self-condensation of ketones may lead to ketols or α,β-unsaturated ketones. Diacetone alcohol is best prepared by condensing acetone using barium oxide catalyst.[145] The best yield (75%)[146, 147] is secured by employing a Soxhlet extractor to separate the catalyst from the ketol as it is formed. Mesityl oxide may be prepared in 79% yield by heating acetone under reflux with a basic ion-exchange resin.[148] (Phorone and isophorone are by-products of acetone self-condensation.) It is also readily prepared by dehydration of diacetone alcohol (iodine catalyst).[149] More vigorous conditions are needed for the self-condensation of symmetrical homologs of acetone such as 3-pentanone and 4-heptanone; catalysts such as sodium ethoxide, aluminum *t*-butoxide, isopropylmagnesium chloride, zinc chloride, and phosphorus oxychloride have been used. Acid catalysts lead to unsaturated ketones. Methylanilinomagnesium bromide in ether-benzene solvent is an excellent catalyst for ketol preparation;[150] it catalyzes self-condensation of 4-heptanone to the ketol in 45% yield and 3-pentanone to the ketol in 60% yield. Attempts to effect self-condensation of diisopropyl and diisobutyl ketones have failed.[151]

$$2\ (n\text{-}C_3H_7)_2CO \xrightarrow[(C_2H_5)_2O]{C_6H_5N(CH_3)MgBr} (n\text{-}C_3H_7)_2COHCH(C_2H_5)COC_3H_7\text{-}n$$

Unsymmetrical aliphatic ketones undergo self-condensation by attack of the carbonyl group on the less hindered α carbon atom (anti-Lieben's

[145] J. Herscovici, T. Bota, and D. Siriteaunu, *Rev. Chim. (Bucharest)*, **15**, 736 (1964) [*C.A.*, **62**, 11679 (1965)].

[146] J. Colonge, *Bull. Soc. Chim. France*, [4] **49**, 441 (1931).

[147] J. B. Conant and N. Tuttle, *Org. Syntheses, Coll. Vol.*, **1**, 199 (1941).

[148] N. B. Lorette, *J. Org. Chem.*, **22**, 346 (1957).

[149] J. B. Conant and N. Tuttle, *Org. Syntheses, Coll. Vol.* **1**, 345 (1941).

[150] V. Grignard and J. Colonge, *Compt. Rend.*, **194**, 929 (1932).

[151] W. Wayne and H. Adkins, *J. Am. Chem. Soc.*, **62**, 3401 (1940).

Rule.). One known exception is 2-butanone which undergoes self-condensation on the ethyl group with acid catalysis.[152–154] All other

$$2\ CH_3COCH_2CH_3 \xrightarrow{\text{Base}} CH_3CH_2C(OH)(CH_3)-CH_2COCH_2CH_3$$

$$2\ CH_3COCH_2CH_3 \xrightarrow{\text{Acid}} CH_3CH_2C(CH_3){=}C(CH_3)COCH_3$$

methyl ketones, including branched ones, undergo primarily 1-condensation (on the methyl group) with acid *or* basic catalysts;[151–158] either ketols or unsaturated ketones may result with base. A very small amount of 3-condensation may also result in certain instances.[152, 156]

$$2\ CH_3(CH_2)_nCOCH_3 \xrightarrow{\text{Acid or base}} CH_3(CH_2)_nC(CH_3){=}CHCO(CH_2)_nCH_3 \quad \text{(or parent } \beta\text{-ketol with base)}$$

$$n > 1$$

The self-condensation of enamines derived from methyl alkyl ketones (generated *in situ* from the ketals and morpholine) ultimately produces α,β-unsaturated ketones resulting from condensation at the methyl group.[60]

$$2\ RCH_2C(CH_3)(OC_2H_5)_2 + \text{morpholine (HN, O)} \xrightarrow[-C_2H_5OH]{\text{Heat, 36–48 hr.}}$$

$$RCH_2C(CH_3){=}CHC(N\text{-morpholino}){=}CHR \xrightarrow{H_3O^{\oplus}} RCH_2C(CH_3){=}CHCOCH_2R$$

152 V. Grignard and J. Colonge, *Compt. Rend.*, **190**, 1349 (1930).

153 J. Colonge and K. Mostafavi, *Bull. Soc. Chim. France*, [5] **5**, 1478 (1938).

154 A. E. Abbott, G. A. R. Kon, and R. D. Satchell, *J. Chem. Soc.*, 2514 (1928).

155 H. Thoms and C. Mannich, *Ber.*, **36**, 2555 (1903).

156 J. Colonge, *Bull. Soc. Chim. France*, [4] **49**, 426 (1931).

157 J. E. Dubois, *Compt. Rend.*, **224**, 1018 (1947).

158 J. E. Dubois and M. Chastrette, *Tetrahedron Letters*, 2229 (1964).

Self-condensation of alicyclic ketones proceeds normally; cyclopentanone, cyclohexanone, and cycloheptanone form ketols or α,β-unsaturated ketones; the ketols derived from cyclopentanone dehydrate most easily. Ketol formation is favored with anilinomagnesium bromide in ether.[159] 2-Cyclobutylidenecyclobutanone has been prepared from cyclobutanone and its enamines.[56b]

$C_6H_5NHMgBr$

OH O

KOH

C_2H_5OH

Acid catalysts favor the monosubstituted unsaturated ketone.[160, 161] With cyclohexanones the double bond in the product may appear in the endocyclic β,γ position as well as the α,β position.[161–164] The use of ethanolic potassium hydroxide as a catalyst leads to mixtures of mono- and bis-condensation products.[165–167a]

$NaOCH_3$, CH_3OH

<10°

2

HCl

Cl O

NaOH, aq.

25°

(See text on p. 22.)

[159] J. Colonge, *Compt. Rend.*, **196,** 1414 (1933).

[160] J. Plesek, *Collection Czech. Chem. Commun.*, **21,** 368 (1956); *Chem. Listy*, **50,** 246 (1956) [*C.A.*, **50,** 7732 (1956)].

[161] W. Hückel, O. Neunhoeffer, A. Gercke, and E. Frank, *Ann.*, **477,** 110 (1930).

[162] O. Wallach, *Ber.*, **40,** 70 (1907).

[163] J. Reese, *Ber.*, **75,** 384 (1942).

[164] E. Wenkert, S. K. Bhattacharya, and E. M. Wilson, *J. Chem. Soc.*, 5617 (1964).

[165] J. Stanek, *Chem. Listy*, **46,** 110 (1952).

[166] J. Plesek, *Collection Czech. Chem. Commun.*, **21,** 375 (1956); *Chem. Listy*, **50,** 252 (1956) [*C.A.*, **50,** 7732 (1956)].

[167] D. Varech, C. Ouannes, and J. Jacques, *Bull. Soc. Chim. France*, 1662 (1965).

[167a] T. A. Favorskaya, A. S. Lozhenitsyna, G. A. Kalabin, and V. M. Vlasov, *Zh. Org. Khim.*, **2,** 739 (1966) [*C.A.*, **65,** 8772 (1966)].

The intermediate β-chloroketone, 2-(1-chlorocyclohexyl)cyclohexanone,[162, 164] prepared from cyclohexanone and hydrogen chloride, may be dehydrohalogenated to the α,β-unsaturated ketone (methanolic sodium methoxide at low temperature, $<10°$) or the endocyclic β,γ isomer (with aqueous sodium hydroxide at room temperature).[163, 164]

Alkyl aryl ketones undergo self-condensation to substituted styryl aryl ketones (dypnones). Aluminum *t*-butoxide[151] and hydrogen bromide[152] are effective catalysts for these condensations. Methylanilinomagnesium bromide in ether produces ketols.[159]

$$2\,ArCOCH_2R \xrightarrow[100°]{Al(OC_4H_9\text{-}t)_3} ArC{=}C(R)COAr + H_2O \quad (\text{ArC bearing } CH_2R)$$

$$2\,ArCOCH_2R \xrightarrow[(C_2H_5)_2O,\ 25°]{C_6H_5N(CH_3)MgBr} ArC(OH)(CH_2R)CH(R)COAr$$

The 1,2-diketones, 2,3-butanedione and 2,3-pentanedione, undergo base-catalyzed self-condensation, in low yield, to p-benzoquinone derivatives.[168–171] The intermediate cyclic ketol has been isolated from 2,3-butanedione after reaction with dilute aqueous sodium hydroxide at low temperature;[168, 170] more concentrated base at higher temperature produces the benzoquinone.

$$2\,CH_3COCOCH_2R \xrightarrow[H_2O]{NaOH} \left[CH_3COCOCH(R)C(OH)(CH_3)COCH_2R\right] \rightarrow$$

R = H, CH_3

$$\text{cyclic ketol (R, OH, CH}_3\text{; HO, CH}_3\text{, R; two C=O)} \xrightarrow[\text{Warm}]{NaOH,\ H_2O} \text{2,5-R}_2\text{-3,6-(CH}_3)_2\text{-}p\text{-benzoquinone}$$

In aqueous acetic acid, 2,3-butanedione undergoes self-condensation to a dihydrofuranone.[171a] Self-condensation has been effected with one

[168] O. Diels, W. M. Blanchard, and H. v.d. Heyden, *Ber.*, **47**, 2355 (1914).

[169] G. Machell, *J. Chem. Soc.*, 683 (1960).

[170] H. v. Pechmann and E. Wedekind, *Ber.*, **28**, 1845 (1895).

[171] H. v. Pechmann, *Ber.*, **21**, 1411 (1888).

[171a] R. Shapiro, J. Hackmann, and R. Wahl, *J. Org. Chem.*, **31**, 2710 (1966).

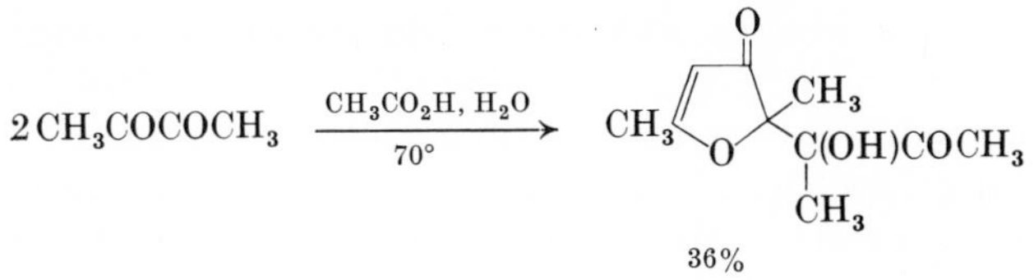

$$2\,CH_3COCOCH_3 \xrightarrow[70^\circ]{CH_3CO_2H,\ H_2O} \text{(furanone: } CH_3,\ O,\ CH_3,\ C(OH)COCH_3,\ CH_3\text{)}\quad 36\%$$

cyclic 1,2-diketone; the ketol formed could not be completely dehydrated to a benzoquinone,[172] presumably because of the strain that would be present in the product.

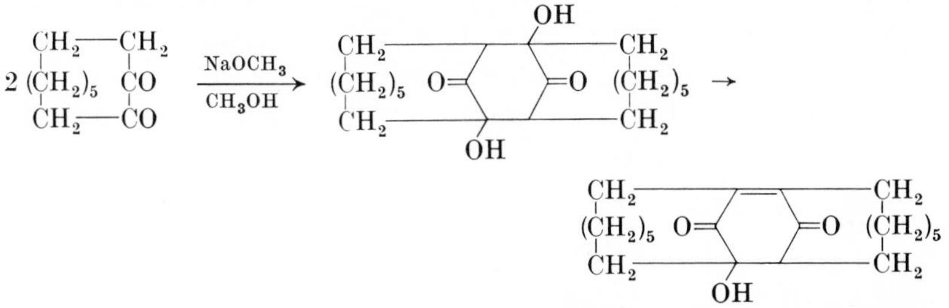

$$2\,(CH_2)_5\begin{matrix}CH_2-CH_2\\ CO\\ CH_2-CO\end{matrix} \xrightarrow[CH_3OH]{NaOCH_3} \text{(bis-ketol, OH, OH, =O, =O, }(CH_2)_5\text{)} \rightarrow \text{(ene-dione, OH, =O, =O, }(CH_2)_5\text{)}$$

Few examples of self-condensation of 1,3-diketones are known. 2,4,6-Heptanetrione was condensed under carefully controlled conditions (aqueous sodium hydroxide and a potassium acid phosphate buffer) to a tetralone (at *p*H 7.1–7.2) or a naphthalene derivative (*p*H 8.2).[173] Attempts to effect self-condensation of other 1,3-diketones have failed.[174]

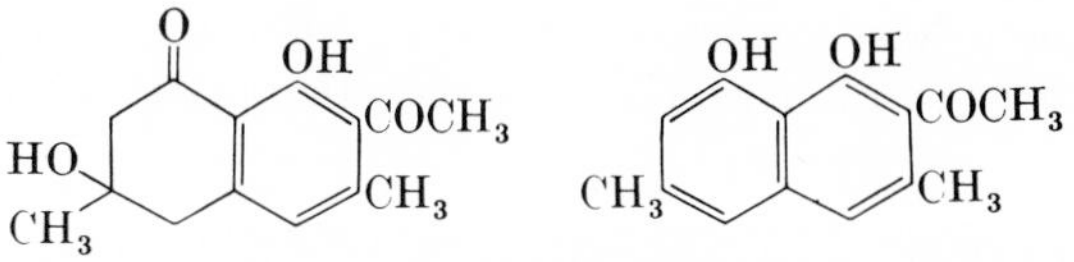

(Structures: tetralone with O, OH, $COCH_3$, CH_3, HO, CH_3; naphthalene with OH, OH, $COCH_3$, CH_3, CH_3)

Diketones having carbonyl groups separated by two or more carbons have not been observed to undergo intermolecular condensation. Those which have been examined readily condense intramolecularly to form cycloalkenones or acyl cycloalkenes (discussed on pp. 49–56).

Mixed Condensation of Ketones

The condensation of two different ketones produces ketols or α,β-unsaturated ketones. The reaction has seldom been applied to two

[172] R. A. Raphael and A. I. Scott, *J. Chem. Soc.*, 4566 (1952).

[173] J. R. Bethell and P. Maitland, *J. Chem. Soc.*, 3751 (1962).

[174] E. E. Blaise, *Compt. Rend.*, **158**, 708 (1914).

different acyclic ketones, and the known procedures appear to be of limited utility for mixed condensations.[174a] Condensation of acetone with 2-butanone gave a mixture of products.[175] Although diisobutyl ketone failed to condense with itself, it has been condensed with acetone to produce a ketol.[176] One should be able to extend this reaction to

$$i\text{-}C_3H_7CH_2COCH_2C_3H_7\text{-}i + CH_3COCH_3 \xrightarrow{i\text{-}C_3H_7MgCl} i\text{-}C_3H_7\overset{(CH_3)_2COH}{\overset{|}{C}}HCOCH_2C_3H_7\text{-}i$$

condensations of other hindered ketones with more reactive ketones employing methylanilinomagnesium bromide as the condensing agent.[177]

Condensation of aliphatic methyl ketones with cyclohexanones occurs by reaction of the cyclic carbonyl group with the methyl group of the methyl ketone.[178–179a] The double bond in the product usually appears

$$\text{cyclohexanone} + CH_3COCH_2CH_3 \xrightarrow{NaOH} \text{1-cyclohexenyl-}CH_2COCH_2CH_3$$

$$\text{cyclohexanone} + \text{1-acetyl-2-methylcyclohexene} \xrightarrow{KOC_4H_9\text{-}t} \text{cyclohexenyl}{=}CH{-}\overset{O}{\overset{\|}{C}}\text{-(2-methylcyclohexenyl)}$$

in cyclohexanone-condensed products in the endocyclic β,γ position rather than the exocyclic α,β position. On the other hand, cyclobutanone[180]

$$\text{cyclobutanone} + CH_3COCH_3 \xrightarrow{\text{Aq.NaOH}} (CH_3)_2C{=}\text{cyclobutanone} \; (20\%) + (CH_3)_2C{=}\text{cyclobutanone}{=}C(CH_3)_2 \; (20\%)$$

$$\text{cyclopentanone} + CH_3COCH_3 \xrightarrow{NaOC_2H_5} (CH_3)_2C{=}\text{cyclopentanone}$$

[174a] S. Jung and P. Cordier, *Compt. Rend.*, Ser. C, **262**, 1793 (1966).
[175] T. Voitila, *Suomen Kemistilehti*, **9B**, 30 (1936) [*C.A.*, **31**, 2582 (1937)].
[176] P. Maroni and J. E. Dubois, *Bull. Soc. Chim. France*, 126 (1955).
[177] A. T. Nielsen, C. Gibbons, and C. Zimmerman, *J. Am. Chem. Soc.*, **73**, 4696 (1951).
[178] G. A. R. Kon, *J. Chem. Soc.*, 1792 (1926).
[179] R. B. Turner and D. M. Voitle, *J. Am. Chem. Soc.*, **72**, 4166 (1950).
[179a] E. A. Brande and O. H. Wheeler, *J. Chem. Soc.*, 329 (1955).
[180] J. P. Sandre and J. M. Conia, *Bull. Soc. Chim. France*, 903 (1962).

and cyclopentanone[181–183] condense with acetone to form isopropylidene derivatives.[184] The relative reactivities of ketones to nucleophilic attack (rates and equilibria) appear to be important here and have been evaluated for the formation of acetals,[185] semicarbazones,[186] cyanohydrins,[187–191] and bisulfite addition compounds,[192, 193] as well as for sodium borohydride reduction.[194, 195] The relative reactivities of ketone enolates may also be pertinent; cf. alkylation rates.[196, 197]

Acetophenone condenses with 2-butanone or 2-pentanone to give mixtures of products.[198]

$$C_6H_5COCH_3 + CH_3COC_3H_7\text{-}n \xrightarrow{C_6H_5N(CH_3)MgBr} \underset{37\%}{C_6H_5\underset{\displaystyle CH_3}{\underset{|}{C}}{=}CHCOC_3H_7\text{-}n} + \underset{32\%}{C_6H_5COCH{=}\underset{\displaystyle CH_3}{\underset{|}{C}}C_3H_7\text{-}n}$$

The condensation of benzils with simple aliphatic ketones leads to the interesting cyclopentenone ketols, "anhydroacetonebenzils **(11)**,"[199–201] which do not dehydrate to stable cyclopentadienones. The ketols

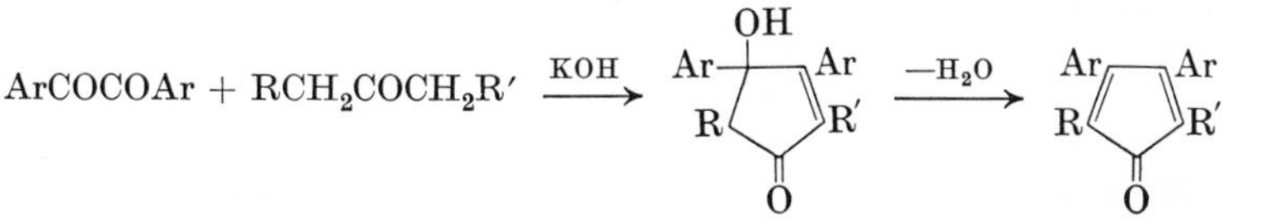

$$ArCOCOAr + RCH_2COCH_2R' \xrightarrow{KOH} \mathbf{11} \xrightarrow{-H_2O} \mathbf{12}$$

11: 4-hydroxy-3,4-Ar,Ar-2,5-R,R'-cyclopent-2-enone (OH on the carbon bearing Ar)

(R, R′ = alkyl, aryl)

12 (stable when R, R′ = aryl only)

[181] G. A. R. Kon and J. H. Nutland, *J. Chem. Soc.*, 3101 (1926).
[182] O. Wallach, *Ann.*, **394**, 362 (1912).
[183] L. Bouveault, *Compt. Rend.*, **130**, 415 (1900).
[184] P. Hudry and P. Cordier, *Compt. Rend.*, **261**, 468 (1965).
[185] J. M. Bell, D. G. Kubler, P. Sartwell, and R. G. Zepp, *J. Org. Chem.*, **30**, 4284 (1965).
[186] J. B. Conant and P. D. Bartlett, *J. Am. Chem. Soc.*, **54**, 2881 (1932).
[187] W. J. Svirbely and J. F. Roth, *J. Am. Chem. Soc.*, **75**, 3106 (1953).
[188] D. P. Evans and J. R. Young, *J. Chem. Soc.*, 1310 (1954).
[189] A. Lapworth and R. H. F. Manske, *J. Chem. Soc.*, 2533 (1928).
[190] A. Lapworth and R. H. F. Manske, *J. Chem. Soc.*, 1976 (1930).
[191] K. L. Servis, L. K. Oliver, and J. D. Roberts, *Tetrahedron*, **21**, 1827 (1965).
[192] K. Shinra, K. Ishikawa, and K. Arai, *J. Chem. Soc. Japan, Pure Chem. Sect.*, **75**, 661 (1954) [*C.A.*, **49**, 5084 (1955)].
[193] M. A. Gubareva, *Zh. Obshch. Khim.*, **17**, 2259 (1947) [*C.A.*, **42**, 4820 (1948)].
[194] H. C. Brown, R. Bernheimer, and K. J. Morgan, *J. Am. Chem. Soc.*, **87**, 1280 (1965).
[195] H. C. Brown, O. H. Wheeler, and K. Ichikawa, *Tetrahedron*, **1**, 214 (1957).
[196] J. M. Conia, *Ann. Chim.* (*Paris*) [12] **8**, 709 (1953).
[197] H. D. Zook and W. L. Rellahan, *J. Am. Chem. Soc.*, **79**, 881 (1957).
[198] V. V. Chelinstev and A. V. Pataraya, *Zh. Obshch. Khim.*, **11**, 461 (1941) [*C.A.*, **35**, 6571 (1941)].
[199] F. R. Japp and C. I. Burton, *J. Chem. Soc.*, **51**, 431 (1887).
[200] F. R. Japp and T. S. Murray, *J. Chem. Soc.*, **71**, 144 (1897).
[201] P. Yates, N. Yoda, W. Brown, and B. Mann, *J. Am. Chem. Soc.*, **80**, 202 (1958).

derived from dibenzyl ketone do dehydrate to produce tetraphenylcyclopentadienones (tetracyclones, **12**).[202–206] The cyclopentadienones derived from dialkyl ketones are unstable and readily form carbonyl

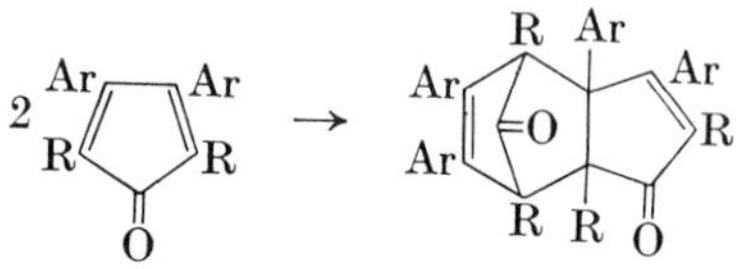

13

bridged dimers such as **13**.[205, 206] Condensation of benzils with cyclohexanones leads to ketols[207] and with acetophenones to β-benzoyl chalcones.[208]

By employing 2 molecular equivalents of potassium amide (in liquid ammonia) with certain 1,3-dicarbonyl compounds to produce a dicarbanion, condensations may be effected at the γ position.[209–212c] Alkylations and acylations also occur at the γ position of the dianion. The dicarbanions are most conveniently prepared from the monoenolate salt of the dicarbonyl compound by treatment with 1 molecular equivalent of

$$(CH_3COCH{=}CHO^{\ominus})Na^{\oplus} \xrightarrow{KNH_2} {}^{\ominus}CH_2COCH{=}CHO^{\ominus}$$

$$(C_6H_5)_2CO + {}^{\ominus}CH_2COCH{=}CHO^{\ominus} \longrightarrow (C_6H_5)_2C(OH)CH_2COCH_2CHO$$

potassium amide.[212] The method has been applied to 1-aryl-1,3,5-hexanetriones (*tris*-anion intermediate) whereby condensation of a carbonyl compound (benzophenone, benzaldehyde) occurs on the terminal methyl group to produce a ketol.[213]

[202] C. G. Henderson and R. H. Corstorphine, *J. Chem. Soc.*, **79,** 1256 (1901).

[203] W. Dilthey, O. Trösken, K. Plum, and W. Schommer, *J. Prakt. Chem.*, [2] **141,** 331 (1934).

[204] J. R. Johnson and O. Grummitt, *Org. Syntheses, Coll. Vol.*, **3,** 806 (1955); L. F. Fieser, *Org. Syntheses*, **46,** 45 (1966).

[205] C. F. H. Allen, *Chem. Rev.*, **37,** 209 (1945); C. F. H. Allen and J. A. Van Allan, *J. Org. Chem.*, **20,** 315 (1955).

[206] M. A. Ogliaruso, M. G. Romanelli, and E. I. Becker, *Chem. Rev.*, **65,** 261 (1965).

[207] C. F. H. Allen and J. A. Van Allan, *J. Org. Chem.*, **16,** 716 (1951).

[208] C. F. H. Allen and H. B. Rosener, *J. Am. Chem. Soc.*, **49,** 2110 (1927).

[209] T. M. Harris and C. R. Hauser, *J. Am. Chem. Soc.*, **84,** 1750 (1962).

[210] J. F. Wolfe, T. M. Harris, and C. R. Hauser, *J. Org. Chem.*, **29,** 3249 (1964).

[211] R. J. Light and C. R. Hauser, *J. Org. Chem.*, **26,** 1716 (1961).

[212] T. M. Harris, S. Boatman, and C. R. Hauser, *J. Am. Chem. Soc.*, **87,** 3186 (1965).

[212a] K. G. Hampton, T. M. Harris, and C. R. Hauser, *J. Org. Chem.*, **31,** 663 (1966).

[212b] S. Boatman and C. R. Hauser, *J. Org. Chem.*, **31,** 1785 (1966).

[212c] T. M. Harris and C. M. Harris, *Org. Reactions*, **17,** in press.

[213] K. G. Hampton, T. M. Harris, C. M. Harris, and C. R. Hauser, *J. Org. Chem.*, **30,** 4263 (1965).

$$(C_6H_5)_2CO + {}^{\ominus}CH_2COCH{=}\underset{O^{\ominus}}{\underset{|}{C}}H{=}\underset{O^{\ominus}}{\underset{|}{C}}Ar \xrightarrow[NH_3]{NaNH_2} (C_6H_5)_2C(OH)CH_2COCH_2COCH_2COAr$$

An unexplored, but potentially useful, synthetic route to α,β-unsaturated ketones might involve condensation of a ketone with a ketone enamine (acetic acid catalyst).[54, 56] The Wittig reaction may be employed for synthesis of α,β-unsaturated ketones.[15]

Condensation of Aldehydes with Acyclic Ketones

The condensation of aldehydes with ketones to yield ketols or α,β-unsaturated ketones is a reaction of broad utility and applicability. A significant limitation results from the greater reactivity of aldehydes, which will undergo self-condensation or will condense at the α carbon atom of ketones.[186–193] Ketones, however, will not condense intermolecularly at the α carbon atom of aldehydes to form aldols with simple base or acid catalysts. By employing certain enamine or imine intermediates this limitation may be circumvented. Condensation of ethyl pyruvate with butanal and higher alkanals[214, 215] in the presence of diethylamine at 4° leads to aldols and probably involves the more readily formed aldehyde enamine. (Pyruvic acid under the same conditions with the same aldehydes forms alkylidene derivatives, $RCH_2CH{=}CHCOCO_2H$.[215a])

$$RCH_2CHO + CH_3COCO_2C_2H_5 \xrightarrow[4^\circ]{(C_2H_5)_2NH} \underset{CHO}{\underset{|}{RCH}}{-}\underset{OH}{\underset{|}{C}}(CH_3)CO_2C_2H_5$$

R = alkyl

Intramolecular condensations of ketonic carbonyl groups on the α carbon atom of aldehydes (via aldehyde enamine intermediates) to produce cycloalkene carboxaldehydes are known and are discussed on p. 57.

An intermolecular condensation of a ketone with an aldehyde enamine has been reported. Acetone reacts with 1-propenylpiperidine (added acetic acid) to produce, ultimately, 2,4,5-trimethyl-2,4-hexadienal in unstated yield;[54] the second molecule of enamine entering the reaction

$$(CH_3)_2CO + 2\,CH_3CH{=}CHN\bigcirc \xrightarrow{CH_3CO_2H} CH_3\underset{CH_3}{\underset{|}{C}}{=}\underset{CH_3}{\underset{|}{C}}CH{=}\underset{CH_3}{\underset{|}{C}}CHO$$

[214] P. Cordier, J. Schreiber, and C. G. Wermuth, *Compt. Rend.*, **250**, 1668, 2587 (1960).
[215] J. Schreiber and C. G. Wermuth, *Bull. Soc. Chim. France*, 2242 (1965).
[215a] C. G. Wermuth *Bull. Soc. Chim. France*, 1435 (1966).

may be attacked by an immonium ion intermediate such as $(CH_3)_2C{=}C(CH_3)CH{=}\overset{\oplus}{N}C_5H_{10}$. This type of reaction would appear to be of much potential synthetic value if it could be extended to other ketones and enamines derived from an aldehyde or ketone, particularly if condensation with only one molecule of enamine could be achieved by proper choice of reaction conditions. Employment of preformed enamines (under mild conditions), rather than those generated *in situ*, appears to be necessary. In each of the few known examples where secondary amine catalysts have been employed in condensations of aldehydes (having α hydrogen atoms) with ketones, the products have been α,β-unsaturated ketones.[216–220]

By employing an aldehyde imine anion, condensation with ketones can be directed to produce α,β-unsaturated aldehydes (a directed Wittig aldol condensation).[221–221b] In this synthesis the imine is metalated with a lithium amide in ether, then treated with a ketone to produce an isolable lithium salt; hydrolysis of this salt in dilute sulfuric acid provides the α,β-unsaturated aldehyde in excellent yield—78% overall in the example shown in the accompanying equations.

$$CH_3CH{=}NC_6H_{11} + LiN(C_3H_7\text{-}i)_2 \xrightarrow{(C_2H_5)_2O} LiCH_2CH{=}NC_6H_{11} + (i\text{-}C_3H_7)_2NH$$

$$(C_6H_5)_2CO + LiCH_2CH{=}NC_6H_{11} \rightarrow$$

$$(C_6H_5)_2C(OLi)CH_2CH{=}NC_6H_{11} \xrightarrow{H^{\oplus}} (C_6H_5)_2C{=}CHCHO$$

The Wittig reaction itself is not applicable to the preparation of α,β-unsaturated aldehydes from ketones and formylalkylidene triphenylphosphoranes.[15, 221b, c]

The entire discussion of aldehyde-ketone condensations that follows applies principally to condensations conducted in protic solvents under equilibrium conditions.

Symmetrical Ketones. Only acetone and 3-pentanone are very reactive in condensations with aldehydes in the presence of ethanolic potassium or sodium hydroxide or ethoxide catalysts. 4-Heptanone and

[216] M. E. McEntee and A. R. Pinder, *J. Chem. Soc.*, 4419 (1957).
[217] G. B. Payne, *J. Org. Chem.*, **24,** 1830 (1959).
[218] K. Eiter, *Ann.*, **658,** 91 (1962).
[219] G. Wermuth, *Compt. Rend.*, **251,** 391 (1960).
[220] B. D. Wilson, *J. Org. Chem.*, **28,** 314 (1963).
[221] G. Wittig and H. D. Frommeld, *Ber.*, **97,** 3548 (1964).
[221a] G. Wittig and P. Suchanek, *Tetrahedron*, **Suppl. 8, Part 1,** 347 (1966).
[221b] G. Wittig, *Record Chem. Prog.* (*Kresge-Hooker Sci. Lib.*), **28,** 45 (1967).
[221c] S. Tripett and D. M. Walker, *J. Chem. Soc.*, 1266 (1961).

$$R'CHO + RCH_2COCH_2R \underset{}{\overset{\text{Base}}{\rightleftharpoons}} R'CHOHCH(R)COCH_2R \xrightarrow{-H_2O} R'CH{=}C(R)COCH_2R$$

R′, R = H, alkyl, aryl

higher *n*-alkanones condense with most monocarboxaldehydes with great difficulty, and yields are poor; no procedure has yet been developed for satisfactorily effecting condensations of this type.[222] However, it appears that several approaches might be applicable to such condensations. For example, a modification of the directed Wittig aldol synthesis[221b] might be applied, namely, addition of an aldehyde to the required ketimine anion. Another approach could make use of a slowly equilibrating lithium enolate (in a solvent such as 1,2-dimethoxyethane, with rigorous exclusion of proton donors),[44] followed by addition of the aldehyde. Alternatively, condensation of an aldehyde with a ketone enamine, or the Wittig reaction,[15] might be employed.

It is interesting that, although most monocarboxaldehydes give very poor yields (usually <20%) of condensation product with 4-heptanone and higher non-methyl *n*-alkanones, *o*-phthalaldehyde is exceptional and reacts readily with these ketones to give excellent yields (73–97%) of 2,7-di-*n*-alkyl-4,5-benzotropones.[223] It is known that intramolecular base-catalyzed enolization is an efficient process;[224] this fact suggests that intramolecular aldol-derived alkoxide-catalyzed enolization of the second α proton may be strongly favored when R is a large group, and may contribute to the success of these condensations. When R is a branched alkyl group, yields are lower, suggesting that in these examples the initial equilibrium favors reactants.

$$C_6H_4(CHO)_2 + RCH_2COCH_2R \underset{C_2H_5OH}{\overset{NaOC_2H_5}{\rightleftharpoons}} \text{[intermediate]} \xrightarrow{-2H_2O} \text{2,7-R}_2\text{-4,5-benzotropone}$$

R = alkyl, aryl

The very unreactive diisopropyl and diisobutyl ketones may be converted into their enolate ions with a strong base such as methylanilinomagnesium bromide. These ions, in contrast to those of *n*-alkanones, do

[222] S. G. Powell and A. T. Nielsen, *J. Am. Chem. Soc.*, **70**, 3627 (1948).
[223] D. Meuche, H. Strauss, and E. Heilbronner, *Helv. Chim. Acta*, **41**, 2220 (1958).
[224] E. T. Harper and M. L. Bender, *J. Am. Chem. Soc.*, **87**, 5625 (1965).

not react with the parent ketone under the reaction conditions and may be condensed with an aldehyde, yielding 62–88% of ketol.[177] Attempts to apply this procedure to more reactive ketones (acetone, 3-pentanone,

$$[(CH_3)_2CH]_2C{=}O \xrightarrow[(C_2H_5)_2O/C_6H_6]{C_6H_5N(CH_3)MgBr} (CH_3)_2CHC(O^{\ominus}){=}C(CH_3)_2 \xrightarrow{RCHO} (CH_3)_2CHC({=}O)C(CH_3)_2CHOHR$$

acetophenone) failed, because self-condensation of both ketone and aldehyde occurred, the aldehyde condensation product ultimately forming a tertiary amine.[177] With 6-undecanone, butanal gave a 15% yield of ketol.

Occasionally, steric hindrance in the aldehyde has been found to prevent condensation. Dehydrocitral (2,2,6-trimethylcyclohexane-1-carboxaldehyde) condenses readily with acetone in the presence of methanolic potassium hydroxide to produce dihydroionone in 55–58% yield.[225] On the other hand, 2,2,6,6-tetramethylcyclohexane-1-carboxaldehyde could not be made to undergo condensation with acetone under a variety of conditions.[225]

Of all acyclic ketones, only acetone readily forms acyclic *bis*-condensation products with monocarboxaldehydes. The products are usually 1,4-pentadien-3-ones, most frequently produced from aromatic aldehydes by employing 2 mole equivalents of aldehyde in the presence of aqueous

$$2\,ArCHO + CH_3COCH_3 \xrightarrow[Aq.\ C_2H_5OH]{NaOH} ArCH{=}CHCOCH{=}CHAr + 2\,H_2O$$

ethanolic sodium hydroxide. 2-Butanone occasionally produces *bis*-condensation products.[226]

The condensation of hydroxyaldehydes with dihydroxyacetone has been employed in sugar syntheses.[35, 131, 132, 227–229a]

Unsymmetrical Ketones. A large number of condensations of aldehydes with various methyl ketones are known, but very few with other unsymmetrical acyclic ketones.[222] α-Ketols of the type RCOCHOHR′ undergo base-catalyzed condensation with formaldehyde and acetaldehyde on the carbon atom bearing the hydroxyl group.[230]

[225] M. de Botton, *Compt. Rend.*, **256,** 2866 (1963).

[226] Y. Kodama, A. Takai, and I. Saikawa, Japan. pat. 19,642 (1964) [*C.A.*, **62,** 10413 (1965)].

[227] R. Schaffer and H. S. Isbell, *J. Org. Chem.*, **27,** 3268 (1962).

[228] R. Schaffer, *J. Org. Chem.*, **29,** 1471 (1964).

[229] J. A. Gascoigne, W. G. Overend, and M. Stacey, *Chem. Ind.* (*London*), 402 (1959).

[229a] C. D. Gutsche, D. Redmore, R. S. Buricks, K. Nowotny, H. Grassner, and C. W. Armbruster, *J. Am. Chem. Soc.*, **89,** 1235 (1967).

[230] J. Colonge and Y. Vaginay, *Bull. Soc. Chim. France*, 3140 (1965).

Ketones of the type $\overset{1}{C}H_3\overset{2}{C}O\overset{3}{C}HRR'$ (R′ = alkyl, aryl, or H; R = alkyl or aryl) may condense with aldehydes at carbon 1 (1-*condensation*) or carbon 3 (3-*condensation*). Four principal factors determine the structure of the product obtained at equilibrium: (1) catalyst; (2) nature of substituents R and R′ in the ketone; (3) structure of the aldehyde; and (4) the solvent, which has received little study. These factors have been reviewed briefly for 2-butanone.[231] Although mixtures of 1- and 3-condensation products are possible and to be expected, few studies of the exact composition of products have been reported; a single substance appears to predominate in most reactions.

Acid catalysts favor 3-condensation except where steric factors prevent it. The most highly branched enol [$CH_3C(OH){=}CHR$] derived from the ketone reacts preferentially with the aldehyde to produce an α,β-unsaturated ketone[42] (see equations, p. 10). It is primarily the direction of enolization which determines the course of the acid-catalyzed condensation. The dehydration of the intermediate ketol is rapid (relative to the condensation step), and ketols seldom result from acid-catalyzed condensations.[30, 40] Anhydrous hydrogen chloride is often employed as acid catalyst; the intermediate β-halo ketone may be dehydrohalogenated by treatment with aqueous or ethanolic alkali hydroxide[232] or carbonate,

$$C_6H_5CHO + CH_3CH_2COCH_3 \xrightarrow{HCl}$$

$$C_6H_5CH(Cl)CH(CH_3)COCH_3 \xrightarrow{\text{Aq. NaOH}} \begin{matrix} C_6H_5 & & CH_3 \\ & C{=}C & \\ H & & COCH_3 \end{matrix}$$

trans

pyridine, or quinoline, or by heating. In the α,β-unsaturated product the larger β-substituent and the α carbonyl group are oriented *trans*.[233–235]

All methyl *n*-alkyl ketones studied are reported to produce 3-condensation with acid catalysts.[236, 237] Examples of acid-catalyzed condensation at carbon atom 3 with ketones of the type $CH_3COCHRR'$ (R, R′ = alkyl or aryl but not hydrogen) are rare. Since the expected intermediate ketol cannot dehydrate and ketol is not a favored product, one would

[231] H. Haeussler and J. Dijkema, *Ber.*, **77**, 601 (1944).

[232] C. Harries and G. H. Müller, *Ber.*, **35**, 966 (1902).

[233] M. E. Kronenberg and E. Havinga, *Rec. Trav. Chim.*, **84**, 17 (1965).

[234] M. E. Kronenberg and E. Havinga, *Rec. Trav. Chim.*, **84**, 979 (1965).

[235] M. E. Kronenberg and E. Havinga, to be published. We are indebted to Prof. Havinga and Dr. Kronenberg for providing us with their data prior to publication.

[236] M. T. Bogert and D. Davidson, *J. Am. Chem. Soc.*, **54**, 334 (1932).

[237] K. Iwamoto and T. Kato, *Sci. Rept. Tohuku Imp. Univ., First Ser.*, **19**, 689 (1930) [*C.A.*, **25**, 2132 (1931)].

predict the major condensation product to be an α,β-unsaturated ketone derived by condensation of the methyl group, RCH=CHCOCHRR′. The only reported exceptions to 3-condensation using an acid catalyst are found with methyl isobutyl ketone which condenses at the methyl group with aromatic aldehydes (benzaldehyde,[238, 239] salicylaldehyde,[239] and 4-methoxybenzaldehyde[240]). Acetaldehyde[241] and chloral[242] are reported to undergo 3-condensation with methyl isobutyl ketone. Condensation of benzaldehyde with 2-heptanone (concentrated hydrochloric acid catalyst) produced 91% 3-condensation and 9% 1-condensation.[243] It is likely that small percentages of 1-condensation accompany many examples of reported 3-condensation. Gas-liquid chromatography and nuclear magnetic resonance spectra would aid in assay of condensation mixtures, but until recently, they have not been extensively employed for this purpose.[63]

Synthesis of β-ketols of known structure (including those derived by 3-condensation from methyl ketones) may be achieved by decarboxylation of a β-keto ester at *p*H 7 (phosphate buffer, 25°, several days) in water or aqueous methanol, presumably to generate the required ketone enol intermediate which condenses with the aldehyde present (Schöpf condensation).[40, 244]

$$\mathrm{RCHO} + \mathrm{R'COCH(R'')CO_2H} \xrightarrow[p\mathrm{H}\ 7]{-\mathrm{CO_2}} \mathrm{RCHOHCH(R'')COR'}$$

The course of the base-catalyzed condensation of aldehydes with unsymmetrical ketones is much more responsive to reaction conditions which influence the condensation (solvent, catalyst, temperature) than is the acid-catalyzed reaction. The reaction sequence is in some ways similar to that of the acid-catalyzed reaction, but relative rates for the steps are different.[40, 42] The rate-determining step with either catalyst is usually the condensation step,[18, 26, 245, 246] but the dehydration step in the base-catalyzed reaction is much slower and more easily reversed (relative to the condensation step) than in the acid-catalyzed reaction[30, 40] (see p. 10). Ketols are frequently produced in the base-catalyzed reaction, especially under mild conditions.

[238] C. V. Gheorghiu and B. Arwentiew, *J. Prakt. Chem.*, [2] **118,** 295 (1928).

[239] I. M. Heilbron and F. Irving, *J. Chem. Soc.*, 936 (1929).

[240] C. V. Gheorghiu and B. Arwentiew, *Bull. Soc. Chim. France*, [4] **47,** 195 (1930).

[241] J. E. Dubois, R. Luft, and F. Weck, *Compt. Rend.*, **234,** 2289 (1952).

[242] J. S. Buck and I. M. Heilbron, *J. Chem. Soc.*, **121,** 1198 (1922).

[243] M. Metayer, *Rec. Trav. Chim.*, **71,** 153 (1952).

[244] C. Schöpf and K. Thierfelder, *Ann.*, **518,** 127 (1935).

[245] G. Kresze and B. Gnauck, *Z. Elektrochem.*, **60,** 174 (1956).

[246] G. Sipos, A. Furka, and T. Szell, *Monatsh. Chem.*, **91,** 643 (1960).

$$CH_3COCH_2R + OH^{\ominus} \rightleftharpoons \underset{\displaystyle\mathbf{14}}{CH_2{=}\underset{O^{\ominus}}{\underset{|}{C}}CH_2R} + H_2O \qquad \text{(Eq. 10a)}$$

$$CH_3COCH_2R + OH^{\ominus} \rightleftharpoons \underset{\displaystyle\mathbf{15}}{CH_3\underset{O^{\ominus}}{\underset{|}{C}}{=}CHR} + H_2O \qquad \text{(Eq. 10b)}$$

$$\mathbf{14} + R'CHO \underset{B}{\overset{BH^{\oplus}}{\rightleftharpoons}} R'CHOHCH_2COCH_2R \qquad \text{(Eq. 11a)}$$

16 (1-Condensation ketol = 1-K)

$$\mathbf{15} + R'CHO \underset{B}{\overset{BH^{\oplus}}{\rightleftharpoons}} R'CHOHCH(R)COCH_3 \qquad \text{(Eq. 11b)}$$

17 (3-Condensation ketol = 3-K)

$$\mathbf{16} \overset{OH^{\ominus}}{\rightleftharpoons} R'CH{=}CHCOCH_2R + H_2O \qquad \text{(Eq. 12a)}$$

18 (1-Condensation unsaturated ketone = 1-U)

$$\mathbf{17} \overset{OH^{\ominus}}{\rightleftharpoons} R'CH{=}C(R)COCH_3 + H_2O \qquad \text{(Eq. 12b)}$$

19 (3-Condensation unsaturated ketone = 3-U)

The composition of the equilibrium mixture of methyl alkyl ketone-derived enolate anions **14** and **15** is affected by the solvent and by the structure of the ketone. It is pertinent to attempt an evaluation of the composition of this mixture in order to interpret the composition of aldol condensation product mixtures. Few systematic studies of such enolate compositions have been made for protic solvents (water, ethanol) which are commonly employed in most aldol condensations.[246a–c] Direct measurements of enolate composition in 1,2-dimethoxyethane (quenching in deuterium oxide-deuterioacetic acid) indicate a stability order $n\text{-}C_4H_9CH{=}\underset{O^{\ominus}}{\underset{|}{C}}CH_3 > n\text{-}C_4H_9CH_2\underset{O^{\ominus}}{\underset{|}{C}}{=}CH_2$;[43, 247] the amount of more highly substituted enolate at equilibrium (58–87%) depends on the cation and the solvent. A more highly branched alkyl group favors the less highly substituted enolate (67–82% **14** when $R = C_3H_7\text{-}i$).[43, 247] The more highly substituted enolate derived from 2-methylcyclopentanone and 2-methylcyclohexanone is the more favored.[43, 247–248c] Alkylation

[246a] C. Rappe, *Acta Chem. Scand.*, **20,** 376, 1721, 2236, 2305 (1966).

[246b] J. Warkentin and O. S. Tee, *Chem. Commun.*, 190 (1966); *J. Am. Chem. Soc.*, **88,** 5540 (1966).

[246c] A. A. Bothner-By and C. Sun, *J. Org. Chem.*, **32,** 492 (1967).

[247] H. O. House, *Record Chem. Progr.*, **28,** 98 (1967).

[248] D. Caine, *J. Org. Chem.*, **29,** 1868 (1964).

[248a] G. Subrahmanyam, S. K. Malhotra, and H. Ringold, *J. Am. Chem. Soc.*, **88,** 1332 (1966).

[248b] H. O. House, W. L. Roelofs, and B. M. Trost, *J. Org. Chem.*, **31,** 646 (1966).

[248c] D. Caine and B. J. L. Huff, *Tetrahedron Letters*, 4695 (1966).

of equilibrated methyl *n*-alkyl ketone enolates favors attack on the more highly substituted enolate,[43, 247, 249–251] as does Michael addition.[14, 252–254] (With non-methyl unsymmetrical acyclic ketones, alkylation occurs preferentially on the less highly substituted enolate,[45, 250] in agreement with the findings for the composition of the enolate.[43, 45]) The enol contents of pure 2-butanone (0.012%) and other methyl alkyl ketones have been found to be greater than that of acetone (0.00015%).[255–257] (The enol contents of the pure liquids parallel the values in alcohol and water.[249, 256, 258, 259]) These considerations all suggest that for methyl *n*-alkyl ketones in all solvents the more highly substituted enolate is the more stable.[246a–c, 260, 261] [The greater rate of proton removal (in water) from isobutyraldehyde relative to acetaldehyde (corrected factor ca. 6.7),[20] combined with the predicted slower C-protonation rate for isobutyraldehyde enolate relative to that of acetaldehyde enolate,[36, 248a] suggests the stability order: $(CH_3)_2C{=}CHO^{\ominus} > CH_2{=}CHO^{\ominus}$.]

The composition of the equilibrium mixture **16** and **17** (1-ketol and 3-ketol) is determined by several factors which are difficult to assess quantitatively but may be expressed in terms of equilibrium constants (Eq. 13). [In weakly basic media an additional equilibrium exists between enolate ion and enol, which favors 3-condensation.[246a–c]] The more

$$\frac{[\text{1-ketol}]}{[\text{3-ketol}]} = \frac{K_{10a}K_{11a}}{K_{11b}K_{10b}} \qquad \text{(Eq. 13)}$$

highly substituted enolate anion **15** leads to the 3-condensation ketol with simple aldehydes and ketones (R, R′ = CH_3, *n*-alkyl); that is, the ratio [1-ketol]/[3-ketol] appears to be determined principally by the relative concentrations of the ketone enolate ions and enols. When R′ (in the aldehyde) and particularly R (in the ketone) both become large, branched, and bulky alkyl groups (*i*-butyl, heptyl, etc.), 1-condensation ketol is

[249] J. M. Conia, *Bull. Soc. Chim. France*, 1392 (1956).
[250] J. M. Conia, *Record Chem. Progr.*, **24,** 43 (1963).
[251] S. K. Malhotra and F. Johnson, *J. Am. Chem. Soc.*, **87,** 5513 (1965).
[252] L. B. Barkley and R. Levine, *J. Am. Chem. Soc.*, **72,** 3699 (1950).
[253] A. D. Campbell, C. L. Carter, and S. N. Slater, *J. Chem. Soc.*, 1741 (1948).
[254] G. R. Zellars and R. Levine, *J. Org. Chem.*, **13,** 911 (1948).
[255] A. Gero, *J. Org. Chem.*, **26,** 3156 (1961); **19,** 1960 (1954).
[256] N. L. Allinger, L. W. Chow, and R. W. Ford, *J. Org. Chem.*, **32,** 1994(1967).
[257] J. E. Dubois and G. Barbier, *Bull. Soc. Chim. France*, 682 (1965).
[258] C. K. Ingold, *Structure and Mechanism in Organic Chemistry*, pp. 554–566, Cornell University Press, Ithaca, New York, 1953.
[259] G. S. Hammond, in *Steric Effects in Organic Chemistry*, ed. by M. S. Newman, pp. 442–454, Wiley, New York, 1956.
[260] H. M. E. Cardwell and A. E. H. Kilner, *J. Chem. Soc.*, 2430 (1951).
[261] H. M. E. Cardwell, *J. Chem. Soc.*, 2442 (1951).

favored for two reasons: bulky groups facilitate cleavage of **17** to reactants,[30, 40, 262, 263] relative to **16**; and the less highly substituted enolate anion becomes more stable as R in the ketone becomes more branched and bulkier.[43, 45]

The composition of the product at the α,β-unsaturated ketone stage (**18** = 1-U and **19** = 3-U) introduces a third pair of equilibria, Eqs. 12a and 12b (actually, two steps are involved in the ketol → unsaturated ketone transformation; see p. 5). It is known that more vigorous reaction conditions favor unsaturated ketone and also often result in an increase in the total percentage of 1-condensation products (1-K and 1-U). It has been established in certain examples that retrogression of **17** to reactants may be favored over dehydration (e.g., **19** is not produced from **17** in basic medium when R′ = C_6H_5, R = CH_3, whereas under the same conditions **16** dehydrates to **18** when R′ = C_6H_5, R = H).[40] In another example, the sodium hydroxide-catalyzed condensation of benzaldehyde with phenylacetone favors 1-condensation.[264] The favored enolate, $C_6H_5CH{=}C(CH_3)O^{\ominus}$, would lead to **20**, but retrogression of **20** to reactants occurs faster than dehydration and **21** is not formed. Less favored ketol **22** dehydrates, effectively irreversibly, to yield the product

$$C_6H_5CHO + CH_3COCH_2C_6H_5 \rightleftharpoons \underset{\mathbf{20}}{C_6H_5CHOHCH(C_6H_5)COCH_3} \xrightarrow[\text{very slow}]{-H_2O} \underset{\mathbf{21}}{C_6H_5CH{=}C(C_6H_5)COCH_3}$$

$$C_6H_5CHO + CH_3COCH_2C_6H_5 \rightleftharpoons \underset{\mathbf{22}}{C_6H_5CHOHCH_2COCH_2C_6H_5} \xrightarrow[\text{fast}]{-H_2O} \underset{\mathbf{23}}{C_6H_5CH{=}CHCOCH_2COC_6H_5}$$

23. In contrast, the piperidine-catalyzed condensation of aldehydes with phenylacetone produces condensation on the methylene group.[265–267] It is known that α-substituents favor β,γ unsaturation in ketones at equilibrium,[142] again suggesting the α,β form **18** (1-U) to be the most stable. It appears likely that $K_{12a}/K_{12b} > 1$. The ratio of unsaturated ketone products is given by Eq. 14.

$$\frac{[1\text{-U}]}{[3\text{-U}]} = \frac{K_{10a}K_{11a}K_{12a}}{K_{10b}K_{11b}K_{12b}} \qquad \text{(Eq. 14)}$$

262 M. R. F. Ashworth and J. E. Dubois, *Bull. Soc. Chim. France*, 147 (1955).
263 C. S. Rondestvedt and M. E. Rowley, *J. Am. Chem. Soc.*, **78**, 3804 (1956).
264 G. Goldschmiedt and G. Knöpfer, *Monatsh.*, **18**, 437 (1897).
265 H. E. Zimmerman, L. Singer, and B. S. Thyagarajan, *J. Am. Chem. Soc.*, **81**, 108 (1959).
266 R. Dickinson, *J. Chem. Soc.*, 2234 (1926).
267 I. M. Heilbron and F. Irving, *J. Chem. Soc.*, 931, 936 (1929).

Table I summarizes the known condensations[141] (solvent water, ethanol, or aqueous ethanol in most cases) of aldehydes with methyl ketones to form ketol and α,β-unsaturated ketone products. Only formaldehyde

TABLE I. SUMMARY OF REPORTED KETOL AND α,β-UNSATURATED KETONE 1:1 PRODUCTS FORMED IN BASE-CATALYZED CONDENSATIONS OF ALDEHYDES WITH METHYL KETONES ($\overset{1}{C}H_3CO\overset{3}{C}HRR'$ and $\overset{1}{C}H_3CO\overset{3}{C}H_2R$)

Principal Product	Aldehyde	Ketone
3-Condensation	Formaldehyde	All
	Acetaldehyde	All except methyl isobutyl and neopentyl ketones
	n-Alkanals, $CH_3(CH_2)_nCHO$*,†	2-Butanone
1- and 3-Condensation mixture	Possible with all; ratio depends on reaction conditions	2-Butanone and 2-pentanone
	Acetaldehyde	Methyl isobutyl ketone
1-Condensation	All having one or no α hydrogen atoms ($ArCHO$, R_3CCHO, R_2CHCHO) except formaldehyde	All except 2-butanone and phenoxyacetone

* Examples studied: $n = 0–7$, with 2-butanone only.

† 3-Methylbutanal produces 3-condensation with 2-butanone.[141]

and acetaldehyde give appreciable amounts of 3-condensation with all ketones; the lower concentration of base required resulting in higher concentration of enol would favor 3-condensation. 2-Butanone is the only ketone reported to yield substantial amounts of 3-condensation with all aldehydes (an exception is phenoxyacetone[268]). Little is known about the products of condensation of n-alkanals above acetaldehyde with methyl n-alkyl ketones, other than 2-butanone. Although acetaldehyde produces ketols derived from 3-condensation with 2-butanone[269] and 2-pentanone[270] under mild conditions, more vigorous conditions lead to a mixture of α,β-unsaturated ketones formed by 1- and 3-condensation.[271] Formaldehyde and acetaldehyde produce 3-condensation ketols with methyl isopropyl ketone, but chloral, 2-ethylbutanal, and various benzaldehydes and furfurals yield the α,β-unsaturated ketones derived

[268] R. Stoermer and R. Wehln, *Ber.*, **35**, 3549 (1902).

[269] J. E. Dubois and R. Luft, *Compt. Rend.*, **238**, 485 (1954).

[270] J. Colonge, *Bull. Soc. Chim. France*, [4] **41**, 325 (1927).

[271] R. Heilmann, G. de Gaudemaris, and P. Arnaud, *Compt. Rend.*, **240**, 1542 (1955).

from 1-condensation;[272, 273] butanal produces a mixture of 1- and 3-condensation ketols (principally 1-condensation) and an α,β-unsaturated ketone (1-condensation) with methyl isopropyl ketone.[272] Aromatic aldehydes produce α,β-unsaturated ketones formed by 1-condensation. However, when the reaction is stopped at the ketol stage, some 3-condensation products may be isolated. For example, *p*-nitrobenzaldehyde and 2-butanone lead to some 3-condensation ketol under mild conditions.[40, 64] Isobutyraldehyde with 2-butanone produces a 55:45 ratio of 1- to 3-condensation unsaturated ketone with aqueous base, but with ethanolic sodium ethoxide the proportion is increased to 85:15;[231,247] with 2-heptanone, isobutyraldehyde produces 1-condensation ketol and a trace of 3-condensation isomer (ethanolic potassium hydroxide).[274] Except as outlined above, the majority of reported examples of base-catalyzed condensations of aldehydes with methyl ketones lead to 1-condensation product, due, in part, to the favored retrogression of 3-condensation ketol to reactants.

Formation of either α,α or α,α′ *bis*-condensation products seldom occurs in reactions of unsymmetrical methyl ketones, with the exception of formaldehyde condensations. 2-Butanone occasionally forms *bis*-condensation products with certain aromatic aldehydes especially in the presence of acid catalysts.[226]

The preceding discussion relates to aldol condensations conducted principally in protic solvents under equilibrating conditions. Use of enolates formed under kinetically controlled conditions would introduce versatility into aldol condensation syntheses, because the enolates derived from methyl ketones would be the less highly substituted isomers **(14)**.[43] Few successful aldol condensations have been reported which employ enolates derived from ketones under non-equilibrating conditions.[177, 213] No examples involving methyl ketones are known. However, what appear to be successful acylations,[8, 43–45] alkylations,[43–45] and carbonations[275] of such enolates have been described. (Enol esters formed by O-acylation are believed to be precursors of C-acylation products.[45]) Another apparently unexplored possibility is the use of preformed acyclic ketone enamines (which have the less substituted structure[52]) in an aldol condensation.[53a, 54]

α,β-Unsaturated ketones are readily obtained by heating ketols (having a hydrogen atom α to the carbonyl group) with an acid catalyst

[272] A. T. Nielsen and E. B. W. Ovist, *J. Am. Chem. Soc.*, **76**, 5156 (1954).

[273] G. B. Marcas, A. M. Municio, and S. Vega, *Anales Real Soc. Espan. Fis. Quim.* (*Madrid*), *Ser. B*, **60**, 639 (1964) [*C.A.*, **63**, 1726 (1965)].

[274] S. G. Powell and F. Hagemann, *J. Am. Chem. Soc.*, **66**, 372 (1944).

[275] R. Levine and C. R. Hauser, *J. Am. Chem. Soc.*, **66**, 1768 (1944).

such as iodine or oxalic acid;[149, 272, 274] yields are excellent except with ketols which dissociate readily on heating (those in which R′ is a large group).

$$\text{RCHOHCHR'COR''} \xrightarrow[\text{Heat}]{\text{H}^{\oplus}} \text{RCH=C(R')COR''} + \text{H}_2\text{O}$$

Condensation of Aldehydes with Alicyclic Ketones

Cyclopentanones and cyclohexanones condense with formaldehyde with basic catalysts to give mono-, unsymmetrical *bis*- and tetra-methylol compounds.[276–278] Acetaldehyde behaves similarly with cyclopentanone.[279] Paraformaldehyde in dimethyl sulfoxide containing boron

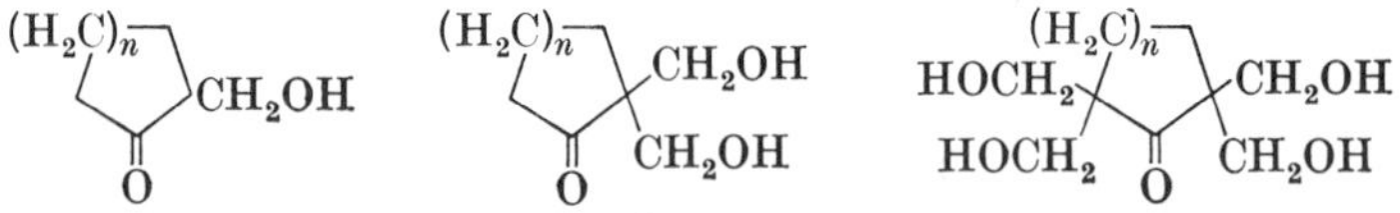

$(H_2C)_n$ CH$_2$OH O; $(H_2C)_n$ CH$_2$OH CH$_2$OH O; $(H_2C)_n$ HOCH$_2$ HOCH$_2$ CH$_2$OH CH$_2$OH O

$n = 1, 2$

trifluoride etherate undergoes α, or vinylogously α, condensation with steroid ketones.[280] α-Methylenecycloalkanones are conveniently prepared by pyrolysis of the Mannich base hydrochlorides.[281] *n*-Alkanals ($>C_2$) yield principally 1:1 condensation products with cycloalkanones, but the yields are low.[282–286] Mild conditions (low temperature,

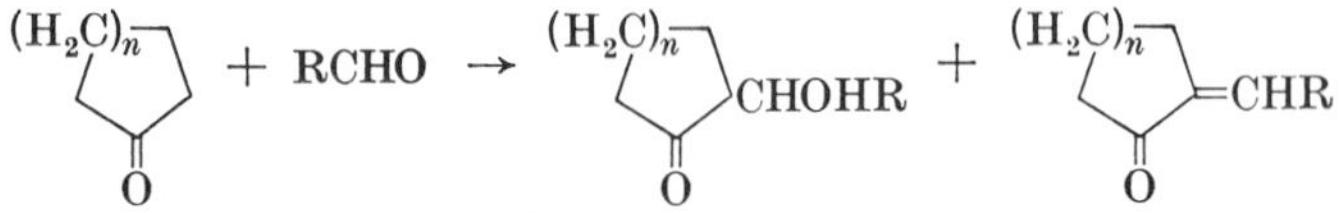

$(H_2C)_n$ (C=O) + RCHO → $(H_2C)_n$ (C=O) CHOHR + $(H_2C)_n$ (C=O) =CHR

R = alkyl, $n > 0$

aqueous ethereal sodium hydroxide) are required to minimize self-condensation of the aldehyde. Chloral[287] and 2-ethylbutanal,[288] which do not readily undergo self-condensation, may be subjected

[276] H. Gault and J. Skoda, *Bull. Soc. Chim. France*, [5] **13**, 308 (1946).

[277] H. Gault and E. Steckl, *Compt. Rend.*, **207**, 475 (1938).

[278] C. Mannich and W. Brose, *Ber.*, **56**, 833 (1923).

[279] J. Skoda, *Bull. Soc. Chim. France*, [5] **13**, 327 (1946).

[280] W. H. W. Lunn, *J. Org. Chem.*, **30**, 2925 (1965).

[281] M. Mühlstädt, L. Zach, and H. Becwar-Reinhardt, *J. Prakt. Chem.*, **29**, 158 (1965).

[282] C. E. Garland and E. E. Reid, *J. Am. Chem. Soc.*, **47**, 2333 (1925).

[283] P. Lambert, G. Durr, and G. Millet, *Compt. Rend.*, **238**, 251 (1954).

[284] I. V. Machinskaya and B. V. Tokarev, *Zh. Obshch. Khim.*, **22**, 1163 (1952) [*C.A.*, **47**, 6360 (1953)].

[285] G. Vavon and V. M. Mitchovitch, *Bull. Soc. Chim. France*, [4] **45**, 961 (1929).

[286] O. B. Edgar and D. H. Johnson, *J. Chem. Soc.*, 3925 (1958).

[287] F. Caujolle, P. Couturier, and C. Dulaurans, *Bull. Soc. Chim. France*, 19 (1950).

[288] H. A. Bruson, U.S. pat. 2,395,453 [*C.A.*, **40**, 3467 (1946)].

to more vigorous reaction conditions to yield 1:1 products. Relatively few symmetrical (α,α') *bis*-condensation products have been reported from condensations of aliphatic aldehydes with cycloalkanones.[279, 286, 289, 290] Enamines derived from cyclopentanone condense with alkanals in boiling benzene to give α-alkylidenecyclopentanones in good yield.[53a]

Many examples of condensations of aromatic aldehydes with alicyclic ketones are known. To obtain 1:1 products an excess of ketone is often, but not always, employed with mild conditions (dilute aqueous sodium hydroxide at room temperature or under reflux[291–297]). Sodium *t*-amyloxide in toluene at temperatures below 0° is an effective catalyst for producing 1:1 condensation products.[295] Ketols or α,β-unsaturated ketones may be obtained, relatively milder conditions favoring ketols.[291] *Ortho*-substituted aromatic aldehydes and aldehydes with electron-releasing groups such as alkyl and methoxyl more readily produce 1:1 products.[291, 293, 298–302] For example, *o*-tolualdehyde condenses with cyclohexanone in boiling aqueous potassium hydroxide to produce 71% of 1:1 α,β-unsaturated ketone.[301] Aqueous, rather than ethanolic, alkali favors 1:1 condensation.[291, 303] Acid catalysts favor formation of a conjugated endocyclic double bond with cyclohexanones *and* cyclopentanones, whereas basic catalysts favor formation of the normal arylidene derivative.[304] A number of reported 1:1 condensation products, obtained in the presence of acid catalysts and described as arylidene cycloalkanones, may, in fact, be the isomers having an endocyclic double bond.

289 A. Roedig and S. Schödel, *Chem. Ber.*, **91,** 320 (1958).

290 R. Mayer, *Chem. Ber.*, **88,** 1853 (1955).

291 J. D. Billimoria, *J. Chem. Soc.*, 1126 (1955).

292 O. Wallach, H. Mallison, and K. von Martius, *Nachr. K. Ges., Wiss. Göttingen*, 399 (1907) [*Chem. Zentr.*, **79,** I, 637 (1908)].

293 H. M. Walton, *J. Org. Chem.*, **22,** 1161 (1957).

294 D. Vorländer and K. Kunze, *Ber.*, **59,** 2078 (1926).

295 G. Vavon and J. M. Conia, *Compt. Rend.*, **234,** 526 (1952).

296 A. R. Poggi, A. Maccioni, and E. Marongiu, *Gazz. Chim. Ital.*, **84,** 528 (1954) [*C.A.*, **50,** 930 (1956)].

297 V. G. Kharchenko, *Uch. Zap. Gos. Saratovsk. Univ.*, **75,** 71 (1962) [*C.A.*, **60,** 485 (1964)].

298 J. Colonge and J. Sibeud, *Bull. Soc. Chim. France*, 786 (1952).

299 W. S. Emerson, G. H. Birum, and R. I. Longley, *J. Am. Chem. Soc.*, **75,** 1312 (1953).

300 R. Poggi and M. Gottlieb, *Gazz. Chim. Ital.*, **64,** 852 (1934) [*C.A.*, **29,** 2152 (1935)].

301 W. S. Rapson and R. G. Shuttleworth, *J. Chem. Soc.*, 636 (1940).

302 R. Baltzly, E. Lorz, P. B. Russell, and F. M. Smith, *J. Am. Chem. Soc.*, **77,** 624 (1955).

303 S. V. Tsukerman, L. A. Kutulya, and V. F. Lavrushin, *Zh. Obshch. Khim.*, **34,** 3597 (1964).

304 A. Hassner and T. C. Mead, *Tetrahedron*, **20,** 2201 (1964).

$$C_6H_5CHO + \text{2-methylcyclohexanone} \xrightarrow{NaOH} C_6H_5CH{=}\text{(2-methylcyclohexanone, 6-ylidene)}$$

$$C_6H_5CHO + \text{2-methylcyclopentanone} \xrightarrow{HCl} C_6H_5CH_2\text{-(2-methylcyclopent-2-enone)}$$

With 2 mole equivalents of aromatic aldehyde and ethanolic sodium hydroxide or ethoxide catalyst, good to excellent yields of *bis*-condensation products are uniformly obtained with cyclopentanone, cyclohexanone, and cycloheptanone.[303, 305] The reactivity of the ketone appears to decrease as ring size increases. Yields with cyclooctanone[306, 307] and higher cycloalkanones[308] are low. Substituted cycloalkanones (exclusive of those substituted in the 2- or 3-position) behave in the same fashion as unsubstituted ones. By starting with a monoarylidene cycloalkanone,

$$2\,ArCHO + (H_2C)_n\text{-cycloalkanone} \xrightarrow[C_2H_5OH]{NaOH} ArCH{=}\text{[(H}_2\text{C)}_n\text{-cycloalkanone]}{=}CHAr$$

Ar = aryl, $n > 0$

condensation with a different aldehyde can lead to unsymmetrical *bis*-arylidene cycloalkanones[303] (Table XIV).

$$ArCH{=}\text{[(H}_2\text{C)}_n\text{-cycloalkanone]} + Ar'CHO \rightarrow ArCH{=}\text{[(H}_2\text{C)}_n\text{-cycloalkanone]}{=}CHAr'$$

Ar, Ar' = aryl, $n > 0$

o-Phthalaldehyde condenses with medium- and large-ring cycloalkanones to yield tricyclic *bis*-ketols which may be dehydrated to cycloheptatrienones.[309]

[305] F. Mattu, *Rend. Seminario Fac. Sci. Univ. Cagliari*, **32**, 230 (1962) [*C.A.*, **63**, 17935 (1965)].

[306] F. Mattu and M. R. Manca-Mura, *Rend. Seminario Fac. Sci. Univ. Cagliari*, **34**, 286 (1964) [*C.A.*, **63**, 13126 (1965)].

[307] E. A. Braude, W. F. Forbes, B. F. Gofton, R. P. Houghton, and E. S. Waight, *J. Chem. Soc.*, 4711 (1957).

[308] E. A. Braude and B. F. Gofton, *J. Chem. Soc.*, 4720 (1957).

[309] E. Kloster-Jensen, N. Tarköy, A. Eschenmoser, and E. Heilbronner, *Helv. Chim. Acta*, **39**, 786 (1956).

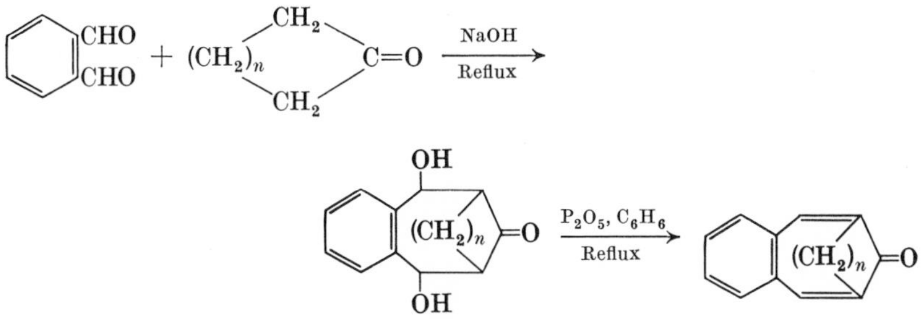

$n = 4\text{–}8, 12$

Nitromalonaldehyde condenses with cycloalkanones to produce *meta*-methylene bridged bicyclic *p*-nitrophenol derivatives;[310–312] when the methylene bridge is too small ($n < 6$), a non-planar cyclohexadienone tautomer is favored.[311] Although cyclobutanone has been condensed

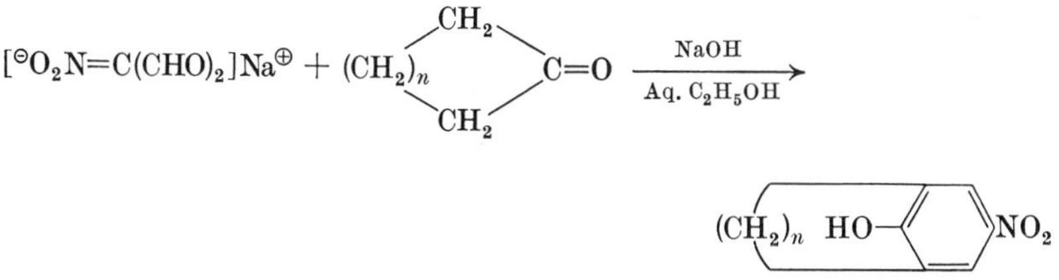

$$[{}^{\ominus}O_2N{=}C(CHO)_2]Na^{\oplus} + (CH_2)_n(CH_2)_2C{=}O \xrightarrow[\text{Aq. } C_2H_5OH]{\text{NaOH}}$$

$n = 6\text{–}18$

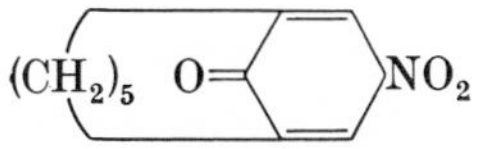

with acetone,[180] no reports of its condensation with aldehydes have yet appeared.

Camphor and other bicyclic ketones produce α,β-unsaturated ketones with basic catalysts. Many aldehydes have been condensed with camphor in the presence of sodium or sodium amide in ether.[313–315] Ethanolic hydrogen chloride-catalyzed condensation of benzaldehyde with nopinone or verbanone caused ring opening to produce a chlorinated *bis*-condensation product, whereas sodium ethoxide led to the normal 1:1 product.[315]

[310] V. Prelog and K. Wiesner, *Helv. Chim. Acta*, **30,** 1465 (1947).

[311] V. Prelog, K. Wiesner, W. Ingold, and O. Häflinger, *Helv. Chim. Acta*, **31,** 1325 (1948).

[312] V. Prelog, K. Wiesner, and O. Häflinger, *Collection Czech. Chem. Commun.*, **15,** 900 (1951).

[313] A. Haller, *Compt. Rend.*, **113,** 22 (1891).

[314] A. Haller and E. Bauer, *Compt. Rend.*, **148,** 1490 (1909).

[315] O. Wallach, *Ann.*, **437,** 187 (1924).

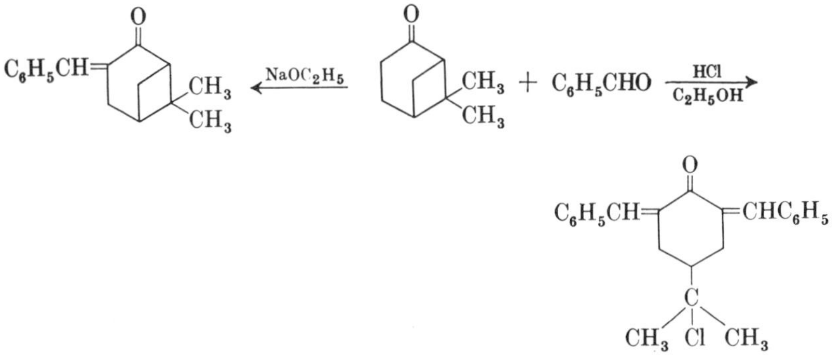

Conjugated cycloalkenones may undergo condensation in the γ-position with basic catalysts. Carvone and menthenone condense with 2 mole equivalents of benzaldehyde.[316–320] 3-Methyl-2-cyclohexen-1-one,[321] 3,5-dimethyl-2-cyclohexen-1-one,[321] isophorone,[321–322a] and piperitone[323] condense with benzaldehyde at the 3-methyl group only (ethanolic sodium ethoxide catalyst). Evidently the most stable, non-cross-conjugated, unsaturated ketone is the favored product of these condensations.

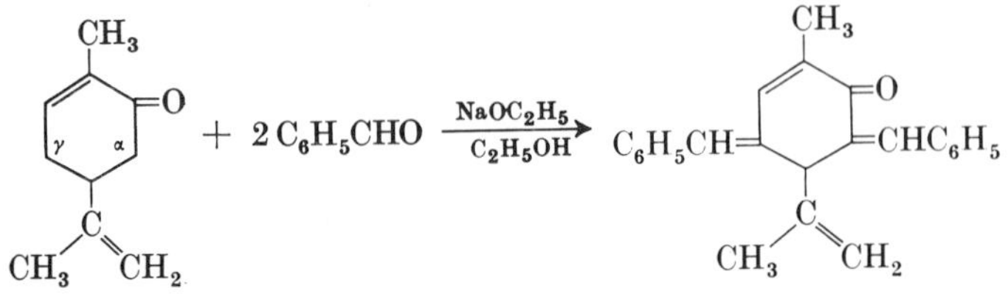

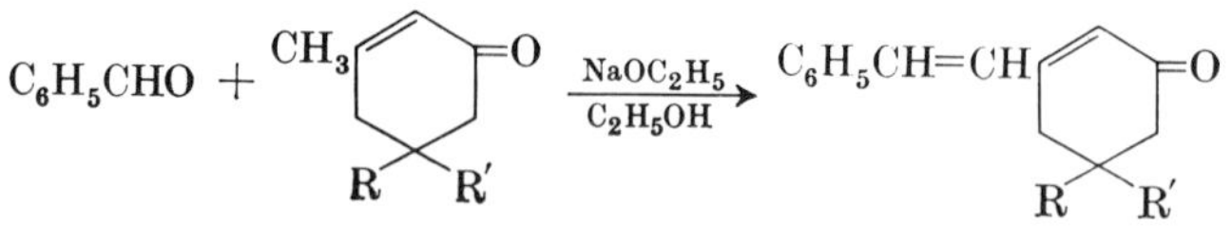

R, R′ = H or alkyl

Anhydroacetonebenzil (3,4-diphenyl-4-hydroxy-2-cyclopenten-1-one), which does not dehydrate to a stable cyclopentadienone (see page 25),

[316] R. E. Christ and R. C. Fuson, *J. Am. Chem. Soc.*, **59**, 893 (1937).

[317] A. Müller, *Ber.*, **54**, 1471 (1921).

[318] O. Wallach, *Ber.*, **29**, 1595 (1896).

[319] O. Wallach, *Ann.*, **305**, 261 (1899).

[320] O. Wallach, *Ann.*, **397**, 211 (1913).

[321] J. M. Conia and V. O'Leary, *Compt. Rend.*, **249**, 1002 (1959).

[322] G. R. Ensor and W. Wilson, *J. Chem. Soc.*, 4068 (1956).

[322a] G. Kabass, Tetrahedron, **22**, 1213 (1966).

[323] J. Dewar, D. R. Morrison, and J. Read, *J. Chem. Soc.*, 1598 (1936).

condenses with aromatic aldehydes in the 5-position[324] with ethanolic potassium hydroxide.

$$\text{4-hydroxy-3,4-diphenyl-2-cyclopentenone} + \text{ArCHO} \xrightarrow[C_2H_5OH]{KOH} \text{5-arylidene derivative}$$

1,2-Cycloalkanediones condense with 2 mole equivalents of aromatic aldehyde in the presence of piperidine to yield *bis*-β,β'-N-piperidyl derivatives, which form *bis*-α,β-unsaturated ketones when heated with methanolic acetic acid.[325–327]

$$(CH_2)_n(CH_2CO)_2 + 2\ \text{ArCHO} \xrightarrow[C_2H_5OH]{\text{Piperidine}} (CH_2)_n[CH(CHArNC_5H_{10})CO]_2 \xrightarrow[CH_3OH]{CH_3CO_2H} (CH_2)_n[C(=CHAr)CO]_2$$

$n = 2\text{–}12$

1,3-Cyclohexanedione[328] and derivatives, such as *dimedon* (5,5-dimethyl-1,3-cyclohexanedione), form 2,2′-*bis* derivatives useful for characterizing aldehydes.[329–333]

$$2\ \text{(1,3-cyclohexanedione)} + \text{RCHO} \xrightarrow{\text{Piperidine}} \text{(1,3-dioxocyclohex-2-yl)}_2\text{CH—R}$$

R = H, alkyl, or aryl

Few condensations of aldehydes with 1,4-cyclohexanedione have been reported;[327, 334] *o*-phthalaldehyde (2 mole equivalents) forms a pentacene derivative in 90% yield.

[324] F. R. Japp and A. Findlay, *J. Chem. Soc.*, **75**, 1017 (1899).
[325] N. J. Leonard, J. C. Little, and A. J. Kresge, *J. Am. Chem. Soc.*, **79**, 6436 (1957).
[326] N. J. Leonard and G. C. Robinson, *J. Am. Chem. Soc.*, **75**, 2714 (1953).
[327] F. Mattu and M. R. Manca, *Chimica* (*Milan*), **33**, 284 (1957) [*C.A.*, **52**, 1934 (1958)].
[328] E. C. Horning and M. G. Horning, *J. Org. Chem.*, **11**, 95 (1946).
[329] F. E. King and D. G. I. Felton, *J. Chem. Soc.*, 1371 (1948).
[330] H. Stetter and U. Milbers, *Chem. Ber.*, **91**, 374 (1958).
[331] V. Kvita and J. Weichet, *Chem. Listy*. **51**, 380 (1957) [*C.A.*, **51**, 9502 (1957)].
[332] D. Vorländer and F. Kalkow, *Ann.*, **309**, 356 (1900).
[333] D. Vorländer and O. Strauss, *Ann.*, **309**, 375 (1900).
[334] F. Mattu, *Chimica* (*Milan*), **38**, 3 (1962) [*C.A.*, **61**, 16025 (1964)].

$$2\ C_6H_4(CHO)_2 + \text{cyclohexane-1,4-dione} \xrightarrow[C_2H_5OH]{KOH} \text{pentacene-6,13-quinone}$$

Non-aromatic heterocyclic ketones (e.g., tetrahydro-γ-pyrone, 4-piperidone) generally appear to behave like their carbocyclic analogs in condensation reactions (Table XIX).[334a]

Condensation of Aldehydes with Alkyl Aryl Ketones

Relatively few aliphatic aldehydes other than formaldehyde have been condensed with alkyl aryl ketones.[177] Yields in the reactions of methyl aryl ketones with aliphatic aldehydes are uniformly poor. No general procedure has been developed for securing good yields in these reactions; the use of enamines or non-equilibrating enolates has not been explored (see p. 37). The Wittig reaction[15] might be applicable.

The condensation of formaldehyde with acetophenone has been studied in some detail;[335, 336] the best yield of the ketol, 3-hydroxy-1-phenyl-1-propanone, was 25% obtained with aqueous sulfuric acid catalyst.[335]

$$C_6H_5COCH_3 + HCHO \xrightarrow[90°,\ 7\ hr.]{H_2SO_4,\ H_2O} C_6H_5COCH_2CH_2OH$$

Basic ion-exchange resin Amberlite IRA 400 at 40–45° gave a 40% yield of ketol from formaldehyde and propiophenone,[337] whereas potassium carbonate gave only 16–19%.[336]

Methylanilinomagnesium bromide, although unsuitable for aromatic methyl ketone condensations (principally because of ketone and aldehyde self-condensation) is a catalyst of general utility for condensing aliphatic and aromatic aldehydes with higher homologs of acetophenone to yield ketols.[177] Ketols have been obtained in the following condensations (yields in parentheses): isobutyraldehyde with propiophenone (83%), hexanal with butyrophenone (40%), and acetaldehyde with valerophenone (50%).[177]

$$\underset{R \neq H}{C_6H_5COCH_2R} + R'CHO \xrightarrow[(C_2H_5)_2O,\ C_6H_6]{C_6H_5N(CH_3)MgBr} C_6H_5COCH(R)CHOHR'$$

[334a] N. Barbulescu and C. Greff, *Rev. Chim.* (*Bucharest*), **17,** 202 (1966) [*C.A.*, **65,** 8869 (1966).

[335] M. G. J. Beets and L. G. Heeringa, *Rec. Trav. Chim.*, **74,** 1085 (1955).

[336] R. C. Fuson, W. E. Ross, and C. H. McKeever, *J. Am. Chem. Soc.*, **60,** 2935 (1938).

[337] S. Yamada, I. Chibata, and H. Matsumae, *Ann. Rept. Gohei Tanabe Co., Ltd.*, **1,** 20 (1956) [*C.A.*, **51,** 6546 (1957)].

In sharp contrast to the paucity of aliphatic aldehyde condensations, a very large number of condensations of substituted benzaldehydes with acetophenone and substituted acetophenones (Table XVII) to yield chalcones are known.[338–341] Many heterocyclic aldehydes have been condensed with acetophenones. The initial studies of Claisen (1881–1889) were extended by v. Kostanecki (1896–1900). Ethanolic sodium hydroxide or sodium ethoxide, at room temperature or below, generally leads to chalcones in good to excellent yields, with few side reactions. Only

$$\mathrm{ArCHO} + \mathrm{CH_3COAr'} \xrightleftharpoons{\mathrm{NaOH,\ C_2H_5OH}} \mathrm{ArCHOHCH_2COAr'}$$

$$\mathrm{ArCHOHCH_2COAr'} \longrightarrow \mathrm{ArCH{=}CHCOAr'} + \mathrm{H_2O}$$

occasionally are ketols obtained. Acid catalysts are seldom employed and, except with certain phenols and phenol esters, appear to offer few advantages. Hydrogen chloride is useful for the condensation of aromatic aldehydes with *n*-alkyl aryl ketones to produce β-chloroketones which, on heating or treatment with base, yield α-alkyl chalcones.[342–345]

$$\mathrm{ArCHO} + \mathrm{RCH_2COAr'} \xrightarrow[\mathrm{R\ =\ alkyl}]{\substack{\mathrm{HCl,\ then}\\ \mathrm{heat\ or\ base}}} \mathrm{ArCH{=}C(R)COAr'} + \mathrm{H_2O}$$

Substituents in the aldehyde or ketone *generally* appear to affect equilibrium yields of chalcone in the same manner for acid- *or* base-catalyzed reactions; electron-withdrawing substituents provide higher yields than electron-releasing ones. Rather poor yields of chalcone are sometimes realized (acid *or* base catalysis) with electron-releasing groups such as alkoxy, amino, and hydroxy.[31, 338, 346–351] Side reactions may ensue with hydroxy compounds. Primary amino groups form Schiff bases, but this difficulty may be circumvented by employing the acetamido

338 M. Vandewalle, *Ind. Chim. Belge*, **26**, 345 (1961); a review of chalcone preparations.

339 F. M. Dean and V. Podimuang, *J. Chem. Soc.*, 3978 (1965).

340 P. Mahanty, S. P. Panda, B. K. Sabata, and M. K. Rout, *Indian J. Chem.*, **3**, 121 (1965).

341 Y. S. Agasimundin, S. D. Jolad, and S. Rajagopal, *Indian J. Chem.*, **3**, 220 (1965).

342 E. P. Kohler, *Am. Chem. J.*, **31**, 642 (1904).

343 L. I. Smith and L. I. Hanson, *J. Am. Chem. Soc.*, **57**, 1326 (1935).

344 R. D. Abell, *J. Chem. Soc.*, **79**, 928 (1901).

345 R. D. Abell, *J. Chem. Soc.*, 2834 (1953).

346 V. Balaiah, L. R. Row, and T. R. Seshadri, *Proc. Indian Acad. Sci.*, **20A**, 274 (1944) [*C.A.*, **39**, 4609 (1945)].

347 J. S. Buck and I. M. Heilbron, *J. Chem. Soc.*, **121**, 1095 (1922).

348 T. A. Geismann and R. O. Clinton, *J. Am. Chem. Soc.*, **68**, 697 (1946).

349 R. E. Lyle and L. P. Paradis, *J. Am. Chem. Soc.*, **77**, 6667 (1955).

350 W. Davey and J. R. Gwilt *J. Chem. Soc.*, 1008 (1957).

351 D. S. Noyce and W. A. Pryor, *J. Am. Chem. Soc.*, **81**, 618 (1959).

derivatives which can subsequently be hydrolyzed. For hydroxy-substituted benzaldehydes the best base-catalyzed procedure appears to be the use of aqueous ethanolic sodium hydroxide at 25–60° for a few hours.[348, 352, 353] The benzoates of hydroxy compounds with hydrogen chloride in ethyl acetate at 0° gave chalcones in good yields.[354–356]

There is a difference in the effect of substituents in benzaldehydes on rates for the condensation step (ketol formation) of acid- and base-catalyzed reactions. For the acid-catalyzed reactions the rate is *mildly* accelerated by electron-releasing substituents.[351, 357, 358] For base-catalyzed reactions the rate is *strongly* accelerated by electron-withdrawing substituents.[31, 349, 356–360] However, yields are determined by the various equilibria involved. The rate of cleavage of the intermediate ketol, relative to the rate of its formation and dehydration to chalcone, is a matter of importance; electron-withdrawing substituents appear to favor dehydration of ketol to chalcone, relative to retrogression in basic or acidic media.[349, 351] Thus the effect of substituents on equilibrium yields of chalcone is roughly the same in all media. Acid catalysts favor irreversible dehydration of ketols to chalcones, and no ketols of this type have been isolated from acid media. Most of the limited number of chalcone precursor ketols that have been isolated have been obtained in alkaline medium, under mild conditions, with reactants having electron-withdrawing groups (nitro, halogen). Several ketols have been prepared from 2-nitrobenzaldehyde and various acetophenones, employing aqueous ethanolic sodium phosphate.[361] Additional quantitative work is needed in this area.

$$o\text{-}O_2NC_6H_4CHO + CH_3COAr \xrightarrow[\text{Aq. } C_2H_5OH]{Na_3PO_4} o\text{-}O_2NC_6H_4CHOHCH_2COAr$$

A variety of aromatic ketones, including 1-indanones, 1-tetralones,[362] aryl acetyl[363] compounds, and benzyl aryl ketones, has been condensed with various aldehydes; yields are generally good and special conditions are not required. Heterocyclic aromatic ketones behave like

[352] T. A. Ellison, *J. Chem. Soc.*, 1720 (1927).
[353] W. Davey and D. J. Tivey, *J. Chem. Soc.*, 1230 (1958).
[354] A. Russell and S. F. Clark, *J. Am. Chem. Soc.*, **61**, 2651 (1939).
[355] A. Russell and J. Todd, *J. Chem. Soc.*, 421 (1937).
[356] R. P. Barnes and J. L. Snead, *J. Am. Chem. Soc.*, **67**, 138 (1945).
[357] T. Széll, A. M. Eastham, and G. Sipos, *Can. J. Chem.*, **42**, 2417 (1964); **43**, 2134 (1965).
[358] S. M. McElvain and R. E. McMahon, *J. Am. Chem. Soc.*, **71**, 901 (1949).
[359] M. Scholtz and L. Huber, *Ber.*, **37**, 390 (1904).
[360] E. A. Walker and J. R. Young, *J. Chem. Soc.*, 2041 (1957).
[361] I. Tanasescu and A. Baciu, *Bull. Soc. Chim. France*, [5] **4**, 1673 (1937).
[362] N. P. Buu-Hoi and G. Saint-Ruf, *Bull. Soc. Chim. France*, 424 (1965).
[363] J. Sam, D. W. Alwan, and K. Aparajithan, *J. Heterocyclic Chem.*, **2**, 366 (1965).

other aromatic ketones,[364–367] except certain nitrogen-containing examples which give best yields with a piperidine catalyst in boiling ethanol.[368] Condensations with most heterocyclic aldehydes appear to proceed normally.[363, 366, 369–372]

Intramolecular Condensations

It is convenient to discuss the intramolecular aldol condensations of dialdehydes, diketones, and ketoaldehydes in one section. The intramolecular reaction is an important tool for synthesis of alicyclic ketones and aldehydes. The reactants may be, and often are, generated *in situ* from such intermediates as alcohols, glycols, acetals, enamines, Mannich bases and quaternary salts thereof, and vinyl chlorides. In addition, the Michael condensation is an important route to 1,5-diketones and δ-ketoaldehydes, which need not be isolated before cyclization (Robinson annelation reaction). As stated in the introduction to this volume, aldol condensations which proceed from, to, or through carbonyl precursors or derivatives have, with certain exceptions, arbitrarily been excluded from the tabular summary of this review. This restriction limits the scope of the following discussion, because carbonyl precursors are often employed in intramolecular aldol condensations and the literature coverage of these particular reactions is incomplete.

Dialdehydes. The intramolecular condensation of alkane α,ω-dialdehydes (chain length $> C_5$) generally leads to alicyclic α,β-unsaturated aldehydes under mild conditions (acid or basic catalysts). Products having rings with five,[373–377] six,[377,378] seven,[379] fifteen,[380] and seventeen[380] members have been prepared by this method. More vigorous

[364] B. Bobarevic, M. Dezelic, and V. Milovic, *Glasnik Hemicara Tehnol. Bosne Hercegovine*, **12,** 111 (1963) [*C.A.*, **63,** 18006 (1965)].

[365] M. Vaysse and P. Pastour, *Compt. Rend.*, **259,** 2865 (1964).

[366] S. V. Tsukerman, V. D. Orlov, V. F. Lavrushin, and Y. K. Yur'ev, *Zh. Org. Khim.*, **1,** 650 (1965).

[367] Mitsubishi Chemical Industries Co., Ltd., Japan. pat. 23,907 (1965) [*C.A.*, **64,** 3499 (1966)].

[368] P. I. Ittyerah and F. G. Mann, *J. Chem. Soc.*, 467 (1958).

[369] J. Koo, *J. Pharm. Sci.*, **53,** 1329 (1964).

[370] A. C. Annigeri and S. Siddappa, *Monatsh. Chem.*, **96,** 625 (1965).

[371] A. C. Annigeri and S. Siddappa, *Indian J. Chem.*, **2,** 413 (1964).

[372] B. S. Tanaseichuk and I. Y. Postovskii, *Zh. Org. Khim.*, **1,** 1279 (1965) [*C.A.*, **63,** 13207 (1965)].

[373] E. B. Reid and J. F. Yost, *J. Am. Chem. Soc.*, **72,** 5232 (1950).

[374] J. B. Brown, H. B. Henbest, and E. R. H. Jones, *J. Chem. Soc.*, 3634 (1950).

[375] Y. Bon and O. Yanemitsu, *Tetrahedron*, **20,** 2877 (1965).

[376] T. P. Kutney, A. By, T. Inaba, and S. Y. Leong, *Tetrahedron Letters*, 2911 (1965).

[377] O. A. Moe, D. T. Warner, and M. I. Buckley, *J. Am. Chem. Soc.*, **73,** 1062 (1951).

[378] J. v. Braun and E. Danziger, *Ber.*, **46,** 103 (1913).

[379] R. Weitzenböck, *Monatsh. Chem.*, **34,** 215 (1913).

[380] M. Stoll and A. Rouvé, *Helv. Chim. Acta*, **20,** 525 (1937).

basic conditions favor formation of lactones or hydroxy acids (see p. 60). The preparation of larger rings (from the corresponding acetals) has been

$$OCH(CH_2)_3CH(C_3H_7\text{-}n)CHO \xrightarrow[115°]{H_2O,\ \text{no catalyst}} \text{cyclopentene-CHO, } C_3H_7\text{-}n$$

$$(C_2H_5O_2C)_2C(CH_2CH_2CHO)_2 \xrightarrow[5\text{–}8°]{NaOC_2H_5,\ C_2H_5OH} \text{cyclohexene } (CO_2C_2H_5)_2\text{, CHO}$$

achieved with acid catalysts; high-dilution conditions were employed to prepare macrocyclic compounds.[380]

$$\text{biphenyl-2,2'-di}(CH_2CH(OCH_3)_2) \xrightarrow[\text{Heat}]{\text{Aq. } H_2SO_4} \text{dibenzocycloheptene-CHO}$$

$$(CH_2)_{12}\begin{cases}CH_2CH(OC_2H_5)_2\\ CH(OC_2H_5)_2\end{cases} \xrightarrow[C_6H_6,\ \text{reflux, high dilution}]{C_6H_5SO_3H} (CH_2)_{12}\begin{cases}CCHO\\ \| \\ CH\end{cases}$$

A cyclobutane aldol is obtained as a major ozonolysis product of 2-hydroxy-6-propyl-1,3,5-triethyl-3-cyclohexene-1-carboxaldehyde; it is derived from the intermediate, 2,4-diethyl-3-propyl-1,5-pentanedial.[381] The buttressing effect of adjacent alkyl groups, as well as poor solvation of the intermediate enolate (or enol), may facilitate this ring closure to a four-membered ring, which is unique.

$$n\text{-}C_3H_7CH\text{—}CH(C_2H_5)CHO,\ C_2H_5CH\text{—}CHO \xrightarrow{H_2O} n\text{-}C_3H_7CH\text{—}C(C_2H_5)(CHO),\ C_2H_5CH\text{—}CHOH$$

Piperidinium acetate catalyzes cyclization of the dialdehyde **24** to the tetracyclic aldehydes **25** and **26**;[382] the mechanism, involving an

[381] A. T. Nielsen, *J. Am. Chem. Soc.*, **79**, 2518 (1957).

[382] R. B. Woodward, F. Sondheimer, D. Taub, K. Heusler, and W. H. McLamore, *J. Am. Chem. Soc.*, **74**, 4223 (1952).

enamine intermediate, has been discussed.[52] The cyclization may also be effected by heating the dialdehyde in water at 145°, whereby the relative amount of **26** in the mixture is increased.[382]

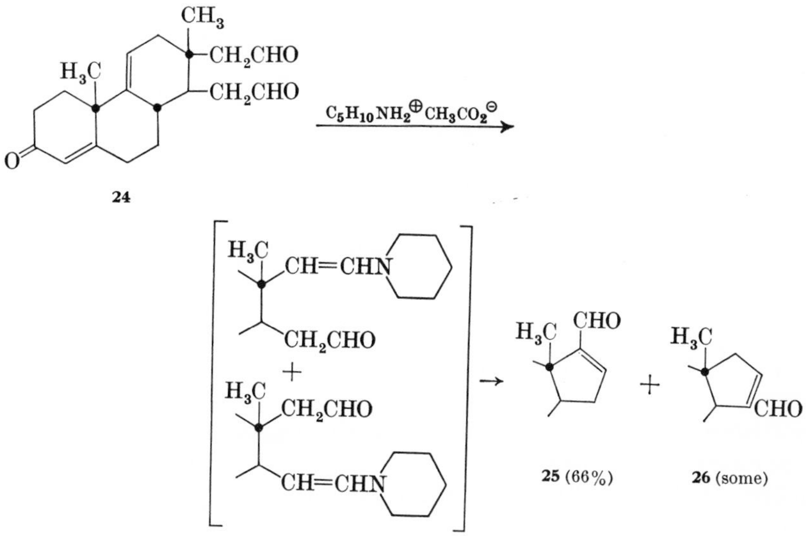
CH3
CH2CHO
CH2CHO
H3C
O
C5H10NH2⊕CH3CO2⊖
H3C
CH=CHN
CH2CHO
+
H3C
CH2CHO
CH=CHN
H3C
CHO
+
H3C
CHO

24 **25** (66%) **26** (some)

Self-condensation of 2-ethyl-2-hexenal leads, by Michael addition, to an intermediate dialdehyde (**27**, not isolated) which is cyclized to a cyclohexene aldol, **28**, existing principally (99%) in the configuration shown.[66, 381, 383]

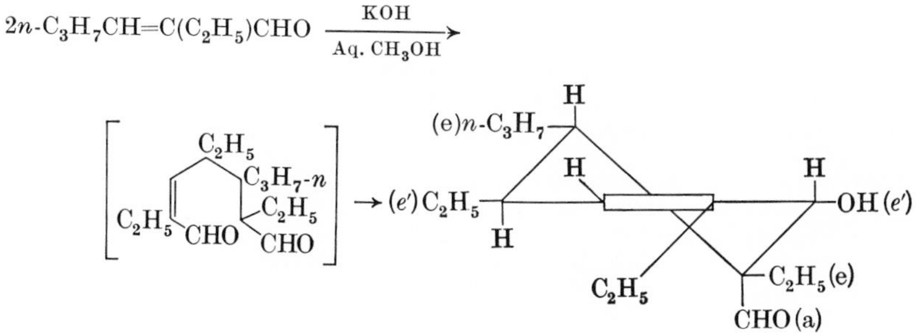
2n-C3H7CH=C(C2H5)CHO
KOH
Aq. CH3OH
C2H5
C3H7-n
C2H5
C2H5
CHO
CHO
(e)n-C3H7
H
H
H
(e′)C2H5
H
OH(e′)
C2H5
C2H5(e)
CHO(a)

27 **28**

Diketones. The intramolecular self-condensation of α,ω-diketones is a frequently employed, useful, and important route to alicyclic β-ketols

[383] A. T. Nielsen, *J. Org. Chem.*, **28**, 2115 (1963).

and α,β-unsaturated ketones. Six-membered rings are easily formed.[384–391] 1,5-Diketones yield cyclohexenones; 1,7-diketones yield acylcyclohexenes.

$$CH_3CO(CH_2)_5COCH_3 \xrightarrow{H_2SO_4} \text{1-acetyl-2-methylcyclohexene } (COCH_3,\ CH_3)$$

$$C_6H_5CH_2COCH(C_6H_5)CH(C_6H_5)CH_2COC_6H_5 \xrightarrow{NaOCH_3} \text{tetraphenylcyclohexenone } (C_6H_5,\ C_6H_5,\ C_6H_5,\ C_6H_5,\ O)$$

$$CH_3CH_2CO(CH_2)_3COCH_2CH_3 \xrightarrow{HCl} \text{cyclohexenone } (CH_2CH_3,\ CH_3,\ O)$$

Although common basic and acidic catalysts (sodium ethoxide, hydrogen chloride) are quite effective in many intramolecular aldol condensations, secondary amine catalysts (pyrrolidine and piperidine) often give outstanding results[392] under mild conditions and also facilitate isolation of β-ketols.[48, 68, 76, 77] The stereochemistry of the ring juncture in decalin derivatives is evidently governed by the size of the ring substituent in the 2-position of the reactant cyclohexanone;[76, 77] sodium ethoxide at low temperature (−10°)[67] leads to the same stereochemistry in the products as does pyrrolidine[76, 77] (see p. 13).

Five-membered[393–397b] and seven-membered[398, 399] rings have been

[384] E. E. Blaise and A. Koehler, *Bull. Soc. Chim. France*, [4] **7**, 655 (1910).

[385] E. E. Blaise, *Compt. Rend.*, **173**, 313 (1921).

[386] P. Yates and J. E. Hyre, *J. Org. Chem.*, **27**, 4101 (1962).

[387] E. Buchta and S. Buchholz, *Ann.*, **688**, 40 (1965).

[388] H. Smith, Brit. pat. 975,594 [*C.A.*, **62**, 9068 (1965)].

[389] G. Nomine, R. Bucourt, J. Tessier, A. Pierdet, G. Costerousse, and J. Mathieu, *Compt. Rend.*, **260**, 4545 (1965).

[390] E. Buchta and H. Kroeger, *Naturwiss.*, **52**, 496 (1965).

[391] E. Buchta and P. Vincke, *Chem. Ber.*, **98**, 208 (1965).

[392] S. Ramachandran and M. S. Newman, *Org. Syntheses*, **41**, 38 (1961).

[393] G. Stork and R. Borch, *J. Am. Chem. Soc.*, **86**, 935, 936 (1964).

[394] R. F. Heck, *J. Org. Chem.*, **30**, 2205 (1965).

[395] J. Kossanyi, *Bull. Soc. Chim. France*, 722 (1965).

[396] P. Doyle, I. R. Maclean, R. D. H. Murray, W. Parker, and R. A. Raphael, *J. Chem. Soc.*, 1344 (1965).

[397] C. P. Chiusoli and G. Bottaccio, *Chim. Ind. (Milan)*, **47**, 165 (1965) [*C.A.*, **63**, 13067 (1965)].

[397a] G. Büchi and H. Wüest, *J. Org. Chem.*, **31**, 977 (1966).

[397b] R. Breslow, W. Vitale, and K. Wendel, *Tetrahedron Letters*, 365 (1965).

[398] A. G. Anderson and J. A. Nelson, *J. Am. Chem. Soc.*, **73**, 232 (1951).

[399] G. Muller and A. Poittevin, Fr. pat. 1,369,321 [*C.A.*, **62**, 7838 (1965)].

made by this intramolecular condensation but no ring closures to cyclobutane derivatives have been reported. 1,4-Diketones yield cyclopentenones; 1,6-diketones yield acylcyclopentenes or cycloheptenones.

$CH_3COCH_2CH_2COCH_2CH_2CH{=}CHCH_2CH_3 \xrightarrow{NaOH}$ 2-methyl-3-($CH_2CH{=}CHCH_2CH_3$)cyclopent-2-enone

Jasmone

$RCO(CH_2)_4COR \xrightarrow[\text{Reflux}]{KOH,\ C_2H_5OH}$ 1-R-2-COR-cyclopentene (70–84%) + 3-R-4-COR-cyclopentene (2–12%)

R = alkyl, aryl

Cyclodecane-1,6-dione $\xrightarrow{Na_2CO_3}$ bicyclic cycloheptenone

A tricyclic β-ketol intermediate incorporating a four-membered ring was found to undergo an irreversible retroaldol condensation.[400]

Tricyclic ketone ($OCOCH_3$) $\xrightarrow{\text{Aq.NaOH}}$ alkoxide intermediate $\xrightarrow{H^{\oplus}}$ bicyclic diketone

Large rings have been prepared by employing high-dilution techniques.[401]

$$CH_3CO(CH_2)_{12}COCH_3 \xrightarrow[(C_2H_5)_2O,\ 35^\circ]{C_6H_5N(CH_3)MgBr} (CH_2)_{12}\text{-bridged } C(CH_3){=}CH{-}C{=}O$$

Numerous condensed ring compounds have been prepared from diketones (Table VI); an interesting example is one involving intramolecular attack at the γ position of an α,β-unsaturated ketone.[402, 403]

[400] H. Hikino and P. de Mayo, *J. Am. Chem. Soc.*, **86**, 3582 (1964).

[401] M. Stoll and A. Rouvé, *Helv. Chim. Acta*, **30**, 2019 (1947).

[402] N. N. Gaidamovich and I. V. Torgov, *Izv. Acad. Nauk SSSR*, 1311 (1964) [*C.A.*, **61,** 8203 (1964)].

[403] C. Sannié, J. J. Panouse, and C. Neuville, *Bull. Soc. Chim. France*, 1435 (1956).

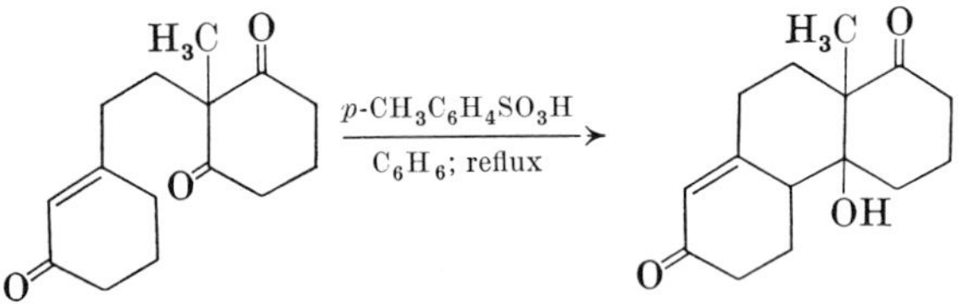

$\xrightarrow[C_6H_6;\ reflux]{p\text{-}CH_3C_6H_4SO_3H}$

Ketols with a bridgehead hydroxyl group have been prepared.[403a, b]

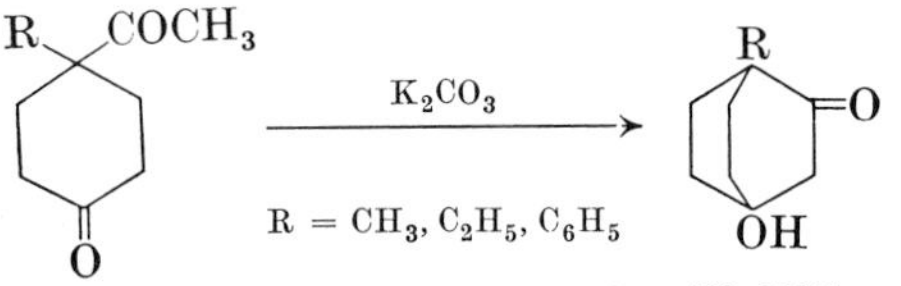

$\xrightarrow{K_2CO_3}$

$R = CH_3, C_2H_5, C_6H_5$

(50–80%)

Certain readily available carboalkoxy diketones are very useful substitutes for the corresponding diketones;[404] the carboalkoxy group may be lost during the condensation.[405]

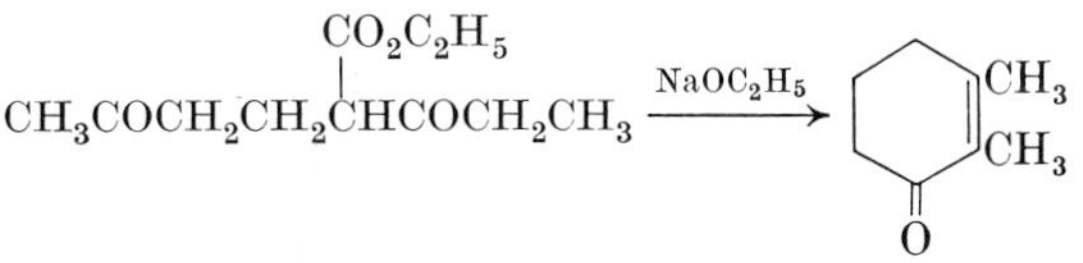

$CH_3COCH_2CH_2CH(CO_2C_2H_5)COCH_2CH_3 \xrightarrow{NaOC_2H_5}$

1,5-Diketones, available by piperidine-catalyzed Knoevenagel condensation from a β-keto ester and an aldehyde, may be cyclized to yield, ultimately, α,β-unsaturated ketones.[406]

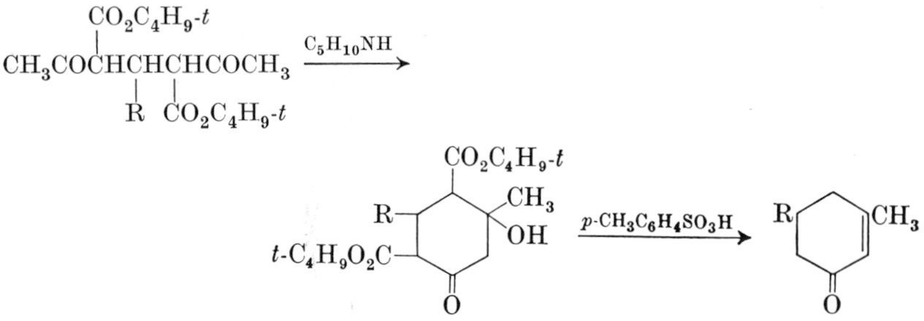

$CH_3COCH(CO_2C_4H_9\text{-}t)CH(R)CH(CO_2C_4H_9\text{-}t)COCH_3 \xrightarrow{C_5H_{10}NH}$

$\xrightarrow{p\text{-}CH_3C_6H_4SO_3H}$

$R = H$, alkyl, aryl

In some reactions, ring closure may be directed by the choice of catalyst.[407] As in the dialdehyde example **24** (p. 49), kinetically controlled aldol

[403a] J. Colonge, P. Francois, and R. Vuillemey, *Bull. Soc. Chim. France*, 1028 (1966).
[403b] J. Colonge and R. Vuillemet, *Bull. Soc. Chim. France*, 2235 (1961).
[404] D. J. Goldsmith and J. A. Hartman, *J. Org. Chem.*, **29**, 3524 (1964).
[405] E. Buchta and G. Satzinger, *Chem. Ber.*, **92**. 468 (1959).
[406] G. Näslund, A. Senning, and S. O. Lawesson, *Acta Chem. Scand.*, **16**, 1329 (1962).
[407] H. Plieninger and T. Suchiro, *Chem. Ber.*, **89**, 2789 (1956).

condensation is favored by piperidine which evidently forms an enamine by reaction with the more reactive, less hindered carbonyl group. Other catalysts appear to favor the thermodynamically more stable condensation product.

$$\text{C}_2\text{H}_5\text{O}_2\text{C}(\text{CH}_3)\text{-ring}(\text{CH}_3)(=\text{O}) \xleftarrow{\text{H}_3\text{PO}_4} \text{CH}_3\text{CO}\overset{\text{CO}_2\text{C}_2\text{H}_5}{\underset{\text{CH}_3}{\text{C}}}\text{CH}_2\text{CH}_2\text{COCH}_3 \xrightarrow[\text{CH}_3\text{CO}_2\text{H}]{\text{C}_5\text{H}_{10}\text{NH}}$$

CH_3 $CO_2C_2H_5$ CH_3 O

1,5-Diketones obtained by Michael additions to α,β-unsaturated ketones may be cyclized to cyclohexenones, a reaction often referred to as the Robinson annelation reaction.[48, 76, 408–409a] For example propiophenone adds to 3-methyl-3-buten-2-one, ultimately forming 4,6-dimethyl-3-phenyl-2-cyclohexen-1-one.[410] Many examples of this type of base-catalyzed reaction (Michael addition followed by aldol condensation leading to an alicyclic ketone) are known.[51, 67, 71, 392, 408, 411–425a]

$$\text{C}_6\text{H}_5\text{COCH}_2\text{CH}_3 + \text{CH}_2{=}\underset{\text{CH}_3}{\text{C}}\text{—COCH}_3 \xrightarrow[\text{2. Aldol}]{\text{1. Michael}}$$

CH_3 C_6H_5 CH_3 O

[408] W. S. Rapson and R. Robinson, *J. Chem. Soc.*, 1285 (1935).

[409] D. J. Baisted and J. S. Whitehurst, *J. Chem. Soc.*, 2340 (1965).

[409a] R. Selvarajan, J. P. John, K. V. Naragan, and S. Swaminathan, *Tetrahedron*, **22**, 949 (1966).

[410] R. Chapurlat and J. Dreux, *Bull. Soc. Chim. France*, 349 (1962).

[411] R. Dickinson, I. M. Heilbron, and F. Irving, *J. Chem. Soc.*, 1888 (1927).

[412] E. R. H. Jones and H. P. Koch, *J. Chem. Soc.*, 393 (1942).

[413] D. Iwanov and T. Iwanow, *Ber.*, **77**, 173 (1944).

[414] H. Meerwein, *Ber.*, **77**, 227 (1944).

[415] J. Colonge and J. Dreux, *Compt. Rend.*, **231**, 1504 (1950).

[416] J. Colonge and J. Dreux, *Bull. Soc. Chim. France*, [5] **19**, 47 (1952).

[417] W. A. Ayer and W. I. Taylor, *J. Chem. Soc.*, 2227 (1955).

[418] E. A. Braude, B. F. Gofton, G. Lowe, and E. S. Waight, *J. Chem. Soc.*, 4054 (1956).

[419] G. Büchi, J. H. Hansen, D. Knutson, and E. Koller, *J. Am. Chem. Soc.*, **80**, 5517 (1958).

[420] E. D. Bergmann and P. Bracha, *J. Org. Chem.*, **26**, 4685 (1961).

[421] W. Bacon, S. Brewis, G. E. Usher, and E. S. Waight, *J. Chem. Soc.*, 2255 (1961).

[422] J. Wiemann and Y. Dubois, *Bull. Soc. Chim. France*, 1813 (1962).

[423] J. Colonge, J. Dreux, and R. Chapurlat, *Compt. Rend.*, **251**, 252 (1960).

[424] N. C. Ross and R. Levine, *J. Org. Chem.*, **29**, 2341 (1964).

[425] J. Wiemann, N. Ronzani, and J. J. Godfroid, *Compt. Rend.*, **256**, 4677 (1963).

[425a] J. J. Beereboom, *J. Org. Chem.*, **31**, 2026 (1966).

The structure of the product may depend on substituents and ring size in certain condensations; in compound **29** acid-catalyzed condensation occurs on the ring methylene group if the ring is large enough; otherwise condensation occurs on the methyl group of the 3-oxobutyl side chain.[409, 426]

CO_2CH_3 / $CCH_2CH_2COCH_3$ / $(CH_2)_n$ CO / CH_2 (**29**) $\xrightarrow[\text{Aq.}CH_3CO_2H]{HCl}$ O= ... CH_2 ... $(CH_2)_n$, $n = 3–5$; or O= ... CH_3 ... $(CH_2)_n$, $n = 5–7, 10, 12$

Evidently this situation holds only when the group in the 2-position is or becomes hydrogen (replacement of CO_2CH_3 by H occurs during the cyclization of **29**), for when an alkyl group is initially present in the 2-position (as in **30**) condensation occurs on the ring methylene group (sulfuric acid catalyst) to yield **31**;[427] in **31** the double bond is not at the bridgehead, however.

32 (CH_3, O=) $\xleftarrow[25°]{KOH, C_2H_5OH}$ **30** (O, CH_3, $CH_2CH_2COCH_3$) $\xrightarrow[25°]{H_2SO_4}$ **31** (CH_3, O=, CH_3)

The structure of the product may depend on reaction conditions; by employing a basic catalyst, condensation of **30** occurs on the methyl group of the 3-oxobutyl side chain to yield an octalone **(32)**.[427] Maintaining reaction conditions of either kinetic or thermodynamic control may determine product composition. It has been established that in cyclization of the diketone **33** the bicyclo[3.3.1]nonane structure **34** (formed by condensation of the more rapidly formed cyclohexanone enolate with the less reactive side chain carbonyl group[190]) is kinetically favored; prolonged reaction (equilibrium conditions) leads to the thermodynamically more stable decalone **35**.[409] Acid-catalyzed cyclization of **33** leads to the octalone **36**,[409] rather than to a bicyclo[3.3.1]non-2-en-9-one as with **30**. (There is no angular 2-alkyl group in **33**.)

Modifications of the Robinson annelation reaction are frequently employed in practice, but are not included in the tabular summary of this

[426] V. Prelog, L. Ruzicka, P. Barman, and L. Frenkiel, *Helv. Chim. Acta*, **31**, 92 (1948).
[427] J. A. Marshall and D. J. Schaeffer, *J. Org. Chem.*, **30**, 3642 (1965).

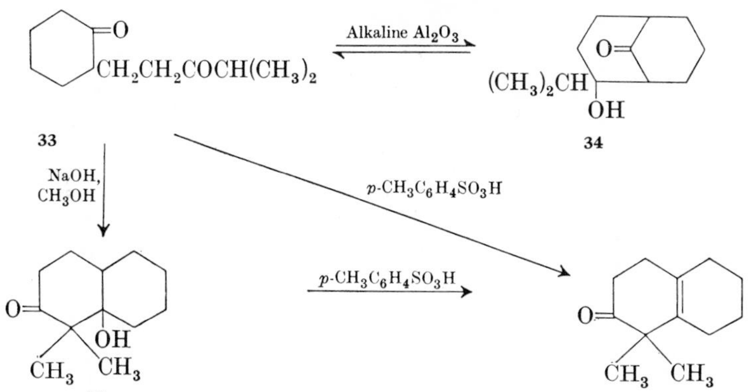

Alkaline Al_2O_3

$CH_2CH_2COCH(CH_3)_2$

$(CH_3)_2CH$ OH

33 34

NaOH, CH_3OH

p-$CH_3C_6H_4SO_3H$

p-$CH_3C_6H_4SO_3H$

OH CH_3 CH_3 CH_3 CH_3

35 36

review. Three of them are mentioned now. (1) The vinyl ketone acceptor may be replaced by a Mannich base[409] or its methiodide,[428–430] thus allowing more vigorous reaction conditions which would polymerize

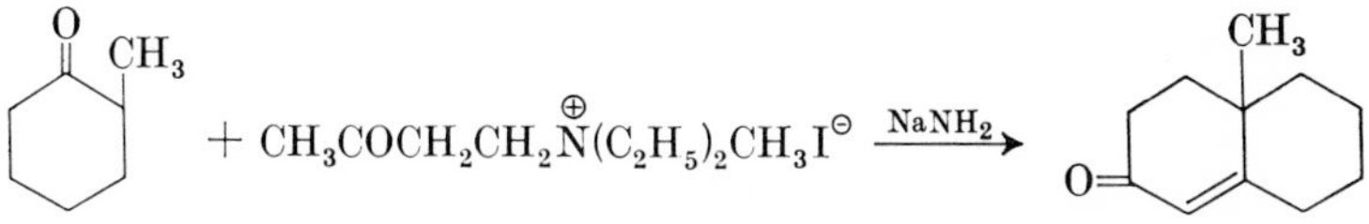

$$+ \; CH_3COCH_2CH_2\overset{\oplus}{N}(C_2H_5)_2CH_3I^{\ominus} \xrightarrow{NaNH_2}$$

32

the vinyl ketone. (2) The Wichterle reaction[427, 431] employs 1,3-dichloro-*cis*-2-butene as a methyl vinyl ketone surrogate to alkylate a ketone (sodium amide catalyst); the 2-(3-chloro-2-butenyl) ketone produced is treated with sulfuric acid to yield a diketone (sometimes isolable[427]) which is then cyclized. (3) Another very useful modification of the

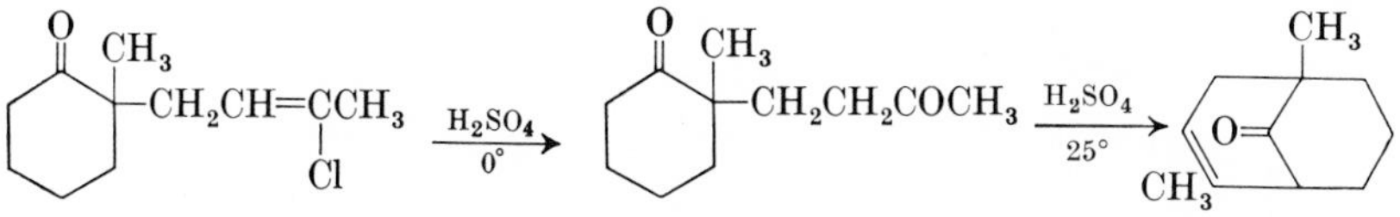

$-CH_2CH{=}C(Cl)CH_3 \xrightarrow[0^\circ]{H_2SO_4} -CH_2CH_2COCH_3 \xrightarrow[25^\circ]{H_2SO_4}$

30 31

reaction is conversion of the addend ketone into an enamine.[52, 53, 432] This allows alkylation at the least substituted position with pyrrolidine-derived enamines[433, 433a] (e.g., the 6-position of 2-alkylcyclohexanones) and also minimizes polymerization of the vinyl ketone.[52]

[428] J. W. Cornforth and R. Robinson, *J. Chem. Soc.*, 1855 (1949).

[429] J. Szmuszkovicz and H. Born, *J. Am. Chem. Soc.*, **75,** 3350 (1953).

[430] E. C. du Feu, F. J. McQuillin, and R. Robinson, *J. Chem. Soc.*, 53 (1937).

[431] O. Wichterle, J. Prochazka and J. Hofmann, *Collection Czech. Chem. Commun.* **13,** 300 (1948).

[432] R. L. Augustine and J. A. Caputo, *Org. Syntheses*, **45,** 80 (1965).

[433] S. K. Malhotra and F. Johnson, *Tetrahedron Letters*, 4027 (1965).

[433a] W. D. Gurowitz and M. A. Joseph, *Tetrahedron Letters*, 4433 (1965).

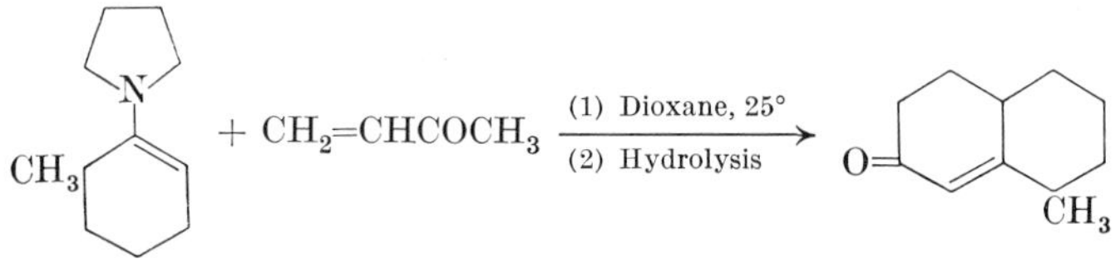

$+ CH_2{=}CHCOCH_3 \xrightarrow[(2)\ Hydrolysis]{(1)\ Dioxane,\ 25°}$

Ketoaldehydes. Ketoaldehydes (sometimes generated *in situ*) have been condensed intramolecularly to form five- or six-membered rings (Table XX).[75b] One example of formation of a seven-membered ring has been reported.[70] The reaction has been employed in steroid syntheses.[434]

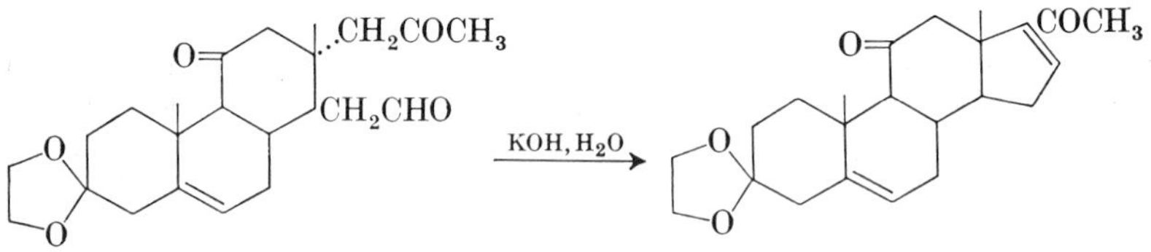

$\xrightarrow{KOH, H_2O}$

It is interesting that condensation of the ketonic carbonyl group occurs on the α carbon atom of the aldehyde in some of these examples producing a cyclopentene-1-carboxaldehyde or the related aldol.[435–443] Cyclization of the ketoaldehyde **37** catalyzed by ethanolic sodium hydroxide led to the five-membered ring aldehyde **38** rather than a seven-membered ring ketone;[443] the result illustrates the dominant

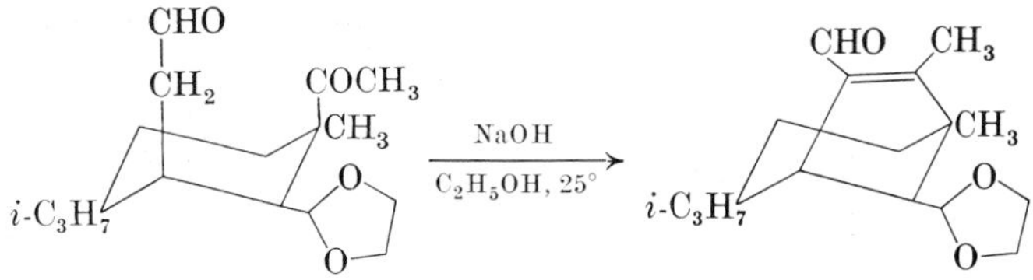

$\xrightarrow[C_2H_5OH,\ 25°]{NaOH}$

37 **38**

[434] G. I. Poos, W. F. Johns. and L. H. Sarett, *J. Am. Chem. Soc.*, **77,** 1026 (1955).

[435] B. Berkoz, E. Denot, and A. Bowers, *Steroids* **1,** 251 (1963) [*C.A.*, **59,** 7588 (1963)].

[436] R. Lalande, J. Moulines, and J. Duboudin, *Compt. Rend.*, **254,** 1087 (1962).

[437] H. E. Eschinazi, U.S. pat. 2.946,823 [*C.A.*, **55,** 1480 (1961)].

[438] N. L. Wendler and H. L. Slates, *J. Am. Chem. Soc.*, **80,** 3937 (1958).

[439] E. J. Corey and S. Nozoe, *J. Am. Chem. Soc.*, **85,** 3527 (1963).

[440] G. H. Whitman and J. A. F. Wickramasinghe, *J. Chem. Soc.*, 1655 (1964).

[441] K. Tanabe and Y. Morisawa, *Chem. Pharm. Bull.* (*Tokyo*), **11,** 536 (1963) [*C.A.*, **59,** 7600 (1963)].

[442] K. Tanabe, R. Hayashi, and R. Takasaki, *Chem. Pharm. Bull.* (*Tokyo*), **9,** 1 (1961) *C.A.*, **60,** 9331 (1964)].

[443] E. J. Corey and S. Nozoe, *J. Am. Chem. Soc.*, **87,** 5728 (1965).

influence of strain and ring size rather than the relative reactivities of aldehyde and ketone carbonyl groups on the structure of the product.[190]

The composition of the product has been shown to depend on the catalyst with 6-oxo-3-isopropylheptanal, piperidinium acetate leading solely to 2-methyl-5-isopropyl-1-cyclopentene-1-carboxaldehyde[436] and aqueous potassium hydroxide or an acid catalyst to 1-acetyl-4-isopropyl-1-cyclopentene as the major product;[436, 437] 6-oxo-3-isopropenylheptanal behaves

$$\text{1-acetyl-4-}(C_3H_7\text{-}i)\text{-cyclopentene} \xleftarrow[\text{or } H^{\oplus}]{KOH, H_2O,} CH_3CO\text{-}CH_2CH_2CH(C_3H_7\text{-}i)CH_2CHO \xrightarrow{\text{Piperidinium acetate}} \text{2-}CH_3\text{-1-}CHO\text{-5-}(C_3H_7\text{-}i)\text{-cyclopentene}$$

similarly.[444–446] This result may be explained by attack of ketone carbonyl on the more rapidly formed aldehyde enamine (kinetic control) with piperidine catalyst and the more reactive aldehyde carbonyl attacking the ketone enolate (thermodynamic control) with hydroxide-ion catalysis.

Michael addition of ketones to α,β-unsaturated aldehydes leads to δ-ketoaldehydes which can be cyclized to cyclohexenones.[447, 448]

$$CH_3COCHRR' + R''CH{=}CHCHO \xrightarrow[CH_3OH]{KOCH_3} CH_3COCRR'{-}CHR''CH_2CHO \rightarrow \text{cyclohexenone (R, R', R'', =O)}$$

Bicyclic bridged ketols (which cannot readily dehydrate) have been prepared from acetals with acid catalysts;[449, 450] more vigorous conditions lead to β,γ-unsaturated ketones.[451]

$$\text{4-}COCH_3\text{-piperidine-N-}CH_2CH(OC_2H_5)_2 \xrightarrow{\text{Aq. HCl}} \text{bicyclic ketol (}COCH_3\text{, OH, N)}$$

444 J. Wolinsky and W. Barker, *J. Am. Chem. Soc.*, **82**, 636 (1960).
445 J. Wolinsky, M. R. Slabaugh, and T. Gibson, *J. Org. Chem.*, **29**, 3740 (1964).
446 J. M. Conia and C. Faget, *Bull. Soc. Chim. France*, 1963 (1964).
447 J. Colonge, J. Dreux, and M. Thiers, *Compt. Rend.*, **243**, 1425 (1956).
448 J. Colonge, J. Dreux, and M. Thiers, *Compt. Rend.*, **244**, 89 (1957).
449 R. W. Guthrie, A. Philipp, Z. Valenta, and K. Wiesner, *Tetrahedron Letters*, 2945 (1965).
450 A. T. Nielsen, *J. Org. Chem.*, **31**, 1053 (1966).
451 E. W. Colvin and W. Parker, *J. Chem. Soc.*, 5764 (1965).

It would be desirable to have rules predicting the course of intramolecular aldol condensations. Potentially two or more routes may be available and may lead to different ring systems or to different arrangements of substituents. It appears that the factors which determine the equilibrium course of intermolecular condensations apply also to intramolecular condensations, with the important restriction that ring size and ring strain in the product are quite crucial. In general, the order of favored ring size formation is: $6 > 5 > 7 \gg 4$. Substitution of endocyclic double bonds by alkyl, aryl, or acyl groups will stabilize the product.

SIDE REACTIONS

Because of the great reactivity of aldehydes and ketones and their derived condensation products, aldol condensation side reactions are ubiquitous. In the following summary some of these reactions (acetal formation, hydride transfer, and additions to α,β-unsaturated ketones) are briefly discussed. Most discussion of mechanism has been omitted.

It is an interesting fact, perhaps not generally recognized, that cyclic acetal formation can occur in basic as well as acidic media when certain reaction conditions and structural requirements combine. The relatively unstable aldehyde trimers, 4-hydroxy-1,3-dioxanes (aldoxans, **39**), are produced from aliphatic aldehydes by mild basic catalysts. For example, propanal with aqueous potassium carbonate at 6–8° leads to **39** (R′ = H, R = CH_3; yield, 94%).[452] The aldoxans readily dissociate into the

3 RR′CHCHO $\xrightarrow[6\text{-}8^\circ]{K_2CO_3}$ [4-hydroxy-1,3-dioxane ring: R, R′, OH at C-4; RR′CH at C-6; CHRR′ at C-2] $\xrightarrow{\text{Heat}}$ RR′CHCHOHC(RR′)CHO + RR′CHCHO

39

R = alkyl or aryl; R′ = alkyl or H

parent aldol and aldehyde (1 mole of each) when strongly heated; for this reason, yields of aldols rarely exceed 66.7%. The reaction has been reported for acetaldehyde, butanal, isobutyraldehyde, and phenylacetaldehyde.[452–458] The reaction is also catalyzed by acids.[459] The

[452] E. Späth, R. Lorenz, and E. Freund, *Ber.*, **76**, 1196 (1943).

[453] F. J. Villani and F. F. Nord, *J. Am. Chem. Soc.*, **68**, 1674 (1946).

[454] F. Urech, *Ber.*, **12**, 191 (1879); **13**, 483 (1880).

[455] J. L. E. Erickson and G. N. Grammer, *J. Am. Chem. Soc.*, **80**, 5466 (1958).

[456] E. R. Alexander and E. N. Marvell, *J. Am. Chem. Soc.*, **72**, 1396 (1950).

[457] R. H. Saunders, M. J. Murray, F. F. Cleveland, and V. T. Komarewsky, *J. Am. Chem. Soc.*, **65**, 1309. 1714 (1943).

[458] R. H. Saunders and M. J. Murray, *J. Am. Chem. Soc.*, **66**, 206 (1944).

[459] J. C. Bevington, *Quart. Rev.* (*London*), **6**, 141 (1952).

bis-1,3-dioxane **40** was obtained in 30% yield from cyclohexanone and formaldehyde with sulfuric acid in acetic acid.[460] The formation of acetals from aldols has been reviewed.[461]

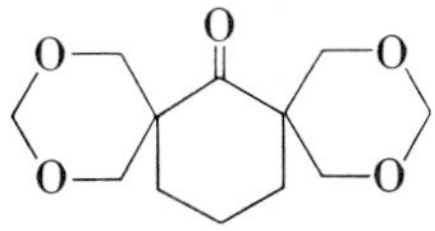

40

Aldols dimerize on standing, often rapidly; acetaldol forms paraldol.[461, 462] The stereochemistry of paraldol has been examined.[462a] Aldol dimerization can be reversed by heat. Distillation of aldols immediately before use is recommended; traces of accumulated acidic impurities

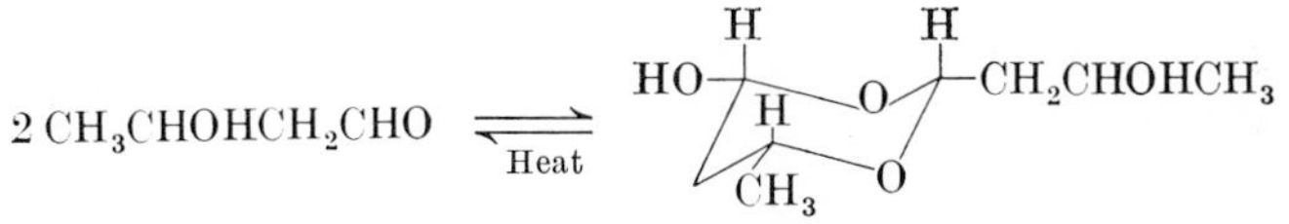

$$2\,CH_3CHOHCH_2CHO \underset{\text{Heat}}{\rightleftharpoons} \text{Paraldol}$$

Paraldol

should be removed prior to this operation to avoid dehydration of the aldol.

Hydride transfer reactions such as the Cannizzaro reaction are encountered with a few reactive aldehydes, notably formaldehyde, under basic conditions.[123, 123a] However, aldol condensation reaction conditions are usually sufficiently mild to obviate extensive hydride transfer as a competing reaction. A molar excess of formaldehyde with basic catalysts, such as aqueous calcium hydroxide, at 25–100°, will produce diols or triols from aliphatic aldehydes;[121, 463–468] acetaldehyde yields pentaerythritol.[469–470b] The reduction of aldols of higher molecular weight is less

$$RCH_2CHO + 3\,HCHO + H_2O \xrightarrow{OH^{\ominus}} RC(CH_2OH)_3 + HCO_2H$$

460 S. Olsen, *Acta Chem. Scand.*, **7**, 1364 (1953).

461 L. N. Owen, *Ann. Rept. Progr. Chem.* (*Chem. Soc. London*), **41**, 139–148 (1945).

462 E. Späth and H. Schmid, *Ber.*, **74**, 859 (1941).

462a M. Vogel and D. Rhum, *J. Org. Chem.*, **31**, 1775 (1966).

463 J. A. Wyler, U.S. pat. 2,468,718 [*C.A.*, **43**, 7505 (1949)].

464 G. Dupont, R. Doulou, and A. Duplessis-Kergomard, *Bull. Soc. Chim. France*, [5] **17**, 314 (1949).

465 O. C. Dermer and P. W. Solomon, *J. Am. Chem. Soc.*, **76**, 1697 (1954).

466 C. M. van Marle and B. Tollens, *Ber.*, **36**, 1341, 1347 (1903).

467 W. Hensinger, Fr. pat. 1,103,113 [*C.A.*, **52**, 19953 (1958)].

468 A. Just, *Monatsh. Chem.*, **17**, 76 (1896).

469 S. F. Marrian, *Chem. Rev.*, **43**, 149 (1948).

470 H. B. J. Schurink, *Org. Syntheses*, *Coll. Vol.*, **1**, 425 (1941).

470a H. C. Da Silva and L. M. Chaves, *Tecnica* (*Lisbon*), **40**, 649 (1965) [*C.A.*, **64**, 17406 (1966)].

470b H. Guenther and H. Mirbach, Ger. pat. 1,220,842 [*C.A.*, **65**, 12108 (1966)].

efficient; nonanal[121] and 2-methylheptanal[471] undergo normal aldol condensation with formaldehyde under conditions whereby lower aldol homologs are reduced. Reduction of ketols by excess formaldehyde can also occur. Although cyclohexanone condenses with 4 mole equivalents of formaldehyde (aqueous calcium hydroxide catalyst) to produce the tetramethylol derivative (40% yield), use of 5.5 mole equivalents of formaldehyde leads to the pentahydroxy compound **41** (73–85%).[472] Some

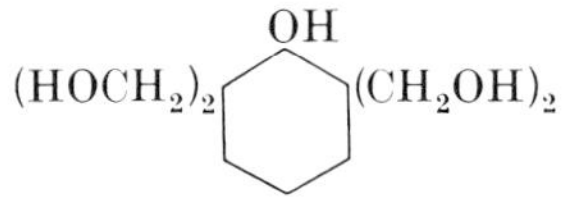

$$\text{OH; } (HOCH_2)_2 \text{ and } (CH_2OH)_2 \text{ on cyclohexane ring}$$

41

examples of reductions of aldols or ketols involving aldehydes other than formaldehyde have been reported (e.g., benzaldehyde, 2-methylpropanal, furfural, 2-methylbutanal, acetaldehyde, and 2-nitrobenzaldehyde).[473–476]

Although alkane α,ω-dicarboxaldehydes produce cycloalkene carboxaldehydes by intramolecular aldol condensation under acidic or mild basic conditions,[373, 374, 377–381] somewhat more vigorous basic conditions employing alcohol solvents (e.g., warm methanolic potassium methoxide) lead to lactones.[477, 478] Warm aqueous alkali leads to hydroxy acids.[477] Similarly, ω-ketoaldehydes may form lactones, hydroxy acids,[477] cycloalkenones, or cycloalkene carboxaldehydes,[434–442] depending on reaction

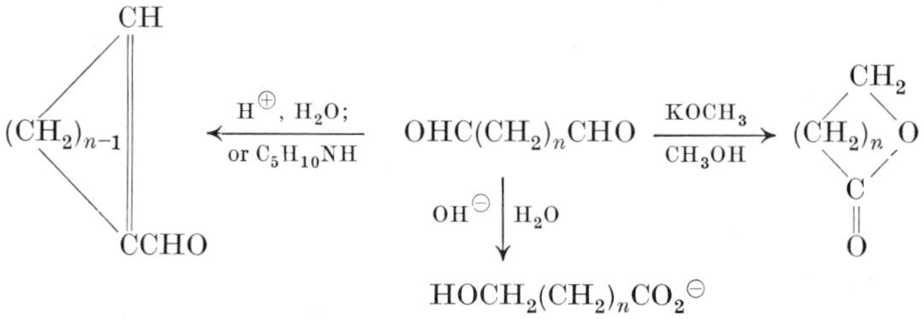

$$(CH_2)_{n-1}\text{[ring: CH}=\text{CCHO]} \xleftarrow[\text{or } C_5H_{10}NH]{H^{\oplus},\ H_2O;} OHC(CH_2)_nCHO \xrightarrow[CH_3OH]{KOCH_3} (CH_2)_n\text{[ring: CH}_2\text{–O–C(=O)]}$$

$$OHC(CH_2)_nCHO \xrightarrow{OH^{\ominus},\ H_2O} HOCH_2(CH_2)_nCO_2^{\ominus}$$

conditions. High concentrations of hydroxide or alkoxide ion favor conversion of the aldehyde carbonyl function into hydroxy or alkoxy alkoxide Cannizzaro intermediates,[123] which undergo facile cyclization

[471] E. Fourneau, G. Benoit, and R. Firmenich, *Bull. Soc. Chim. France*, [4] **47**, 858 (1930).

[472] H. Wittcoff, *Org. Syntheses, Coll. Vol.*, **4**, 907 (1963).

[473] R. Cornubert and H. Le Bihan, *Bull. Soc. Chim. France*, [4] **41**, 1077 (1927).

[474] G. Lindauer, *Monatsh. Chem.*, **21**, 72 (1900).

[475] A. Franke and R. Stern, *Monatsh. Chem.*, **49**, 21 (1928).

[476] A. Baeyer and V. Drewsen, *Ber.*, **15**, 2856 (1882).

[477] A. Meerwein, *Ber.*, **53**, 1829 (1920).

[478] R. H. Hall, *J. Chem. Soc.*, 4303 (1954).

to lactones (ring opening occurs in aqueous base). Enamine intermediates and acid catalysts lead to aldol condensation products.

The condensation of aliphatic aldehydes to esters (Tischenko condensation) is most readily effected by aluminum alkoxides at room temperature.[104, 479] Alkoxides such as $Mg[Al(OC_2H_5)_4]_2$ and calcium ethoxide cause formation of glycol esters.[104, 453]

$$2\ RCH_2CHO \xrightarrow{Al(OC_3H_7\text{-}i)_3} RCH_2CO_2CH_2CH_2R$$

$$3\ RCH_2CHO \xrightarrow{Ca(OC_2H_5)_2} RCH_2CO_2CH_2CH(R)CHOHCH_2R$$

When aliphatic aldehydes are subjected to vigorous, base-catalyzed reaction conditions (relatively high temperatures, high catalyst concentration), numerous reactions (including Cannizzaro reaction) other than aldol condensation may ensue and some of these have been studied. Resin-like polymers result from very reactive aldehydes such as acetaldehyde and certain α,β-unsaturated aldehydes.[457, 480–482] Butanal undergoes self-condensation with concentrated aqueous alcoholic potassium or sodium hydroxide (25–110°) to produce several products including butanol, 2-ethyl-2-hexen-1-ol, butyric acid, 2-ethyl-2-hexenoic acid, the aldol **42,** the glycol **43,** and the lactones **44** and **45,** some of these **(42–44)** in yields up to 40–50% depending on reaction conditions.[66, 381,383, 483–485] Products **43–45** are derived from **42** and butanal or 2-ethyl-2-hexenal by

C_2H_5, C_3H_7-n, C_2H_5, C_2H_5, OH, CHO

42

C_2H_5, C_3H_7-n, C_2H_5, C_2H_5, OH, CH_2OH

43

C_2H_5, n-C_3H_7, $CH(C_2H_5)CH_2OH$, C_2H_5, O, O

44

C_2H_5, n-C_3H_7, C_2H_5, O, O

45

[479] I. Lin and A. R. Day, *J. Am. Chem. Soc.*, **74,** 5133 (1952).

[480] A. Michael and A. Kopp, *Am. Chem. J.*, **5,** 182 (1883).

[481] J. Furukawa and T. Saegusa, *Polymerization of Aldehydes and Oxides,* pp. 43–123, Interscience, New York, 1963.

[482] E. E. Degering and T. Stoudt, *J. Polymer Sci.*, **7,** 653 (1951).

[483] M. Häusermann, *Helv. Chim. Acta,* **34,** 1482 (1951).

[484] P. Y. Blanc, A. Perret, and F. Teppa, *Helv. Chim. Acta,* **47,** 567, 725 (1964).

[485] R. H. Hall and K. H. W. Tuerck, Brit. pat. 608,985 [*C.A.*, **44,** 4493 (1950)].

reactions which include hydride transfer.[381, 383] Compounds of undetermined structure, composed of four aldehyde units, have been reported to form with strong base from other aliphatic aldehydes including propanal,[383, 485–487] pentanal,[488–491] 2-hexenal,[492] hexanal,[492] and heptanal.[483, 487–496] It is likely that some of these products are homologs of **42–45**.

The self-condensation of ketones can lead to various side reactions.[497–506] Isophorone is produced from phorone (from acetone) with basic catalysts by an intramolecular Michael addition;[502] homoisophorones are formed similarly from 2-butanone.[503] Homoisophorone formation is also acid-catalyzed.[504] "Kostanecki's compound,"[497] formed by self-condensation of phenoxyacetone with sodium or sodium amide in xylene, has been shown to be 1,2,3-tribenzoylpropene.[498–501]

$$(CH_3)_2C{=}CHCOCH{=}C(CH_3)_2 \rightarrow \text{Isophorone (3,5,5-trimethyl-2-cyclohexen-1-one)}$$

Phorone → Isophorone

$$3\ C_6H_5COCH_2OC_6H_5 \xrightarrow{NaNH_2,\ xylene} C_6H_5COCH_2C(COC_6H_5){=}CHCOC_6H_5$$

Kostanecki's compound

1,3,5-Trisubstituted benzenes are obtained from methyl ketones, particularly with acid catalysts. Acetone yields mesitylene,[505] and

[486] J. Ducasse, *Bull. Soc. Chim. France*, [5] **11**, 333 (1944).
[487] A. v. Lenz, *Monatsh. Chem.*, **24**, 155 (1903).
[488] F. Gäss and C. Hell, *Ber.*, **8**, 369 (1875).
[489] G. Bruylants, *Ber.*, **8**, 414 (1875).
[490] A. Borodin, *Ber.*, **5**, 480 (1872).
[491] A. Borodin, *Ber.*, **6**, 982 (1873).
[492] H. Anselm and E. Nickl, Ger. pat. 752,482 [*C.A.*, **50**, 2658 (1956)].
[493] W. H. Perkin, Jr., *J. Chem. Soc.*, **43**, 45 (1883).
[494] W. H. Perkin, Jr., *J. Chem. Soc.*, **43**, 79 (1883).
[495] W. H. Perkin, Jr., *Ber.*, **15**, 2802 (1882).
[496] H. G. Kuivila, S. C. Slack, and P. K. Siiteri, *J. Am. Chem. Soc.*, **73**, 123 (1951).
[497] S. v. Kostanecki and J. Tambor, *Ber.*, **35**, 1679 (1902).
[498] R. E. Lutz and F. S. Palmer, *J. Am. Chem. Soc.*, **57**, 1947 (1935).
[499] P. Yates, D. G. Farnum, and B. H. Stout, *Chem. Ind.* (*London*), 821 (1956).
[500] P. F. Devitt, E. M. Philbin, and T. S. Wheeler, *Chem. Ind.* (*London*), 822 (1956).
[501] P. F. Devitt, E. M. Philbin, and T. S. Wheeler, *J. Chem. Soc.*, 510 (1958).
[502] K. Hess and K. Munderloh, *Ber.*, **51**, 377 (1918).
[503] C. Porlezza and V. Gatti, *Gazz. Chim. Ital.*, **55**, 224 (1925).
[504] F. B. Ekeley and W. W. Howe, *J. Am. Chem. Soc.*, **45**, 1917 (1923).
[505] R. Adams and R. W. Hufferd, *Org. Syntheses, Coll. Vol.*, **1**, 341 (1941).
[506] A. F. Odell and C. W. Hines, *J. Am. Chem. Soc.*, **35**, 81 (1913).

$$3\,CH_3COR \rightarrow$$ (1,3,5-trisubstituted benzene, R, R, R)

R = alkyl or aryl

acetophenone yields 1,3,5-triphenylbenzene.[506] Such reactions may also be catalyzed by bases[507] and can occur with cycloalkanones. Cyclohexanone with sodium methoxide or acid catalysts forms the dodecahydrotriphenylene **45a**; cyclopentanone and 1-indanone behave analogously.[160, 508, 509]

3 (cyclohexanone) $\xrightarrow{NaOCH_3 \text{ or } H^{\oplus}}$ (dodecahydrotriphenylene)

45a

Side reactions involving the aldehyde are encountered in aldehyde-ketone condensations. An important side reaction is self-condensation of aliphatic aldehydes (particularly those having an α-methylene group) to aldols, since this reaction may proceed more rapidly than condensation with the ketone. Slow addition of the aldehyde to a large molar excess of ketone containing the catalyst may circumvent this difficulty; or, for ketones which do not undergo self-condensation readily (diisopropyl ketone), the ketone anion may be formed irreversibly with a strong base before adding the aldehyde.[177] Only occasionally does the separation of aldehyde self-condensation by-products create difficulty (butanal-acetone condensation, for example).[510–512] Side reactions of ketone self-condensation are important only with acetone (to yield mesityl oxide) in condensation with unreactive aldehydes such as pivaldehyde which do not undergo self-condensation.[511]

Retrograde aldol condensation may present difficulties. An attempt to condense propanal with a large excess of 3-methyl-2-butanone led only to propionaldol and 2-methyl-2-pentenal; no aldehyde-ketone condensation products could be isolated under conditions whereby acetaldehyde and

[507] P. D. Bartlett, M. Roha, and M. Stiles, *J. Am. Chem. Soc.*, **76**, 2349 (1954).

[508] J. Hausmann, *Ber.*, **22**, 2019 (1889).

[509] H. Stobbe and F. Zschoch, *Ber.*, **60**, 457 (1927).

[510] E. N. Eccott and R. P. Linstead, *J. Chem. Soc.*, 905 (1930).

[511] R. Heilmann, G. de Gaudemaris, P. Arnaud, and G. Scheuerbrandt, *Bull. Soc. Chim. France*, 112 (1957).

[512] G. de Gaudemaris and P. Arnaud, *Compt. Rend.*, **241**, 1311 (1955).

butanal readily produce good yields of ketols with this ketone.[272] An attempt to condense 3-(3,4-methylenedioxyphenyl)-2-propenal with 3,3-dimethyl-2-butanone led to a piperonal condensation product (73% yield) due to cleavage of the propenal.[513] An attempt to condense 2-ethyl-2-hexenal with 4-heptanone (aqueous ethanolic potassium hydroxide catalyst) led to 5-ethyl-5-nonen-4-one derived from butanal and the ketone.[222, 514] The formation of *bis*-styryl ketones from styryl alkyl ketones with basic catalysts is a similar reaction.[515]

$$n\text{-}C_3H_7CH{=}C(C_2H_5)CHO + (n\text{-}C_3H_7)_2CO \xrightarrow[\text{Aq. } C_2H_5OH]{\text{KOH}} n\text{-}C_3H_7CH{=}C(C_2H_5)COC_3H_7\text{-}n$$

The position of the double bond in unsaturated carbonyl compounds obtained as condensation products is pertinent. α,β-Unsaturation predominates in the reaction equilibria in most examples. β,γ-Unsaturation is favored by alkyl substituents in the α position and, to a greater extent, in the γ position.[80, 142, 274, 516–518] Acid catalysts favor formation of α,β-unsaturated ketones with endocyclic double bonds in condensations of benzaldehyde with cyclohexanones *and* cyclopentanones; with basic catalysts the product has an exocyclic double bond.[304] The mechanism of the base-catalyzed $\alpha,\beta \rightleftharpoons \beta,\gamma$ interconversion has been studied.[94]

α,β-Unsaturated ketones undergo many additions to the olefinic double bond which account for side reactions. Addition of alkoxide ion may lead to ethers when alcoholic or aqueous alcoholic base is employed.[513, 519] The presence of these products can complicate the separation of liquid ketols by distillation, as the mixture of 1-cyclopropyl-3-ethoxy-4-methyl-1-pentanone and the related ketol.[520] Secondary amine catalysts may undergo a similar addition reaction leading to β-aminoketones.[325, 327, 521]

$$\triangleright COCH{=}CHCH(CH_3)_2 \xrightarrow{^{\ominus}OC_2H_5} \triangleright COCH_2\underset{\displaystyle OC_2H_5}{\underset{|}{C}}HCH(CH_3)_2$$

513 S. G. Powell and W. J. Wasserman, *J. Am. Chem. Soc.*, **79**, 1934 (1957).
514 E. G. Fischer and K. Löwenberg, *Ann.*, **494**, 263 (1932).
515 I. M. Heilbron and J. S. Buck, *J. Chem. Soc.*, **119**, 1500 (1921).
516 S. K. Malhotra and H. J. Ringold, *J. Am. Chem. Soc.*, **85**, 1538 (1963).
517 P. Maroni and J. E. Dubois, *Bull. Soc. Chim. France*, 126 (1955).
518 K. G. Lewis and G. J. Williams, *Tetrahedron Letters*, 4573 (1965).
519 R. C. Fuson and C. H. McKeever, *J. Am. Chem. Soc.*, **62**, 999 (1940).
520 A. T. Nielsen, D. W. Moore, and K. Highberg, *J. Org. Chem.*, **26**, 3691 (1961).
521 W. Dilthey and N. Nagel, *J. Prakt. Chem.*, [2] **130**, 147 (1931).

1-Alkyl-1-(2-tetrahydropyranyl)-2-propanones are reported to result from the condensation of 5-hydroxypentanal with methyl ketones.[522] Hydrogen chloride adds to α,β-unsaturated ketones forming β-chloroketones;

$$HOCH_2(CH_2)_3CHO + RCH_2COCH_3 \xrightarrow{NaOH} \text{(2-tetrahydropyranyl)}CH(R)COCH_3$$

however, these products are ordinarily quite easily reconverted to α,β-unsaturated ketones by heating or treatment with bases such as quinoline, potassium carbonate, potassium hydroxide, or sodium ethoxide (ethanol or water solvent).

$$RCH{=}CHCOR \underset{\text{Base or heat}}{\overset{HCl}{\rightleftharpoons}} RCHClCH_2COR$$

The formation of γ-pyrones, frequently encountered in base-catalyzed condensations of benzaldehyde (2 mole equivalents) with certain aliphatic ketones (2-butanone, 3-pentanone) is a result of intramolecular alkoxide addition in a *bis*-condensation product.[523, 524]

$$ArCH(O^{\ominus})CH(R)COC(R'){=}CHAr \xrightarrow[\text{Aq. } C_2H_5OH]{NaOH} \text{2,6-diaryl-3-R-5-R'-tetrahydro-}\gamma\text{-pyrone}$$

Ar = aryl; R, R′ = alkyl

The base-catalyzed formation of chromanones (flavanones) from 2-hydroxyphenyl styryl ketones is another example of this type of reaction.[525–529] γ-Pyrone formation may be acid-catalyzed as in the

$$o\text{-}(O^{\ominus})C_6H_4COC(R){=}CHAr \xrightarrow{BH} \text{2-Ar-3-R-chromanone}$$

condensation of benzaldehyde with 2,6-dialkylcyclohexanones; **46** ($R = CH_3$, $Ar = C_6H_5$) was formed in 88% yield (hydrogen chloride

[522] J. Colonge and P. Corbet, *Bull. Soc. Chim. France*, 283 (1960).
[523] F. R. Japp and W. Maitland, *J. Chem. Soc.*, **85,** 1488 (1904).
[524] D. Vorländer and F. Kalkow, *Ber.*, **30,** 2268 (1897).
[525] S. v. Kostanecki, R. Levi, and J. Tambor, *Ber.*, **32,** 326 (1899).
[526] S. v. Kostanecki and S. Oderfeld, *Ber.*, **32,** 1926 (1899).
[527] L. Reichel, W. Burkhart, and K. Müller, *Ann.*, **550,** 146 (1942).
[528] C. C. Patel and N. M. Shah, *J. Indian Chem. Soc.*, **31,** 867 (1954).
[529] R. Neu, *Arch. Pharm.*, **295,** 1 (1962) [*C.A.*, **57,** 7153 (1962)].

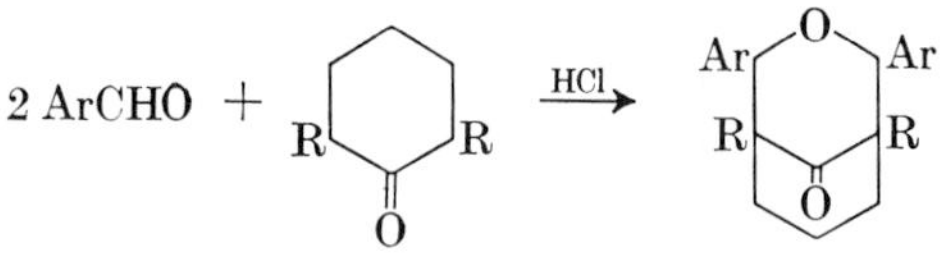

$$2\,ArCHO + \text{2,6-R}_2\text{-cyclohexanone} \xrightarrow{HCl} \mathbf{46}$$

46

catalyst).[530–532] The hydrogen chloride-catalyzed condensation of nitro-2-hydroxyacetophenones with an excess of aromatic aldehyde leads to 3-arylideneflavanones.[533, 534]

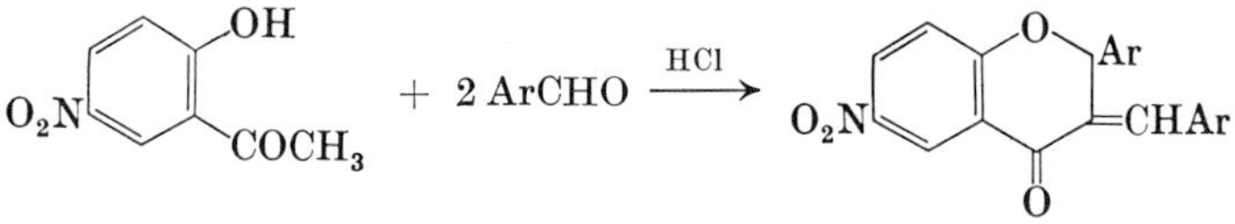

$$O_2N\text{-}C_6H_3(OH)COCH_3 + 2\,ArCHO \xrightarrow{HCl} O_2N\text{-flavanone (2-Ar, 3=CHAr)}$$

Michael additions to α,β-unsaturated ketones are the basis for many side reactions. 1,5-Diketones result from the addition of ketones, a reaction favored by use of more concentrated base and excess ketone. The diketones **(47)** frequently are formed from aromatic ketones and aromatic aldehydes,[526, 532, 535–549] less frequently (except formaldehyde) from aliphatic aldehydes.[547] Substitution in the acetophenone inhibits this side reaction.[549] Benzaldehyde and desoxybenzoin may yield benzamaron (**47,** Ar, Ar′, Ar″, R = C_6H_5).[546, 548] Cycloalkanones also form 1,5-diketones **(48),**[550–551a] which may undergo intramolecular aldol

[530] A. Haller and R. Cornubert, *Bull. Soc. Chim. France*, [4] **39,** 1621 (1926).

[531] A. Haller and R. Cornubert, *Bull. Soc. Chim. France*, [4] **41,** 367 (1927).

[532] R. Cornubert and C. Borrel, *Compt. Rend.*, **183,** 294 (1926).

[533] T. Széll and R. E. M. Unyi, *J. Org. Chem.*, **28,** 1146 (1963).

[534] A. C. Annigeri and S. Siddappa, *Indian J. Chem.*, **1,** 484 (1963).

[535] M. N. Tilichenko and V. G. Kharchenko, *Zh. Obshch. Khim.*, **32,** 1192 (1962) [*C.A.* **58,** 1414 (1963)].

[536] W. Feuerstein and S. v. Kostanecki, *Ber.*, **32,** 315 (1899).

[537] S. v. Kostanecki and L. Podrajansky, *Ber.*, **29,** 2248 (1896).

[538] S. v. Kostanecki and G. Rossbach, *Ber.*, **29,** 2245 (1896).

[539] J. Blumstein and S. v. Kostanecki, *Ber.*, **33,** 1478 (1900).

[540] W. Dilthey, *J. Prakt. Chem.*, [2] **101,** 177 (1921).

[541] W. Dilthey and R. Taucher, *Ber.*, **53,** 252 (1920).

[542] A. Cornelson and S. v. Kostanecki, *Ber.*, **29,** 240 (1896).

[543] W. B. Black and R. E. Lutz, *J. Am. Chem. Soc.*, **77,** 5134 (1955).

[544] H. v. Lendenfeld, *Monatsh. Chem.*, **27,** 969 (1906).

[545] C. S. Marvel, L. E. Coleman, and G. P. Scott, *J. Org. Chem.*, **20,** 1785 (1955).

[546] F. Klingemann, *Ann.*, **275,** 50 (1893).

[547] M. N. Tilichenko, *Zh. Obshch. Khim.*, **25,** 2503 (1955) [*C.A.*, **50,** 9327 (1956)].

[548] F. Klingemann, *Ber.*, **26,** 818 (1893).

[549] L. C. Raiford and L. K. Tanzer, *J. Org. Chem.*, **6,** 722 (1941).

[550] J. Colonge, J. Dreux, and H. Delplace, *Bull. Soc. Chim. France*, 1635 (1956).

[551] M. N. Tilichenko, *Zh. Obshch. Khim.*, **10,** 718 (1940) [*C.A.*, **35,** 2471 (1940)].

[551a] A. Polsky, J. Huet, and J. Dreux, *Compt. Rend., Ser. C*, **262,** 1543 (1966).

$$\mathrm{ArCOCH_2R + Ar'CHO \rightarrow ArCOC(R){=}CHAr' + H_2O}$$

$$\mathrm{ArCOC(R){=}CHAr' + Ar''COCH_2R \rightarrow ArCOCH(R)CH(Ar')CH(R)COAr''}$$

47

Ar, Ar′, Ar″ = aryl
R = aryl, alkyl, or H

condensation to yield tricyclic ketols, **49**.[552–555a] The latter have also been obtained from cyclohexanone and a primary alcohol, RCH_2OH.[556] 1,3-Diketones[328–332] and desoxybenzoins[557] readily form 1,5-diketones, especially with formaldehyde.[558]

2 cyclohexanone + RCHO $\xrightarrow{OH^{\ominus}}$ bis(2-oxocyclohexyl)CH–R (**48**) $\xrightarrow{OH^{\ominus}}$ **49**

R = aryl, alkyl, or H

The base-catalyzed self-condensation of α,β-unsaturated ketones proceeds initially as a Michael condensation to form as the primary product an acyclic monoolefinic 1,5- or 1,7-diketone. As the final product there results either (A) the acyclic monoolefinic diketone, (B) a cyclic saturated diketone, (C) a cyclic monoolefinic ketol, or (D) a cyclic diolefinic monoketone; products B–D arise from A by Michael or aldol condensations. Numerous examples of these various possibilities are known;[322a, 520, 559–560b]

[552] M. N. Tilichenko, *Uch. Zap. Saratovsk. Gos. Univ.*, **75**, 60 (1962) [*C.A.*, **60**, 419 (1964)].

[553] N. Barbulescu, *Rev. Chim. (Bucharest)*, **7**, 45 (1958) [*C.A.*, **51**, 3470 (1957)].

[554] J. Pithia, M. N. Tilichenko, and V. G. Kharchenko, *Zh. Obshch. Khim.*, **34**, 1936 (1964) [*C.A.*, **61**, 8163 (1964)].

[555] J. Pithia, J. Plesek, and M. Horak, *Collection Czech. Chem. Commun.* **26**, 1209 (1961).

[555a] L. Ivan and N. Barbulescu, *Analele Univ. Bucuresti Ser. Stiint. Nat.*, **12**, 155 (1963) [*C.A.* **65**, 2145 (1966)].

[556] P. Munk and J. Plesek, *Chem. Listy*, **51**, 633 (1957) [*C.A.*, **51**, 11261 (1957)].

[557] L. Mehr, E. I. Becker, and P. E. Spoerri, *J. Am. Chem. Soc.*, **77**, 984 (1955).

[558] M. Scholtz, *Ber.*, **35**, 2295 (1897).

[559] J. Wiemann and J. Dupayrat, *Bull. Soc. Chim. France*, 209 (1961).

[560] B. Furth and J. Wiemann, *Bull. Soc. Chim. France*, 1819 (1965).

[560a] G. Kabas and H. C. Rutz, *Tetrahedron*, **22**, 1219 (1966).

[560b] O. Samuel and R. Setton, *Bull. Soc. Chim. France*, 1201 (1966).

e.g., **50** (type A) from cyclopropyl methyl ketone and isobutyraldehyde,[520] **51** (type A) from benzaldehyde and cyclohexanone,[561–563] and **52** (type C) from mesityl oxide.[418]

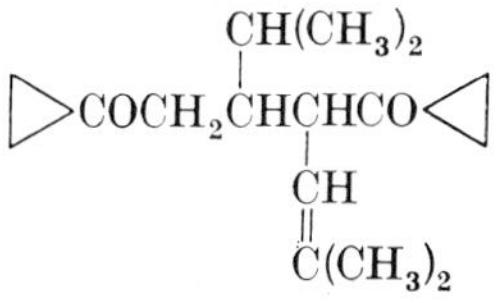

50

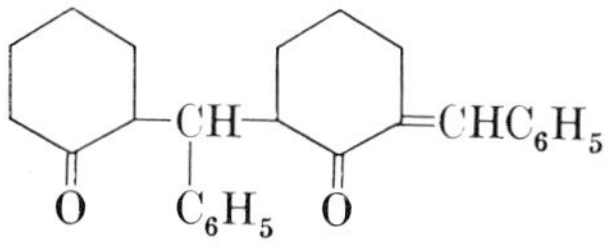

51

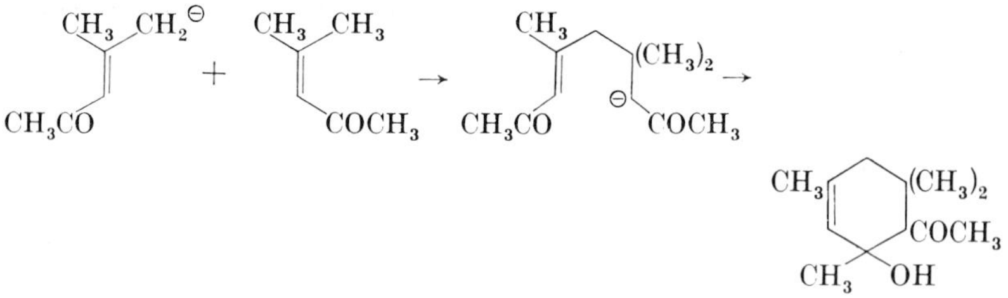

52

Self-condensation of styryl alkyl ketones of the type

$$ArCH{=}CHCOCH_2R$$

with ethanolic sodium ethoxide leads to 3,5-diaryl-4-alkanoyl-2-alkylcyclohexanones (**53,** 10–20% yield).[564] Electron-releasing groups in the *meta* or *para* position of the aryl group facilitate the reaction; R may be any alkyl group except methyl. Styryl methyl or ethyl ketones may undergo self-condensation to styryl aryl cyclohexenones, **54**.[411, 564a]

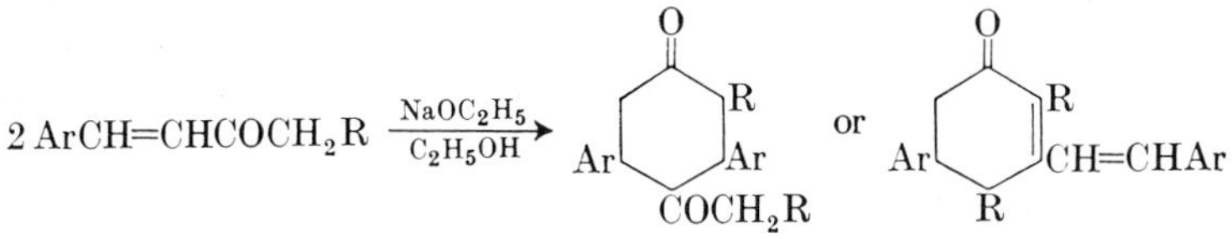

$$2\,ArCH{=}CHCOCH_2R \xrightarrow[C_2H_5OH]{NaOC_2H_5}$$

53
R = alkyl except CH_3

or

54
R = H or CH_3

Acid catalysts may produce side reactions with styryl ketones. The reactions involve oxidation and lead to pyrylium salts. The process with

[561] M. N. Tilichenko and V. G. Kharachenko, *Zh. Obshch. Khim.*, **29,** 1909 (1959) [*C.A.*, **54,** 9783 (1960)].

[562] M. N. Tilichenko and V. G. Kharachenko, *Dokl. Akad. Nauk SSSR*, **110,** 226 (1956) [*C.A.*, **51,** 5037 (1957)].

[563] M. N. Tilichenko, V. G. Kharachenko and T. I. Krupina, *Zh. Obshch. Khim.*, **34,** 2721 (1964) [*C.A.*, **61,** 14637 (1964)].

[564] A. T. Nielsen and H. J. Dubin, *J. Org. Chem.*, **28,** 2120 (1963).

[564a] A. T. Nielsen and co-workers, forthcoming publication.

chalcones is favored by the presence of electron-releasing groups such as methoxy or dimethylamino.[565–567] The reaction may result by using 2 mole equivalents of the ketone and 1 of the aldehyde with acid catalysts (with basic catalysts an intermediate 1,5-diketone is formed which may subsequently be cyclized by acid catalysts). The pyrylium cations may be isolated as their chloroferrates or perchlorates.

$$\text{ArCH{=}CHCOAr} + \text{CH}_3\text{COAr} \xrightarrow[2.\ X^{\ominus}]{1.\ H^{\oplus}} \text{2,4,6-triarylpyrylium } X^{\ominus}$$

$X^{\ominus} = FeCl_4^{\ominus}, ClO_4^{\ominus}$
Ar = aryl with electron-releasing substituents

Benzopyrylium salts result from 2-hydroxy aromatic aldehydes and aromatic ketones.[568–573] Spiro compounds are obtained from 2-hydroxy

$$o\text{-HOC}_6\text{H}_4\text{CHO} + \text{RCH}_2\text{COAr} \xrightarrow[2.\ X^{\ominus}]{1.\ H^{\oplus}} \text{3-R-2-Ar-benzopyrylium } X^{\ominus}$$

aromatic aldehydes (2 mole equivalents) and alkyl ketones (including dibenzyl ketone) with hydrogen chloride catalyst.[239, 574, 575]

$$2\ o\text{-HOC}_6\text{H}_4\text{CHO} + \text{RCH}_2\text{COCH}_2\text{R}' \xrightarrow{\text{HCl}} \text{3-R-3'-R'-2,2'-spirobichromene}$$

R, R′ = alkyl or aryl

EXPERIMENTAL CONDITIONS

Reaction conditions are extremely important. A great variety of experimental procedures has been employed and it is quite difficult to make generalizations of broad applicability. In a series of papers Dubois

565 W. Dilthey, *J. Prakt. Chem.*, [2] **95,** 107 (1917).

566 W. Dilthey, C. Berres, E. Holterhoff, and H. Wübken, *J. Prakt. Chem.*, [2] **114,** 179 (1926).

567 R. Wizinger, S. Losinger, and P. Ulrich, *Helv. Chim. Acta*, **39,** 5 (1956).

568 H. Decker and T. von Fellenberg, *Ber.*, **40,** 3815 (1907).

569 H. Decker and T. von Fellenberg, *Ann.*, **364,** 1 (1909).

570 J. W. Gramshaw, A. W. Johnson, and T. J. King, *J. Chem. Soc.*, 4040 (1958).

571 W. H. Perkin, Jr., R. Robinson, and M. R. Turner, *J. Chem. Soc.*, **93,** 1085 (1908).

572 S. G. Sastry and B. N. Ghosh, *J. Chem. Soc.*, **107,** 1442 (1915).

573 A. Robertson and R. Robinson, *J. Chem. Soc.*, 1951 (1926).

574 W. Dilthey and H. Wübken, *Ber.*, **61,** 963 (1928).

575 R. Dickinson and I. M. Heilbron, *J. Chem. Soc.*, 1699 (1927).

reported a detailed study of experimental conditions influencing the aldol condensation.[269, 576–581] Condensations of aldehydes with 2-butanone[582, 583] and chalcone formation[338] have been reviewed. The reaction conditions (catalyst, solvent, temperature, reaction time, and ratio of reactants) are summarized in the following discussion, which applies principally to conditions such that the reaction course is determined by thermodynamic considerations.

Catalyst. The choice of catalyst and its concentration are extremely important factors in determining a successful condensation. The matter of catalyst concentration has often been overlooked. Dubois' experiments clearly demonstrate the signal importance of this factor, particularly in base-catalyzed reactions of simple aliphatic aldehydes and ketones.[269, 576–581] For example, in the condensation of acetaldehyde with acetone to yield the ketol, 4-hydroxy-2-pentanone, a maximum yield of 87% could be obtained under appropriate conditions at a particular concentration of potassium hydroxide in ethanol; a very slight increase or decrease of catalyst concentration from the optimum value produced an abrupt decrease in yield of ketol.[577] The rates of the forward and reverse reactions and their relative magnitudes are important in determining the required catalyst concentration for a particular condensation. Excess base favors the retrograde process of cleaving aldols to reactants, but if the base concentration is too low the forward rate is too slow. Fortunately, most aldol condensations are not extremely sensitive to catalyst concentration. Often the reverse process is not rapid, as when an α,β-unsaturated carbonyl compound, rather than an aldol, is the product and only a small amount of catalyst is needed to secure maximum yield. Excess base favors dehydration and is to be avoided if aldols are desired. In certain procedures, salts of weak acids such as sodium carbonate, sodium phosphate, or potassium cyanide serve to provide a buffered control of *p*H and optimum conditions. Acid-catalyzed condensations are generally not so sensitive to catalyst concentration because these reactions are largely irreversible and, with few exceptions, lead to α,β-unsaturated carbonyl compounds.

Although a wide variety of materials have been employed as catalysts in the aldol condensation, frequent use is limited to a few. In the following

[576] J. E. Dubois, *Bull. Soc. Chim. France*, [5] **16**, 66 (1949).

[577] J. E. Dubois, *Ann. Chim. (Paris)*, [12] **6**, 406 (1951).

[578] J. E. Dubois, *Compt. Rend.*, **235**, 296 (1952).

[579] J. E. Dubois and R. Luft, *Bull. Soc. Chim. France*, 1148 (1954).

[580] J. E. Dubois and R. Luft, *Bull. Soc. Chim. France*, 1153 (1954).

[581] J. E. Dubois, *Bull. Soc. Chim. France*, 272 (1955).

[582] M. G. J. Beets and H. van Essen, *Rec. Trav. Chim.*, **77**, 1138 (1958).

[583] H. Haeussler and C. Brugger, *Ber.*, **77**, 152 (1944).

discussion, catalysts have been classified by chemical type and approximate base strength.

Primary and secondary, but not tertiary, amines are of value as catalysts for reactions with very reactive aldehydes that are sensitive to stronger bases, and for reactions with ketones of high enol content such as 1,3-diketones. The amines can react to form imine or enamine intermediates. Pyrrolidine and piperidine are quite effective for intramolecular condensations.[48, 68, 76, 382, 406, 436] Added acid, usually acetic, often facilitates the condensation (cf. p. 8). Pyrrolidinium and piperidinium acetates have been employed in self-condensation or mixed condensation reactions of α,β-unsaturated aldehydes leading to polyenals,[105–111, 115, 514, 584–587] and in condensation of aldehydes with 1,3-diketones.[216, 217, 328–332, 588–596] These catalysts have also been employed in condensations of aldehydes with 1,2-diketones,[325] desoxybenzoins,[597, 598] and tetrahydro-γ-pyrone.[599] In the condensation of 4-dimethylaminobenzaldehyde with certain 1,2,3,4-tetrahydro-1-methyl-4-oxoquinolines, piperidine was effective where sodium hydroxide failed, although the situation was reversed in the condensation of benzaldehyde with these ketones; i.e., sodium hydroxide was the most effective catalyst.[368] In the condensation of aromatic aldehydes with phenylacetone, piperidine caused condensation at the methylene group,[265–267] whereas sodium hydroxide produced condensation at the methyl group (see discussion p. 35).[264] Piperidine was effective at high temperature (175–200°) for the rapid condensation of indole-3-carboxaldehydes with acetophenone.[600] Diethylamine effected condensation of *o*-phthalaldehyde with phenylacetone and with 2,4-pentanedione to yield the corresponding *bis*-ketols in excellent yields; potassium hydroxide was ineffective for this purpose.[601] Other amines have found

[584] G. W. Seymour and V. S. Salvin, U.S. pat. 2,408.127 [*C.A.*, **41**, 772 (1947)].

[585] E. Barraclough, J. W. Batty, I. M. Heilbron, and W. E. Jones, *J. Chem. Soc.*, 1549 (1939).

[586] R. Kuhn and C. J. O. Morris, *Ber.*, **70**, 853 (1937).

[587] P. Karrer and A. Rüegger, *Helv. Chim. Acta*, **23**, 284 (1940).

[588] E. Knoevenagel, *Ber.*, **37**, 4461 (1904).

[589] E. Hase and G. Stjernvall, *Suomen Kemistilehti*, **21B**, 42 (1948) [*C.A.*, **42**, 8176 (1948)].

[590] H. Midorikawa, *Bull. Chem. Soc. Japan*, **27**, 149 (1954).

[591] P. T. Mora and T. Széki, *J. Am. Chem. Soc.*, **72**, 3009 (1950).

[592] H. Schlenk, *Chem. Ber.*, **81**, 175 (1948).

[593] H. Schlenk, *Chem. Ber.*, **85**, 901 (1952).

[594] N. A. Sörensen, E. Samuelsen, and F. R. Oxaal, *Acta Chem. Scand.*, **1**, 458 (1947).

[595] R. H. Wiley, C. H. Jarboe, and H. G. Ellert, *J. Am. Chem. Soc.*, **77**, 5102 (1955).

[596] P. Delest and R. Pallaud, *Compt. Rend.*, **245**, 2056 (1957).

[597] A. Dornow and F. Boberg, *Ann.*, **578**, 122 (1952).

[598] K. Matsumura, *J. Am. Chem. Soc.*, **57**, 496 (1935).

[599] N. J. Leonard and D. Choudhury, *J. Am. Chem. Soc.*, **79**, 156 (1957).

[600] R. B. Van Order and H. G. Lindwall, *J. Org. Chem.*, **10**, 128 (1945).

[601] W. Davey and H. Gottfried, *J. Org. Chem.*, **26**, 3699 (1961).

limited use as catalysts for the aldol condensation (dimethylamine, morpholine, pyridine,[50] triethylamine[482]) and generally are less effective than piperidine and pyrrolidine.

Alkali and alkaline earth metal hydroxides in dilute solution are very effective catalysts for the aldol condensation and have found the widest use. A study has been made of the effect of the nature of the cation on catalytic activity of hydroxides.[582] Sodium hydroxide in aqueous or aqueous ethanolic solution is the catalyst most frequently employed; potassium hydroxide is equally effective. By selecting a suitable catalyst concentration and solvent, excellent yields of condensation product may often be obtained. For condensations involving very reactive or base-sensitive aldehydes such as formaldehyde or α,β-unsaturated aldehydes the alkaline earth hydroxides, calcium or barium hydroxide, are valuable catalysts.[278, 514, 602] Barium oxide is effective for preparing diacetone alcohol from acetone (Soxhlet apparatus, 75% yield),[147] but 2-butanone under the same conditions affords the homologous ketol in only 11% yield.[146] Quaternary ammonium hydroxides such as Triton-B have found only limited use as aldol condensation catalysts.[582]

Alkoxides of the alkali and alkaline earth metals and of magnesium and aluminum are frequently employed as catalysts. These strong bases are often effective when sodium hydroxide fails. Sodium ethoxide in ethanol has been widely employed and is very efficient, especially for preparation of most chalcones; sodium methoxide, potassium methoxide, and lithium ethoxide[582] have been used effectively, but less frequently. Other alkoxides have been employed, principally in aprotic solvents such as ether, benzene, or toluene. Excellent solvents less commonly employed are tetrahydrofuran, 1,2-dimethoxyethane, and dimethylformamide.[603] Sodium *t*-amyloxide (soluble in benzene or toluene) rapidly condenses aromatic aldehydes with cycloalkanones at 0°,[295] and sodium pentoxide has been employed for self-condensation of 3-methylbutanal at 75°.[604] Potassium *t*-butoxide in boiling benzene has been employed for self-condensation of cycloalkanones to α,β-unsaturated ketones[605] and for the cyclization of certain diketones.[606] Solvent-free potassium *t*-butoxide is less soluble in benzene than is sodium *t*-amyloxide; it is soluble in 1,2-dimethoxyethane, whereas the solvated material is not. Although aluminum alkoxides

[602] A. Kuzin, *Zh. Obshch. Khim.*, **8,** 592 (1938) [*C.A.*, **33,** 1271 (1939)].

[603] A. J. Parker, *Quart. Rev.* (*London*), **16,** 163 (1962); A. J. Parker, *Advan. Org. Chem.*, **5,** 1–46 (1965).

[604] V. I. Lyubomilov, *Zh. Obshch. Khim.*, **26,** 2738 (1956) [*C.A.*, **51,** 7293 (1957)].

[605] W. Rosenfelder and D. Ginsburg, *J. Chem. Soc.*, 2955 (1954).

[606] P. Wieland, K. Heusler, H. Ueberwasser, and A. Wettstein, *Helv. Chim. Acta*, **41,** 74 (1958).

(particularly isopropoxide) generally produce Tischenko reactions (ester formation) with aldehydes,[104, 479] aluminum *t*-butoxide in boiling benzene has been found effective for condensation of high-molecular-weight polyenals with ketones[607, 608] or for the self-condensation of certain ketones.[151, 609] In a novel procedure an allylic alcohol is oxidized to the corresponding aldehyde with aluminum isopropoxide in the presence of acetone; an aldol condensation follows to yield an α,β-unsaturated ketone.[610] Sodium phenoxide, employed as a catalyst in only one instance, was claimed to be

$$CH_2{=}CHCH_2CH_2C(CH_3){=}CHCH_2OH + CH_3COCH_3 \xrightarrow{Al(OC_3H_7\text{-}i)_3} CH_2{=}CHCH_2CH_2C(CH_3){=}CHCH{=}CHCOCH_3$$

very effective in condensing citral with acetone (81% yield of unsaturated ketone).[611, 612] Butoxymagnesium bromide and iodide have been employed for aldehyde self-condensations but are not too effective.[613] Alkali metal alkoxyborohydrides, with some exceptions, are ineffective catalysts.[614] Sodium metal was employed in condensations of aldehydes with camphor in the early work of Haller,[313] but its use is now generally avoided because of the competing reductions.

Alkali metal salts of weak acids have been employed effectively as catalysts. They provide the constant *p*H needed for condensations sensitive to hydroxide-ion concentration. Potassium and sodium carbonate are useful as mild catalysts for many aldehyde-aldehyde and aldehyde-ketone condensations for which a stronger base is undesirable; i.e., with formaldehyde, acetaldehyde, and α,β-unsaturated aldehydes. Sodium phosphate was found suitable for condensation of nitrobenzaldehydes with acetophenones to produce ketols.[361] Although potassium cyanide has been employed effectively for some condensations (the preparation of aldol itself[6] and the condensation of acetaldehyde with methyl isopropyl ketone[615]), it is seldom used; its catalytic effect appears to resemble that of sodium hydroxide very closely.[582] Potassium cyanide condenses aromatic aldehydes with acetophenones to yield complex cyano compounds[616] and causes self-condensation of aromatic aldehydes to

[607] C. K. Warren and B. C. L. Weedon, *J. Chem. Soc.*, 3972 (1958).

[608] C. K. Warren and B. C. L. Weedon, *J. Chem. Soc.*, 3986 (1958).

[609] K. Billig, Ger. pat. 639,291 [*C.A.*, **31**, 1431 (1939)].

[610] R. Helg, F. Zobrist, A. Lauchenauer, K. Brack, A. Caliezi, D. Stauffacher, E. Zweifel, and H. Schinz, *Helv. Chim. Acta*, **39**, 1269 (1956).

[611] Z. Arnold and K. Hejno, Czech. pat. 85,207 [*C.A.*, **50**, 10781 (1956)].

[612] E. I. Kozlov, M. T. Yanotovskii, and G. I. Samokhvalov, *Zh. Obshch. Khim.*, **34**, 2748 (1964) [*C.A.*, **61**, 14716 (1964)].

[613] V. Grignard and M. Fluchaire, *Ann. Chim.* (*Paris*), [10] **9**, 1 (1928).

[614] G. Hesse and M. Maurer, *Ann.*, **658**, 21 (1962).

[615] J. Salkind, *J. Russ. Phys. Chem. Soc.*, **37**, 484 (1905); [*Chem. Zentr.*, **76**, II, 752 (1905)].

[616] R. B. Davis, *J. Org. Chem.*, **24**, 880 (1959).

benzoins (the benzoin condensation).[617] The bases, sodium amide, sodium hydride, and calcium hydride, have been used occasionally and have value for condensations involving hindered aldehydes which do not readily undergo self-condensation.[143, 314, 514, 618–622] They are polymeric solids insoluble in solvents with which they do not react. They are useful for preparing certain enolates in aprotic solvents; the reaction may be followed by measuring the hydrogen evolved. Enolates of methyl ketones usually cannot be prepared without concomitant aldol condensation.[247] Certain soaps have been used effectively for self-condensation of aliphatic aldehydes.[623–626] Sodium sulfite condensed 3,5-dimethyl-2,6-octadienal with acetone to 6,8-dimethyl-3,5,9-undecatrien-2-one in 67% yield.[627] Alumina at room temperature has been used.[404, 441] Barium nitride is reported to be a powerful catalyst for effecting self-condensation of ketones.[627a] Several other salts have been employed at times as aldol condensation catalysts, but none appears to offer distinct advantages.

Aminomagnesium halides and some Grignard reagents are very effective aldol condensation agents in certain cases. Methylanilinomagnesium bromide provides high yields of ketols by self-condensation of most ketones.[150, 159] Certain hindered ketones (non-methyl) for which this reagent cannot effect self-condensation are converted to their enolate anions, which may be treated with a different ketone or aldehyde to prepare a wide variety of ketols in good yield.[177, 198] Certain Grignard reagents such as isopropyl- and *t*-butyl-magnesium bromide have been employed for self-condensation of hindered ketones to ketols,[157] but in general these catalysts appear to be less effective than methylanilinomagnesium bromide. Diethylzinc is an ineffective catalyst.[628] Lithium amides [e.g., $(C_2H_5)_2$-NLi][221, 629] are very strong bases and appear to be of potential value as aldol condensation catalysts.

Use of ion-exchange resins (basic types most frequently) as aldol

617 W. S. Ide and J. S. Buck, *Org. Reactions*, **4**, 269–304 (1948).

618 N. Wolff, *Ann. Chim.* (*Paris*), [9] **20**, 82 (1923).

619 C. Daesslé and H. Schinz, *Helv. Chim. Acta*, **40**, 2270 (1957).

620 H. Favre and H. Schinz, *Helv. Chim. Acta*, **41**, 1368 (1958).

621 L. Ruzicka and H. Schinz, *Helv. Chim. Acta*, **23**, 959 (1940).

622 I. M. Heilbron, W. E. Jones, and J. W. Batty, Brit. pat. 510,540 [*C.A.*, **34**, 5092 (1940)].

623 R. A. Reck and R. P. Arthur, U.S. pat. 2,982,784 [*C.A.*, **55**, 20962 (1961)].

624 R. M. Cole, U. S. pat. 2,545,261 [*C.A.*, **45**, 6654 (1951)].

625 F. A. Biribauer, C. S. Carlson, and C. E. Morrell, U.S. pat. 2,684,385 [*C.A.*, **49**, 11002 (1955)].

626 J. Kollar and J. G. Schulz, U.S. pat. 3,060,236 [*C.A.*, **58**, 5517 (1963)].

627 A. A. Petrov, N. A. Razumova, and M. L. Genusov, *Zh. Obshch. Khim.*, **28**, 1128 (1958) [*C.A.*, **52**, 20027 (1958)].

627a Y. Okamoto and J. C. Goswani, *Bull. Chem. Soc. Japan*, **39**, 2778 (1966).

628 R. Rieth and F. Beilstein, *Ann.*, **126**, 241 (1863).

629 G. Wittig and H. D. Frommeld, *Chem. Ber.*, **97**, 3541 (1964).

condensation catalysts has been studied often for self-condensation of aldehydes,[100, 101, 630–632] but seldom for ketone-ketone or aldehyde-ketone condensations.[337, 633, 634] Acetone provides a 79% yield of mesityl oxide when heated under reflux with Dowex 50H; good yields of aldols may be obtained from aliphatic aldehydes with ion-exchange catalysts when other catalysts fail.[100] In many other reactions, however, yields obtained with ion-exchange catalysts are somewhat inferior to those obtained with more conventional catalysts. In use, resin catalysts are often deactivated rapidly and cannot be reused.[100]

Acid catalysts are less frequently employed for aldol condensations. Ordinarily α,β-unsaturated carbonyl products rather than aldols, are formed. A remarkable exception is phosphorus oxychloride which is reported to produce aldols from alkanals.[102, 103] In general, yields are lower with acid catalysts and the products are more difficult to purify. With aldehyde-aldehyde condensations, acid catalysts usually lead to resin formation or other side reactions, and, except for intramolecular condensation of some α,ω-dialdehydes[373–381] and ketoaldehydes[449, 450] (formed *in situ* from acetal precursors), have found little practical use. In the self-condensation of 2-butanone, and in condensations of aldehydes with methyl ketones, acid catalysts permit condensation at the methylene carbon, whereas basic catalysts cause condensation at the methyl group. Hydrogen chloride is the acid catalyst most often employed, although hydrogen bromide and hydrogen iodide appear to be more effective;[152, 156] about two-thirds of 1 mole equivalent of hydrogen halide is needed to attain maximum yields.[152] Frequently β-halo ketones result. They can be dehydrohalogenated to α,β-unsaturated ketones by treatment with ethanolic sodium hydroxide, but this treatment may cause some polymerization of the product. However, milder base treatment (pyridine, quinoline or potassium carbonate) in water or ethanol is often sufficient to effect dehydrohalogenation. Perchloro aldehydes such as chloral and trichloropropenal undergo condensations with ketones to yield ketols or α,β-unsaturated ketones in excellent yields with acid catalysts such as acetic or sulfuric acid.[289, 635–639]

[630] G. Durr, *Compt. Rend.*, **235,** 1314 (1952).

[631] I. G. Farbenind. A.–G., Brit. pat. 349,556 [*C.A.*, **26,** 5430 (1932)].

[632] G. Durr and R. L. Farhi, *Compt. Rend.*, **245,** 75 (1957)].

[633] B. Tomek and J. Cvrtnik, Czech. pat. 100,648 [*C.A.*, **58,** 4429 (1963)].

[634] Rheinpreussen A.–G. für Bergbau und Chemie, Fr. pat. 1,383,548 [*C.A.*, **62,** 9013 (1965)].

[635] F. L. Breusch and H. Keskin, *Arch. Biochem.*, **18,** 305 (1948).

[636] W. Koenigs, *Ber.*, **25,** 792 (1892).

[637] L. K. Zakharkin and L. P. Sorokina, *Izv. Akad. Nauk SSSR, Otd. Khim. Nauk*, 1445 (1958) [*C.A.*, **53,** 8130 (1959)].

[638] L. K. Zakharkin and L. P. Sorokina, *Izv. Akad. Nauk SSSR, Otd. Khim. Nauk,* 287 (1962) [*C.A.*. **57,** 12417 (1962)].

[639] A. Roedig, R. Manger, and S. Schödel, *Chem. Ber.*, **93,** 2294 (1960).

Acid catalysts are of value for preparing certain hydroxy-substituted chalcones.[640, 641] Condensation of 2-nitrobenzaldehyde with tetralones gives excellent yields of α,β-unsaturated ketones with sulfuric acid catalysis.[642–644] In the reaction of 5-chloro- and 4-chloro-2-nitrobenzaldehyde with 4,4-dimethyl-1-tetralone, aqueous acetic acid and iron powder produced 90–94% yields of the corresponding amino α,β-unsaturated ketones by aldol condensation and reduction of the nitro group.[645] This procedure may be of general value in preparing condensation products containing primary amino groups since condensations involving aminobenzaldehydes often give very poor yields[350] and condensations with primary amino ketones may lead to Schiff bases.[352, 359] Nitrofuran carboxaldehydes and, to a lesser extent, nitrothiophene carboxaldehydes and ketones undergo resinification with base, and sulfuric acid in acetic acid has been advantageously employed in aldol condensations involving these substances.[646, 647] Sulfuric acid has occasionally been used for cyclization of α,ω-diketones,[384, 648] as has p-toluenesulfonic acid,[403] but other catalysts, such as sodium carbonate and sodium methoxide, are often better. Seldom used, and generally less effective acid catalysts, are nitric acid, boron trifluoride, phosphorus oxychloride, phosphorus pentoxide, zinc chloride, ferric chloride, aluminum chloride, and acetic anhydride. Acetic acid is often used as a solvent when acid catalysts are employed.

Solvent. The choice of solvent generally depends on solubility of reactants and catalyst. Ethanol, aqueous ethanol, or water is most commonly employed; sodium ethoxide is used in ethanol. The heterogeneous mixture, aqueous sodium hydroxide-ether, has been found excellent for condensation reactions of reactive aliphatic aldehydes.[649] Strong bases generally require aprotic solvents;[609] methylanilinomagnesium bromide is employed in ether-benzene, and sodium amide is used as a suspension in ether, benzene, or toluene. Solvents less often exploited, such as tetrahydrofuran, 1,2-dimethoxyethane, dimethylformamide, and dimethyl sulfoxide[649a] might offer advantages in some reactions. Hydrogen

[640] J. F. Miguel, *Bull. Soc. Chim. France*, 1369 (1961).

[641] T. Széll, *Chem. Ber.*, **92**, 1672 (1959).

[642] V. L. Bell and N. H. Cromwell, *J. Org. Chem.*, **23**, 789 (1958).

[643] A. Hassner and N. H. Cromwell, *J. Am. Chem. Soc.*, **80**, 893 (1958).

[644] J. L. Adelfang and N. H. Cromwell, *J. Org. Chem.*, **26**, 2368 (1961).

[645] N. H. Cromwell and V. L. Bell, *J. Org. Chem.*, **24**, 1077 (1959).

[646] V. F. Lavrushin, S. V. Tsukerman, and A. I. Artemenko, *Zh. Obshch. Khim.*, **32**, 1324, 1329 (1962) [*C.A.*, **58**, 1422, 5603 (1963)].

[647] S. V. Tsukerman, V. M. Nikitchenko, and V. F. Lavrushin, *Zh. Obshch. Khim.*, **32**, 2324 (1962) [*C.A.*, **58**, 7896 (1963)].

[648] W. G. Dauben and J. W. McFarland, *J. Am. Chem. Soc.*, **82**, 4245 (1960).

[649] V. Grignard and M. Dubien, *Ann. Chim. (Paris)*, [10] **2**, 282 (1924).

[649a] B. Wesslen, *Acta. Chem. Scaud.*, **21**, 713 (1967).

chloride is often used in the absence of a solvent and less frequently with methanol, ethanol, ethyl acetate, ether, or benzene. Ether or benzene, rather than an alcohol, would be preferred to avoid side reactions such as acetal formation. Acetic acid has been used as solvent for sulfuric acid and hydrogen chloride catalysis. The effect of solvent on the kinetics of the base-catalyzed cleavage of tertiary β-ketols has been studied.[650, 651] The selection of solvent may be exploited to determine the ratio of 1-condensation to 3-condensation products derived from condensation reactions of methyl ketones.[231, 248]

Temperature and Reaction Time. The aldol condensation is reversible and slightly exothermic, and best yields are generally obtained at 5–25° (however, low temperatures also favor aldehyde trimer formation; see p. 58). Rates are sufficiently rapid at room temperature that maximum yields are usually reached within 12 to 24 hours. For reactions which are more rapid or produce an unstable product, temperatures between 0 and 5°, or a short reaction time at room temperature, may be employed. Acid-catalyzed reactions usually require a low temperature; hydrogen chloride-catalyzed reactions are often conducted at 0–5°. For base-catalyzed reactions, reflux temperatures of solvents such as ethanol, ether, or benzene may be used to reduce the reaction time. Condensations of indole-3-carboxaldehydes with acetophenones have been accomplished at 175–200° (in the absence of added catalyst) within 5 minutes, giving good yields of α,β-unsaturated ketones.[600] If the reaction is essentially irreversible, as is chalcone formation, and the product is stable, the higher temperature does not result in appreciably lower yields. Excessively high temperatures are generally unnecessary and are avoided, because side reactions are more likely to result. High temperatures favor the formation of α,β-unsaturated carbonyl compounds rather than aldols or ketols and favor 1-condensation over 3-condensation in reactions of aldehydes with methyl alkyl ketones.

Ratio of Reactants. In mixed condensations the ratio of reactants may determine products. Generally, stoichiometric quantities are employed. An important exception involves condensations of certain reactive aliphatic aldehydes with ketones where a large molar excess of ketone is often used to minimize self-condensation of the aldehyde and to secure best yields of 1:1 condensation products; the excess ketone does not undergo self-condensation under the reaction conditions and is recovered. Self-condensation of the aldehyde may also be minimized by very slow addition of the aldehyde to an excess of ketone containing the catalyst. In condensations of alicyclic ketones with aldehydes an excess

[650] J. Barthel and J. E. Dubois, *Z. Phys. Chem.*, **32**, 296 (1962).

[651] J. Barthel and J. E. Dubois, *Compt. Rend.*, **248**, 1649 (1959).

of ketone may be required to minimize formation of *bis*-condensation products;[297] with other ketones (except acetone) it is usually difficult to obtain other than 1:1 products.

An excess of aldehyde may be of value in condensations with unreactive ketones if the aldehyde does not readily condense with itself (e.g., aromatic aldehydes). Because formaldehyde does not readily undergo self-condensation and is very reactive, polymethylol compounds are easily produced with most active methylene compounds; however, an excess of formaldehyde can lead to reduction of carbonyl groups.

EXPERIMENTAL PROCEDURES

The best yields in aldol condensations are obtained when all reactants and solvents are carefully purified immediately before use. Although traces of peroxides do not inhibit the aldol condensation, larger amounts may cause lower yields.[381] Liquid aldehydes readily accumulate carboxylic and peroxidic impurities on storage; the impurities may be removed by careful washing with dilute aqueous sodium hydroxide or carbonate, followed by drying and distillation in a nitrogen atmosphere. Although liquid ketones are much less susceptible to oxidation than aldehydes, these reactants, too, should be carefully distilled before use. Melting points of solids should be checked as an indication of purity and samples recrystallized if necessary. Use of a nitrogen atmosphere during the reaction is of value, particularly in reactions of strongly basic enolates (prepared in aprotic solvents) since these intermediates react with oxygen and carbon dioxide; the inert atmosphere is also of value when oxidizable reactants or products are involved (α,β-unsaturated aldehydes or ketones, dialdehydes), and when reaction time is prolonged. However, there are routine condensations (many chalcone preparations, for example) which do not require an inert atmosphere.

In working up reaction mixtures conventional procedures are followed, but special care should be taken if products are to be distilled. Because many aldols and ketols readily dissociate to reactants when heated, isolation of these substances by distillation at a low temperature under reduced pressure in a nitrogen atmosphere is recommended. Complete removal of acidic or basic substances (including those in solvents used) prior to distillation is necessary for efficient recovery of liquid aldols or ketols because these impurities catalyze dissociation or dehydration. Thorough washing of the product (conveniently in purified ether solution) with saturated aqueous sodium bicarbonate will remove most undesirable impurities. α,β-Unsaturated aldehydes and ketones are readily oxidized, especially when heated in air, and should be distilled immediately after work-up, under reduced pressure, preferably in a nitrogen atmosphere. Aldols dimerize on storage and should be redistilled before use. Redistillation of stored products may result in their decomposition (dealdolization, dehydration, or polymerization) unless accumulated acidic or peroxidic impurities are removed by washing with dilute aqueous alkali.

A few special procedures have been developed for certain reactions to avoid reversal of the condensation and to secure higher yields of product. One example is use of a column reactor in which formaldehyde and sodium hydroxide are added simultaneously to the vapor of refluxing ketone in a tube; a non-volatile acid is contained in the liquid ketone to neutralize the basic catalyst.[652] The use of barium oxide in a Soxhlet extractor leads to diacetone alcohol in good yield from acetone.[146, 147] A cation-exchange resin has been employed for continuous removal of sodium hydroxide catalyst after condensation of acetone to diacetone alcohol.[653]

Several aldol condensation procedures are to be found in volumes of *Organic Syntheses*. Aldehydes condensed with acetone to yield α,β-unsaturated methyl ketones, $RCH{=}CHCOCH_3$, are as follows (yields in parentheses): citral (45–49),[654] benzaldehyde (65–78),[655] 4-methoxybenzaldehyde (83),[655] and furfural (60–66).[656] Dibenzalacetone (90–94%) has been prepared from benzaldehyde and acetone.[657] Benzaldehyde has also been condensed with acetophenone to form chalcone (85%)[658] and with pinacolone to form benzalpinacolone (88–93%).[659]

Also described in *Organic Syntheses* are procedures for the self-condensation of acetone to diacetone alcohol (71%)[147] employing a Soxhlet apparatus; the dehydration of this ketol to mesityl oxide;[149] the sulfuric acid-catalyzed condensation of acetone to mesitylene (13–17%);[505] the preparation of dypnone in 77–82% yield from acetophenone;[660] and the preparation of tetraphenylcyclopentadienone (tetracyclone) from dibenzyl ketone and benzil (91–96%).[204]

Three procedures for intramolecular condensation of diketones are described in *Organic Syntheses*. 4-Carbethoxy-3,5-dimethyl-2-cyclohexen-1-one has been prepared by cyclization (sulfuric acid-acetic acid) of

[652] J. T. Hays, G. F. Hager, H. M. Engelmann, and H. M. Spurlin, *J. Am. Chem. Soc.*, **73**, 5369 (1951).

[653] D. C. Buttle, Brit. pat. 917,782 [*C.A.*, **59**, 1491 (1963)].

[654] A. Russell and R. L. Kenyon, *Org. Syntheses, Coll. Vol.*, **3**, 747 (1955).

[655] N. L. Drake and P. Allen, *Org. Syntheses, Coll. Vol.*, **1**, 77 (1941).

[656] C. J. Leuck and L. Cejka, *Org. Syntheses, Coll. Vol.*, **1**, 283 (1941).

[657] C. R. Conard and M. A. Dolliver, *Org. Syntheses, Coll. Vol.*, **2**, 167 (1943).

[658] E. P. Kohler and H. M. Chadwell, *Org. Syntheses, Coll. Vol.*, **1**, 78 (1941).

[659] G. A. Hill and G. M. Bramann, *Org. Syntheses, Coll. Vol.*, **1**, 81 (1941).

[660] W. Wayne and H. Adkins, *Org. Syntheses, Coll. Vol.*, **3**, 367 (1955).

ethylidene *bis*-acetoacetic ester (prepared from acetaldehyde and acetoacetic ester); the overall yield of cyclized product based on acetaldehyde is 47–50%.[661] Cyclization of 2-(3-oxobutyl)-2-methyl-1,3-cyclohexanedione to 1,6-dioxo-8a-methyl-1,2,3,4,6,7,8,8a-octahydronaphthalene occurs in 75% yield with pyrrolidine catalyst.[392] $\Delta^{1(9)}$-2-Octalone has been prepared by reaction of 1-morpholino-1-cyclohexene with methyl vinyl ketone to generate the diketone enamine *in situ*, followed by ring closure.[432]

Aldol condensation followed by Canizzaro reaction is illustrated by the condensation of formaldehyde with acetaldehyde (to yield pentaerythritol, 55–57%)[470] and with cyclohexanone (to yield 2,2,6,6-tetramethylolcyclohexanol, 73–85%).[472]

Following are representative aldol condensation procedures not found in *Organic Syntheses.*

Aldol (Self-Condensation of Acetaldehyde).[662] Acetaldehyde (1 kg., 22.7 moles) contained in a 2-l. flask is cooled below 5°. Over a 20-minute period 25 ml. of 10% aqueous sodium hydroxide is added dropwise with vigorous stirring while the temperature of the reaction mixture is maintained at 4–5°. After the mixture has been stirred for 1 hour at 4–5°, it is made slightly acidic with tartaric acid. The mixture is filtered to remove sodium tartrate (addition of diethyl ether facilitates the filtration). The filtrate is distilled under reduced pressure from an oil bath by raising the temperature slowly. Acetaldehyde and aldol slowly distil together as the aldoxan dissociates (see p. 58). Redistillation of the aldol fraction yields about 500 g. (50%) of aldol, b.p. 72°/12 mm. Aldol should be distilled immediately before use since it dimerizes to paraldol on standing.[461, 462]

2-Ethyl-2-hexenal (Self-Condensation of Butanal).[483] To 750 ml. of aqueous 1 *M* sodium hydroxide at 80° is added, dropwise and with vigorous stirring during 1.5 hours, 2520 g. (35 moles) of freshly distilled *n*-butanal. A large efficient reflux condenser is required for rapid addition of the aldehyde; the solution temperature rises to 93°. After the addition of aldehyde is complete, the mixture is heated under reflux for 1 hour. It is then cooled to room temperature, the organic layer is separated and, without further treatment, distilled through a 150-cm. Vigreux column to yield 1880 g. (86%) of pure 2-ethyl-2-hexenal, b.p. 59.5–60°/10 mm.; n_D^{18} 1.4556.*

[661] E. C. Horning, M. O. Denekas, and R. E. Field, *Org. Syntheses, Coll. Vol.*, **3,** 317 (1955).

[662] L. P. Kyriakides, *J. Am. Chem. Soc.*, **36,** 530 (1914).

* The following measurements were made on the 2-ethyl-2-hexenal. Gas-liquid chromatograph of the product (10 ft. × ¼ in. column, 20% Apiezon L on Chromosorb W, 150°, flow rate 60 cc./minute) indicated a purity of 97–99% (single major peak with retention time of about 8 minutes). A single isomer is also indicated by the n.m.r. spectrum; vinyl

2-Cyclohexylidenecyclohexanone (Self-Condensation of Cyclohexanone).[163, 164] *A. 2-(1-Chlorocyclohexyl)cyclohexanone.*[162, 164] Dry hydrogen chloride is bubbled through pure cyclohexanone (40 g.) contained in a flask cooled by a water bath. After 15 hours the gas flow is stopped and the crystalline 2-(1-chlorocyclohexyl)cyclohexanone (32 g., 71%) is collected by filtration. Wallach reports m.p. 41–43°.[162]

B. 2-Cyclohexylidenecyclohexanone.[163, 164] To 2-(1-chlorocyclohexyl)-cyclohexanone (70 g., 0.326 mole) in 100 ml. of diethyl ether is added, dropwise and with vigorous stirring, a cooled solution of 7.5 g. (0.326 g.-atom) of sodium in 150 g. of methanol while the temperature is kept below 10°. After the reaction is complete, water is added and the ether solution separated and washed until it is neutral. Evaporation of the ethereal solution under reduced pressure leaves a crystalline residue which is crystallized from methanol-water (4:1) to furnish 51.0 g. (88%) of 2-cyclohexylidenecyclohexanone, m.p. 56–57°.

Dehydrochlorination with aqueous sodium hydroxide at 25° yields principally 2-(1-cyclohexenyl)cyclohexanone.[164]

3-Methyl-4-phenyl-*trans*-3-buten-2-one (Condensation of Benzaldehyde with 2-Butanone).[233–235] A mixture of 2-butanone (25 g., 0.35 mole) and benzaldehyde (35 g., 0.33 mole) is saturated with dry hydrogen chloride with cooling. After it has stood for 2 hours, the brown reaction mixture, is shaken with dilute aqueous sodium hydroxide, washed with water, and extracted with diethyl ether. After the ether solution has been dried and concentrated under reduced pressure, the crude product is crystallized from petroleum ether (b.p. 60–80°) to yield 45 g. (85%) of 3-methyl-4-phenyl-*trans*-3-buten-2-one, m.p. 37–38°. Harries and Müller report m.p. 38° after distilling the crude product at b.p. 127–130°/12 mm. and recrystallizing the distillate from petroleum ether.[232]

2-Furfurylidenecyclopentanone (Condensation of Furfural with Cyclopentanone).[297] A mixture of freshly distilled furfural (32 g., 0.33 mole), cyclopentanone (28.0 g., 0.33 mole), diethyl ether (150 ml.) and 0.1 N sodium hydroxide solution (300 ml.) is stirred with external cooling to moderate the exothermic reaction. After about 30 minutes, yellow crystalline material (presumably 2,5-difurfurylidenecyclopentanone) begins to separate in rapidly increasing amounts. After the mixture has been stirred for a total of 45 minutes it is filtered with suction. The solid on the funnel and the aqueous layer from the filtrate are extracted

---

triplet (J = 7 c.p.s.) centered at τ 4.42 (Varian A-60 instrument; measurement in carbon tetrachloride); infrared bands (liquid film) at 1685 (C=O) and 1643 (C=C) cm.$^{-1}$ (Perkin Elmer Model 621 grating spectrophotometer). These results suggest that the β-propyl and carboxaldehyde groups are *trans* (cf. tiglaldehyde[87]).[88, 89]

with ether. The combined etheral solutions are washed twice with water and concentrated on the steam bath. The residue is distilled under reduced pressure. After a fore-run of starting material, the product is collected as a yellow oil, b.p. 154°/15 mm., which readily crystallizes (55 g., 60% yield). Recrystallization from diisopropyl ether furnishes 2-furfurylidencyclopentanone melting at 59–60°.

2,4-Dimethyl-3-hydroxy-1-phenyl-1-pentanone (Condensation of 2-Methylpropanal with Propiophenone).[177] Methylanilinomagnesium bromide is prepared by adding, with cooling and stirring, a solution of 33.8 g. (0.315 mole) of freshly distilled N-methylaniline in 100 ml. of dry benzene to an ether solution of ethylmagnesium bromide [prepared from 8.0 g. (0.33 mole) of magnesium turnings, 40 g. (0.364 mole) of ethyl bromide, and 80 ml. of diethyl ether under a nitrogen atmosphere].

To the freshly prepared solution of methylanilinomagnesium bromide is added, during 30 minutes, a solution of 48.8 g. (0.364 mole) of dry propiophenone in 100 ml. of dry benzene while the temperature is kept at 15–20°. The resulting mixture is allowed to stand at 25° for 2 hours and a solution of 15.8 g. (0.291 mole) of freshly distilled 2-methylpropanal in 20 ml. of dry benzene is then added during 15 minutes keeping the temperature at −10°. After the reaction mixture has stood for 2½ hours at −10° to 0°, 300 ml. of aqueous 3 *M* hydrochloric acid is added and the organic layer separated, washed with five portions of 6 *M* hydrochloric acid and finally with water. The organic solution is dried with anhydrous sodium sulfate and the solvents are removed by distillation at 15–30 mm. from a hot-water bath (60–70°). The residue is distilled to separate, after a fore-run of recovered propiophenone, 37.5 g. (83%) of 2,4-dimethyl-3-hydroxy-1-phenyl-1-pentanone, b.p. 101–107/0.4 mm.

3-Ferrocenyl-1-phenyl-2-propen-1-one (Condensation of Ferrocene Carboxaldehyde with Acetophenone).[663] To a stirred solution of 2.56 g. (0.064 mole) of sodium hydroxide in 20 ml. of water (cooled to 15°) are added, successively, solutions of 6.0 g. (0.05 mole) of acetophenone in 10 ml. of 95% ethanol and 10.8 g. (0.05 mole) of ferrocene carboxaldehyde in 30 ml. of 95% ethanol. The mixture is stirred at room temperature for 3 hours and allowed to stand overnight. The thick purple suspension is filtered and the collected solid washed thoroughly with water, followed by a small portion of ice-cold 95% ethanol. After drying, 14.5 g. (92%) of 3-ferrocenyl-1-phenyl-2-propen-1-one (purple solid), m.p. 123–126°, is obtained. Recrystallization from 95% ethanol gives the pure unsaturated ketone; deep purple needles, m.p. 126–128°.

4-Keto-1,2,3,4,5,6,7,8-octahydroazulene (Intramolecular Self-Condensation of 1,6-Cyclodecanedione).[398] 1,6-Cyclodecanedione (50

[663] C. R. Hauser and J. K. Lindsay, *J. Org. Chem.*, **22**, 906 (1957).

g., 0.3 mole) is heated under reflux in 500 ml. of aqueous 5% sodium carbonate for 1 hour. The product is then steam-distilled and the distillate simultaneously extracted with chloroform in an apparatus described by Vogel.[664] The chloroform is removed and the residue distilled. The fraction (43 g., 96%) boiling at 126–128°/15 mm. is collected.

***cis*-9-Acetoxy-10-hydroxy-5-methyldecalin-1,6-dione [Intramolecular Condensation of 2-Acetoxy-2-(3-oxopentyl)cyclohexane-1,3-dione].**[665] To a stirred solution of 42.3 g. (0.167 mole) of 2-acetoxy-2-(3-oxopentyl)cyclohexane-1,3-dione[665] in 850 ml. of dry benzene is added 33.3 ml. (0.40 mole) of pyrrolidine, and then 25 ml. (0.44 mole) of glacial acetic acid is added dropwise (noticeable heating occurs). The yellow solution is stirred at room temperature for 72 hours. The entire reaction mixture is poured onto a column packed with 900 g. of acid-washed alumina. Elution with a 2:3 acetone-hexane mixture affords 28.42 g. (67%) of nearly pure ketol, m.p. 216–218°, after the crude fractions have been washed with ether. Recrystallization from acetone and from an acetone-hexane mixture affords the pure product, m.p. 217–217.5°.

β-Phenylcinnamaldehyde (Condensation of Benzophenone with Acetaldehyde).[221] *A. Ethylidenecyclohexylamine.* Acetaldehyde (44.1 g., 1 mole) is slowly added dropwise, with vigorous stirring, to 99.2 g. (1 mole) of cyclohexylamine, keeping the temperature at −20°. After the mixture has stood for 1 hour at −20°, it is treated with 5 g. of sodium sulfate. The organic layer is separated at room temperature and treated with anhydrous sodium carbonate. The dried product is rapidly distilled under reduced pressure through a fractionating column. The ethylidenecyclohexylamine fraction is redistilled just before use; b.p. 47–48°/12 mm. Tiollais reports b.p. 54°/18 mm., n_D^{15} 1.4647, and 76% yield by a similar procedure.[666]

B. 3,3-Diphenyl-3-hydroxypropylidenecyclohexylamine. A nitrogen atmosphere is employed throughout this procedure. Ethylidenecyclohexylamine (3.0 g., 0.025 mole) in 20 ml. of absolute diethyl ether is added to a cold (0°) ether solution of lithium diisopropylamide (0.025 mole) prepared from 2.53 g. (0.025 mole) of diisopropylamine in 25 ml. of ether at room temperature by addition of 0.025 mole of methyllithium[667] in 24 ml. of ether.[221, 629] The solution is left for 10 minutes at 0°, cooled to −70°, and treated, dropwise, with a solution of 4.55 g.

[664] A. I. Vogel, *A Textbook of Practical Organic Chemistry*, 3rd ed., p. 224, Longmans, Green, New York, 1956.

[665] T. A. Spencer, K. K. Schmiegel, and W. W. Schmiegel, *J. Org. Chem.*, **30**, 1626 (1965).

[666] R. Tiollais, *Bull. Soc. Chim. France* [5] **14**, 708 (1947).

[667] A solution of methyllithium in diethyl ether (0.5 *M*) is available from Foote Mineral Co., Exton, Pennsylvania, U.S.A.

(0.025 mole) of benzophenone in 25 ml. of ether. The mixture is allowed to stand for 24 hours at room temperature, cooled to 0°, and then treated with water. From the ether solution is isolated 7.05 g. (92%) of 3,3-diphenyl-3-hydroxypropylidenecyclohexylamine, m.p. 127–128°.

C. β-Phenylcinnamaldehyde. 3,3-Diphenyl-3-hydroxypropylidenecyclohexylamine (1.54 g., 0.005 mole) and oxalic acid (10.0 g.) are added to water and the mixture is steam-distilled. From the distillate is isolated 1.03 g. (99%) of slightly impure β-phenylcinnamaldehyde, m.p. 42–44°. Recrystallization from petroleum ether (b.p. 40°) furnishes 0.89 g. (85%) of pure product, m.p. 46–47°.

TABULAR SURVEY

The following tables summarize data in the literature through August 1966. The general arrangement is explained by the titles of the tables which are grouped: II–V, condensations of aldehydes only; VI and VII, condensations of ketones only; VIII–XX, condensations of aldehydes with ketones. Procedures leading to no aldol condensation and/or side reactions have for the most part been omitted. Aldol condensations which proceded from, to, or through carbonyl precursors or derivatives, with certain exceptions, have been excluded.

Reactants. Entries in each table are arranged by molecular formula. Radical prefixes in names appear in alphabetical order; in abbreviated structural formulas they appear in numerical order. The sequence for substances having the same molecular formula follows Beilstein; acyclic (unbranched, branched), alicyclic by ring size, etc. An aldehyde or ketone is designated aromatic if the carbonyl group is attached directly to the aromatic ring. A heterocyclic ketone has a carbonyl group attached directly to the heterocyclic ring system or has a keto group within the heterocyclic ring.

In mixed condensations of aldehydes with aldehydes (Tables III–V) and ketones with ketones (VII), entries are found under the reactant having the lowest carbon content, except in Table IV where entries are in order of molecular formula of the aliphatic aldehyde. The aldehyde-ketone condensations (Tables VII–XIX) are arranged by ketone.

Catalyst. The catalyst(s) listed give(s) the product indicated in the yield stated. Where more than one catalyst is listed for a particular condensation the order of arrangement is as follows (incomplete list): (1) bases–metals, alkali hydroxides, alkaline earth hydroxides, alkali metal alkoxides, salts, alkoxides of metals other than alkali, aminomagnesium halides, amines; (2) ion-exchange resins (basic, acidic);

and (3) acids: acidic salts, acid anhydrides, acid halides, acids. Occasionally, after the catalyst leading to the product cited, other catalysts are listed which yield the same product in lower or unstated yield. When different reaction conditions (solvent, temperature, reaction time) produce different products with the same catalyst, the conditions are given with the appropriate product.

Product. The product indicated by name or structural formula is obtained with the catalyst(s) and reaction conditions cited. Products designated by a molecular formula are of unknown or unassigned structure. In the aldehyde-ketone tables the condensation product formula is abbreviated, R being employed to indicate the carboxaldehyde substituent in RCHO.

Yield. The yield is listed in parentheses after the product and refers to product formed with the catalyst and conditions cited, and in most instances is the highest value reported. When several catalysts giving the same product are listed, a range of yields may be reported. A dash indicates yield not stated or unavailable from data in reference cited.

References. The first reference cited refers to the catalyst and conditions listed leading to the highest yield stated. The remaining references are listed in numerical order and refer to other preparations of the same product by aldol condensation, possibly under quite different conditions, but in lower or unstated yield.

TABLE II. SELF-CONDENSATION OF ALIPHATIC ALDEHYDES

Aldehyde	Catalyst	Product(s) (Yield, %)	Refs.
CH_3CHO*	NaOH; $Na_2B_4O_7$	$CH_3CHOHCH_2CHO$ (50–75)	662, 668
	$POCl_3$	$CH_3CHOHCH_2CHO$ (27)	103
	$SOCl_2$; H_2SO_4	$CH_3CH{=}CHCHO$ (43–61)	103, 669–671
C_2H_5CHO	KOH	$C_2H_5CHOHCH(CH_3)CHO$ (—), $C_2H_5CH{=}C(CH_3)CHO$ (—)	738, 137, 696, 739
	$NaOCH_3$, CH_3CO_2Na, soap, or $K[B(OCH_3)_4]$	$C_2H_5CH{=}C(CH_3)CHO$ (59–83)	614, 483, 624, 716, 740, 741
	K_2CO_3	1,3-dioxane ring (O, O) bearing C_2H_5, CH_3, OH and C_2H_5 (94)	452
	$Al(OC_3H_7\text{-}n)_3$	$C_2H_5CO_2C_3H_7\text{-}n$ (—), $C_2H_5CHOHCH(CH_3)CO_2C_3H_7\text{-}n$ (—)	742
	Ion-exchange resin	$C_2H_5CH{=}C(CH_3)CHO$ (—)	732
	$POCl_3$	$C_2H_5CHOHCH(CH_3)CHO$ (31), $C_2H_5CH{=}C(CH_3)CHO$ (—)	103, 102
CH_3OCH_2CHO	K_2CO_3	$CH_3OCH_2CHOHCH(OCH_3)CHO$ (10)	743
$HOCH_2CHOHCHO$	NaOH	Fructose (41), sorbose (40)	35, 744–746
$CH_3CH{=}CHCHO$	NaHg	$n\text{-}C_3H_7CHOHCH(CH_3)CH_2CH_2OH$ (50), tetrahydrofuran ring (O) bearing CH_3, $CH_2CH{=}CH_2$ and OH (40), cyclopentene ring bearing CHO, CH_3, CH_3 (Low)	747
	K_2CO_3	$CH_3CH{=}CHCHOHCH_2CH{=}CHCHO$ (—)	748, 722, 749
	C_2H_5OMgCl	cyclohexadiene ring bearing CHO and CH_3 (—)	750
	$(CH_2)_5NH$ or morpholine, CH_3CO_2H	$CH_3(CH{=}CH)_3CHO$ (—), $CH_3(CH{=}CH)_5CHO$ (—)	751, 752, 109, 110, 482, 729, 753–759

	HCl	CHO, CH_3, O, CH_3 (ring structure) (—)	760–762
n-C_3H_7CHO	NaOH; also KOH, $NaOC_2H_5$, $(CH_3)_2C_6H_3SO_3Na$	n-$C_3H_7CH{=}C(C_2H_5)CHO$ (65–97), n-$C_3H_7CHOHCH(C_2H_5)CHO$ (—)	483, 104, 138, 139, 222, 635, 680, 696, 731, 763–771
	KOH, H_2O, $(C_2H_5)_2O$; also $Ca(OH)_2$, n-C_4H_9OMgI	n-$C_3H_7CHOHCH(C_2H_5)CHO$ (75)	772, 613, 773
	K_2CO_3, H_2O, 10°	n-C_3H_7, C_2H_5, OH, O, O, C_3H_7-n (ring structure) (73)	452–454
	Ion-exchange resin, Wolfatit L 150	n-$C_3H_7CHOHCH(C_2H_5)CHO$ (58–79)	774, 100, 631, 775
	Ion-exchange resin, Amberlite IR-4B	n-$C_3H_7CH{=}C(C_2H_5)CHO$ (14–54)	630, 101, 632, 776
	$POCl_3$; also $SOCl_2$, H_2SO_4	n-$C_3H_7CH{=}C(C_2H_5)CHO$ (53)	103, 102, 777–779
i-C_3H_7CHO	NaOH; also KOH, $Mg[Al(OC_4H_9\text{-}n)_3]_2$, $POCl_3$	$(CH_3)_2CHCHOHC(CH_3)_2CHO$ (85)	780, 102, 103, 457, 767, 781–794
	K_2CO_3, H_2O, 20°	i-C_3H_7, $(CH_3)_2$, OH, O, O, C_3H_7-i (ring structure) (75)	452
	NaCN	i-$C_3H_7CHOHC(CH_3)_2CHOHCN$ (—)	791
$CH_3CHOHCH_2CHO$	CaO; also NaOH, K_2CO_3	$CH_3(CHOHCH_2)_nCHO$ (—)	795, 684, 722, 729, 796

Note: References 668–2359 are on pp. 403–438.

* Acetaldehyde is also converted to aldol and/or 2-butenal in mostly unspecified yield by the following catalysts. The numbers following each catalyst are the pertinent references. MgHg, 672; NaOH or KOH, 480, 631, 673–711; $Ba(OH)_2$, 712, 713; MgO, 714–716; $NaOC_2H_5$, 700–702; K_2CO_3, 480, 717–722; KCN, 6; HCO_2K, 480; CH_3CO_2Na, 480, 723, 724; Na_2SO_3, 691, 725; ROMgI (R = alkyl), 613; morpholine, 726; amino acids, amines, 482, 609, 727–729; $ZnCl_2$, 480, 724, 730; ion-exchange resin Amberlite, 101, 630, 632, 731–734; silica gel, 735; HCl, 1, 736, 737.

TABLE II. Self-Condensation of Aliphatic Aldehydes (*Continued*)

Aldehyde	Catalyst	Product(s) (Yield, %)	Refs.
$(CH_3)_2C{=}CHCHO$	NaOH; also $Ba(OH)_2$	$(CH_3)_2C{=}CHCH{=}CHC(CH_3){=}CHCHO$ (—), 3,7,11-trimethyl-2,4,6,8,10-dodecapentaenal (—), $C_{15}H_{20}O$ (—)	105, 106
	$NaNH_2$	CH_3, $(CH_3)_2$, CHO (—)	105
	$(CH_2)_5NH$, CH_3CO_2H	$(CH_3)_2C{=}CHCH{=}CHC(CH_3){=}CHCHO$ (14.5)	115
n-C_4H_9CHO	KOH	$C_{20}H_{38}O_3$ (—)	489–491
	K_2CO_3; also $Ca(OH)_2$, $Zn(C_2H_5)_2$, HCl	n-$C_4H_9CH{=}C(C_3H_7$-$n)CHO$ (—), $C_{15}H_{28}O$ (—), $C_{20}H_{30}O$ (—)	488, 290, 628, 767, 797–800
$C_2H_5CH(CH_3)CHO$	K_2CO_3	$C_2H_5CH(CH_3)CHOHC(CH_3)(C_2H_5)CHO$ (—)	801
i-$C_3H_7CH_2CHO$	$NaOC_5H_{11}$-n; also NaOH, MgI_2, leucine	i-$C_3H_7CH_2CH{=}C(C_3H_7$-$i)CHO$ (66)	604, 802–804
	KOH; also $Ba(OH)_2$	i-$C_3H_7CH_2CH{=}C(C_3H_7$-$i)CHO$ (17), i-$C_3H_7CH_2CHOHCH(C_3H_7$-$i)CHO$ (—)	805, 797, 806–812
$CH_3CH(OCH_3)CH_2CHO$	K_2CO_3	$CH_3CH(OCH_3)CH_2CHOHC(CHO){=}CHCH_3$ (30)	813
O, CHO	NaOH or $(C_6H_5CO_2)_2Cu$	O, CHOH, CHO, O (59–76)	814–816
n-$C_3H_7CH{=}CHCHO$	NaOH	High-mol.-wt. acid, alcohol, lactone (—)	492
$CH_2{=}CH(CH_2)_3CHO$†	KOC_4H_9-t	$CH_2{=}CH(CH_2)_3CH{=}C(CHO)CH_2CH_2$-$CH{=}CH_2$ (—)	817
$OHC(CH_2)_4CHO$	KOH; also CH_3CO_2H, H_3PO_4	CHO (58–62)	374, 818–820
$HN[(CH_2)_2CHO]_2$	HCl	CHO, $\oplus$NH_2 $Cl^{\ominus}$ (75)	821

	HCl, C_6H_5COCl, C_5H_5N	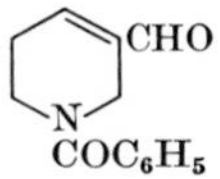 1-benzoyl-1,2,5,6-tetrahydropyridine-3-carbaldehyde (N–COC_6H_5, CHO) (100)	821
n-$C_5H_{11}CHO$	Ion-exchange resin, Amberlite IR-4B	n-$C_5H_{11}CH{=}C(C_4H_9\text{-}n)CHO$ (16–60)	101, 630
	Ion-exchange resins	n-$C_5H_{11}CHOHCH(C_4H_9\text{-}n)CHO$ (50–66)	100
 3-cyclohexene-1-carbaldehyde (CHO)	NaOH	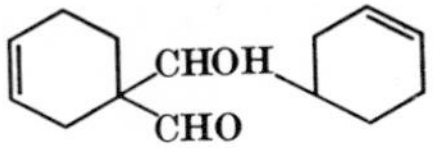 1-[CHOH(3-cyclohexenyl)]-3-cyclohexene-1-carbaldehyde (CHOH, CHO) (—)	822
$OHC(CH_2)_5CHO$	H_2SO_4	 1-cyclohexene-1-carbaldehyde (CHO) (—)	378
n-$C_6H_{13}CHO$	KOH; also K_2CO_3, $(i\text{-}C_3H_7O)_3B$, $Zn(C_2H_5)_2$, $ZnCl_2$, $HCONH_2$	n-$C_6H_{13}CH{=}C(C_5H_{11}\text{-}n)CHO$ (80), $C_{28}H_{50}O$ (—), $C_{28}H_{54}O_3$ (—)	493, 489–496, 628, 823, 824
	Ion-exchange resin, Amberlite	n-$C_6H_{13}CH{=}C(C_5H_{11}\text{-}n)CHO$ (16–57)	101, 632, 825
	$POCl_3$; also $SOCl_2$, HNO_3	n-$C_6H_{13}CHOHCH(C_5H_{11}\text{-}n)CHO$ (41–46), n-$C_6H_{13}CH{=}C(C_5H_{11}\text{-}n)CHO$ (—)	103, 102, 792, 793, 826
$C_6H_5CH_2CHO$	$(CH_2)_5NH$, CH_3CO_2H	$C_6H_5CH_2CH{=}C(C_6H_5)CHO$ (35)	827
	KOH	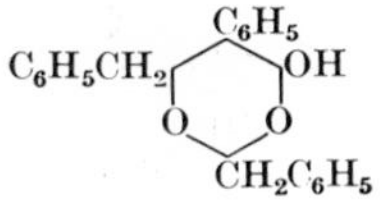 1,3-dioxane ring with $C_6H_5CH_2$, C_6H_5, OH, $CH_2C_6H_5$ (27.5)	455
n-$C_3H_7CH{=}C(C_2H_5)CHO$	NaOH	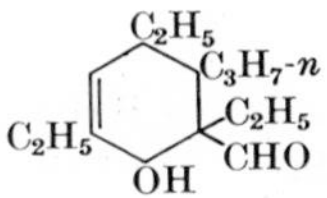 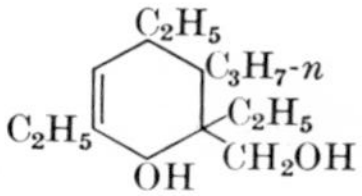 cyclohexene ring with C_2H_5, C_2H_5, C_3H_7-n, C_2H_5, CHO, OH (45), cyclohexene ring with C_2H_5, C_2H_5, C_3H_7-n, C_2H_5, CH_2OH, OH (16–25)	381, 66, 383, 483–485, 492

(See also p.90)

Note: References 668–2359 are on pp. 403–438.

† Formed *in situ* from cyclohexane-*cis*-(or *trans*)-1,3-diol mono-4-bromobenzenesulfonate.

TABLE II. SELF-CONDENSATION OF ALIPHATIC ALDEHYDES (*Continued*)

Aldehyde	Catalyst	Product(s) (Yield, %)	Refs.
n-$C_3H_7CH{=}C(C_2H_5)CHO$ (*contd.*)	NaOH (*contd.*)	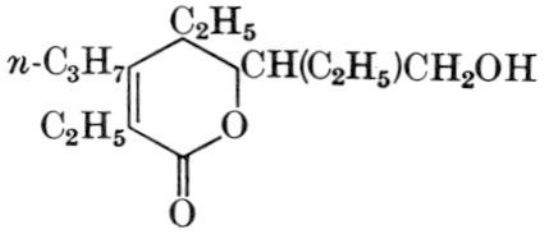 n-C_3H_7, C_2H_5, C_2H_5, $CH(C_2H_5)CH_2OH$, O, O (30–56),	381, 66, 383, 483–485, 492
		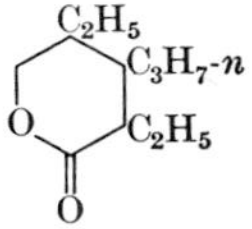 C_2H_5, C_3H_7-n, C_2H_5, O, O (9–36),	
		n-$C_3H_7CO_2H$ (5), n-$C_3H_7CH{=}C(C_2H_5)CO_2H$ (1.5)	
n-$C_7H_{15}CHO$	$NaOC_2H_5$; also fatty acid metal salt, MgI_2, ion-exchange resin, Amberlite, IR-4B	n-$C_7H_{15}CH{=}C(C_6H_{13}$-$n)CHO$ (36–79)	104, 101, 623, 803, 2349
i-$C_3H_7(CH_2)_4CHO$	$NaNH_2$	C_{16} alcohol (—), C_{16} diol (—)	828
	$[i$-$C_3H_7(CH_2)_4CO_2]_2Mg$	i-$C_3H_7(CH_2)_4CH{=}C(CHO)(CH_2)_3C_3H_7$-$i$ (98)	626, 2349
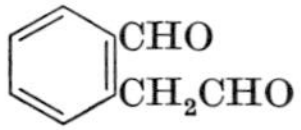 CHO, CH_2CHO (benzene ring)	NaOH	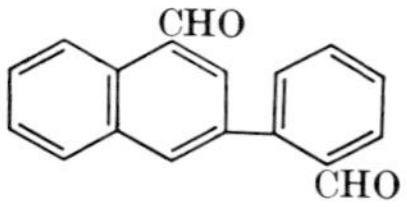 CHO, CHO (46)	829, 830
$C_6H_5CH_2CH_2CHO$	K_2CO_3	$C_6H_5CH_2CH_2CH{=}C(CHO)CH_2C_6H_5$ (26)	831
$OHC(CH_2)_3CH(C_3H_7$-$n)CHO$	None, H_2O, 115°	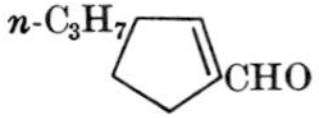 n-C_3H_7, CHO (62)	818
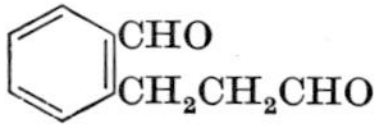 CHO, CH_2CH_2CHO (benzene ring)	H_2SO_4	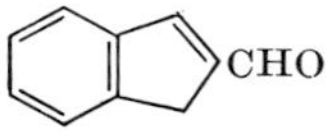 CHO (—)	832

$(CH_3)_2C{=}CHCH_2CH_2$- $C(CH_3){=}CHCHO$, citral	$(CH_2)_5NH$	$C_{20}H_{30}O$ (—)	833
$(CH_3)_2C{=}CHCH_2CH_2$- $CH(CH_3)CH_2CHO$, citronellal	KOH	$C_{20}H_{32}O_2$ (64)	834
CH_3, CH_3, CH_3, CHO (cyclohexene)	"	Polymer (—)	835
$CH_3CO(CH_2)_2CH[C(CH_3){=}CH_2]$- CH_2CHO	$(CH_2)_5NH$, CH_3CO_2H	CH_3, CHO, $C(CH_3){=}CH_2$ (cyclopentene) (59)	836
n-$C_9H_{19}CHO$	Fatty acid metal salt	n-$C_9H_{19}CH{=}C(CHO)C_8H_{17}$-$n$ (70)	623
$OHC(CH_2)_2C(CO_2C_2H_5)_2$- $(CH_2)_2CHO$	$NaOC_2H_5$	$(C_2H_5O_2C)_2$, CHO (cyclohexene) (15)	377
$OHCCH(C_3H_7$-$n)CH_2$- $CH(C_3H_7$-$n)CH_2CHO$‡	None, heat	n-C_3H_7, CHO, C_3H_7-n (cyclopentene) (80)	818
$OHCCH(C_2H_5)CH(C_3H_7$-$n)$- $CH(C_2H_5)CHO$§	"	C_2H_5, n-C_3H_7, CHO, C_2H_5, OH (cyclobutane) (30)	381
$CO_2C_2H_5$, CH_2CH_2CHO, O (cyclohexanone)	H_2SO_4	$CO_2C_2H_5$, I (—); $CO_2C_2H_5$ II (—) I:II=4.7:1	451

Note: References 668–2359 are on pp. 403–438.

‡ Formed *in situ* by lead tetracetate oxidation of 3,5-di-n-propylcyclohexane-1,2-diol.

§ Formed *in situ* as an ozonolysis product of C_2H_5, C_3H_7-n, C_2H_5, CHO, C_2H_5, OH (cyclohexene).

TABLE II. Self-Condensation of Aliphatic Aldehydes (*Continued*)

Aldehyde	Catalyst	Product(s) (Yield, %)	Refs.
$OHC(CH_2)_2C(CO_2C_2H_5)_2$-$(CH_2)_2CHO$	$NaOC_2H_5$	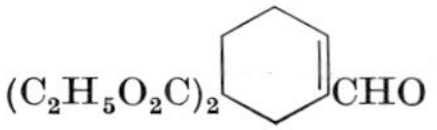 $(C_2H_5O_2C)_2$ cyclohexene CHO (22)	377
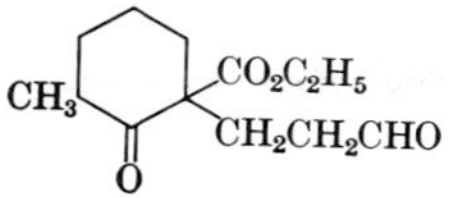 CH_3, $CO_2C_2H_5$, CH_2CH_2CHO, O (cyclohexanone)	H_2SO_4	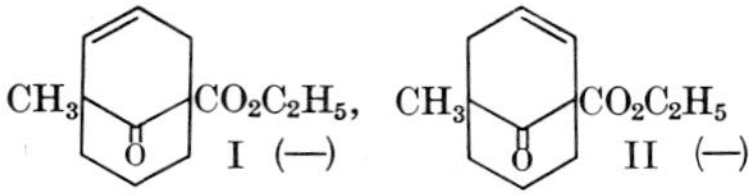 CH_3, $CO_2C_2H_5$, I (—); CH_3, $CO_2C_2H_5$, II (—); I:II = 2.8:1	451, 2318, 2358
$OHCCH_2CH(C_4H_9\text{-}sec)CH_2$-$CH(C_4H_9\text{-}sec)CHO$ ‖	None	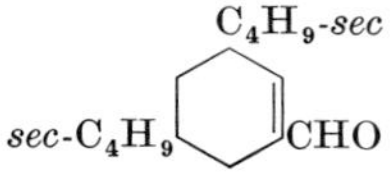 C_4H_9-*sec*, *sec*-C_4H_9, CHO (—)	373
$OHCCH(CH_3)CH_2C(CO_2C_2H_5)_2$-$CH_2CH(CH_3)CHO$	$NaOC_2H_5$	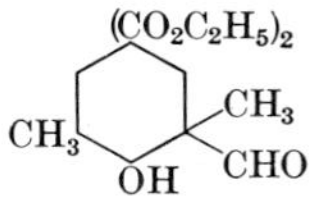 $(CO_2C_2H_5)_2$, CH_3, CH_3, CHO, OH (13)	377
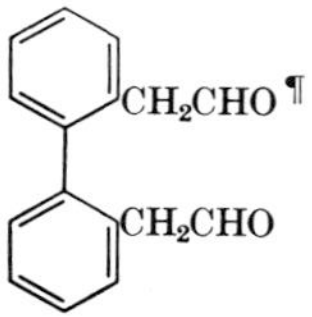 CH_2CHO ¶, CH_2CHO (biphenyl)	H_2SO_4	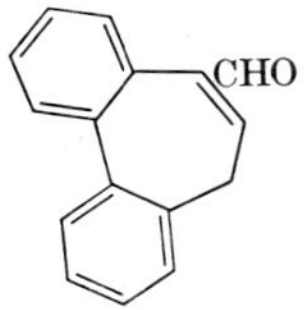 CHO (dibenzocycloheptene) (—)	379
$OHC(CH_2)_{14}CHO$	$C_6H_5SO_3H$	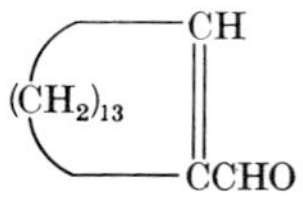 $(CH_2)_{13}$, CH, CCHO (—)	380

	$(C_2H_5)_3N$, CH_3CO_2H		(—)	837, 838
$OHC(CH_2)_{16}CHO$	$C_6H_5SO_3H$		(—)	380
n-$C_{17}H_{35}CHO$	$(n$-$C_{17}H_{35}CO_2)Co$	n-$C_{17}H_{35}CH{=}C(C_{16}H_{33}$-$n)CHO$	(81)	623
	None	or	(—)	375
	NaOH		(—)	376

Note: References 668–2359 are on pp. 403–438.

‖ Formed *in situ* by lead tetracetate oxidation of 3,5-di-*sec*-butylcyclohexane-1,2-diol.

¶ Formed *in situ* from the tetramethyl acetal.

TABLE III. MIXED CONDENSATION OF ALIPHATIC ALDEHYDES

Reactants	Catalyst	Product(s) (Yield, %)	Refs.
CH_2O and			
CH_3CHO	$Ca(OH)_2$; also NaOH, KOH, H_2SO_4	$C(CH_2OH)_4$ (79–80), $HC(CH_2OH)_3$ (70)	839, 117, 118, 722, 840–863
	MgO	$C(CH_2OH)_4$ (90)	864
	K_2CO_3	$HOCH_2CH_2CHO$ (—), $CH_3CHOHCH_2CHO$ (—)	865–867
	CsOH; also Al_2O_3, SiO_2, or MnO_2, 300°	$CH_2{=}CHCHO$ (62)	125, 126, 868–884
$HOCH_2CHO$	$Ca(OH)_2$	$HOCH_2CHOHCHO$ (75)	602
C_2H_5CHO	NaOH; also $Ca(OH)_2$	$CH_3C(CH_2OH)_3$ (94)	885, 463, 886–894
	KOH	$HOCH_2CH(CH_3)CHO$ (—), $CH_2{=}C(CH_3)CHO$ (—), $C_2H_5CHOHC(CH_3)(CH_2OH)CHO$ (—)	895
	K_2CO_3	$CH_3C(CH_2OH)_3$ (—), $CH_3C(CHO)(CH_2OH)_2$ (—)	895
	Na_2O, SiO_2, 275°	$CH_2{=}C(CH_3)CHO$ (46)	125, 126, 880
	BF_3	$CH_2{=}C(CH_3)CHO$ (—), $C_2H_5CH{=}C(CH_3)CHO$ (—)	896
	H_2SO_4; also H_2SO_4, CH_3CO_2H	$CH_2{=}C(CH_3)CHO$ (60), $C_2H_5CH{=}C(CH_3)CHO$ (—), $CH_3C(CH_2OH)_3$ (—)	897, 460, 898
CH_3OCH_2CHO	CaO	$CH_3OC(CH_2OH)_3$ (15)	465
$HOCH_2CH_2CHO$	$Ca(OH)_2$	$C(CH_2OH)_4$ (63)	899, 722
$CH_3CH{=}CHCHO$	K_2CO_3	Polymer (—)	900–902
	NaO_2CCH_3	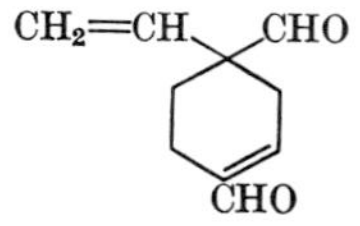 $CH_2{=}CH$, CHO, CHO (—)	900
	C_5H_5N	$C_{10}H_{12}O_2$ (—)	903
	H_2SO_4	$CH_3(CH{=}CH)_3CHO$ (—)	501

n-C_3H_7CHO	NaOH or $Ca(OH)_2$	$C_2H_5C(CH_2OH)_3$ (80–90)	887, 463, 886, 888, 894, 905–908
	K_2CO_3	$C_2H_5C(CH_2OH)_2CHO$ (90)	122
	Na_2O, silica gel	$CH_2{=}C(C_2H_5)CHO$ (49)	125, 120, 880, 909
	H_2SO_4	$CH_2{=}C(C_2H_5)CHO$ (40)	898
i-C_3H_7CHO	$Ca(OH)_2$; also NaOH, KOH	$HOCH_2C(CH_3)_2CH_2OH$ (90)	463, 468, 471, 890, 894, 910–917, 925
	K_2CO_3	$HOCH_2C(CH_3)_2CHO$ (40)	918, 471, 919–924
$C_2H_5OCH_2CHO$	CaO	$C_2H_5OC(CH_2OH)_3$ (24)	465
	$(C_2H_5)_2NH{\cdot}HCl$	$CH_2{=}C(OC_2H_5)CHO$ (67)	904
$CH_3CHOHCH_2CHO$	$Ca(OH)_2$ or $Pb(OH)_2$	$C(CH_2OH)_4$ (—)	926
	Na_2CO_3	$CH_3CHOHC(CH_2OH)_2CHO$ (91)	927
$CH_2{=}CH(CH_2)_2CHO$	NaOH	$CH_2{=}CHCH_2C(CH_2OH)_3$ (73)	928
$OHC(CH_2)_3CHO$	"	$(HOCH_2)_2$ / O / $(CH_2OH)_2$ (33)	929
n-C_4H_9CHO	None, H_2O, 100°	n-$C_3H_7C(CHO){=}CH_2$ (65)	127, 125, 880
$C_2H_5CH(CH_3)CHO$	KOH	$C_2H_5C(CH_3)(CH_2OH)CHO$ (60)	930
	"	$C_2H_5C(CH_3)(CH_2OH)_2$ (55–87)	931, 471, 932, 933
i-C_4H_9CHO	NaOH	i-$C_3H_7C(CH_2OH)_3$ (63)	934, 466, 935
	K_2CO_3	i-$C_3H_7CH(CH_2OH)CHO$ (52)	120
n-$C_3H_7OCH_2CHO$	CaO	n-$C_3H_7OC(CH_2OH)_3$ (10)	465
$CH_3CH(OCH_3)CH_2CHO$	K_2CO_3	$CH_3CH{=}C(CH_2OH)CHO$ (15)	813, 900
	C_5H_5N	$CH_2{=}CH$, CHO, CHO (—),	900
		$CH_3CH{=}C(CH_2OH)CHO$ (—)	

Note: References 668–2359 are on pp. 403–438.

TABLE III. MIXED CONDENSATION OF ALIPHATIC ALDEHYDES (*Continued*)

Reactants	Catalyst	Product(s) (Yield, %)	Refs.
CH_2O (***contd.***) and $HOCH_2(CH_2)_3CHO$	NaOH	$HOCH_2(CH_2)_2C(CH_2OH)_3$ (75)	936, 937
	,,	O, $(CH_2OH)_2$, OH (—)	937
CHO	KOH	$(CH_2OH)_2$ (65)	938
n-$C_5H_{11}CHO$	NaOH	n-$C_4H_9C(CH_2OH)_3$ (70)	934, 465, 939
	K_2CO_3	n-$C_4H_9C(CH_2OH)_2CHO$ (70)	122
	None, H_2O, 100°	n-$C_3H_7C(CHO){=}CH_2$ (65)	127
n-$C_3H_7CH(CH_3)CHO$	KOH	n-$C_3H_7C(CH_3)(CH_2OH)_2$ (—)	940
i-$C_3H_7CH(CH_3)CHO$	,,	i-$C_3H_7C(CH_3)(CH_2OH)_2$ (46)	931
$(C_2H_5)_2CHCHO$	,,	$(C_2H_5)_2C(CH_2OH)_2$ (90)	913
CHO	,,	$(CH_2OH)_2$ (91)	941, 835
$CH_2{=}CH(CH_2)_2CH(CH_3)CHO$	K_2CO_3	$CH_2{=}CH(CH_2)_2C(CH_3)(CH_2OH)_2$ (—)	471
OH, $(CH_3)_2$, O, O, O, CHO	NaOH	OH, $(CH_3)_2$, O, O, O, $(CH_2OH)_2$ (90)	942
n-$C_6H_{13}CHO$	KOH	n-$C_5H_{11}C(CH_2OH)_2CHO$ (—)	121
	$Ca(OH)_2$	n-$C_5H_{11}C(CH_2OH)_3$ (60)	464, 466, 939, 943
sec-$C_4H_9CH(CH_3)CHO$	NaOH	sec-$C_4H_9C(CH_3)(CH_2OH)_2$ (41)	944, 945
CH_3, CHO	KOH	CH_3, $(CH_2OH)_2$ (50–60)	835

$n\text{-}C_3H_7CH{=}C(C_2H_5)CHO$	"	$n\text{-}C_3H_7CH{=}C(CH_2OH)C_2H_5$ (48)	946
$C_2H_5O_2C(CH_2)_3CHO$	CaO	$(HOCH_2)_3C(CH_2)_2CO_2H$ (17)	947
$n\text{-}C_7H_{15}CHO$	"	$n\text{-}C_6H_{13}C(CH_2OH)_3$ (—)	465
$n\text{-}C_5H_{11}CH(CH_3)CHO$	K_2CO_3	$n\text{-}C_5H_{11}C(CH_3)(CH_2OH)CHO$ (80)	471
$C_6H_5CH(CH_3)CHO$	KOH; also K_2CO_3	$C_6H_5C(CH_3)(CH_2OH)_2$ (85)	948, 932
	K_2CO_3	$C_6H_5C(CH_3)(CH_2OH)CHO$ (—)	471
CHO	$Ca(OH)_2$	$(CH_2OH)_2$ (—)	948
$n\text{-}C_8H_{17}CHO$	KOH	$n\text{-}C_7H_{15}C(CH_2OH)_2CHO$ (—)	121
	"	$n\text{-}C_7H_{15}C(CH_2OH)_3$ (65)	939
$i\text{-}C_3H_7(CH_2)_3CH(CH_3)CHO$	K_2CO_3	$i\text{-}C_3H_7(CH_2)_3C(CH_3)(CH_2OH)_2$ (80–90)	471
CHO, CH_3	NaOH	$(CH_2OH)_2$, CH_3 (—)	948
CH_3, CH_3, CH_3, CHO	KOH	CH_3, CH_3, CH_3, $(CH_2OH)_2$ (50–60)	835
$(CH_3)_2C{=}CHCH_2CH_2CH(CH_3)\text{-}CH_2CHO$	"	$(CH_3)_2C{=}CHCH_2CH_2CH(CH_3)C(CH_2OH)_3$ (5)	465
$n\text{-}C_9H_{19}CHO$	"	$n\text{-}C_8H_{17}C(CH_2OH)_3$ (47)	939, 465
$n\text{-}C_{10}H_{21}CHO$	"	$n\text{-}C_9H_{19}C(CH_2OH)_3$ (—)	939
$n\text{-}C_8H_{17}CH(CH_3)CHO$	K_2CO_3	$n\text{-}C_8H_{17}C(CH_3)(CH_2OH)_2$ (80–90)	471
$n\text{-}C_9H_{19}CH(CH_3)CHO$	"	$n\text{-}C_9H_{19}C(CH_3)(CH_2OH)_2$ (80–90)	471
$n\text{-}C_{12}H_{25}CHO$	KOH	$n\text{-}C_{11}H_{23}C(CH_2OH)_3$ (—)	939
$(C_6H_5)_2CHCHO$	NaOH	$(C_6H_5)_2C(CH_2OH)_2$ (25)	949
$C_6H_5COCH(C_6H_5)CHO$	KOH	$C_6H_5COC(C_6H_5)(CH_2OH)_2$ (60)	949
CH_3, CH_3, C_6H_5, CHO	"	CH_3, CH_3, C_6H_5, $(CH_2OH)_2$ (50–60)	835

Note: References 668–2359 are on pp. 403–438.

TABLE III. MIXED CONDENSATION OF ALIPHATIC ALDEHYDES (*Continued*)

Reactants	Catalyst	Product(s) (Yield, %)	Refs.
CH_2O (***contd.***) and			
HO, O, H_3C, H_3C, CHO*	H_2SO_4	HO, O, H_3C, H_3C, $(CH_2OH)_2$ (58)	124
O, O, O, H_3C, H_3C, CH_2CHO	$HCON(CH_3)_2$	O, O, H_3C, H_3C, $CH(CH_2OH)_2$ (20)	124
		O, O, H_3C, H_3C, $C(CH_2OH)_3$ (8)	
		O, O, O, H_3C, H_3C, $CH_2{=}CCHO$ (60)	

Reactants	Catalyst	Product (Yield, %)	Refs.
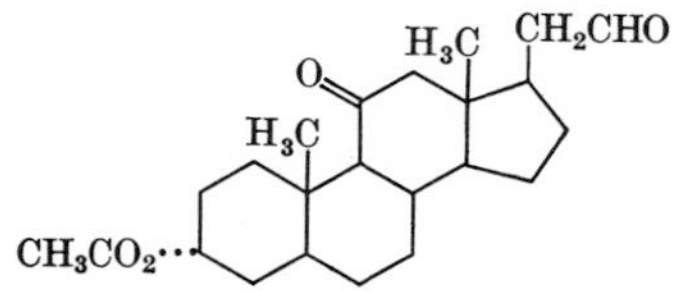 CH_3CO_2, H_3C, O, H_3C, CH_2CHO (steroid structure)	CaO, $HCON(CH_3)_2$	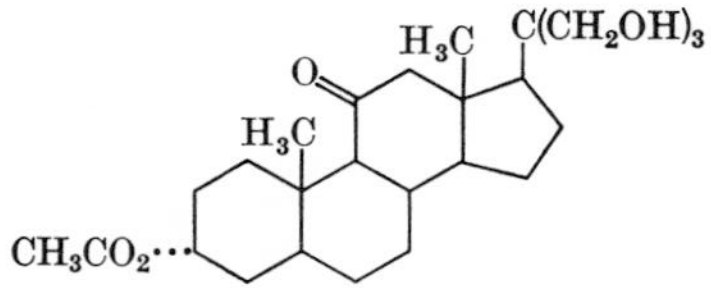 CH_3CO_2, H_3C, O, H_3C, $C(CH_2OH)_3$ (steroid structure) (—)	950
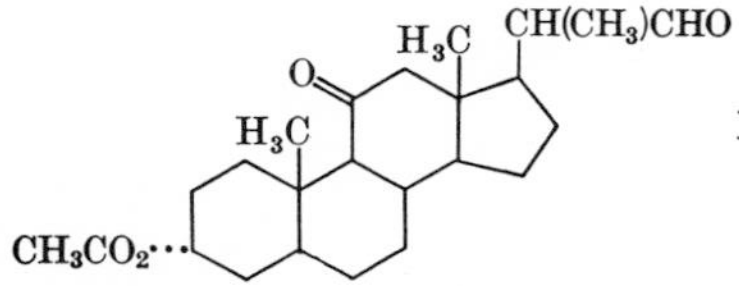 CH_3CO_2, H_3C, O, H_3C, $CH(CH_3)CHO$ (steroid structure)	KOH	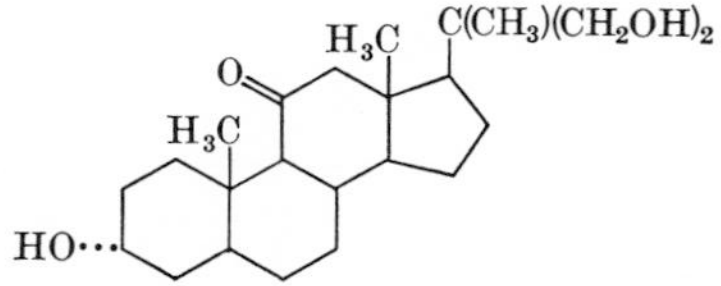 HO, H_3C, O, H_3C, $C(CH_3)(CH_2OH)_2$ (steroid structure) (—)	951, 124
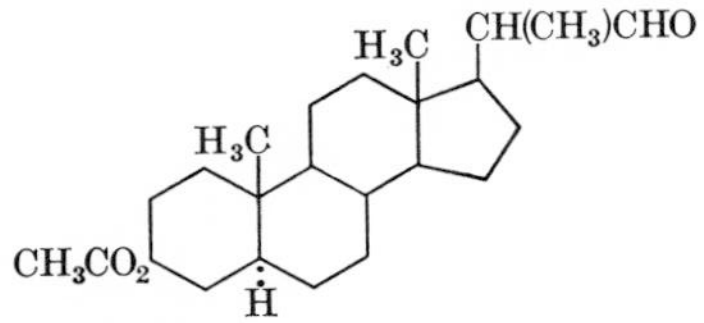 CH_3CO_2, H_3C, H, H_3C, $CH(CH_3)CHO$ (steroid structure)		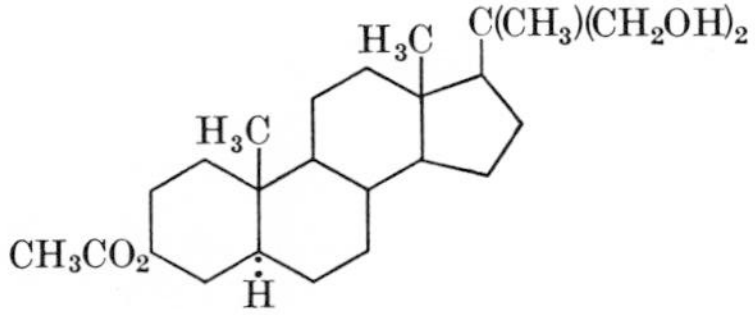 CH_3CO_2, H_3C, H, H_3C, $C(CH_3)(CH_2OH)_2$ (steroid structure) (—)	952
$(CHO)_2$ and CH_3CHO	HCl or $(CH_2)_5NH$, CH_3CO_2H	Polymer (—)	953
i-C_3H_7CHO	KOH	$OHCC(CH_3)_2CHOHCHOHC(CH_3)_2CHO$ (—), $OHCCHOHC(CH_3)_2CHOHC(CH_3)_2CHO$ (—)	954
CCl_3CHO **and** CH_3CHO	CH_3CO_2H	$C_6H_9Cl_3O_3$ (—), $C_4H_5Cl_3O_3$ (—)	636
CH_3CHO **and** C_2H_5CHO	NaOH	$CH_3CH{=}C(CH_3)CHO$ (53)	138, 139
	KOH	$CH_3CHOHC(CH_3)CHO$ (—)	137, 955
	H_2SO_4, CH_3CO_2H	$CH_3CHOHC(CH_3)(CH_2OH)CHOHCH_3$ (—)	460
$CH_3CHClCCl_2CHO$	CH_3CO_2H	$CH_3CHClCCl_2CHOHCH(CHO)CHOHCH_3$ (—)	636

Note: References 668–2359 are on pp. 403–438.

* This compound was formed *in situ* from its dimethyl acetal.

TABLE III. Mixed Condensation of Aliphatic Aldehydes (*Continued*)

Reactants	Catalyst	Product(s) (Yield, %)	Refs.
CH_3CHO (***contd.***) **and**			
$CH_3CH{=}CHCHO$	$(CH_2)_5NH$, CH_3CO_2H	$CH_3(CH{=}CH)_2CHO$ (—)	751, 956, 957
	,,	$CH_3(CH{=}CH)_2CHO$ (—), $CH_3(CH{=}CH)_3$-CHO (—)	958, 107, 956, 957, 959–963
n-C_3H_7CHO	NaOH; also ion-exchange resin	$CH_3CH{=}C(C_2H_5)CHO$ (41)	138, 776
	KOH	$CH_3CHOHCH(C_2H_5)CHO$ (61)	662
	$Ni_3(PO_4)_2$ or $Cd_3(PO_4)_2$ on silica gel at 275°; also K_2CO_3	$CH_3CH{=}CHCHO$ (—), $CH_2{=}CHCH_2CHO$ (—), $CH_3CH{=}C(C_2H_5)CHO$ (—), $CH_3(CH_2)_2CH{=}CHCHO$ (—), n-$C_3H_7CH{=}C(C_2H_5)CHO$ (—)	691, 963
i-C_3H_7CHO	KOH	$CH_3CHOHC(CH_3)_2CHO$ (40)	964, 918
$C_2H_5OCH_2CHO$	K_2CO_3	$C_2H_5OCH_2CHOHCH_2CHO$ (—)	965
i-C_4H_9CHO	CH_3CO_2Na	Unsaturated aldehydes (—)	966
$C_6H_5CH_2CHO$	,,	$CH_3CH{=}C(C_6H_5)CHO$ (44)	967
$CH_3(CH{=}CH)_3CHO$	$(CH_2)_5NH$, CH_3CO_2H	$CH_3(CH{=}CH)_4CHO$ (—)	958
[cyclohexyl]CH_2CHO	$NaOC_2H_5$	[cyclohexyl]$CH_2CH{=}CHCHO$ (—)	968
2-$O_2NC_6H_4CH{=}CHCHO$	NaOH	2-$O_2NC_6H_4(CH{=}CH)_2CHO$ (—)	969
$C_6H_5CH{=}CHCHO$	NaOH; also $NaNH_2$	$C_6H_5(CH{=}CH)_2CHO$ (20)	970, 585
4-$O_2NC_6H_4CH{=}C(CH_3)CHO$	NaOH	4-$O_2NC_6H_4CH{=}C(CH_3)CH{=}CHCHO$ (—)	971
$C_6H_5CH{=}C(CH_3)CHO$	,,	$C_6H_5CH{=}C(CH_3)CH{=}CHCHO$ (21)	972
$CH_2{=}C(CH_3)$[cyclohexenyl]CHO	,,	$CH_2{=}C(CH_3)$[cyclohexenyl]$CH{=}CHCHO$ (—)	973
Citral	$NaNH_2$	Citrylideneacetaldehyde (9)	585, 622
	$(CH_2)_5NH$, CH_3CO_2H	Citrylideneacetaldehyde (—), α- and β-citrylidenecrotonaldehyde (—)	585, 958
$C_6H_5(CH{=}CH)_2CHO$	NaOH	$C_6H_5(CH{=}CH)_3CHO$ (—)	970
CH_3[cyclohexenyl]$CH(CH_3)CH_2CHO$	KOH	CH_3[cyclohexenyl]$CH(CH_3)CH_2CH{=}CHCHO$ (—)	144

n-$C_{10}H_{21}CHO$	$NaNH_2$	n-$C_{10}H_{21}CHOHCH_2CHO$ (41), n-$C_9H_{19}CH(CHOHCH_3)CHO$ (19)	143
$HOCH_2CHO$ and $HOCH_2CHOHCHO$	$Ca(OH)_2$	Ribose, arabinose, xylose (—)	129, 974, 975
C_2H_5CHO and			
n-C_3H_7CHO	NaOH	$C_2H_5CH{=}C(C_2H_5)CHO$ (42)	138
i-C_3H_7CHO	KOH	$C_2H_5CHOHC(CH_3)_2CHO$ (—)	133
$C_6H_5CH{=}CHCHO$	NaOH	$C_6H_5CH{=}CHCH{=}C(CH_3)CHO$ (60)	972, 976
CHO (ring) $CH_2CH_2CH{=}C(CH_3)_2$	KOH	$(CH_3)_2C{=}CHCH_2CH_2$ (ring) $CH{=}C(CH_3)CHO$ (—)	977
$CH_3CH{=}CHCHO$ and			
i-C_4H_9CHO	—	Liquid, b.p. 85–120°/12 mm. (8)	978
$CH_3(CH{=}CH)_2CHO$	$(CH_2)_5NH$, CH_3CO_2H	$CH_3(CH{=}CH)_6CHO$ (—), $CH_3(CH{=}CH)_8CHO$ (—)	753
$CH_3(CH{=}CH)_3CHO$	,,	$CH_3(CH{=}CH)_5CHO$ (7)	754
$C_6H_5CH{=}CHCHO$	,,	$C_6H_5(CH{=}CH)_5CHO$ (—), $C_6H_5(CH{=}CH)_7CHO$ (—)	979
$(CH_3)_2C{=}CHCH_2CH_2$-$C(CH_3){=}CHCHO$, citral	$NaNH_2$	α-Citrylidenecrotonaldehyde (—)	585, 622
	$(CH_2)_5NH$, CH_3CO_2H	α- and β-Citrylidenecrotonaldehyde (12–19)	585, 980
$C_6H_5(CH{=}CH)_2CHO$	,,	$C_6H_5(CH{=}CH)_4CHO$ (—), $C_6H_5(CH{=}CH)_6CHO$ (—)	116
n-$C_{10}H_{21}CHO$	,,	$CH_3CH{=}C(C_9H_{19}$-$n)CHO$ (—), n-$C_{10}H_{21}(CH{=}CH)_4CHO$ (—)	143
$CH_3(CH{=}CH)_5CHO$	,,	$CH_3(CH{=}CH)_7CHO$ (—)	753
$C_6H_5(CH{=}CH)_5CHO$	,,	$C_6H_5(CH{=}CH)_7CHO$ (—)	116
n-C_3H_7CHO and			
i-C_3H_7CHO	NaOH	n-$C_3H_7CH{=}C(C_2H_5)CHO$, i-$C_3H_7CH{=}C(C_2H_5)CHO$ (Total, 88)	767, 136, 731
$C_6H_5CH{=}CHCHO$	,,	$C_6H_5CH{=}CHCH{=}C(C_2H_5)CHO$ (55)	976, 972
$CH_3C(OH)(CO_2C_2H_5)$-$CH(C_2H_5)CHO$	$(C_2H_5)_2NH$	$CH_3C(OH)(CO_2C_2H_5)CH(C_2H_5)CH{=}C(C_2H_5)CHO$ (20–30)	981

Note: References 668–2359 are on pp. 403–438.

TABLE III. Mixed Condensation of Aliphatic Aldehydes (*Continued*)

Reactants	Catalyst	Product(s) (Yield, %)	Refs.
i-C_3H_7CHO **and**			
i-C_4H_9CHO	KOH	i-$C_4H_9CHOHC(CH_3)_2CHO$ (—)	134, 135
$C_2H_5CH{=}C(CH_3)CHO$	”	$C_2H_5CH{=}C(CH_3)CHOHC(CH_3)_2CHO$ (—)	982
$C_6H_5CH{=}CHCHO$	”	$C_{13}H_{16}O_2$ (—)	983
$(CH_3)_2C{=}CHCHO$ **and**			
$(CH_3)_2C{=}CHCH_2CH_2$-$C(CH_3){=}CHCHO$, citral	$NaNH_2$	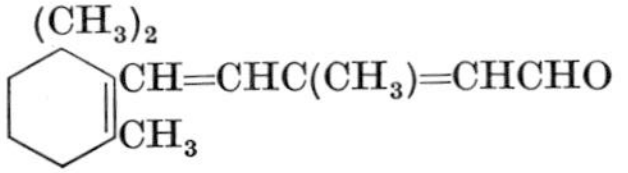 $(CH_3)_2$ [cyclohexene ring] $CH{=}CHC(CH_3){=}CHCHO$, CH_3 (—)	585, 622
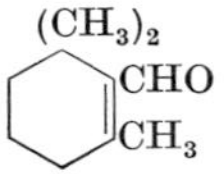 $(CH_3)_2$ [cyclohexene ring] CHO, CH_3	$(CH_2)_5NH$, CH_3CO_2H	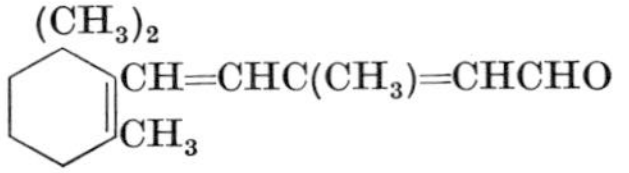 $(CH_3)_2$ [cyclohexene ring] $CH{=}CHC(CH_3){=}CHCHO$, CH_3 (—)	984
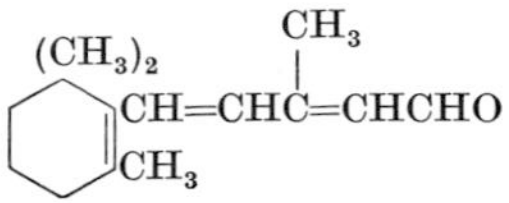 $(CH_3)_2$ [cyclohexene ring] $CH{=}CHC(CH_3){=}CHCHO$, CH_3	”	Vitamin A aldehyde (—)	586, 984
n-C_4H_9CHO **and** i-C_3H_7CHO	NaOH	n-$C_4H_9CH{=}C(C_3H_7$-$n)CHO$, i-$C_3H_7CH{=}C(C_3H_7$-$n)CHO$ (Total, 90)	767
n-$C_7H_{15}CHO$ **and**			
$C_6H_5CH{=}CHCHO$	”	$C_6H_5CH{=}CHCH{=}C(C_6H_{13}$-$n)CHO$ (54)	972
n-$C_{11}H_{23}CHO$	$(C_{17}H_{35}CO_2)_2Fe$	n-$C_7H_{15}CH{=}C(C_6H_{13}$-$n)CHO$ (—), n-$C_{11}H_{23}CH{=}C(C_6H_{13}$-$n)CHO$ (—), n-$C_7H_{15}CH{=}C(C_{10}H_{21}$-$n)CHO$ (—), n-$C_{11}H_{23}CH{=}C(C_{10}H_{21}$-$n)CHO$ (—)	623

Note: References 668–2359 are on pp. 403–438.

TABLE IV. CONDENSATION OF CARBOCYCLIC AROMATIC ALDEHYDES WITH ALDEHYDES

Reactants	Catalyst	Product(s) (Yield, %)	Refs.
CH_2ClCHO **and** $4\text{-}O_2NC_6H_4CHO$	KOH	$4\text{-}O_2NC_6H_4CH{=}CClCHO$ (—)	985
CH_3CHO **and**			
$2\text{-}ClC_6H_4CHO$	"	$2\text{-}ClC_6H_4CH{=}CHCHO$ (—)	986
$2\text{-}O_2NC_6H_4CHO$	NaOH	$2\text{-}O_2NC_6H_4CH{=}CHCHO$ (—)	987
	$Ba(OH)_2$	$2\text{-}O_2NC_6H_4CHOHCH_2CH_2OH$ (—)	476
$3\text{-}O_2NC_6H_4CHO$	NaOH, 5 min.	$3\text{-}O_2NC_6H_4CHOHCH_2CHO$ (—)	988
	NaOH, 12 hr.	$3\text{-}O_2N_6H_4CH{=}CHCHO$ (50)	989
$4\text{-}O_2NC_6H_4CHO$	NaOH, H_2O	$4\text{-}O_2NC_6H_4CHOHCH_2CHO$ (—)	990
	KOH, CH_3OH	$4\text{-}O_2NC_6H_4CH{=}CHCHO$ (70)	991
C_6H_5CHO	NaOH, $POCl_3$, or HCl	$C_6H_5CH{=}CHCHO$ (42–90)	992, 103, 779, 993–998
	NaOH	$C_6H_5(CH{=}CH)_2CHO$ (15), $C_6H_5(CH{=}CH)_3CHO$ (—)	999, 1000
	Ion-exchange resins IR-4B and IR-120	$CH_3CH{=}CHCHO$ (9–13)	101
$2\text{-}H_2NC_6H_4CHO$	NaOH	Quinoline (—)	1001
$C_6H_4(CHO)_2\text{-}1,4$	KOH	$C_6H_4(CH{=}CHCHO)_2\text{-}1,4$	1002, 1003
$3,4\text{-}(OCH_2O)C_6H_3CHO$	NaOH	$3,4\text{-}(OCH_2O)C_6H_3CH{=}CHCHO$ (—)	1004
$2\text{-}CH_3C_6H_4CHO$	"	$2\text{-}CH_3C_6H_4CH{=}CHCHO$ (—)	1005
$4\text{-}CH_3C_6H_4CHO$	"	$4\text{-}CH_3C_6H_4CH{=}CHCHO$ (25–30)	1006
$4\text{-}CH_3OC_6H_4CHO$	"	$4\text{-}CH_3OC_6H_4CH{=}CHCHO$ (—)	1006
$2\text{-}C_2H_5C_6H_4CHO$	"	$2\text{-}C_2H_5C_6H_4CH{=}CHCHO$ (—)	1005
$2,3\text{-}(CH_3O)_2C_6H_3CHO$	KOH	$2,3\text{-}(CH_3O)_2C_6H_3CH{=}CHCHO$ (33)	1007
$2,4\text{-}(CH_3O)_2C_6H_3CHO$	NaOH	$2,4\text{-}(CH_3O)_2C_6H_3CH{=}CHCHO$ (77)	1008
$3,4\text{-}(CH_3O)_2C_6H_3CHO$	KOH	$3,4\text{-}(CH_3O)_2C_6H_3CH{=}CHCHO$ (22)	1007
$4\text{-}(CH_3OCH_2O)C_6H_4CHO$	"	$4\text{-}(CH_3OCH_2O)C_6H_4CH{=}CHCHO$ (37)	1009
$4\text{-}(CH_3)_2NC_6H_4CHO$	H_2SO_4	$4\text{-}(CH_3)_2NC_6H_4CH{=}CHCHO$ (—), $4\text{-}(CH_3)_2NC_6H_4(CH{=}CH)_2CHO$ (—), $4\text{-}(CH_3)_2NC_6H_4(CH{=}CH)_3CHO$ (6), $4\text{-}(CH_3)_2NC_6H_4(CH{=}CH)_5CHO$ (—)	1010
$2\text{-}CH_3OCH_2O\text{-}3\text{-}CH_3OC_6H_3CHO$	KOH	$2\text{-}CH_3OCH_2O\text{-}3\text{-}CH_3OC_6H_3CH{=}CHCHO$ (49)	1009
$3\text{-}CH_3O\text{-}4\text{-}(CH_3OCH_2O)C_6H_3CHO$	"	$3\text{-}CH_3O\text{-}4\text{-}(CH_3OCH_2O)C_6H_3CH{=}CHCHO$ (26–45)	1009

Note: References 668–2359 are on pp. 403–438.

TABLE IV. CONDENSATION OF CARBOCYCLIC AROMATIC ALDEHYDES WITH ALDEHYDES (*Continued*)

Reactants	Catalyst	Product(s) (Yield, %)	Refs.
CH_3CHO (***contd.***) **and**			
4-$(CH_3OCH_2O)C_6H_4CHO$	NaOH	4-$(CH_3OCH_2O)C_6H_4CH{=}CHCHO$ (—)	1011
3,5-$(CH_3O)_2$-4-(CH_3OCH_2O)-C_6H_2CHO	KOH	3,5-$(CH_3O)_2$-4-$(CH_3OCH_2O)C_6H_2$-$CH{=}CHCHO$ (25–28)	1012
C_2H_5CHO **and**			
C_6H_5CHO	NaOH	$C_6H_5CH{=}C(CH_3)CHO$ (—), $C_{10}H_{12}O_2$ (—)	1013, 967, 976, 1011, 1014, 2330
3,4-$(CH_2O_2)C_6H_3CHO$	KOH	3,4-$(CH_2O_2)C_6H_3CH{=}C(CH_3)CHO$ (75)	1015
4-*i*-$C_3H_7C_6H_4CHO$	NaOH	4-*i*-$C_3H_7C_6H_4CH{=}C(CH_3)CHO$ (80)	1016
$CH_3CH{=}CHCHO$ **and** C_6H_5CHO	”	$C_6H_5(CH{=}CH)_2CHO$ (11)	1017
	$(CH_2)_5NH$, CH_3CO_2H	$C_6H_5(CH{=}CH)_2CHO$ (—), $C_6H_5(CH{=}CH)_3CHO$ (—), $C_6H_5(CH{=}CH)_5CHO$ (—)	116
n-C_3H_7CHO **and** C_6H_5CHO	KOH	$C_6H_5CH{=}C(C_2H_5)CHO$ (58)	1018, 976
	NaOH	$C_6H_5CH{=}C(C_2H_5)CHO$ (27–53), *n*-$C_3H_7CH{=}C(C_2H_5)CHO$ (43–57)	998, 993
	Ion-exchange resins IR-4B and IR-120	$C_6H_5CH{=}C(C_2H_5)CHO$ (5–16), *n*-$C_3H_7CH{=}C(C_2H_5)CHO$ (34–52)	101
i-C_3H_7CHO **and**			
C_6H_5CHO	KOH	$C_6H_5CHOHC(CH_3)_2CHOHC(CH_3)_2CHO$ (—)	1019
4-HOC_6H_4CHO	”	*i*-$C_3H_7CHOHC(CH_3)_2CHO$ (—)	1020
3-$C_2H_5OC_6H_4CHO$	”	3-$C_2H_5OC_6H_4CHOHC(CH_3)_2CHO$ (—)	1021
4-$C_2H_5OC_6H_4CHO$	”	4-$C_2H_5OC_6H_4CHOHC(CH_3)_2CHO$ (—)	1020
$C_2H_5CH(CH_3)CHO$ **and** C_6H_5CHO	”	$C_6H_5CHOHC(CH_3)(C_2H_5)CH_2OH$ (—)	475
i-C_4H_9CHO **and**			
C_6H_5CHO	”	$C_6H_5CH{=}C(C_3H_7$-*i*$)CHO$ (—)	976

2-HOC_6H_4CHO	HCl	2-Acetoxy-3-isopropylbenzopyrylium chloride (C_3H_7-*i*, $OCOCH_3$, $Cl^{\ominus}$) (—)	772
n-$C_6H_{13}CHO$ **and** C_6H_5CHO	NaOH	$C_6H_5CH{=}C(C_5H_{11}$-*n*$)CHO$ (—)	1022, 901
	Ion-exchange resins IR-4B and IR-120	$C_6H_5CH{=}C(C_5H_{11}$-*n*$)CHO$ (2–13), *n*-$C_6H_{13}CH{=}C(C_5H_{11}$-*n*$)CHO$ (27–49)	101
	NaOH	$C_6H_5CH{=}C(C_5H_{11}$-*n*$)CHO$, *n*-$C_6H_{13}CH{=}C(C_5H_{11}$-*n*$)CHO$ (Total, 14–28)	998, 993
	$POCl_3$	$C_6H_5CH{=}C(C_5H_{11}$-*n*$)CHO$ (73)	103, 779
2-HOC_6H_4CHO	KOH	*n*-$C_6H_{13}CH{=}C(C_5H_{11}$-*n*$)CHO$ (—)	1023
3,4-$(CH_2O_2)C_6H_3CHO$	,,	3,4-$(CH_2O_2)C_6H_3CH{=}C(C_5H_{11}$-*n*$)CHO$ (17)	1023
	$(CH_2)_5NH$	3,4-$(CH_2O_2)C_6H_3CH{=}C(C_5H_{11}$-*n*$)CHO$ (—)	1024
4-$CH_3OC_6H_4CHO$	KOH	4-$CH_3OC_6H_4CH{=}C(C_5H_{11}$-*n*$)CHO$ (14)	1023, 976
3-CH_3O-4-HOC_6H_3CHO	,,	*n*-$C_6H_{13}CH{=}C(C_5H_{11}$-*n*$)CHO$ (60)	1023
2,3-$(CH_3O)_2C_6H_3CHO$	$(CH_2)_5NH$	2,3-$(CH_3O)_2C_6H_3CH{=}C(C_5H_{11}$-*n*$)CHO$ (—)	1024
$C_6H_5CH_2CHO$ **and** C_6H_5CHO	KOH	$C_6H_5CH{=}C(C_6H_5)CHO$ (69)	967, 88
n-$C_7H_{15}CHO$ **and** C_6H_5CHO	NaOH	$C_6H_5CH{=}C(C_6H_{13}$-*n*$)CHO$ (—)	1026
4-Methylcyclohex-3-enyl-$CH(CH_3)CH_2CHO$ **and** C_6H_5CHO	KOH	4-Methylcyclohex-3-enyl-$CH(CH_3)C(CHO){=}CHC_6H_5$ (—)	144

Note: References 668–2359 are on pp. 403–438.

TABLE V. CONDENSATION OF HETEROCYCLIC ALDEHYDES WITH ALDEHYDES

Reactants	Catalyst	Product(s) (Yield, %)	Refs.
CH_2O and			
C_4H_3O–CHO (furan ring, O)	NaOH	Resin (—)	1028, 1029
C_4H_7O–CHO (tetrahydrofuran ring, O)	”	$C_4H_7O(CH_2OH)_2$ (90)	1025, 1027
C_5H_7O–CHO (dihydropyran ring, O)	”	$C_5H_7O(CH_2OH)_2$ (83)	1027
$C_4H_3O(CH{=}CH)_2CHO$ (furan ring, O)	”	Resin (—)	1028
CH_2ClCHO and C_4H_3S–CHO (thiophene ring, S)	”	C_4H_3S–CH=CClCHO (42)	1030
CH_3CHO and			
Br–C_4H_2O–CHO	”	Br–C_4H_2O–CH=CHCHO (36)	1031
Cl–C_4H_2S–CHO	”	Cl–C_4H_2S–CH=CHCHO (42)	1030, 1032
Cl–C_4H_2O–CHO	”	Cl–C_4H_2O–CH=CHCHO (36)	1031
I–C_4H_2O–CHO	KOH	I–C_4H_2O–CH=CHCHO (52)	1031

$O_2N(C_4H_2S)CHO$	"	$O_2N(C_4H_2S)CH{=}CHCHO$ (49)	1033, 1032
$O_2N(C_4H_2O)CHO$	$(CH_2)_5NH$, CH_3CO_2H	$O_2N(C_4H_2O)CH{=}CHCHO$ (49)	1034
$(C_4H_3S)CHO$	NaOH	$(C_4H_3S)CH{=}CHCHO$ (54)	1035
	$(CH_2)_5NH$, CH_3CO_2H	$(C_4H_3S)CH{=}CHCHO$ (37), $(C_4H_3S)(CH{=}CH)_2CHO$ (13)	1036
$(C_4H_3O)CHO$	NaOH; also Na_2CO_3, $NaNO_3$	$(C_4H_3O)CH{=}CHCHO$ (60–91), $(C_4H_3O)(CH{=}CH)_2CHO$ (34), $(C_4H_3O)(CH{=}CH)_3CHO$ (—)	1037, 978, 996, 1028, 1035–1047
	Ion-exchange resin Amberlite IRA-400	$(C_4H_3O)CH{=}CHCHO$ (8)	1048, 101
$CH_3(C_4H_2O)CHO$	NaOH	$CH_3(C_4H_2O)CH{=}CHCHO$ (24)	2329
$(C_5H_4N)CHO$	NaOH; also H_2SO_4	$(C_5H_4N)CH{=}CHCHO$ (5)	1121

Note: References 668–2359 are on pp. 403–438.

TABLE V. CONDENSATION OF HETEROCYCLIC ALDEHYDES WITH ALDEHYDES (*Continued*)

Reactants	Catalyst	Product(s) (Yield, %)	Refs.
CH_3CHO (***contd.***) **and**			
(3-CH_3-2-thienyl)CHO	NaOH	(3-CH_3-2-thienyl)CH=CHCHO (39)	1030
O_2N-(2-furyl)CH=CHCHO	$(CH_2)_5NH$, CH_3CO_2H	O_2N-(2-furyl)$(CH{=}CH)_2CHO$ (—),	1049
		O_2N-(2-furyl)$(CH{=}CH)_3CHO$ (—)	
(2-furyl)CH=CHCHO	$FeCl_3$	Resin (—)	1050
C_2H_5CHO **and**			
O_2N-(2-furyl)CHO	KOH	O_2N-(2-furyl)$CH{=}C(CH_3)CHO$ (60)	1033
(2-furyl)CHO	NaOH; also ion-exchange resin Amberlite IRA-400	(2-furyl)$CH{=}C(CH_3)CHO$ (72)	1043, 625, 632, 972, 1045, 1051
$CH_3CH{=}CHCHO$ **and**			
(2-thienyl)CHO	$(CH_2)_5NH$, CH_3CO_2H	(2-thienyl)$(CH{=}CH)_2CHO$ (36),	1036
		(2-thienyl)$(CH{=}CH)_4CHO$ (16)	
(2-furyl)CHO	NaOH	(2-furyl)$(CH{=}CH)_2CHO$ (—)	978, 1028

Reactants	Catalyst	Product (Yield, %)	References
$2\text{-}C_4H_3O$CH=CHCHO	$(CH_2)_5NH$, CH_3CO_2H	$2\text{-}C_4H_3O(CH{=}CH)_3CHO$ (—), $2\text{-}C_4H_3O(CH{=}CH)_5CHO$ (—)	116
$2\text{-}C_4H_3O(CH{=}CH)_2CHO$	NaOH	Resin (—)	1028
	$(CH_2)_5NH$, CH_3CO_2H	$2\text{-}C_4H_3O(CH{=}CH)_4CHO$ (—), $2\text{-}C_4H_3O(CH{=}CH)_6CHO$ (—)	116
$2\text{-}C_4H_3O(CH{=}CH)_5CHO$	,,	$2\text{-}C_4H_3O(CH{=}CH)_7CHO$ (—)	116
***n*-C_3H_7CHO and**			
$5\text{-}O_2N\text{-}2\text{-}C_4H_2O$CHO	KOH	$5\text{-}O_2N\text{-}2\text{-}C_4H_2O$CH=C($C_2H_5$)CHO (42)	1033
$2\text{-}C_4H_3O$CHO	NaOH; also Na_2CO_3, $NaNO_3$, $n\text{-}C_4H_9OMgBr$	$2\text{-}C_4H_3O$CH=C(C_2H_5)CHO (60–70)	1045, 613, 1040, 1043, 1047
	Ion-exchange resin Amberlite IRA-400	$2\text{-}C_4H_3O$CH=C(C_2H_5)CHO (22)	101
i-C_3H_7CHO **and** $2\text{-}C_4H_3O$CHO	KOH	$2\text{-}C_4H_3O$CHOHC$(CH_3)_2CH_2OH$ (60)	474, 1052

Note: References 668–2359 are on pp. 403–438.

TABLE V. CONDENSATION OF HETEROCYCLIC ALDEHYDES WITH ALDEHYDES (*Continued*)

Reactants	Catalyst	Product(s) (Yield, %)	Refs.
$n\text{-}C_4H_9CHO$ and 2-furyl-CHO	NaOH	2-furyl-$CH{=}C(C_3H_7\text{-}n)CHO$ (—)	1043, 1047
$C_2H_5CH(CH_3)CHO$ and 2-furyl-CHO	"	2-furyl-$CHOHC(CH_3)(CH_2OH)C_2H_5$ (—)	1052
$i\text{-}C_3H_7CH_2CHO$ and 2-furyl-CHO	"	2-furyl-$CH{=}C(C_3H_7\text{-}i)CHO$ (60–70)	1045, 978
$n\text{-}C_5H_{11}CHO$ and 2-furyl-CHO	Ion exchange resin Amberlite IRA-400	2-furyl-$CH{=}C(C_4H_9\text{-}n)CHO$ (36)	1053, 101, 1048
$n\text{-}C_3H_7CH(CH_3)CHO$ and 2-furyl-CHO	NaOH	2-furyl-$CHOHC(CH_3)(CH_2OH)C_3H_7\text{-}n$ (—)	1052
$(C_2H_5)_2CHCHO$ and 2-furyl-CHO	"	2-furyl-$CHOHC(C_2H_5)_2CH_2OH$ (—)	1052
$n\text{-}C_6H_{13}CHO$ and 2-furyl-CHO	Ion-exchange resin Amberlite IRA-400; also Na_2CO_3	2-furyl-$CH{=}C(C_5H_{11}\text{-}n)CHO$ (37)	1048, 101, 1043, 1047

$C_6H_5CH_2CHO$ **and** 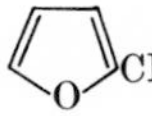CHO	NaOH	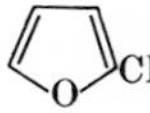$CH{=}C(C_6H_5)CHO$ (73)	972
n-$C_7H_{15}CHO$ **and** 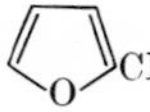CHO	Ion-exchange resin Amberlite IRA-400	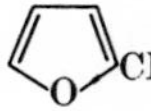$CH{=}C(C_6H_{13}$-$n)CHO$ (—)	1048, 101
n-$C_4H_9CH(C_2H_5)CHO$ **and** 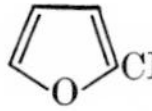CHO	NaOH	$CHOHC(C_2H_5)(CH_2OH)C_4H_9$-$n$ (—)	1052
n-$C_9H_{19}CHO$ **and** 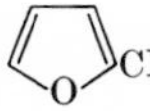CHO	Ion-exchange resin Amberlite IRA-400	$CH{=}C(C_8H_{17}$-$n)CHO$ (—)	1048, 101

Note: References 668–2359 are on pp. 403–438.

TABLE VI. SELF-CONDENSATION OF KETONES

Ketone	Catalyst	Product(s) (Yield, %)	Refs.
CH_3COCH_3*	BaO or CaO; Al_2O_3, Fe_2O_3, various temps.	$(CH_3)_2C(OH)CH_2COCH_3$ (75), $(CH_3)_2C{=}CHCOCH_3$ (—), $(CH_3)_2C{=}CHCOCH{=}C(CH_3)_2$ 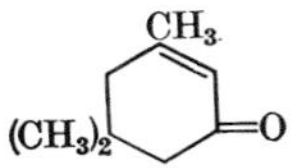 (—), [structure: CH_3, $(CH_3)_2$, O] (62), [structure: CH_3, CH_3, CH_3] (—) 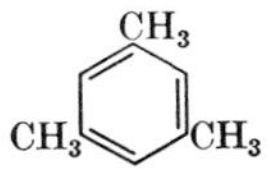	146, 145, 1054–1067
	$Al(OC_4H_9\text{-}t)_3$; also C_2H_5OMgBr, $i\text{-}C_3H_7OMgBr$	$(CH_3)_2C{=}CHCOCH_3$ (37), $(CH_3)_2C{=}CHCOCH{=}C(CH_3)_2$ (19)	151, 1068, 1069
	CaC_2	$(CH_3)_2C{=}CHCOCH_3$ (48)	1070, 502, 1067, 1071, 1072
	Ion-exchange resin Dowex 50HH	$(CH_3)_2C{=}CHCOCH_3$ (79)	148, 1073
	H_2SO_4	Mesitylene, durene, isodurene, pentamethylbenzene, $C_{14}H_{20}$, $C_{15}H_{18}$, $C_{16}H_{20}$ (—)	1074–1079, 2, 1072
$CH_3COCOCH_3$	KOH	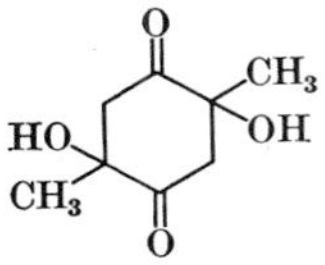 [structure: O, CH_3, OH, HO, CH_3, O] (—)	168, 169
$CH_3COC_2H_5$	NaOH	$CH_3COC(OH)(CH_3)CH_2COCOCH_3$ (—)	170, 171
	$NaOCH_3$; also NaOH, KOH, or BaO	$C_2H_5COCH{=}C(CH_3)C_2H_5$ (29), $C_2H_5COCH_2C(OH)(CH_3)C_2H_5$ (2–11)	146, 156, 504, 1094, 1122–1125
	$Al(OC_4H_9\text{-}t)_3$	$C_2H_5COCH_2C(OH)(CH_3)C_2H_5$ (72)	151, 146, 157, 613, 1126, 1127

	$C_6H_5N(CH_3)MgBr$	$C_2H_5COCH_2C(OH)(CH_3)C_2H_5$ (60–67)	150, 1128, 1129
	Alkali and alkaline earth metals and their hydrides, carbides, and amides; also Al_2O_3	$C_2H_5COCH{=}C(CH_3)C_2H_5$ (13), [ring structure: CH_3, CH_3, C_2H_5, CH_3, O] (—), $C_2H_5C(CH_3){=}CHCOC(CH_3){=}C(CH_3)$-$C_2H_5$ (—), $C_2H_5COCH_2C(OH)(CH_3)C_2H_5$ (—), 1,3,5-$(C_2H_5)_3C_6H_3$ (—)	1130, 154, 503, 1058, 1060, 1119, 1131–1136, 2325
	Ion-exchange resin, Dowex-50	$C_8H_{14}O$ (—)	148
	HCl or HBr	$CH_3COC(CH_3){=}C(CH_3)C_2H_5$ (46–67)	152, 154, 156, 1058, 1122, 1138
	H_2SO_4	$CH_3COC(CH_3){=}C(CH_3)C_2H_5$ (12), $C_2H_5C(CH_3){=}CHCOC(CH_3){=}C(CH_3)$-$C_2H_5$ (13), 1,3,5-$(C_2H_5)_3C_6H_3$ (—)	504, 154, 1139
	$POCl_3$ or $ZnCl_2$	$C_8H_{14}O$ (48)	153
$CH_3COCH{=}CHCH_3$	n-C_4H_9OMgBr	$CH_3CH{=}CHCOCH{=}C(CH_3)$-$CH{=}CHCH_3$ (—)	613
▷$COCH_3$	KOH	▷$COCH{=}C(CH_3)$◁ (14), $C_{15}H_{22}O_2$ (—)	1140

Note: References 668–2359 are on pp. 403–438.

* Self-condensation of acetone is also brought about by the following catalysts. The numbers following each catalyst are the pertinent references. Alkali or alkaline earth metals or amalgams, 1080–1086, 2324; NaOH or KOH, 653, 694, 1072, 1087–1096; $Ba(OH)_2$ or $Ca(OH)_2$, 662, 694, 1097–1099; Fe_2O_3, ThO_2, ZnO, or PbO, 1055–1061, 1117–1120; $NaOC_2H_5$, 1062, 1100–1102; C_2H_5OMgI, 613, 649, 1103–1105; $ZnCl_2$, $AlCl_3$, BF_3, or HCl, 156, 1057, 1072, 1105–1116.

TABLE VI. Self-Condensation of Ketones (*Continued*)

Ketone	Catalyst	Product(s) (Yield, %)	Refs.
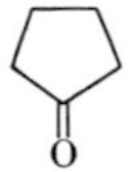	KOH; also NaOH, $NaOC_2H_5$	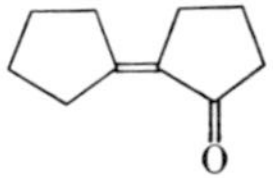 (12–50),	160, 165, 167, 181, 605, 1073, 1141–1150, 2323, 2348
		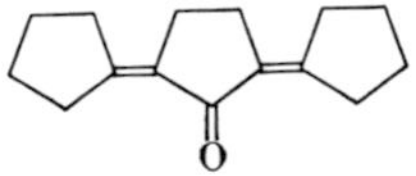 (12–81), $C_{20}H_{22}O$ (2)	
	CaH_2 or CaC_2; also Al_2O_3	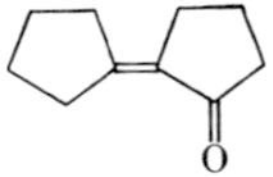 (—), $C_{15}H_{20}O$ (—)	1151, 1152
	sec-C_4H_9MgBr; also $C_6H_5NHMgBr$, *i*-C_3H_7MgCl	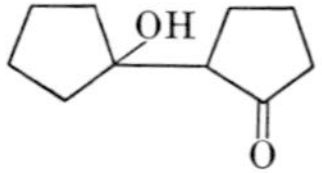 (42)	1153, 159, 1128, 1154, 1155
	HCl; also P_2O_5, $AlCl_3$, or H_2SO_4	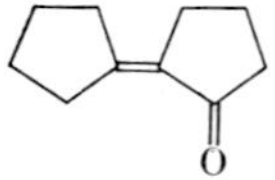 (54), 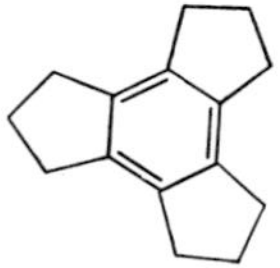 (8)	825, 160, 826, 1109, 1146, 1156, 1157
$CH_3COCOC_2H_5$	NaOH	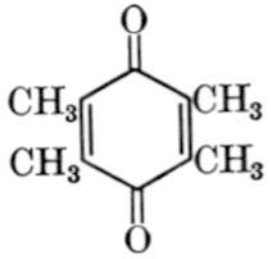 (10)	171
$CH_3COCH_2COCH_3$	KOH	No condensation	174

$CH_3COC_3H_7$-*n*	NaOH, *n*-C_4H_9OMgBr, or BaO	*n*-$C_3H_7COCH_2C(OH)(CH_3)C_3H_7$-*n* (3–57)	1124, 146, 157, 613
	HCl, HBr, or HI	*n*-$C_3H_7COCH{=}C(CH_3)C_3H_7$-*n* (27–74)	152, 156
$(C_2H_5)_2CO$	$NaOC_2H_5$, or $Al(OC_4H_9\text{-}t)_3$	$C_2H_5COC(CH_3){=}C(C_2H_5)_2$ (21–40)	1158, 151, 503, 1124, 1159
	$C_6H_5N(CH_3)MgBr$; also *i*-C_3H_7MgCl, *n*-C_4H_9OMgBr	$C_2H_5COCH(CH_3)C(OH)(C_2H_5)_2$ (60)	1128, 150, 157, 613, 1126, 1161
	Ion-exchange resin Dowex-50	$C_2H_5COC(CH_3){=}C(C_2H_5)_2$ (10)	148
	HBr, HCl, $POCl_3$, $ZnCl_2$, or $AlCl_3$	$C_2H_5COC(CH_3){=}C(C_2H_5)_2$ (12–52)	833, 153, 156, 830, 1160, 1162, 1163
$CH_3COC_3H_7$-*i*	*i*-C_3H_7MgCl; also NH_4Cl, $C_6H_5N(CH_3)MgBr$, or $Al(OC_4H_9\text{-}t)_3$	*i*-$C_3H_7COCH_2C(OH)(CH_3)C_3H_7$-*i* (70)	1153, 150, 151, 157, 1164, 1165
	t-$C_5H_{11}MgCl$, or $Zn(CH_3)_2$	*i*-$C_3H_7COCH{=}C(CH_3)C_3H_7$-*i* (36)	1164, 1069
$CH_3COC(OH)(CH_3)_2$	KOH; K_2CO_3	$(CH_3)_2C(OH)C(CH_3){=}CH\text{-}COC(OH)(CH_3)_2$ (52)	1166, 1085, 1167
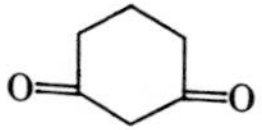	$NaHCO_3$, *p*H 6	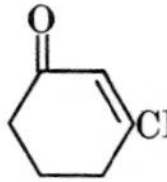 $CH_2CO(CH_2)_3CO_2H$ (49)	1168, 1169
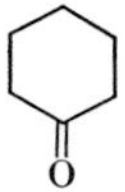	KOH; also $C_6H_5N(CH_3)MgBr$	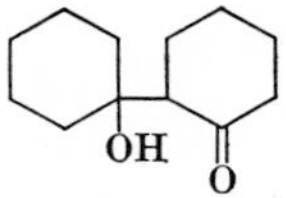 OH (22), 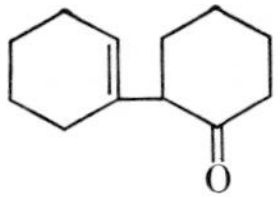 (12), 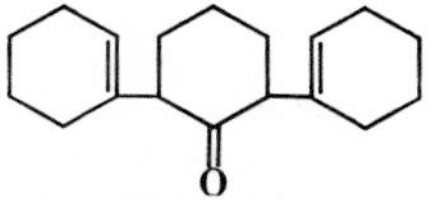 (22)	165, 166, 557, 716, 1129, 1141, 1151, 1171-1176, 2353

Note: References 668–2359 are on pp. 403–438.

TABLE VI. SELF-CONDENSATION OF KETONES (*Continued*)

Ketone	Catalyst	Product(s) (Yield, %)	Refs.
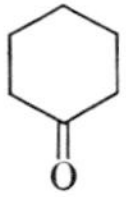 O (*contd.*)	NaOH	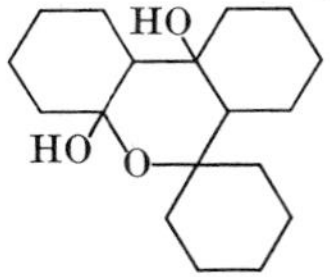 HO HO O (40)	1177, 1137
	$NaOCH_3$	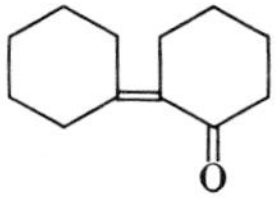 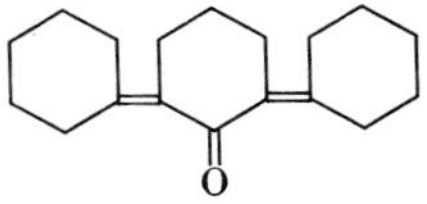 O (—), O (—),	1178, 1146
		 (—)	
		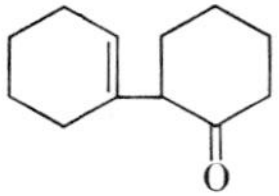 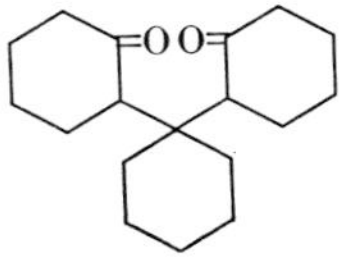 O (—), O O (—)	1179
		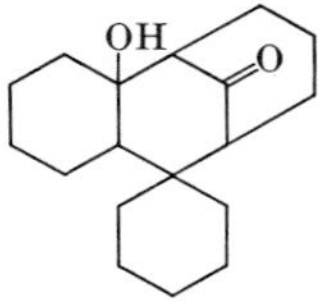 OH O (40)	1180

Catalyst / Conditions	Product (Yield, %)	Refs.
$Al(OC_4H_9\text{-}t)_3$; also $i\text{-}C_3H_7MgCl$, $C_6H_5N(CH_3)$-MgBr	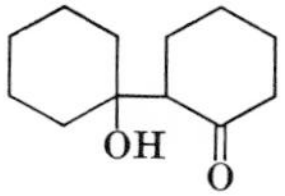 (78)	151, 159, 1153
C_5H_5N, C_6H_5COCl; also aliphatic amines, ion-exchange resin Amberlite IR-120	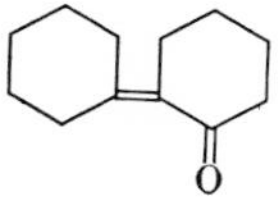 (20–77)	1177, 1098, 1181
Ion-exchange resin Dowex-50	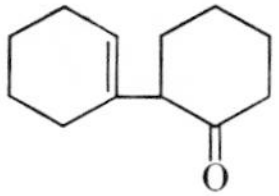 (54)	148, 181, 1182
$ZnCl_2$; also BF_3, $AlCl_3$-C_6H_5-$N(CH_3)_2$, Al_2O_3	2-Cyclohexylidenecyclohexanone (19), 2,6-dicyclohexylidenecyclohexanone (30), dodecahydrotriphenylene (7)	1183, 1109, 1178
HCl	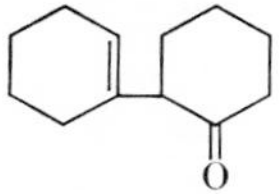 (63)	1184, 162–164, 1176, 1185
H_2SO_4, 30°	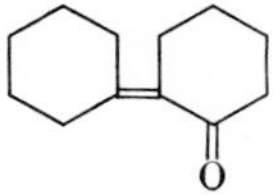 (83–93), 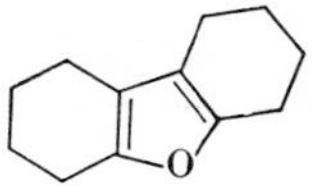 (6), 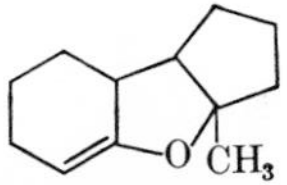 (3)	1186–1188, 1175, 1177

Note: References 668–2359 are on pp. 403–438.

TABLE VI. SELF-CONDENSATION OF KETONES (*Continued*)

Ketone	Catalyst	Product(s) (Yield, %)	Refs.
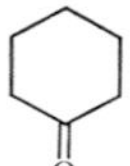 (*contd.*)	H_2SO_4, rfx.	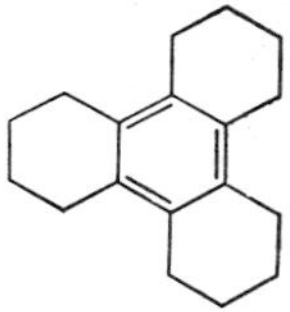 (9)	1189, 624, 1175, 1178
$CH_3COCH{=}C(CH_3)_2$	Li	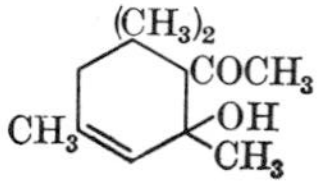 (—)	1190, 418, 1191
	$Ba(OH)_2$ or ion-exchange resins	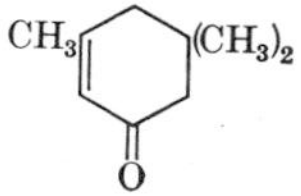 (—), xylitones (—)	1192–1194
	$NaOC_5H_{11}$-*t*, BaO, MgO, ion-exchange resins, or CaC_2	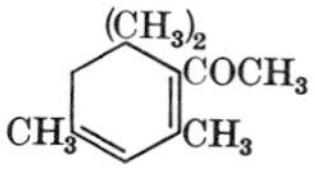 (—), 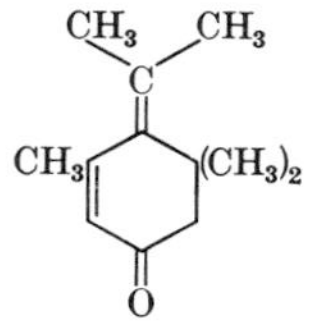 (—), xylitones (—), isoxylitones (—)	1190, 1131, 1132, 1193, 1195–1197
	$C_6H_5N(CH_3)MgBr$	$(CH_3)_2C{=}CHC(OH)(CH_3)CH_2CO$-$CH{=}C(CH_3)_2$ (67)	198
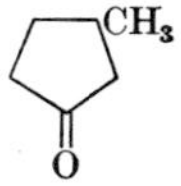	CaH_2 or HCl	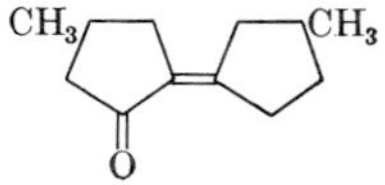 (—)	1151, 1146

$CH_3COCH_2CH_2COCH_3$	NaOH	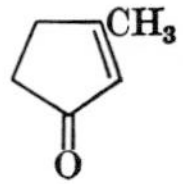 CH_3, O (42)	1198, 174
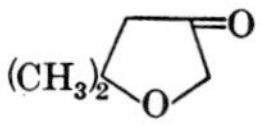 $(CH_3)_2$, O, =O	K	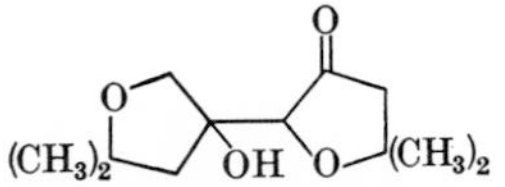 O, $(CH_3)_2$, OH, O, $(CH_3)_2$, O (50)	1199
$CH_3COC_4H_9$-*n*	$Al(OC_4H_9$-*t*$)_3$; also *n*-C_4H_9OMgBr	*n*-$C_4H_9COCH_2C(OH)(CH_3)C_4H_9$-*n* (73)	151, 613
	HCl	*n*-$C_4H_9COCH{=}C(CH_3)C_4H_9$-*n* (22)	156
$CH_3COC_4H_9$-*sec*	$C_6H_5N(CH_3)MgBr$	*sec*-$C_4H_9COCH_2C(OH)(CH_3)C_4H_9$-*sec* (55)	1200, 1128
	HBr	*sec*-$C_4H_9COCH{=}C(CH_3)C_4H_9$-*sec* (35–40)	152, 156
$CH_3COC_4H_9$-*i*	$C_6H_5N(R)MgBr$ (R=H or alkyl); also $Al(OC_4H_9$-*t*$)_3$ or *i*-C_3H_7MgCl	*i*-$C_4H_9COCH_2C(OH)(CH_3)C_4H_9$-*i* (61–77)	1200, 151, 157, 1128, 1201
	HCl or CaC_2	*i*-$C_4H_9COCH{=}C(CH_3)C_4H_9$-*i* (20)	156, 1131, 1132
$C_2H_5COC_3H_7$-*i*	*i*-C_3H_7MgCl	*i*-$C_3H_7COCH(CH_3)C(OH)(C_2H_5)C_3H_7$-*i* (60)	157, 1126
$CH_3COC_4H_9$-*t*	$C_6H_5N(R)MgBr$ (R = CH_3, C_2H_5, C_6H_5); also CH_3MgI, *i*-C_3H_7MgCl, *t*-C_4H_9MgCl, or $Al(OC_4H_9$-*t*$)_3$	*t*-$C_4H_9COCH_2C(OH)(CH_3)C_4H_9$-*t* (60–70)	150, 151, 157, 1126, 1128, 1153, 1161, 1200, 1202–1204
	KOC_4H_9-*t*	*t*-$C_4H_9COCH{=}C(CH_3)C_4H_9$-*t* (70)	507
	HCl, HBr, or HI	No condensation	152, 156
$CH_3COC(OH)(CH_3)C_2H_5$	KOH	$C_2H_5C(OH)(CH_3)COCH{=}C(CH_3)$-$C(OH)(CH_3)C_2H_5$ (30)	1205
$CH_3COCH_2COCH_2COCH_3$	NaOH, KH_2PO_4, *p*H 7.1–7.2	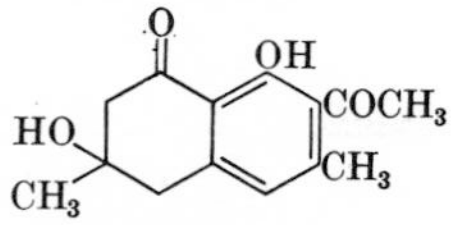 O, OH, $COCH_3$, HO, CH_3, CH_3 (20)	173

Note: References 668–2359 are on pp. 403–438.

TABLE VI. SELF-CONDENSATION OF KETONES (*Continued*)

Ketone	Catalyst	Product(s) (Yield, %)	Refs.
$CH_3COCH_2COCH_2COCH_3$ (*contd.*)	NaOH, KH_2PO_4, *p*H 8.2	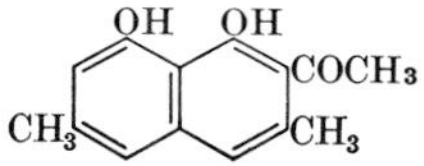 (5)	173
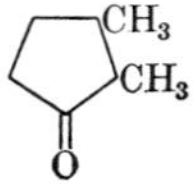	KOH	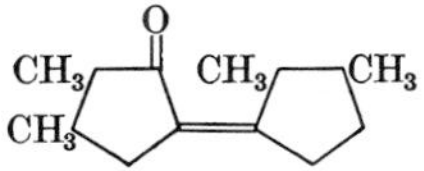 (60)	1206
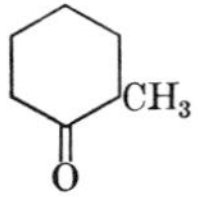	CaH_2	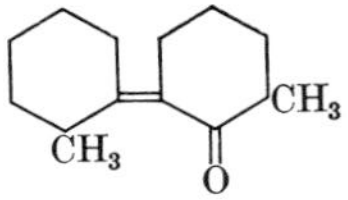 (—)	1207, 530, 1173, 1208
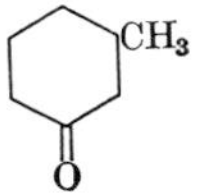	”	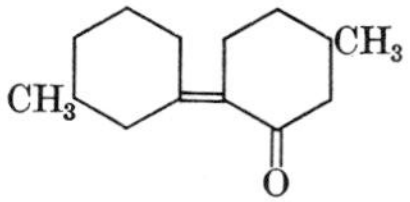 (Very good)	1151, 1146, 1173
	HCl	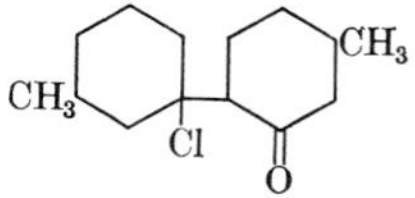 (—)	1209, 1146
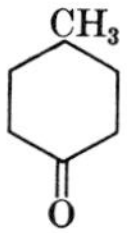	CaH_2	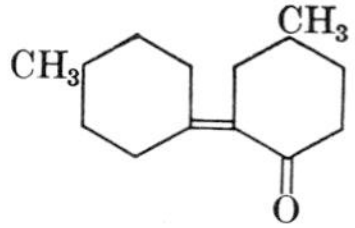 (—)	1207, 1173

Reactant	Catalyst	Product (Yield, %)	Refs.
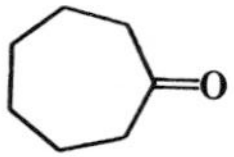	KOC_4H_9-*t*; also CaH_2, $AlCl_3$, or BF_3	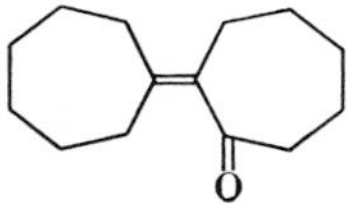 (50)	605, 1109, 1210
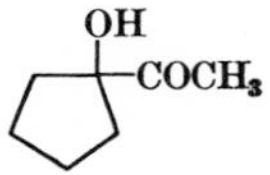 OH, $COCH_3$	KOH	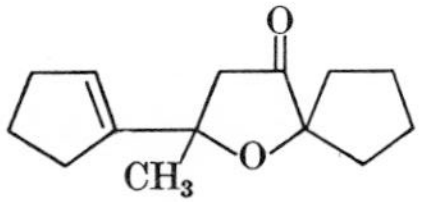 O, O, CH_3 (14)	1211
$CH_3COC_5H_{11}$-*n*	*i*-C_3H_7MgCl	*n*-$C_5H_{11}COCH_2C(OH)(CH_3)C_5H_{11}$-*n* (50)	157, 1126
	HCl	$C_{14}H_{26}O$ (14)	156
	$Al(OC_4H_9$-*t*$)_3$	No condensation	151
n-$C_3H_7COC_3H_7$-*n*	$C_6H_5N(CH_3)MgBr$	*n*-$C_3H_7COCH(C_2H_5)C(OH)(C_3H_7$-*n*$)_2$ (45)	150, 1128, 1200
	HBr or HCl	*n*-$C_3H_7COC(C_2H_5){=}C(C_3H_7$-*n*$)_2$ (30)	156
i-$C_3H_7COC_3H_7$-*i*	$Al(OC_4H_9$-*t*$)_3$ or $C_6H_5N(CH_3)MgBr$	No condensation	151, 177
4-$ClC_6H_4COCH_2Cl$	C_2H_5MgCl	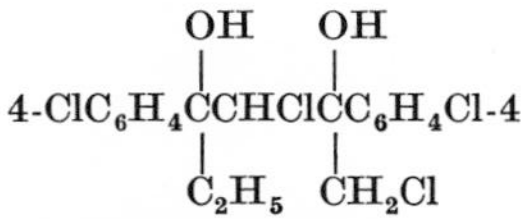 4-$ClC_6H_4C(OH)(C_2H_5)CHClC(OH)(CH_2Cl)C_6H_4Cl$-4 (—)	1212
$C_6H_5COCH_3$	$NaOC_2H_5$, C_2H_5OH, 130–140°	3,5-$(C_6H_5)_2C_6H_3CH_3$ (—), 2-CH_3-4,6$(C_6H_5)_2C_6H_2COC_6H_5$ (—)	1213–1215
	$Al(OC_4H_9$-*t*$)_3$; also ion exchange resins, $AlCl_3$, PCl_5, CaH_2, $ZnCl_2$, HBr	$C_6H_5COCH{=}C(CH_3)C_6H_5$ (82)	151, 148, 152, 1093, 1216–1219
	$C_6H_5N(CH_3)MgBr$	$C_6H_5COCH_2C(OH)(CH_3)C_6H_5$ (25)	1128, 159
	$K_2S_2O_7$, H_2SO_4; also $ZnCl_2$, BF_3, $ArSO_3H$	1,3,5-$(C_6H_5)_3C_6H_3$ (68–85)	506, 1119, 1215, 1220–1224

Note: References 668–2359 are on pp. 403–438.

TABLE VI. Self-Condensation of Ketones (*Continued*)

Ketone	Catalyst	Product(s) (Yield, %)	Refs.
$C_6H_5COCH_3$ (*contd.*)	$AlCl_3$	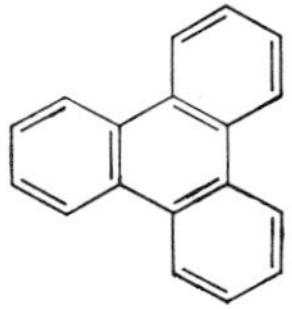 (50)	1109
	BF_3	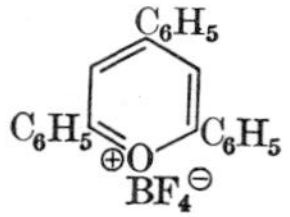 C_6H_5, C_6H_5, C_6H_5, $\oplus O$, $BF_4^{\ominus}$ (22)	1225
	HCl; also SiO_2 or Al_2O_3	$C_6H_5COCH{=}C(CH_3)C_6H_5$ (21–45), 1,3,5-$(C_6H_5)_3C_6H_3$ (60)	1226, 156, 1113, 1119, 1215, 1223, 1227–1231
$CH_3CO(CH_2)_2CO(CH_2)_2CO_2H$	KOH	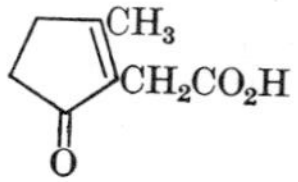 CH_3, CH_2CO_2H, O (97)	1232
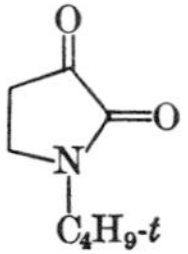 O, O, N, C_4H_9-*t*	None, heat	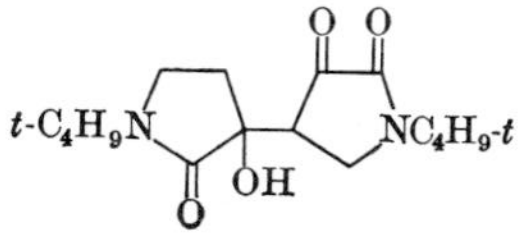 *t*-C_4H_9N, O, OH, O O, NC_4H_9-*t* (Quant.)	1233
	C_5H_5N	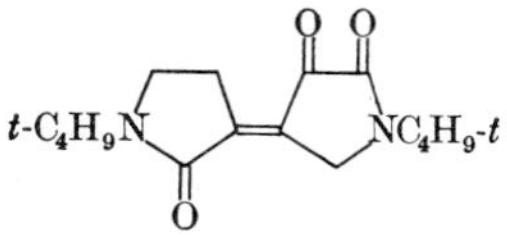 *t*-C_4H_9N, O, O O, NC_4H_9-*t* (96)	1233

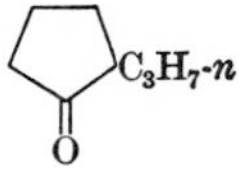 C_3H_7-*n* (cyclopentanone)	KOH	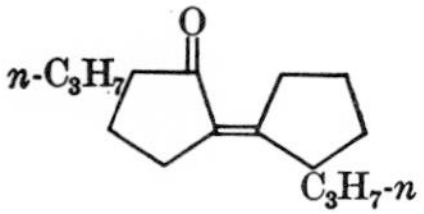 *n*-C_3H_7 … C_3H_7-*n* (83)	1206
$CH_3CO(CH_2)_4COCH_3$	KOH	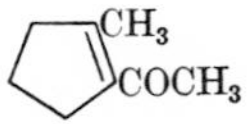 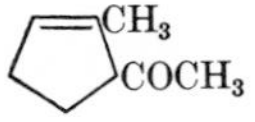 CH_3, $COCH_3$ (83), CH_3, $COCH_3$ (8)	395
$CH_3COCH_2C(CH_3)_2COCH_3$	NaOH	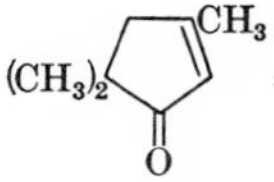 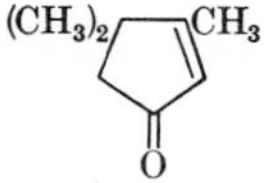 $(CH_3)_2$, CH_3, O , $(CH_3)_2$, CH_3, O (Total, 70)	1234
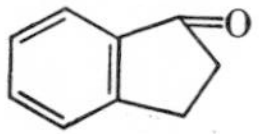 (indanone, O)	$Al(OC_4H_9\text{-}t)_3$; also H_2SO_4, ion-exchange resins, $AlCl_3$, HCl	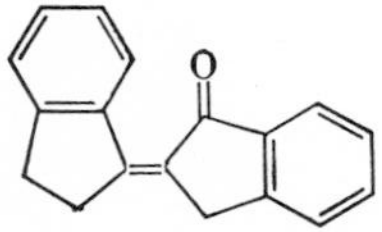 O (48), 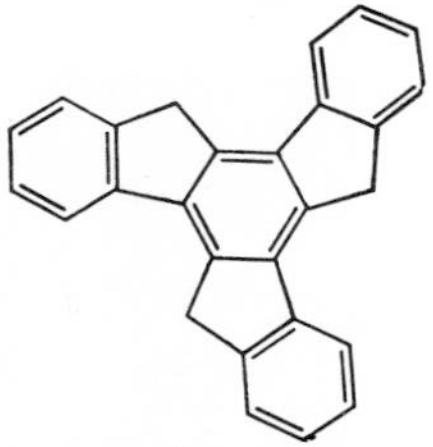	151, 508, 509 1109, 1235, 1236
		(—)	
$C_6H_5COC_2H_5$	$Al(OC_4H_9\text{-}t)_3$; also HBr	$C_6H_5COC(CH_3){=}C(C_2H_5)C_6H_5$ (22)	151, 152, 156
4-$CH_3C_6H_4COCH_3$	$C_6H_5N(CH_3)MgBr$	4-$CH_3C_6H_4COCH{=}C(CH_3)C_6H_4CH_3$-4 (53)	1237

Note: References 668–2359 are on pp. 403–438.

TABLE VI. SELF-CONDENSATION OF KETONES (*Continued*)

Ketone	Catalyst	Product(s) (Yield, %)	Refs.
$C_6H_5CH_2COCH_3$	HCl	$C_6H_5CH_2C(CH_3){=}CHCOCH_2C_6H_5$ (50)†	156
	70% H_2SO_4, rfx.	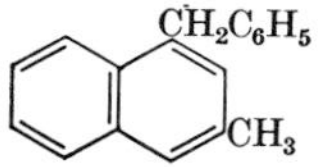 [structure: naphthalene with $CH_2C_6H_5$ and CH_3] (18)	1238
4-$CH_3OC_6H_4COCH_3$	BF_3	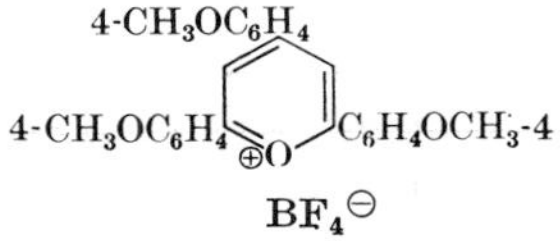 [structure: pyrylium with 4-$CH_3OC_6H_4$, 4-$CH_3OC_6H_4$, $C_6H_4OCH_3$-4] $BF_4^{\ominus}$ (10)	1225
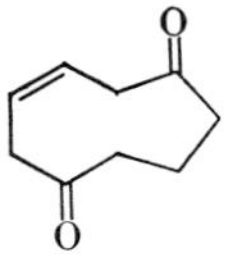 [structure: cyclic enedione]	CH_3CO_2Na	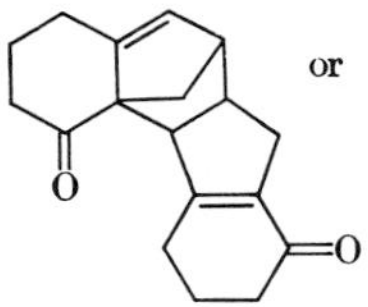 [structure] or [structure] (—) 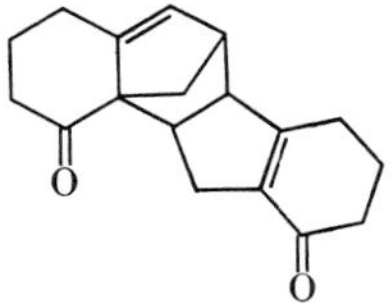	1239
$(CH_3)_2C{=}CHCOCH{=}C(CH_3)_2$	$Al(OC_4H_9\text{-}t)_3$	$C_{18}H_{26}O$ (21)	151
$CH_3CO(CH_2)_2CO(CH_2)_2CH{=}CH_2$	—	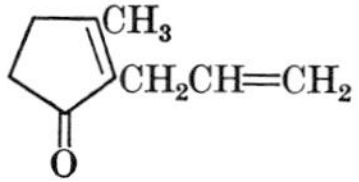 [structure: cyclopentenone with CH_3 and $CH_2CH{=}CH_2$] (93)	393
 CH_3CO[cyclopentane]$COCH_3$	$NaOC_2H_5$	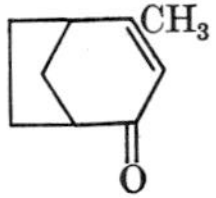 [structure: bicyclic enone with CH_3] (—)	1240, 1241

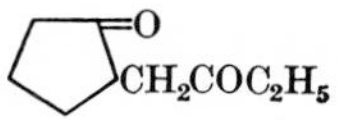 $CH_2COC_2H_5$, O	KOH	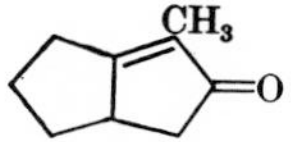 CH_3, O (—)	1242
$CH_3COCH_2CH(CO_2C_2H_5)COCH_3$	NaOH	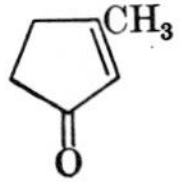 CH_3, O (60)	1198
$CH_3COCH_2CH(CO_2CH_3)COC_2H_5$	$NaOCH_3$	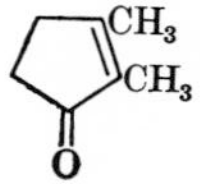 CH_3, CH_3, O (—)	1243
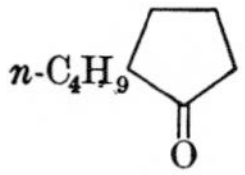 n-C_4H_9, O	KOH	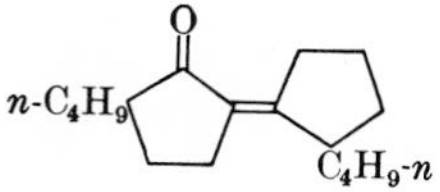 n-C_4H_9, O, C_4H_9-n (78)	1206
$CH_3CO(CH_2)_5COCH_3$	H_2SO_4	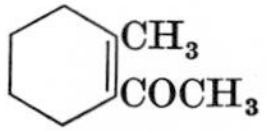 CH_3, $COCH_3$ (85)	384, 1244, 1245
$C_2H_5CO(CH_2)_3COC_2H_5$	HCl	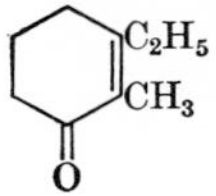 C_2H_5, CH_3, O (—)	385
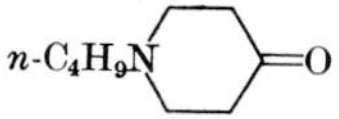 n-C_4H_9N, O	$C_6H_{11}MgCl$	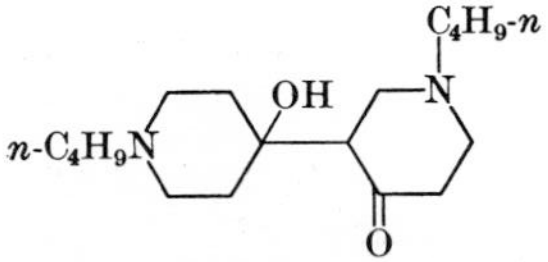 n-C_4H_9N, OH, N, C_4H_9-n, O (60)	1246

Note: References 668–2359 are on pp. 403–438.

† The structure of this product was not established.

TABLE VI. SELF-CONDENSATION OF KETONES (*Continued*)

Ketone	Catalyst	Product(s) (Yield, %)	Refs.
$CH_3COC_7H_{15}$-*n*	HCl	*n*-$C_7H_{15}C(CH_3)$=$CHCOC_7H_{15}$-*n* (—)	155
i-$C_4H_9COC_4H_9$-*i*	$Al(OC_5H_9$-*t*$)_3$ or $C_6H_5N(CH_3)MgBr$	No condensation	151, 177
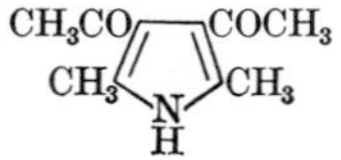 CH_3CO $COCH_3$ CH_3 CH_3 N H	NaH, NaOH, or $NaOC_4H_9$-*t*	No condensation	1247
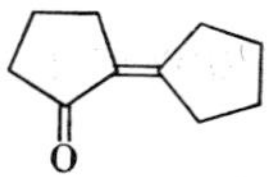 O	4-$CH_3C_6H_4SO_3H$	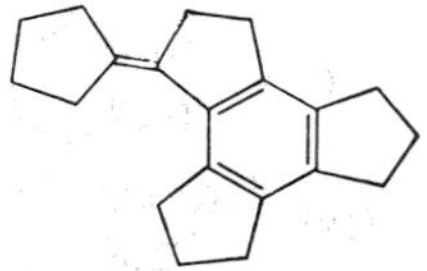 (—)	160
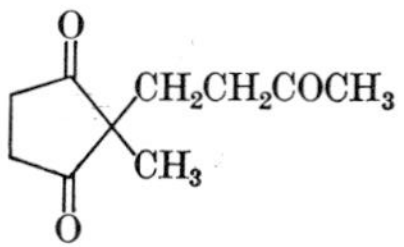 O $CH_2CH_2COCH_3$ CH_3 O	$(CH_2)_5NH$, CH_3CO_2H	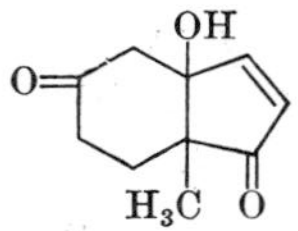 OH O= H_3C O (—)	1248, 1249
	$(CO_2H)_2$; also H_3BO_3	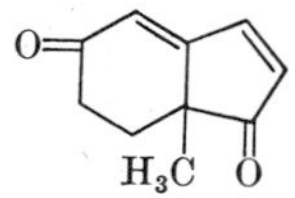 O= H_3C O (—)	1248, 1249
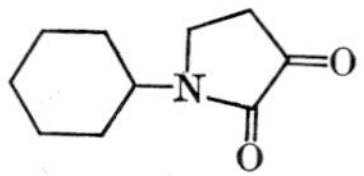 N =O O	C_5H_5N	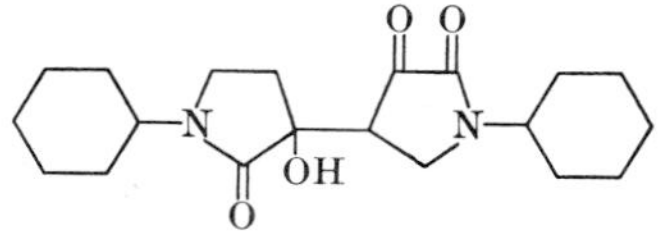 O O N OH N O (96)	1233

Reactant	Catalyst	Product (Yield, %)	Refs.
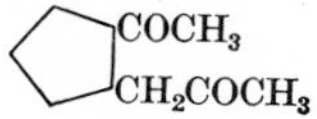 $COCH_3$, CH_2COCH_3	NaOH	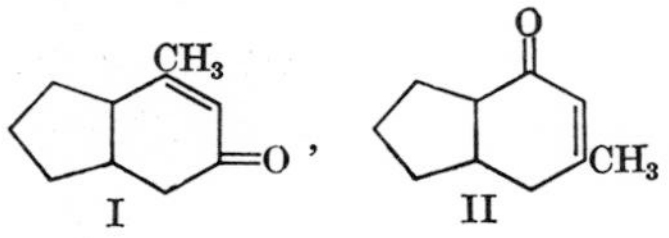 CH_3, O, I ; O, CH_3, II I:II = 3:2	1250
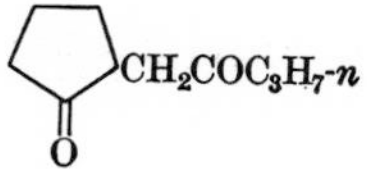 $CH_2COC_3H_7$-*n*, O	KOH	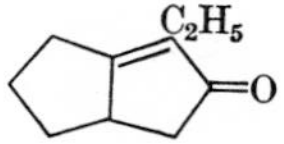 C_2H_5, O (85)	1242
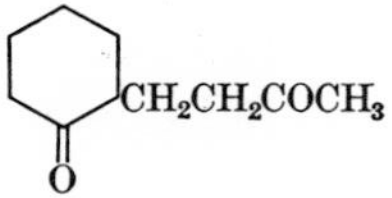 $CH_2CH_2COCH_3$, O	4-$CH_3C_6H_4SO_3H$	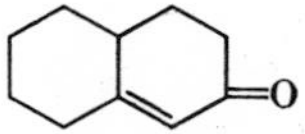 O (90)	1251, 1249
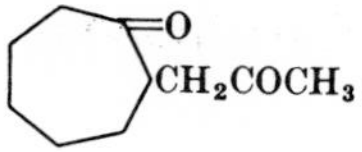 O, CH_2COCH_3	KOH	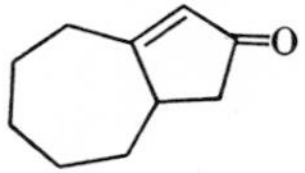 O (72)	1252
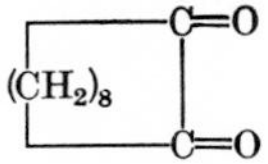 $(CH_2)_8$, C=O, C=O	$NaOCH_3$	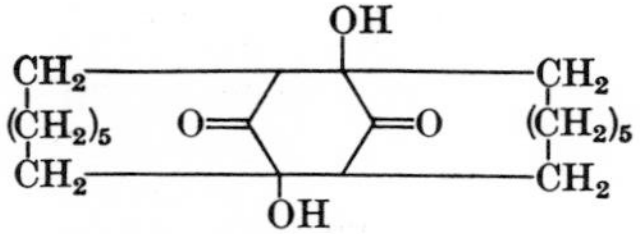 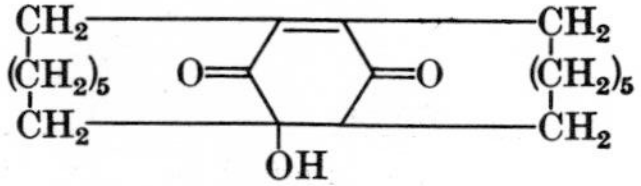 OH, CH_2, $(CH_2)_5$, CH_2, O, O, OH, CH_2, $(CH_2)_5$, CH_2 (—), CH_2, $(CH_2)_5$, CH_2, O, O, OH, CH_2, $(CH_2)_5$, CH_2 (—)	172

Note: References 668–2359 are on pp. 403–438.

TABLE VI. SELF-CONDENSATION OF KETONES (*Continued*)

Ketone	Catalyst	Product(s) (Yield, %)	Refs.
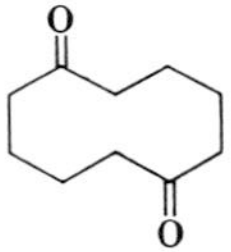 O, O (cyclodecane-1,6-dione structure)	Na_2CO_3; also H_2SO_4	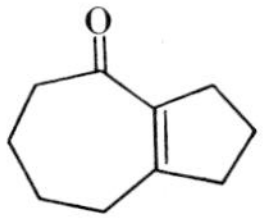 O (bicyclic enone structure) (96)	398, 1253–1255
$CH_3COCH(CO_2C_2H_5)CH(CH_3)COCH_3$	None	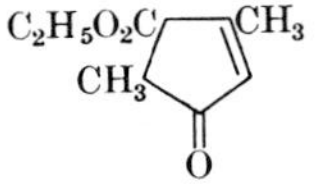 $C_2H_5O_2C$, CH_3, CH_3, O (—)	1256
$CH_3COCH(CO_2C_2H_5)CH_2COC_2H_5$	$NaOC_2H_5$	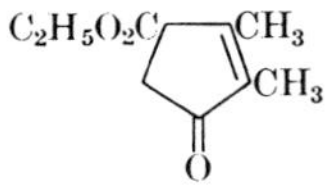 $C_2H_5O_2C$, CH_3, CH_3, O (—)	1257
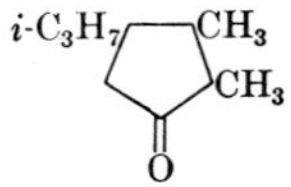 $i\text{-}C_3H_7$, CH_3, CH_3, O	CaH_2	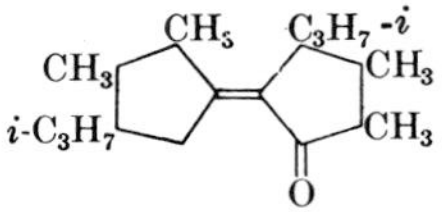 CH_3, CH_3, $i\text{-}C_3H_7$, $C_3H_7\text{-}i$, CH_3, CH_3, O (—)	1258
$C_2H_5CO(CH_2)_4COC_2H_5$	KOH	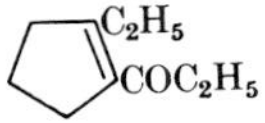 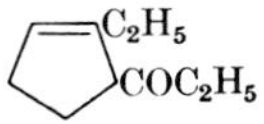 C_2H_5, COC_2H_5 (79), C_2H_5, COC_2H_5 (7)	395
$4\text{-}BrC_6H_4CO(CH_2)_2COCH_3$	NaOH	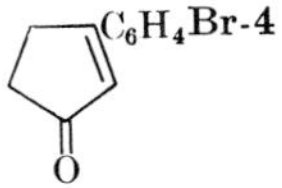 $C_6H_4Br\text{-}4$, O (40–55)	394
$C_6H_5CO(CH_2)_2COCH_3$	"	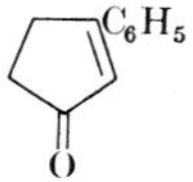 C_6H_5, O (50)	1259, 174

Reactant	Catalyst	Product (Yield, %)	Refs.
$CH_3CO(CH_2)_2CO(CH_2)_2CH{=}CHCH{=}CH_2$	"	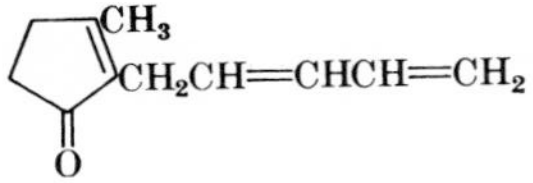 CH_3, $CH_2CH{=}CHCH{=}CH_2$, O (77)	397a, 393, 1260
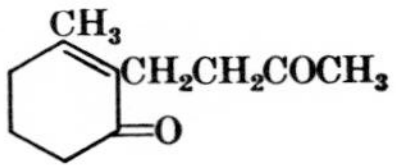 CH_3, $CH_2CH_2COCH_3$, O	$NaOCH_3$	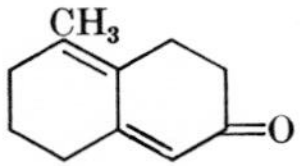 CH_3, O (—)	1261
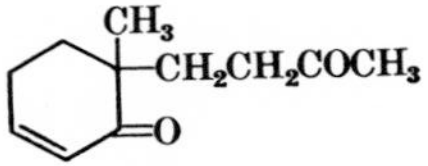 CH_3, $CH_2CH_2COCH_3$, O	KOH	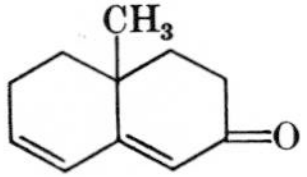 CH_3, O (89)	427
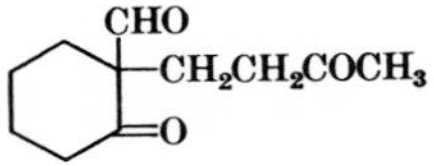 CHO, $CH_2CH_2COCH_3$, O	KOH; also $4\text{-}CH_3C_6H_4SO_3H$	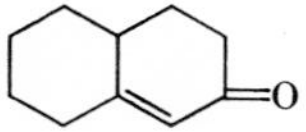 O (56)	1251
	$4\text{-}CH_3C_6H_4SO_3H$	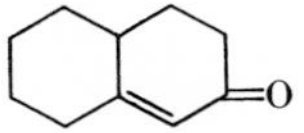 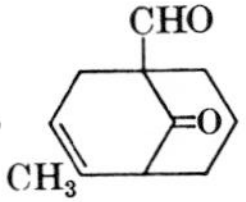 O, CHO, O, CH_3 (Total, 61)	1251
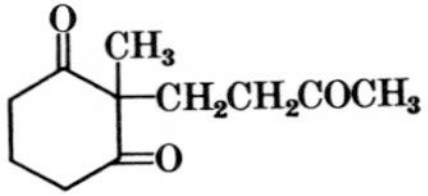 O, CH_3, $CH_2CH_2COCH_3$, O	$Al(OC_4H_9\text{-}t)_3$	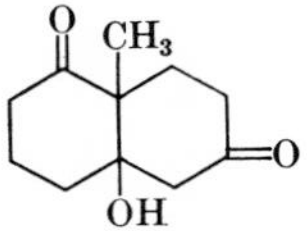 O, CH_3, O, OH (7)	1147, 1262
	$(CH_2)_4NH$	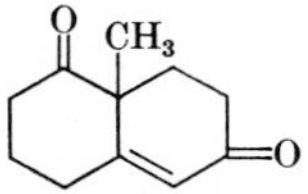 O, CH_3, O (75)	76

Note: References 668–2359 are on pp. 403–438.

TABLE VI. SELF-CONDENSATION OF KETONES (*Continued*)

Ketone	Catalyst	Product(s) (Yield, %)	Refs.
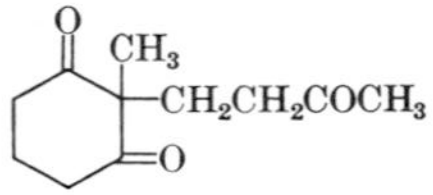 2-CH_3-2-$CH_2CH_2COCH_3$-cyclohexane-1,3-dione (*contd.*)	NaOH	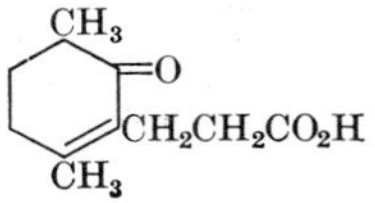 CH_3, =O, $CH_2CH_2CO_2H$, CH_3 (cyclohexenone) (—)	1262
$CH_3CO(CH_2)_2CO(CH_2)_2CH{=}CHC_2H_5$	,,	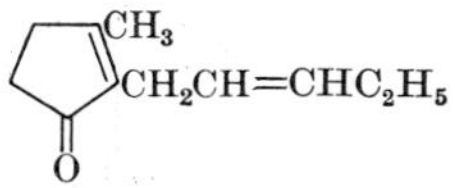 CH_3, $CH_2CH{=}CHC_2H_5$ (cyclopentenone) (80)	393, 1260, 1263
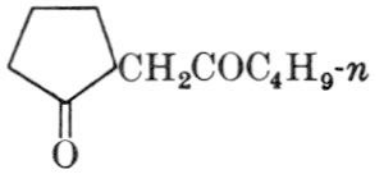 2-$CH_2COC_4H_9$-*n*-cyclopentanone	KOH	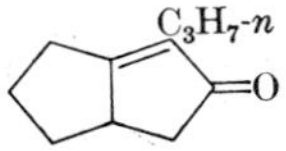 C_3H_7-*n*, =O (bicyclic enone) (73)	1242
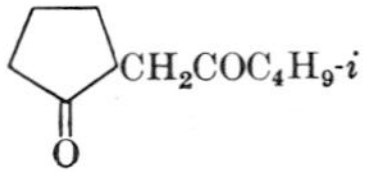 2-$CH_2COC_4H_9$-*i*-cyclopentanone	,,	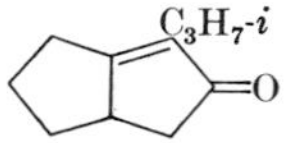 C_3H_7-*i*, =O (bicyclic enone) (73)	1242
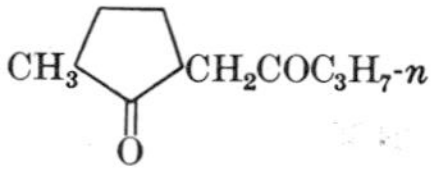 CH_3, $CH_2COC_3H_7$-*n* (cyclopentanone)	NaOH	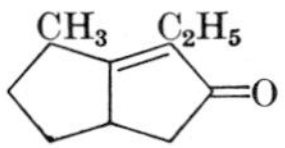 CH_3, C_2H_5, =O (bicyclic enone) (—)	397
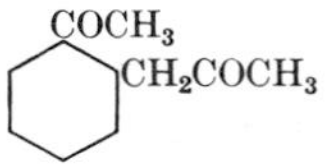 1-$COCH_3$-2-CH_2COCH_3-cyclohexane	HCl	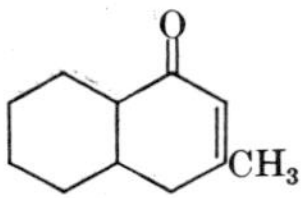 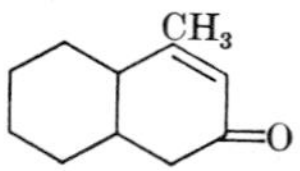 O, CH_3 (octalone) (59), CH_3, =O (octalone) (36)	1250, 1264

Reactant	Catalyst	Product (Yield, %)	Refs.
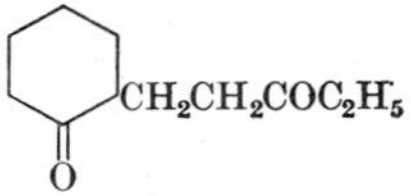 $CH_2CH_2COC_2H_5$, O	KOH	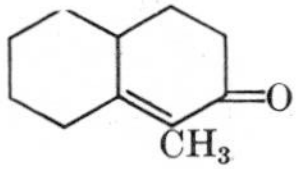 O, CH_3 (90)	1265
$CH_3CO(CH_2)_2CH(CO_2C_2H_5)COC_2H_5$	$NaOC_2H_5$	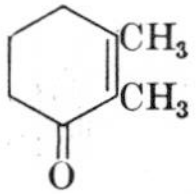 CH_3, CH_3, O (97)	405
$CH_3COCH(CO_2C_2H_5)(CH_2)_3COCH_3$	HCl	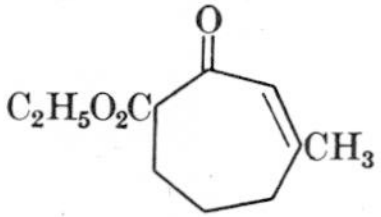 O, $C_2H_5O_2C$, CH_3 (60)	1266
$CH_3COC(CO_2C_2H_5)(CH_3)(CH_2)_2COCH_3$	$(CH_2)_5NH$, CH_3CO_2H	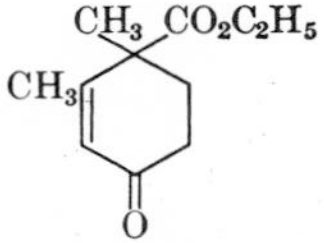 CH_3, $CO_2C_2H_5$, CH_3, O (—)	407
	H_3PO_4	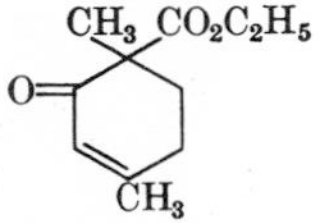 CH_3, $CO_2C_2H_5$, O, CH_3 (—)	407
$CH_3COC(CH_3)_2CH(CO_2C_2H_5)COCH_3$	KOH	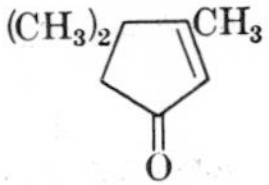 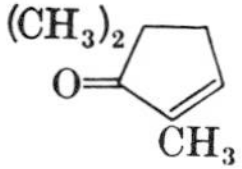 $(CH_3)_2$, CH_3, O (—), $(CH_3)_2$, O, CH_3 (—)	1267
$CH_3CO(CH_2)_2COC_6H_{13}$-$n$	NaOH	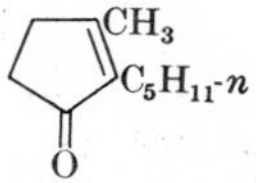 CH_3, C_5H_{11}-n, O (92)	1232, 1268, 393

Note: References 668–2359 are on pp. 403–438.

TABLE VI. Self-Condensation of Ketones (*Continued*)

Ketone	Catalyst	Product(s) (Yield, %)	Refs.
$CH_3CO(CH_2)_3COC_5H_{11}$-*n*	NaOH	CH_3, C_4H_9-*n*, O (—)	393
$C_2H_5CO(CH_2)_5COC_2H_5$	H_2SO_4	No condensation	384
O	KOH	O (45),	166
		O (—),	
		O (35)	
CH_3, CH_3, $COCH_3$, $COCH_3$	$NaOCH_3$	O, CH_3, CH_3, CH_3 (—)	1256
O, CH_3, $CH{=}CHCOC_2H_5$	$NaOC_2H_5$; also H_2SO_4	CH_3, O, CH_3 (79)	2336

Reactant	Catalyst	Product (Yield, %)	Refs.
O; CH_3; $CH_2CH_2CH_2COCH_3$	KOC_4H_9-*t*	HO; O; CH_3 (78)	1269
O; CH_3; $CH_2CH_2COCH_3$; O	$(CH_2)_4NH$	H_3C; O; O; OH (16), CH_3; O; O; CH_3; OH (16)	48
$CH_3COCHOHCH_2CO(CH_2)_2(CH{=}CH)_2CH_3$	NaOH	HO; CH_3; $CH_2(CH{=}CH)_2CH_3$; O (30)	1270
$CO_2C_2H_5$; $CH_2CH_2COCH_3$; O	H_2SO_4	$CO_2C_2H_5$; O; CH_3; OH (75)	648
$CO_2C_2H_5$; CH_2COCH_3; O	KOH	O (73)	1271
CO_2CH_3; $CH_2CH_2COCH_3$; O	HCl, CH_3CO_2H	O (—)	426, 2311
CH_3; O; $CH_2COCH_2CH_2CO_2H$	KOH	CH_3; CH_2CO_2H; O (75)	1272

Note: References 668–2359 are on pp. 403–438.

TABLE VI. SELF-CONDENSATION OF KETONES (*Continued*)

Ketone	Catalyst	Product(s) (Yield, %)	Refs.
$CH_3CO(CH_2)_2CO(CH_2)_2CH{=}CHC_3H_7$-*n*	NaOH	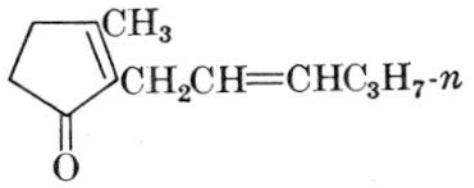 CH_3, $CH_2CH{=}CHC_3H_7$-*n*, O (77)	1260
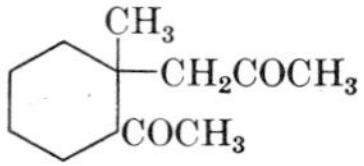 CH_3, CH_2COCH_3, $COCH_3$	,,	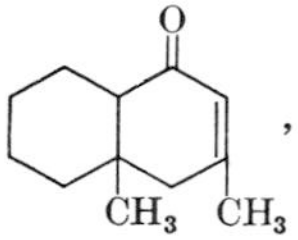 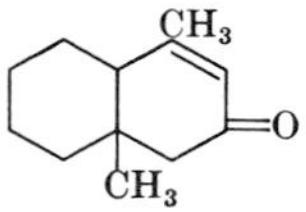 O, CH_3 CH_3 I , CH_3, =O, CH_3 II (—) (I:II = 4:1)	1250
$CH_3COCH_2CH(CO_2CH_3)COC_5H_{11}$-*n*	,,	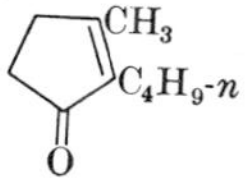 CH_3, C_4H_9-*n*, O (68)	1232
	$NaOCH_3$	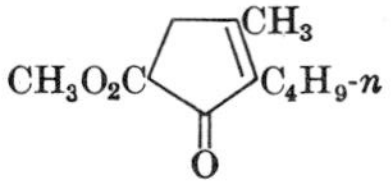 CH_3O_2C, CH_3, C_4H_9-*n*, O (59)	
$CH_3COCH(CO_2C_2H_5)C(CH_3)_2CH_2COCH_3$	$NaOC_2H_5$	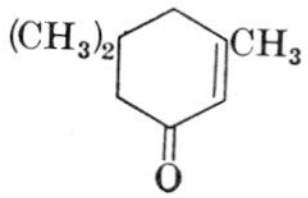 $(CH_3)_2$, CH_3, O (—)	1273
$CH_3CO(CH_2)_2COC_7H_{15}$-*n*	$Na_2B_4O_7$	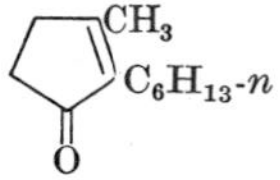 CH_3, C_6H_{13}-*n*, O (—)	1232

Reactant	Catalyst	Product (Yield, %)	Refs.
n-$C_3H_7CO(CH_2)_4COC_3H_7$-n	KOH	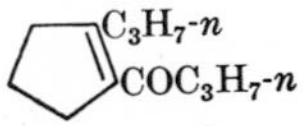 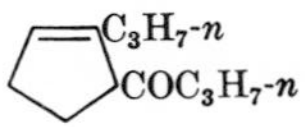 C_3H_7-n, COC_3H_7-n (80), C_3H_7-n, COC_3H_7-n (8)	395
$CH_3COCH_2C(CH_3)_2C(CH_3)_2CH_2COCH_3$	None	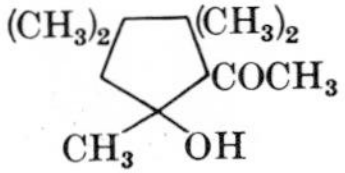 $(CH_3)_2$, $(CH_3)_2$, $COCH_3$, CH_3, OH (—)	69
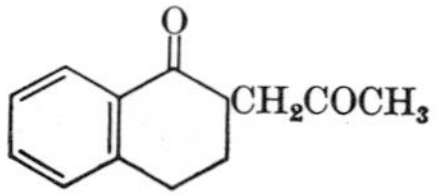 CH_2COCH_3	KOH	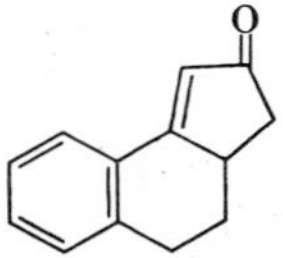 (80)	1274
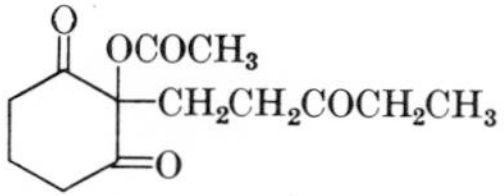 $OCOCH_3$, $CH_2CH_2COCH_2CH_3$	$(CH_2)_4NH$, CH_3CO_2H	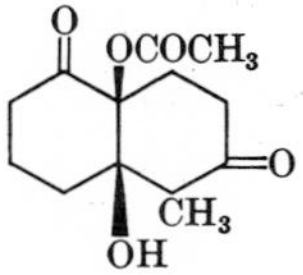 $OCOCH_3$, CH_3, OH (—)	665
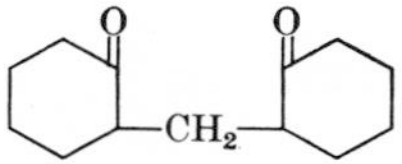 CH_2	KOH	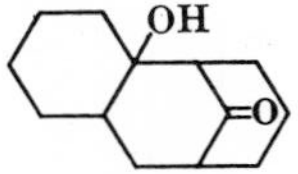 OH (22)	555
$CH_3COCH_2CH(CO_2CH_3)CO$-$(CH_2)_2CH{=}CHC_2H_5$	NaOH	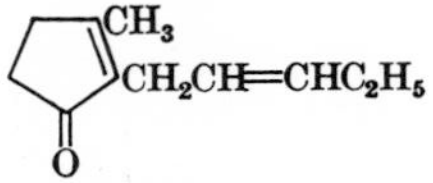 CH_3, $CH_2CH{=}CHC_2H_5$ (73)	393, 1263

Note: References 668–2359 are on pp. 403–438.

TABLE VI. SELF-CONDENSATION OF KETONES (*Continued*)

Ketone	Catalyst	Product(s) (Yield, %)	Refs.
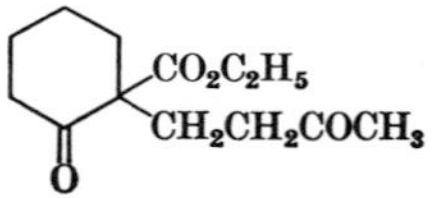 2-oxocyclohexyl with $CO_2C_2H_5$, $CH_2CH_2COCH_3$	$NaOC_2H_5$	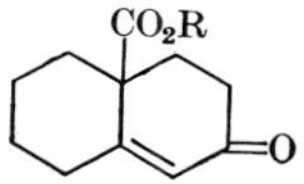 CO_2R; $R = H$ (15), $R = C_2H_5$ (66)	1276
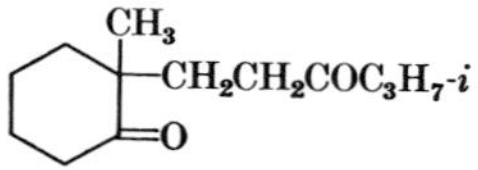 2-oxocyclohexyl with CH_3, $CH_2CH_2COC_3H_7$-*i*	$NaOCH_3$	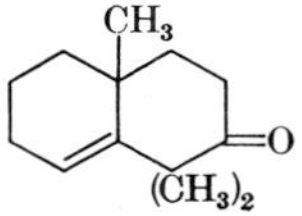 CH_3, $(CH_3)_2$ (—)	1277
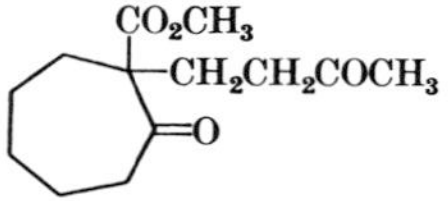 2-oxocycloheptyl with CO_2CH_3, $CH_2CH_2COCH_3$	HCl, CH_3CO_2H	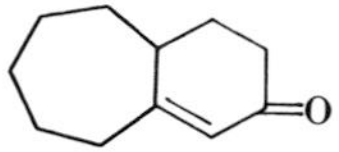 (—)	426
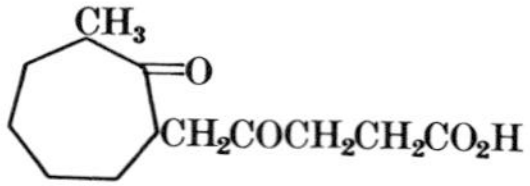 2-oxocycloheptyl with CH_3, $CH_2COCH_2CH_2CO_2H$	KOH	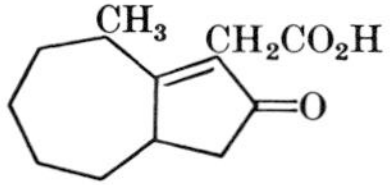 CH_3, CH_2CO_2H (80)	1272
$CH_3COCH_2CH_2C(CO_2C_2H_5)(CH_3)$-$COCO_2C_2H_5$	HCl	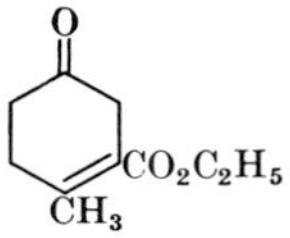 $CO_2C_2H_5$, CH_3 (—)	404
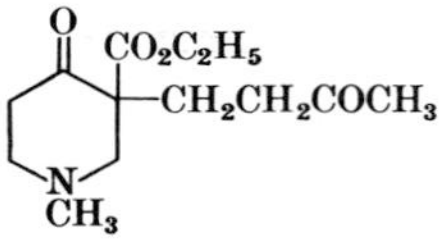 4-oxo-1-methylpiperidyl with $CO_2C_2H_5$, $CH_2CH_2COCH_3$	KOH	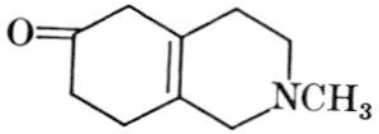 NCH_3 (20)	1278

Reactant	Conditions	Product (Yield, %)	Refs.
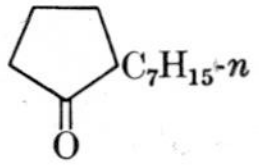 C_7H_{15}-*n*, O	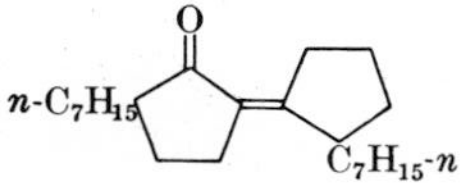 ,,	*n*-C_7H_{15}, O, C_7H_{15}-*n* (50)	1206
$CH_3CO(CH_2)_2CO(CH_2)_2CH{=}CHC_4H_9$-*i*	NaOH 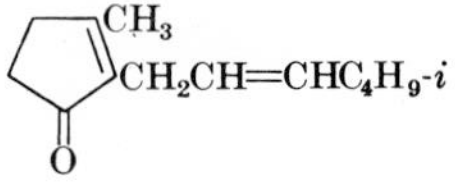	CH_3, $CH_2CH{=}CHC_4H_9$-*i*, O (82)	1260
O, $(CH_3)_2$, O, CH_3	,,	$(CH_3)_2$, O, CH_3 (—)	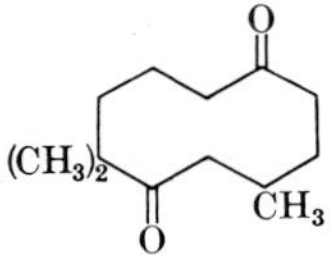 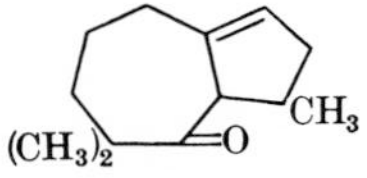 1279
$CH_3COC(CO_2C_2H_5)(C_3H_7$-*i*$)(CH_2)_2COCH_3$	$C_6H_5N(C_2H_5)_2$, 140°	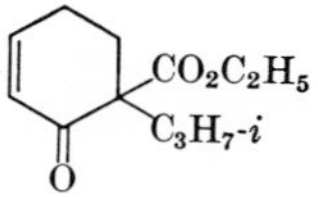 $CO_2C_2H_5$, C_3H_7-*i*, O (43)	1280
$CH_3COCH_2CH(CO_2C_2H_5)COC_5H_{11}$-*n*	NaOH	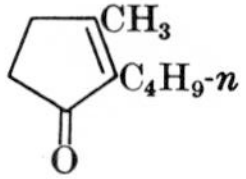 CH_3, C_4H_9-*n*, O (68)	1232
$C_6H_5CH_2COC_6H_5$	Polyphosphoric acid, 150° 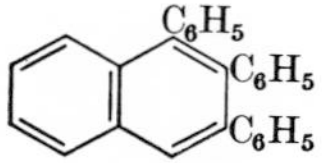	C_6H_5, C_6H_5, C_6H_5 (15)	1281
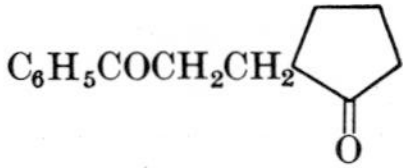 $C_6H_5COCH_2CH_2$, O	4-$CH_3C_6H_4SO_3H$	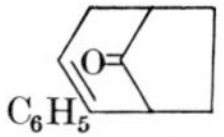 O, C_6H_5 (—)	1282

Note: References 668–2359 are on pp. 403–438.

TABLE VI. SELF-CONDENSATION OF KETONES (*Continued*)

Ketone	Catalyst	Product(s) (Yield, %)	Refs.
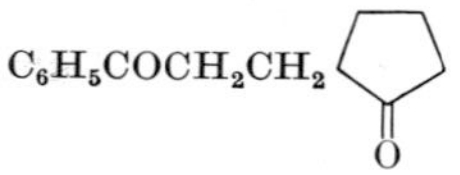 $C_6H_5COCH_2CH_2$-(cyclopentanone) (*contd.*)	HCl, CH_3CO_2H	 C_6H_5-(cycloheptene)-CO_2H (—)	1282
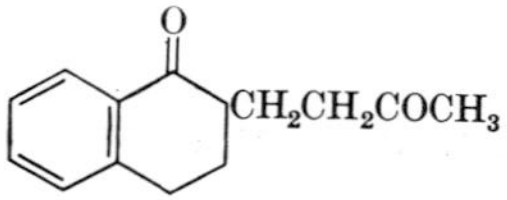 2-($CH_2CH_2COCH_3$)-1-tetralone	KOC_2H_5	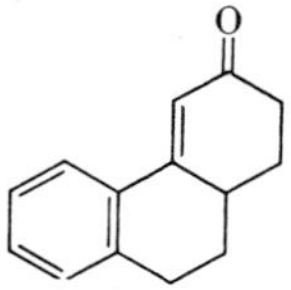 (tricyclic enone, O) (73)	1142
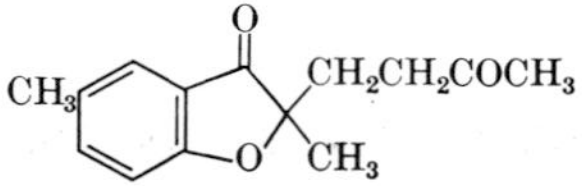 CH_3-benzofuranone, 2-$CH_2CH_2COCH_3$, 2-CH_3	KOH	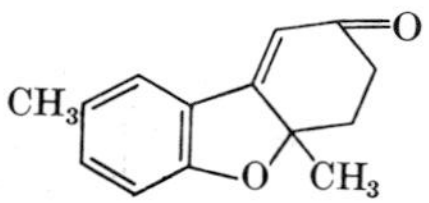 CH_3-(tetracyclic enone, O, CH_3) (—)	1283
$C_6H_5COCH_2CH(COCH_3)CO_2C_2H_5$	NaOH	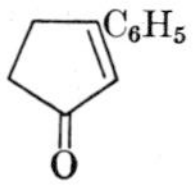 C_6H_5-cyclopentenone (—)	1284
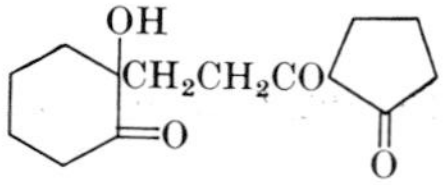 OH, CH_2CH_2CO-(cyclopentanone), cyclohexanone	KOH	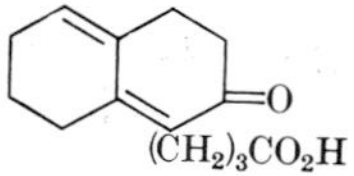 (bicyclic dienone), $(CH_2)_3CO_2H$ (—)	1285
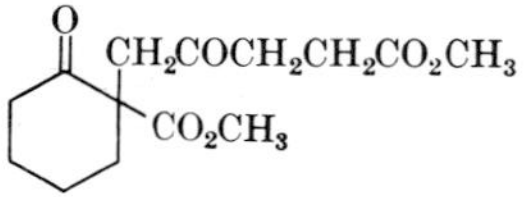 2-($CH_2COCH_2CH_2CO_2CH_3$)-2-(CO_2CH_3)cyclohexanone	KOC_4H_9-*t*	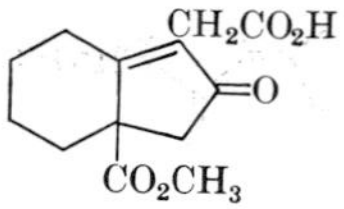 CH_2CO_2H, (bicyclic enone), CO_2CH_3 (—)	1286

Reactant	Catalyst	Product (Yield, %)		Refs.
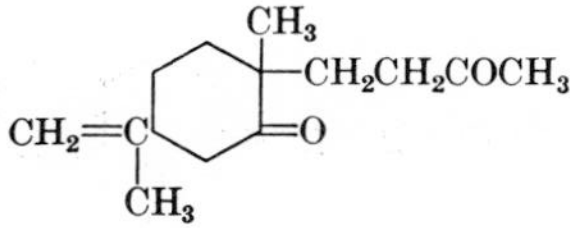 CH_3, $CH_2CH_2COCH_3$, $CH_2{=}C$, CH_3, $=O$	$NaOC_2H_5$	CH_3, CH_2, C, CH_3, $=O$	(—)	1288
CH_3, $=O$, $CH_2COC_2H_5$	KOH	CH_3, CH_3, $=O$	(57)	1275
$(CH_3)_2$, $COCH_3$, $CH_2CH_2COCH_3$, O	,,	$(H_3C)_2$, $COCH_3$, $=O$	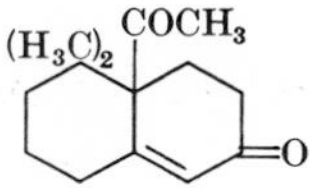 (40)	1289
$C_2H_5COCH_2CH_2C(CH_3)(CO_2C_2H_5)$- $COCO_2C_2H_5$	,,	O, CH_3, CO_2H, CH_3	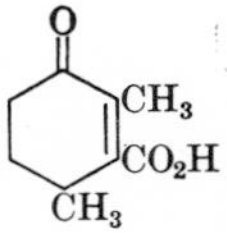 (10)	404
	$NaOCH_3$	$CO_2C_2H_5$, CH_3, $CO_2C_2H_5$, $O{=}$, CH_3	(—)	404
	HCl	O, CH_3, $CO_2C_2H_5$, CH_3 , O, CH_3, CO_2H, CH_3 , O, CH_3, $CO_2C_2H_5$, CH_3 $CO_2C_2H_5$	(—)	404

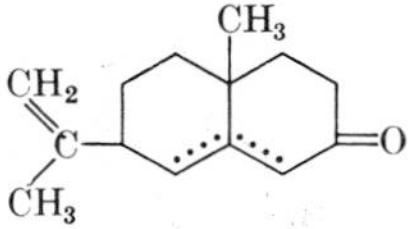

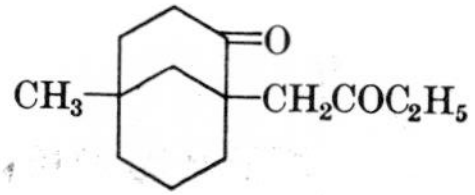

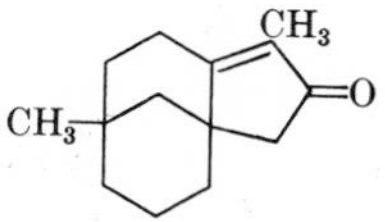

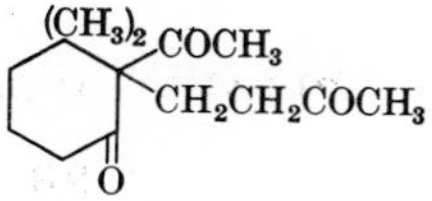

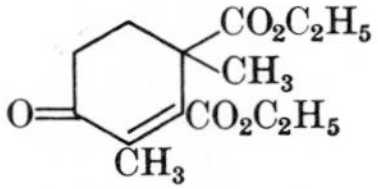

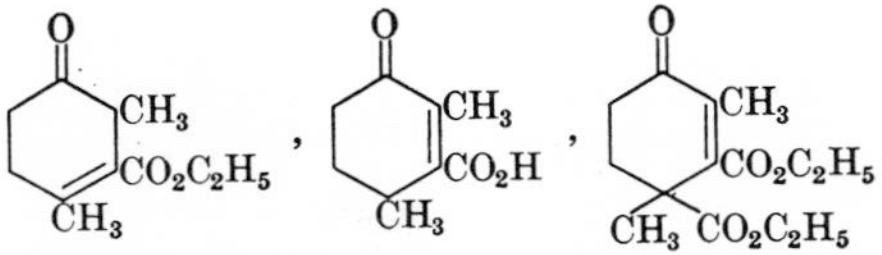

Note: References 668–2359 are on pp. 403–438.

TABLE VI. SELF-CONDENSATION OF KETONES (*Continued*)

Ketone	Catalyst	Product(s) (Yield, %)	Refs.
n-$C_4H_9CO(CH_2)_4COC_4H_9$-n	KOH	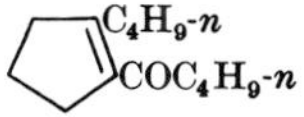 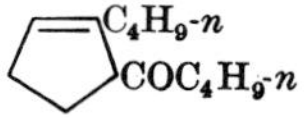 C_4H_9-n, COC_4H_9-n (79), C_4H_9-n, COC_4H_9-n (7)	395
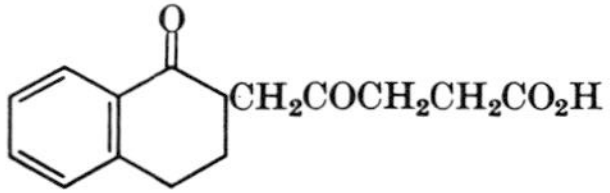 O, $CH_2COCH_2CH_2CO_2H$	KOC_4H_9-t	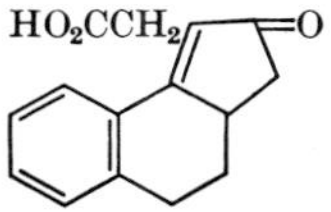 HO_2CCH_2, O (51)	1290
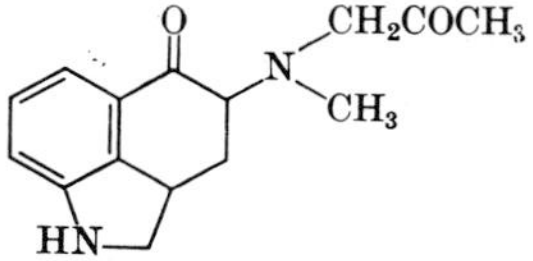 O, CH_2COCH_3, N, CH_3, HN	$NaOCH_3$	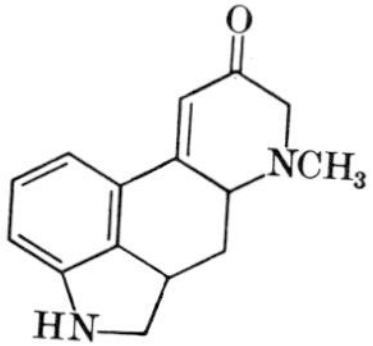 O, NCH_3, HN (69)	1291
$(C_6H_5)_2CHCOCH_3$	C_6H_5MgBr or $(CH_2)_5CHMgBr$	$(C_6H_5)_2CHC(CH_3){=}CHCOCH(C_6H_5)_2$ (30)	1292
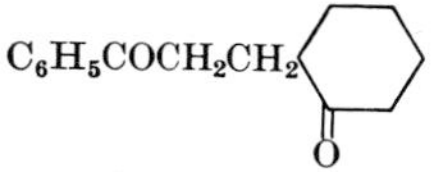 $C_6H_5COCH_2CH_2$, O	–	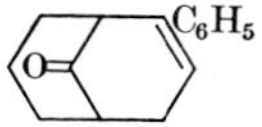 O, C_6H_5 (—)	1293
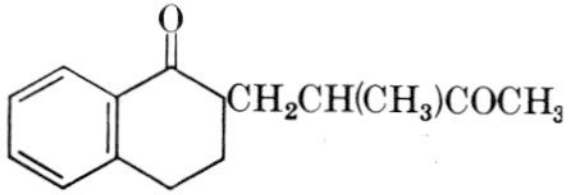 O, $CH_2CH(CH_3)COCH_3$	KOH	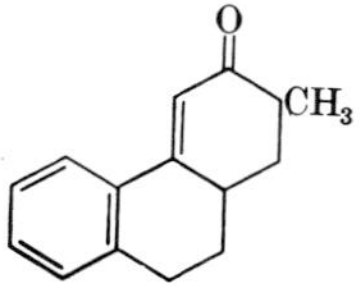 O, CH_3 (—)	1294

Reactant	Catalyst	Product (Yield, %)		Refs.
	H_3PO_4	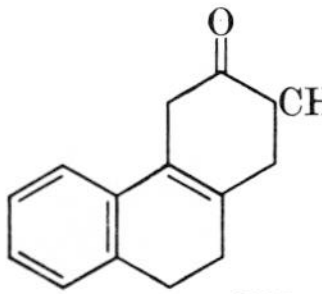 O; CH_3	(—)	1294
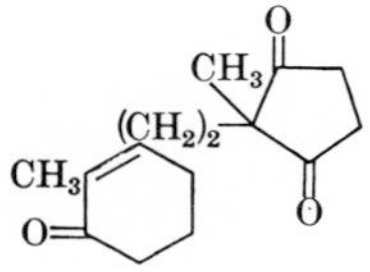 CH_3; O; $(CH_2)_2$; CH_3; O; O	$4\text{-}CH_3C_6H_4SO_3H$	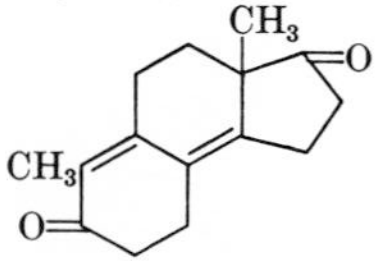 CH_3; O; CH_3; O	(76)	403
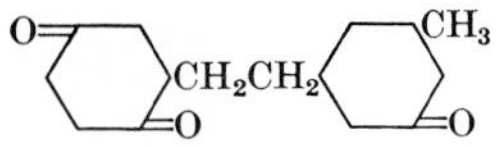 O; CH_2CH_2; O; CH_3; O	HCl	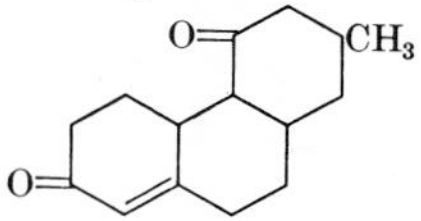 O; CH_3; O	(70)	1295
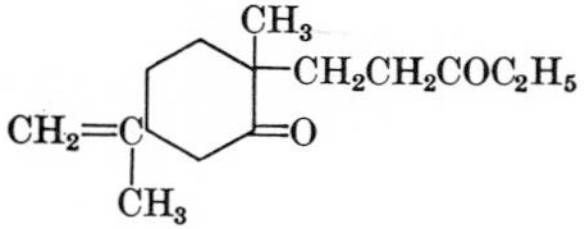 $CH_2{=}C$; CH_3; CH_3; $CH_2CH_2COC_2H_5$; O	$NaOC_2H_5$ or H_2SO_4	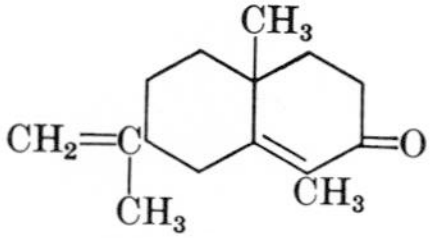 $CH_2{=}C$; CH_3; CH_3; O; CH_3	(—)	1288
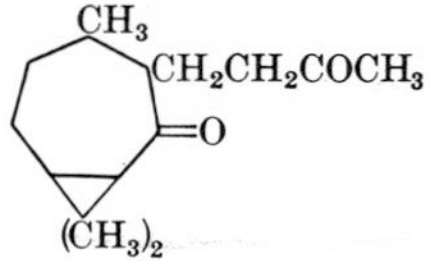 CH_3; $CH_2CH_2COCH_3$; O; $(CH_3)_2$	KOH	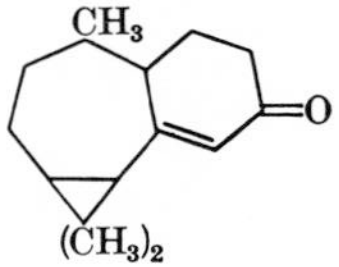 CH_3; O; $(CH_3)_2$	(—)	1296
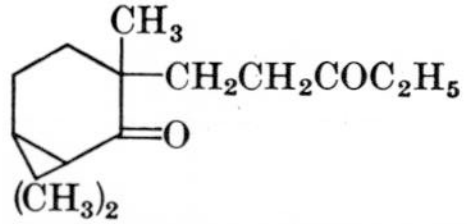 CH_3; $CH_2CH_2COC_2H_5$; O; $(CH_3)_2$	Various catalysts	No condensation		1297

Note: References 668–2359 are on pp. 403–438.

TABLE VI. SELF-CONDENSATION OF KETONES (*Continued*)

Ketone	Catalyst	Product(s) (Yield, %)	Refs.
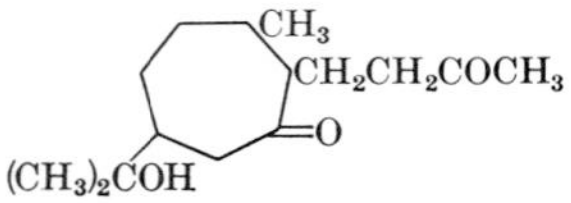 CH_3, $CH_2CH_2COCH_3$, O, $(CH_3)_2COH$	KOH	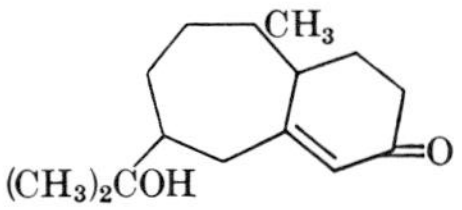 CH_3, O, $(CH_3)_2COH$ (—),	1296, 1298
		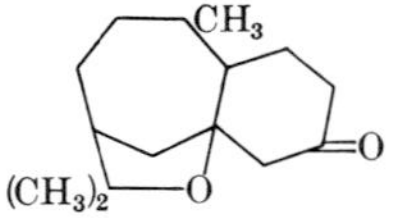 CH_3, O, $(CH_3)_2$, O (—)	
$CH_3CO(CH_2)_2CO(CH_2)_3CH(OC_5H_{11}\text{-}i)CH_3$	HBr	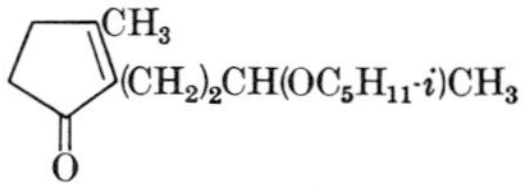 CH_3, $(CH_2)_2CH(OC_5H_{11}\text{-}i)CH_3$, O (—)	1232
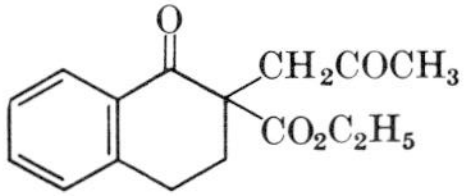 O, CH_2COCH_3, $CO_2C_2H_5$	KOH	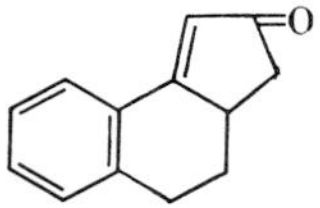 O (51)	1274
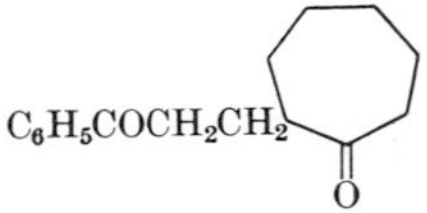 $C_6H_5COCH_2CH_2$, O	4-$CH_3C_6H_4SO_3H$	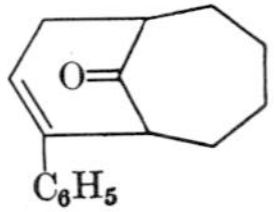 O, C_6H_5 (—)	1282
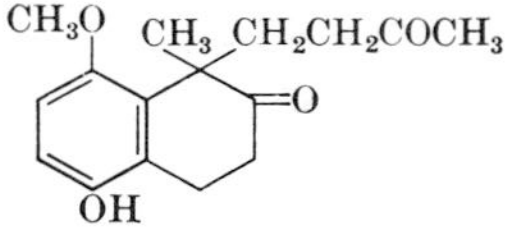 CH_3O, CH_3, $CH_2CH_2COCH_3$, O, OH	HCl	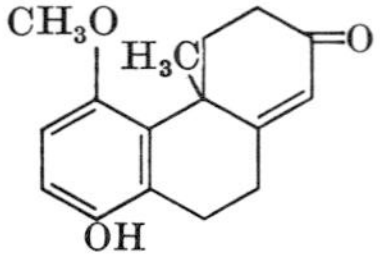 CH_3O, H_3C, O, OH (47)	1299

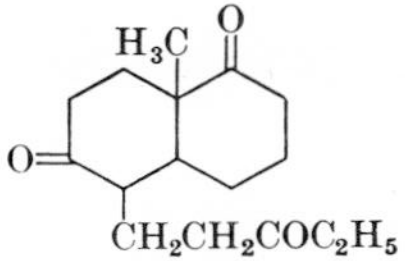 H_3C, O, O, $CH_2CH_2COC_2H_5$	”	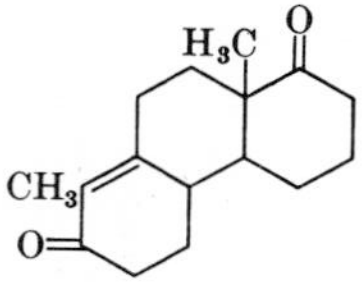 H_3C, O, CH_3, O	(—)	1300
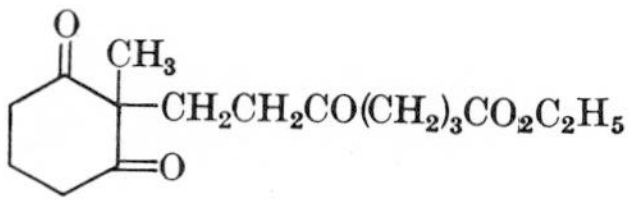 O, CH_3, $CH_2CH_2CO(CH_2)_3CO_2C_2H_5$, O	$(C_2H_5)_3N$, $C_6H_5CO_2H$	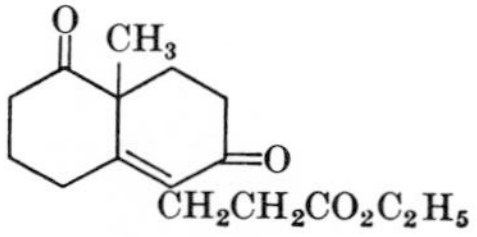 O, CH_3, O, $CH_2CH_2CO_2C_2H_5$	(—)	1300
$CH_3COC(C_2H_5)(CO_2C_2H_5)(CH_2)_2$-$CH(CO_2C_2H_5)COCH_3$	KOH	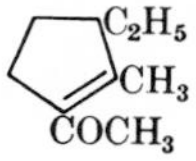 C_2H_5, CH_3, $COCH_3$	(—)	1301
$CH_3CO(CH_2)_{12}COCH_3$	$C_6H_5N(CH_3)MgBr$	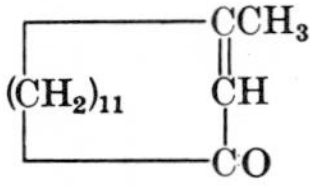 $(CH_2)_{11}$, CCH_3, CH, CO	(—)	1302
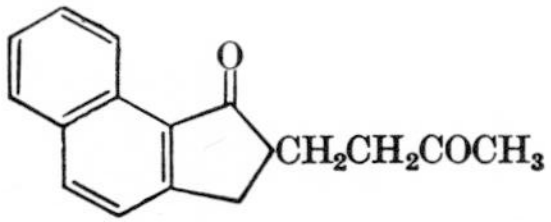 O, $CH_2CH_2COCH_3$	HCl	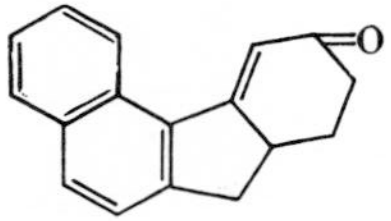 O	(—)	1303
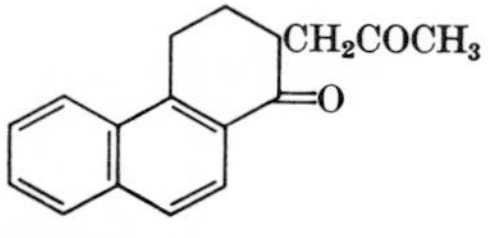 CH_2COCH_3, O	KOH	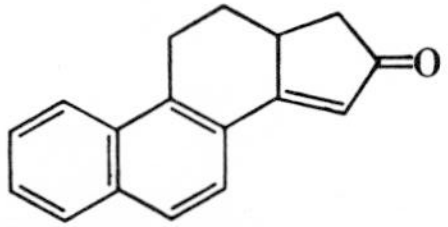 O	(90)	1304

Note: References 668–2359 are on pp. 403–438.

TABLE VI. SELF-CONDENSATION OF KETONES (*Continued*)

Ketone	Catalyst	Product(s) (Yield, %)	Refs.
$CH_3COCH(COCH_3)CH(C_6H_5)$- $CH(COCH_3)COCH_3$	KOH	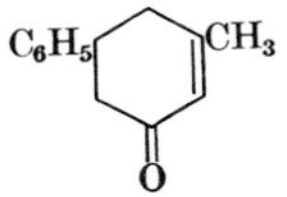 C_6H_5, CH_3, O (—)	46
	HCl	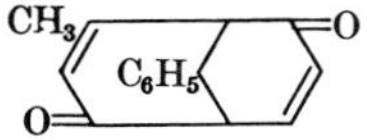 CH_3, C_6H_5, O, O (—)	46
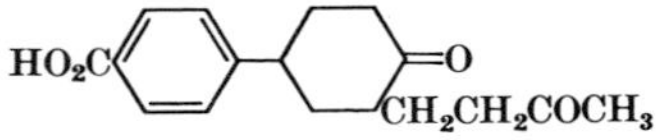 HO_2C, O, $CH_2CH_2COCH_3$	NaOH	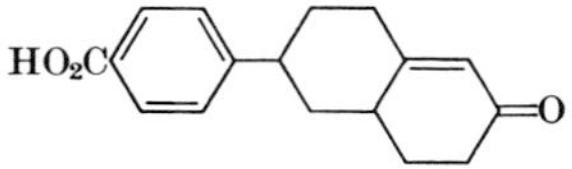 HO_2C, O (72)	1305
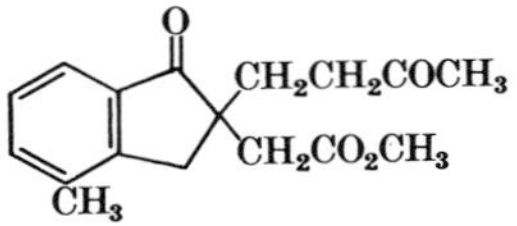 O, $CH_2CH_2COCH_3$, $CH_2CO_2CH_3$, CH_3	$NaOCH_3$	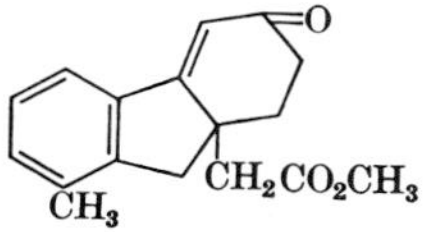 O, $CH_2CO_2CH_3$, CH_3 (33)	1306
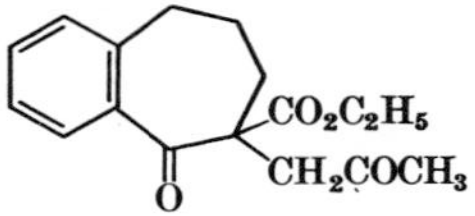 $CO_2C_2H_5$, CH_2COCH_3, O	KOH	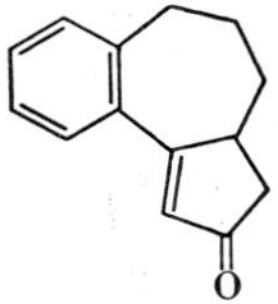 O (61)	1307
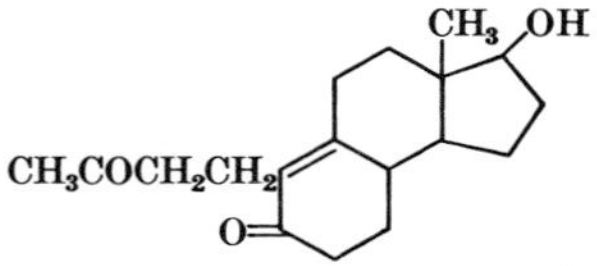 CH_3, OH, $CH_3COCH_2CH_2$, O	HCl	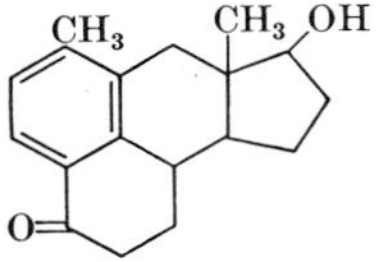 CH_3, CH_3, OH, O (—)	1308

Reactant	Catalyst	Product (Yield, %)	Refs.
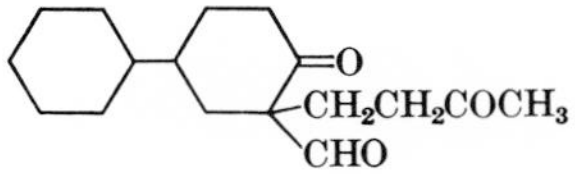 =O, $CH_2CH_2COCH_3$, CHO	HCl; also KOH	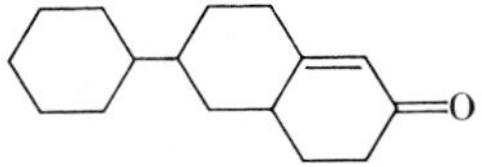 =O (82)	1309
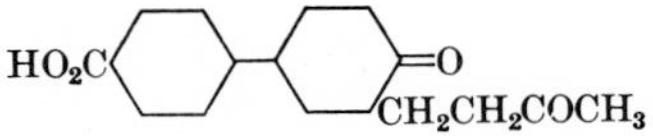 HO_2C, =O, $CH_2CH_2COCH_3$	NaOH	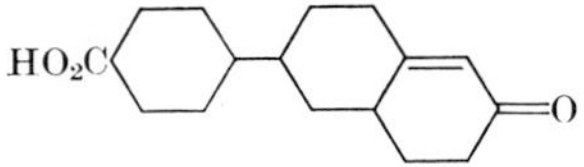 HO_2C, =O (83–90)	1305
$C_6H_5CO(CH_2)_4COC_6H_5$	KOH	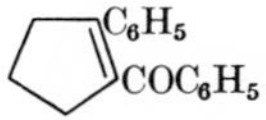 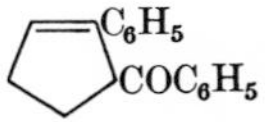 C_6H_5, COC_6H_5 (79), C_6H_5, COC_6H_5 (12)	395
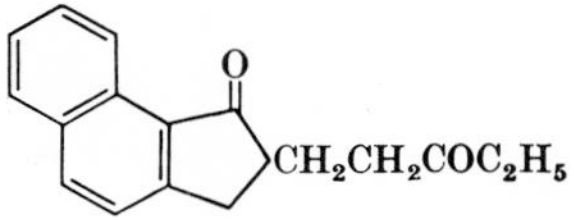 O, $CH_2CH_2COC_2H_5$	HCl	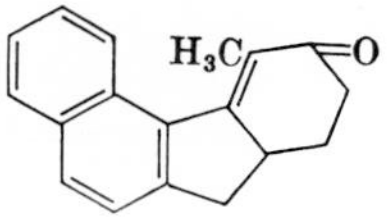 H_3C, O (—)	1303
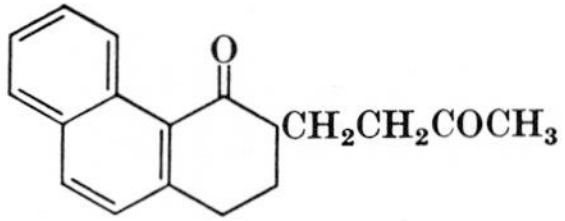 O, $CH_2CH_2COCH_3$	KOH	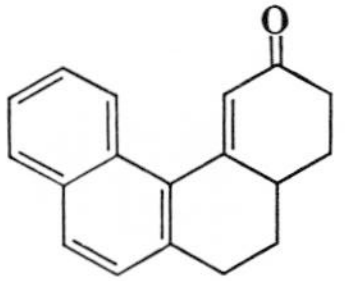 O (91)	1310
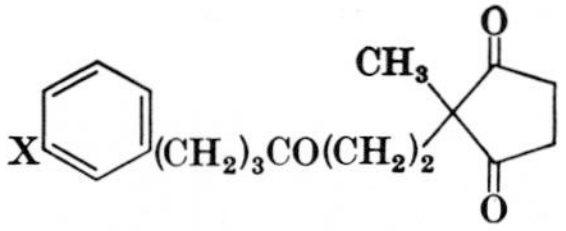 X, $(CH_2)_3CO(CH_2)_2$, CH_3, O, O X = H, O_2N, CH_3O	$(C_2H_5)_3N$, $C_6H_5CO_2H$, $4\text{-}CH_3C_6H_4SO_3H$, or $Al(OC_4H_9\text{-}t)_3$	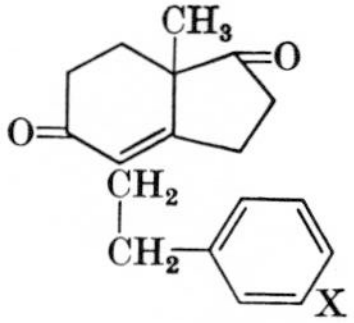 CH_3, O, O=, CH_2, CH_2, X (—) X = H, NO_2, OCH_3	388

Note: References 668–2359 are on pp. 403–438.

TABLE VI. SELF-CONDENSATION OF KETONES (*Continued*)

Ketone	Catalyst	Product(s) (Yield, %)	Refs.
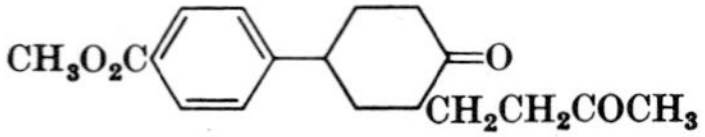 CH_3O_2C, O, $CH_2CH_2COCH_3$	KOH	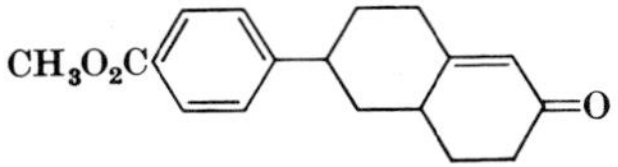 CH_3O_2C, O (50)	1305
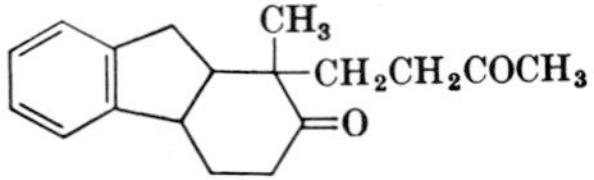 CH_3, $CH_2CH_2COCH_3$, O	NaOH	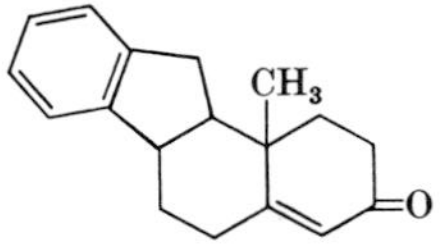 CH_3, O (—)	1311
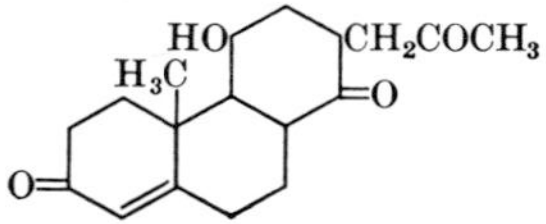 HO, H_3C, CH_2COCH_3, O, O	KOC_4H_9-*t*	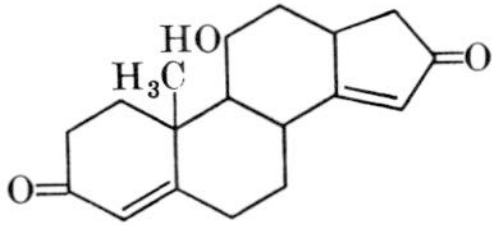 HO, H_3C, O, O (—)	606
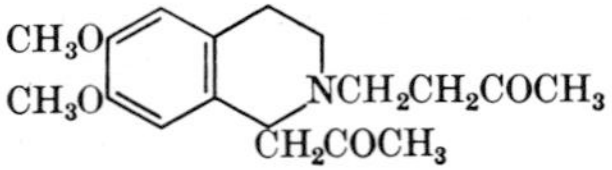 CH_3O, CH_3O, $NCH_2CH_2COCH_3$, CH_2COCH_3	HCl or $NaOCH_3$	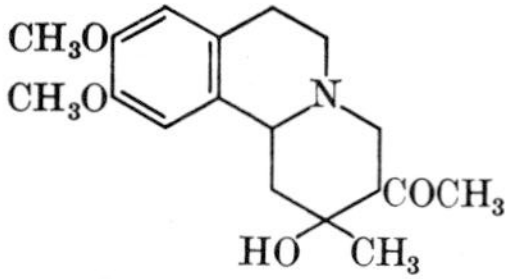 CH_3O, CH_3O, N, $COCH_3$, HO, CH_3 (—)	1312
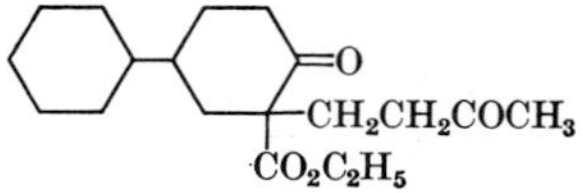 O, $CH_2CH_2COCH_3$, $CO_2C_2H_5$	KOH	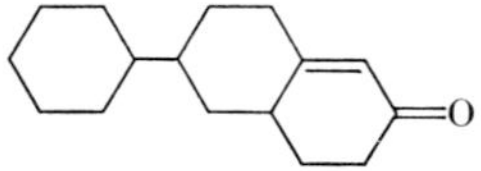 O (48)	1309
	$NaOCH_3$	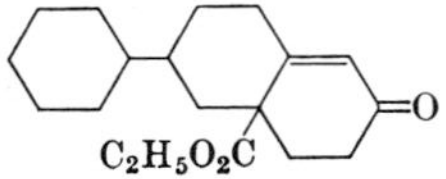 O, $C_2H_5O_2C$ (17)	1309

Reactant	Catalyst	Product (Yield, %)	Refs.
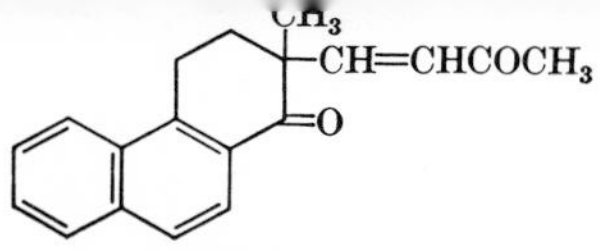 CH_3, $CH{=}CHCOCH_3$, O	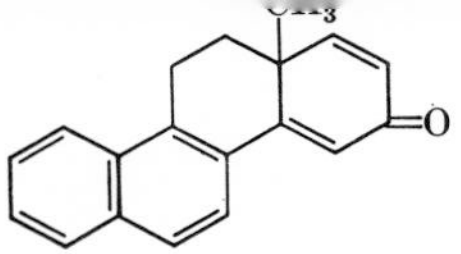 KOH	CH_3, O (37–40)	1313
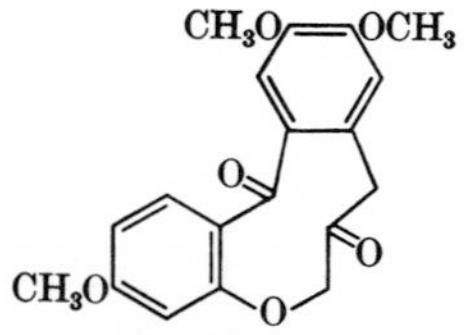 CH_3O, OCH_3, O, O, CH_3O, O	CH_3CO_2Na, $(CH_3CO)_2O$	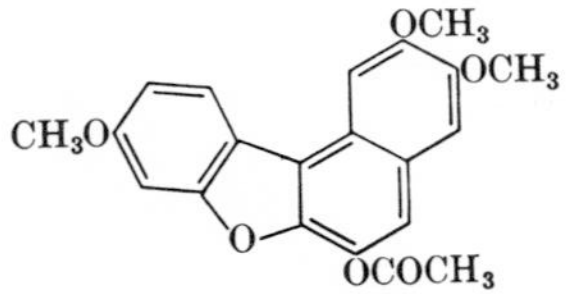 OCH_3, OCH_3, CH_3O, O, $OCOCH_3$ (70)	1314
$C_6H_5CO(CH_2)_5COC_6H_5$	H_2SO_4	No condensation	1244
	P_2O_5	$C_{19}H_{18}O$ (—) •	1244
$COCH_3$, $COCH_3$, CH_3O	$NaOCH_3$	O, CH_3, CH_3O (—) or	1315
		CH_3, O, CH_3O (—)	
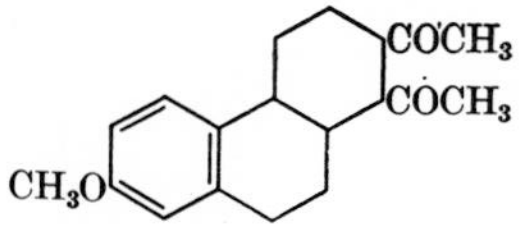 $COCH_3$, $COCH_3$, CH_3O	$NaOC_2H_5$ 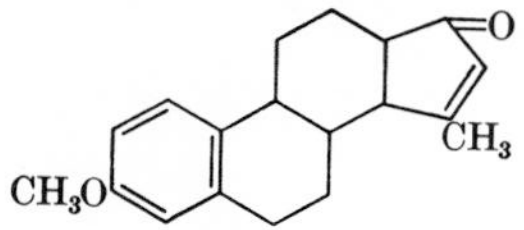	O, CH_3, CH_3O (—) or	1315
		CH_3, O, CH_3O (—)	

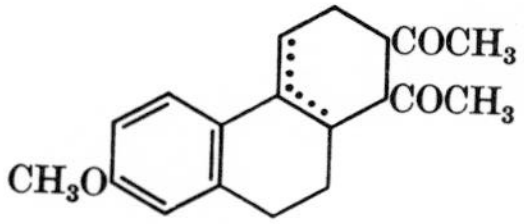

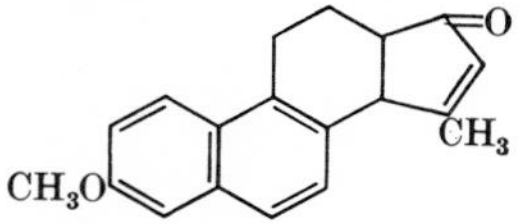

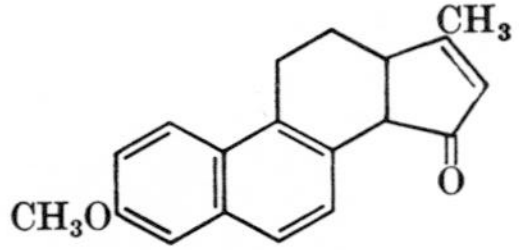

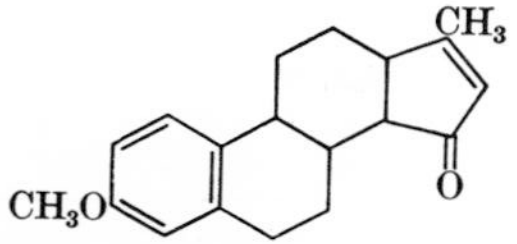

Note: References 668–2359 are on pp. 403–438.

TABLE VI. SELF-CONDENSATION OF KETONES (*Continued*)

Ketone	Catalyst	Product(s) (Yield, %)	Refs.
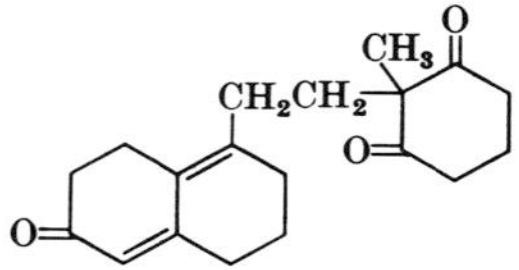	$4\text{-}CH_3C_6H_4SO_3H$	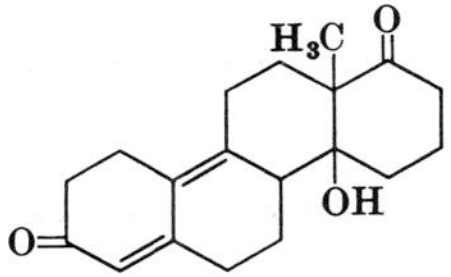 (—)	402
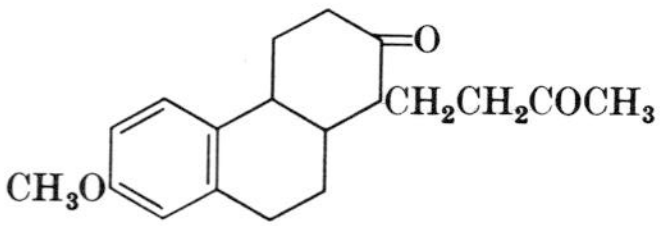 $CH_2CH_2COCH_3$	KOH, 25°	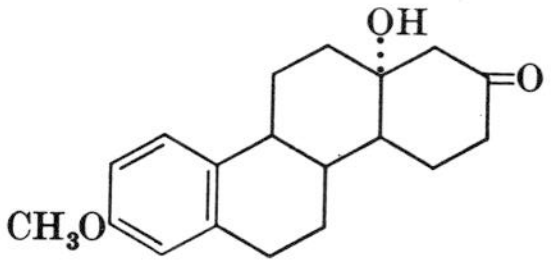 (88)	1316
	KOH, heat	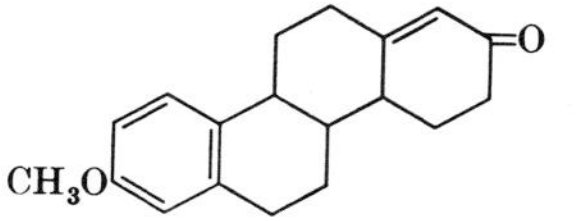 (—)	1316
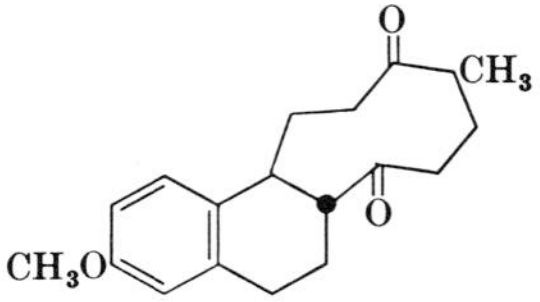	KOH	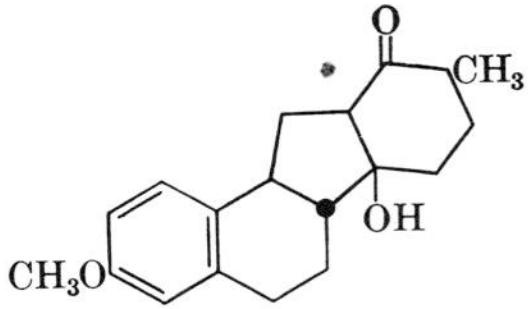 (29),	1316
		(32)	

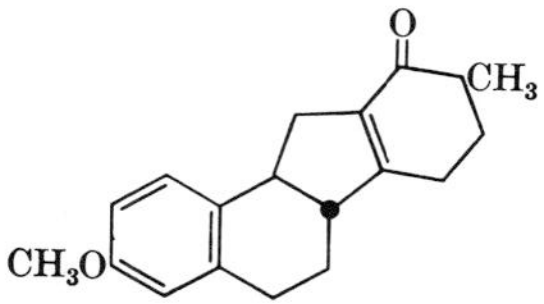

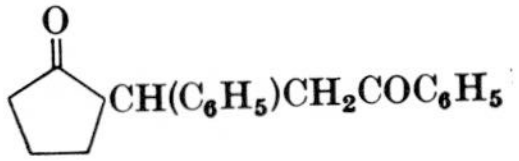

Reactant	Catalyst	Product (Yield, %)		Refs.
$CH(C_6H_5)CH_2COC_6H_5$	HCl	C_6H_5, C_6H_5 †	(—)	1317
$CH(CO_2C_2H_5)COCH_3$	KOH		(—)	1318
$CH(CO_2C_2H_5)COCH_3$	,,		(84)	1304
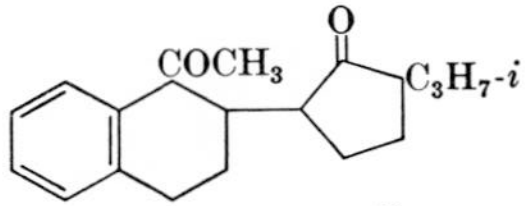 $COCH_3$, C_3H_7-*i*	$NaOC_2H_5$	C_3H_7-*i*	(—)	1319
CH_3O, $(CH_2)_3COCH_3$	$NaOCH_3$	CH_3O, $COCH_3$	(—)	1320
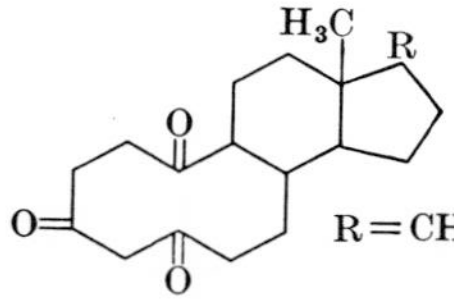 H_3C, R; $R=CH_3CO_2$, CH_3CO	CH_3CO_2H	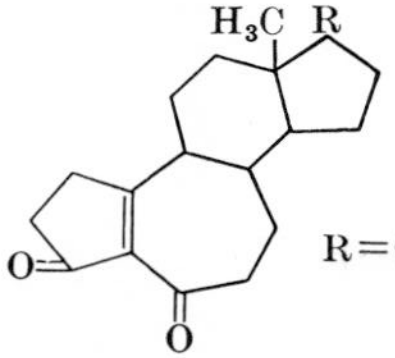 H_3C, R; $R=CH_3CO_2$, CH_3CO	(—)	1321, 399, 1322

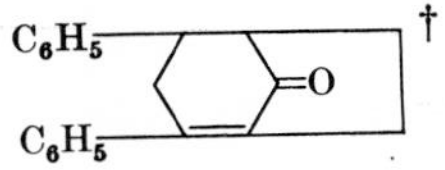

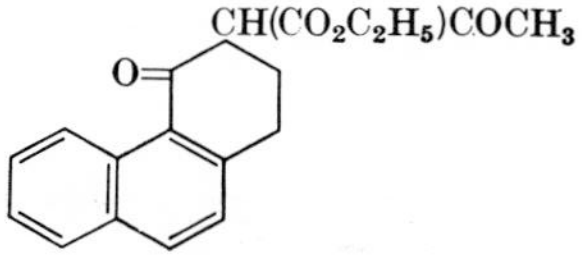

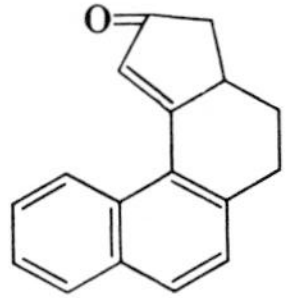

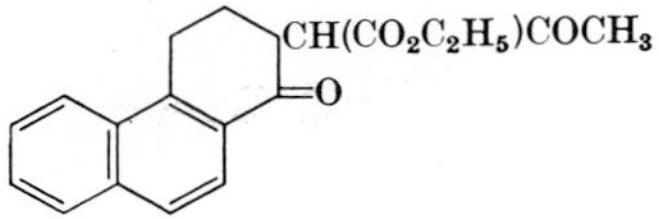

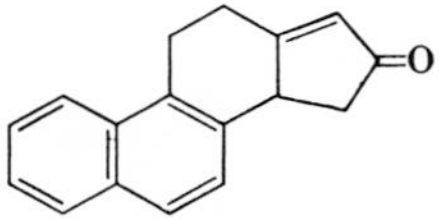

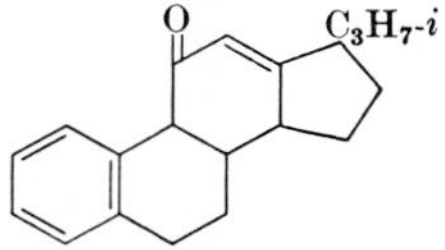

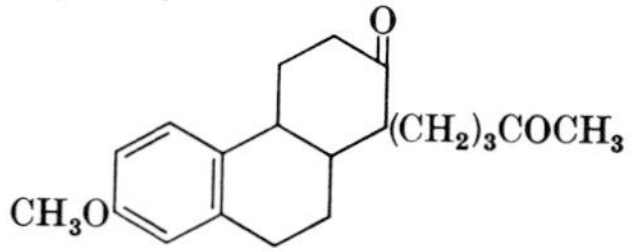

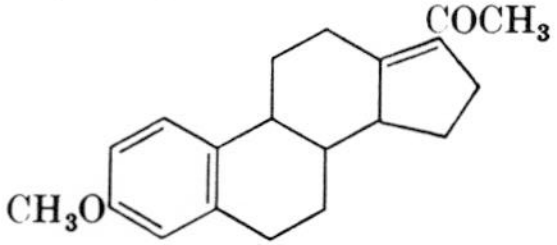

Note: References 668–2359 are on pp. 403–438.

† The structure of this product was not established.

TABLE VI. SELF-CONDENSATION OF KETONES (*Continued*)

Ketone	Catalyst	Product(s) (Yield, %)	Refs.
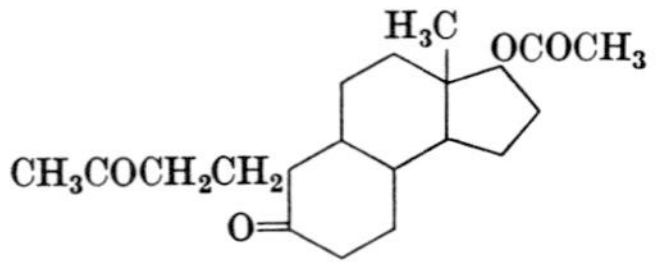 H_3C, $OCOCH_3$, $CH_3COCH_2CH_2$, O	KOH	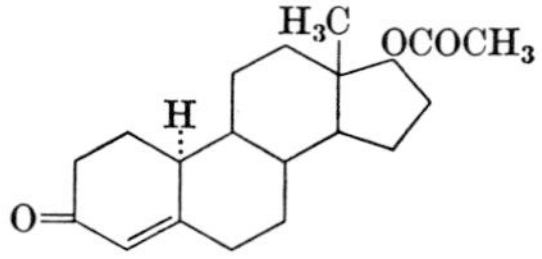 H_3C, $OCOCH_3$, H, O (—)	1323
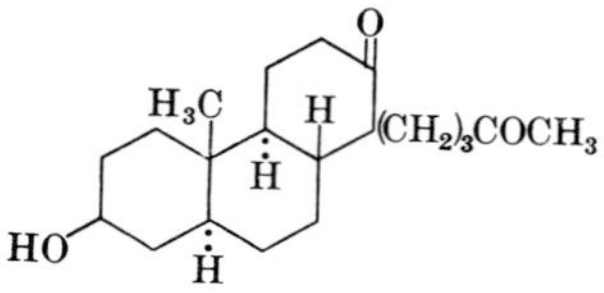 O, H_3C, H, $(CH_2)_3COCH_3$, H, HO, H	NaOH	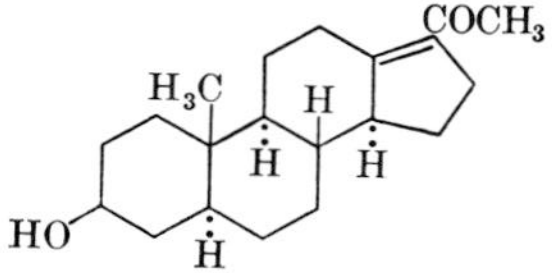 $COCH_3$, H_3C, H, H, H, HO, H (—)	1324, 1325
$(C_6H_5)_3CCOCH_3$	$(C_6H_5)_3CNa$	$(C_6H_5)_3CCOCH_2COCH_3$ (92)	1326
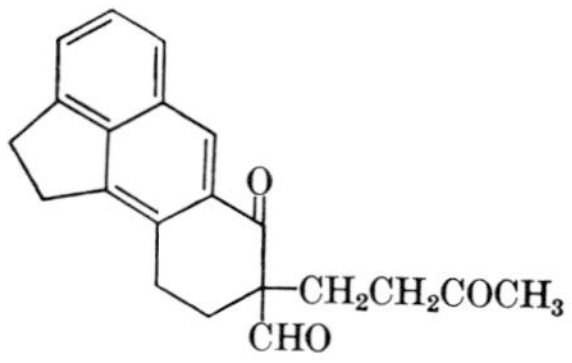 O, $CH_2CH_2COCH_3$, CHO	KOH	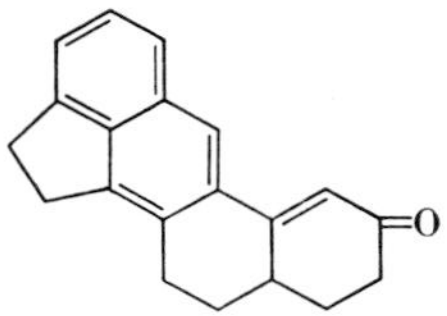 O (90)	1327
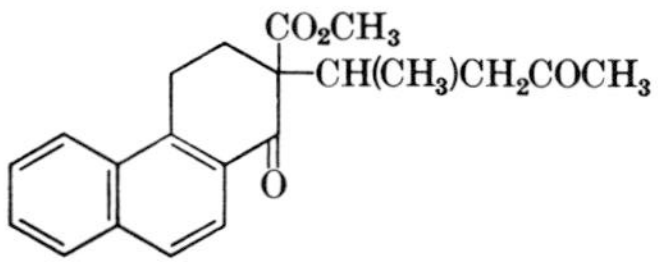 CO_2CH_3, $CH(CH_3)CH_2COCH_3$, O	"	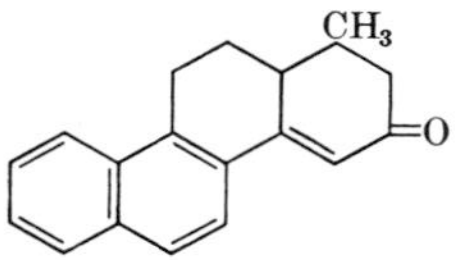 CH_3, O (50)	1313, 1328

Reactant	Catalyst	Product (Yield, %)		Refs.
CO_2CH_3, $CH_2CH_2COCH_2CH_3$, O	HCl, CH_3CO_2H; also KOH	O, CH_3	(69)	1310
CH_3, CH_3, $COCH_3$, $CH_3CO(CH_2)_2$, O	$NaOC_5H_{11}$-*t*	CH_3, CH_3, $COCH_3$, O	(—)	389
O, $CH_2CH_2COC_2H_5$	KOH	O, CH_3	(72)	1329
CH_3, O, $CH_2CH_2COCH_3$, CHO	"	CH_3, O	(88)	1327

Note: References 668–2359 are on pp. 403–438.

TABLE VI. SELF-CONDENSATION OF KETONES (*Continued*)

Ketone	Catalyst	Product(s) (Yield, %)	Refs.
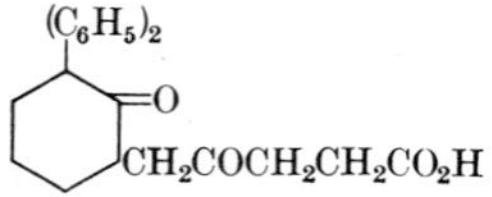 $(C_6H_5)_2$; $CH_2COCH_2CH_2CO_2H$	KOH	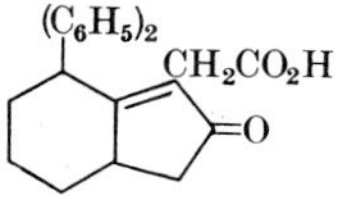 $(C_6H_5)_2$; CH_2CO_2H (72)	1272
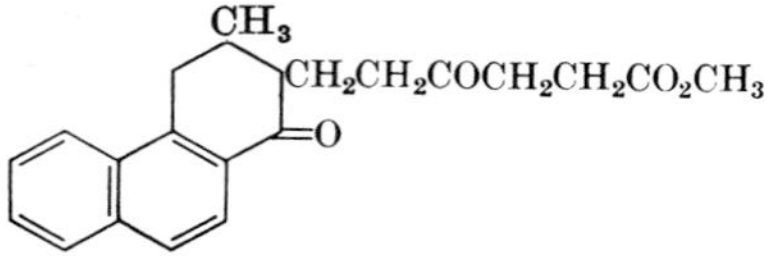 CH_3; $CH_2CH_2COCH_2CH_2CO_2CH_3$	$NaOCH_3$	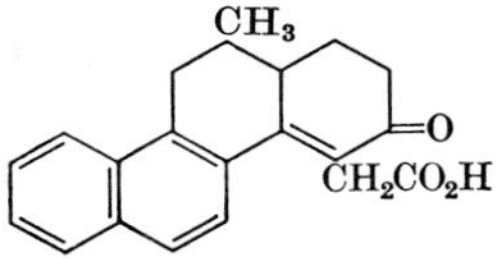 CH_3; CH_2CO_2H (88)	387
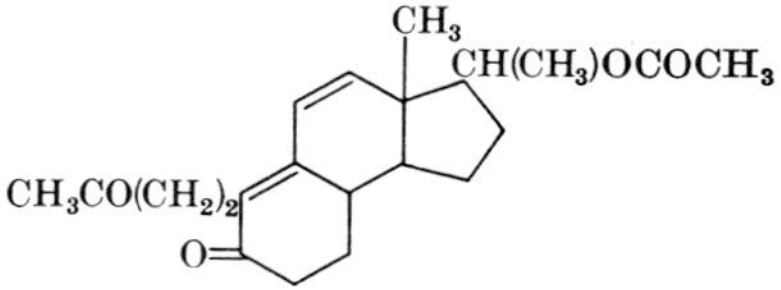 CH_3; $CH(CH_3)OCOCH_3$; $CH_3CO(CH_2)_2$	$NaOC_5H_{11}$-*t*	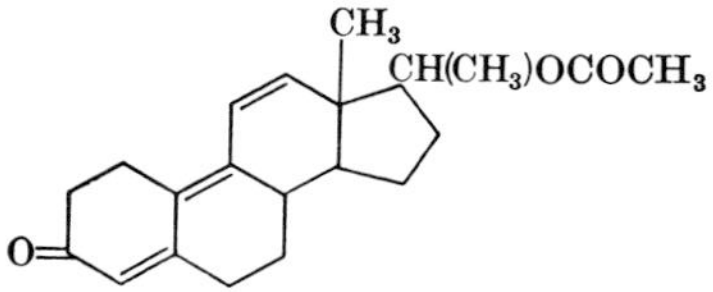 CH_3; $CH(CH_3)OCOCH_3$ (—)	389
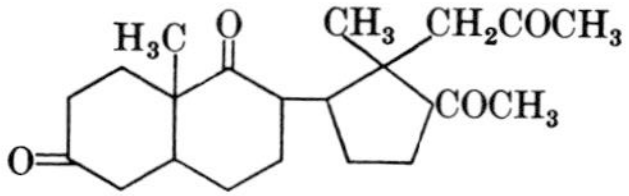 H_3C; CH_3; CH_2COCH_3; $COCH_3$	$NaOCH_3$	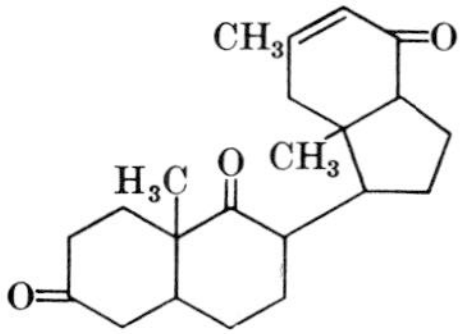 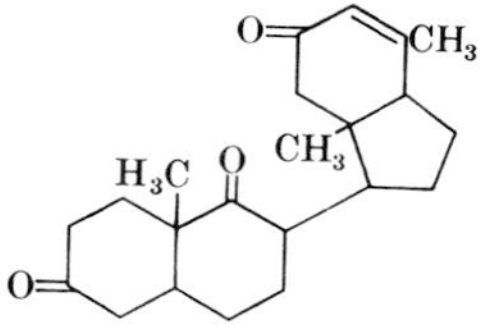 CH_3; H_3C; CH_3, CH_3; CH_3; H_3C (Total, 35)	1330

Reactant	Catalyst	Product (Yield, %)		Refs.
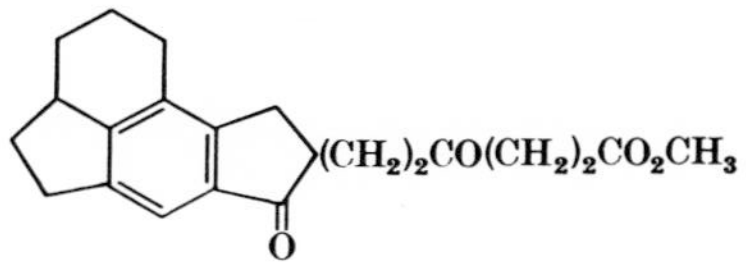 $(CH_2)_2CO(CH_2)_2CO_2CH_3$	$NaOCH_3$, C_5H_5N	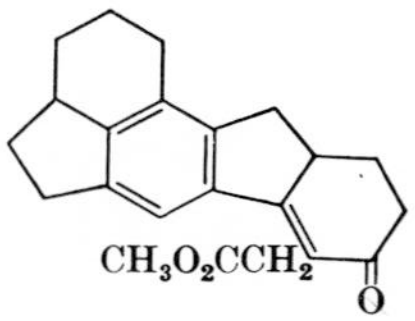 $CH_3O_2CCH_2$	(—)	390
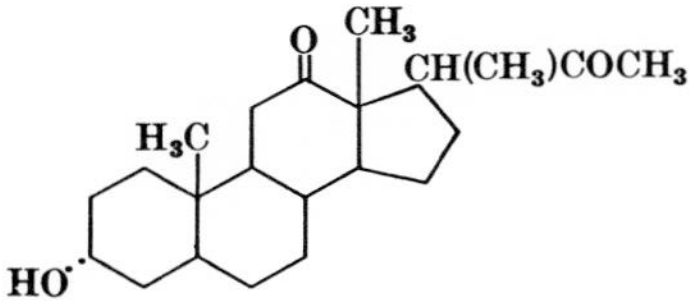 CH_3, $CH(CH_3)COCH_3$, H_3C, HO	NaH	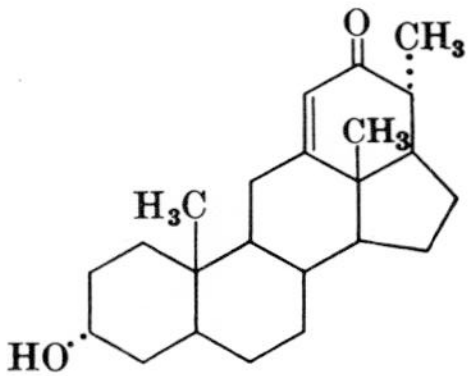 O, CH_3, CH_3, H_3C, HO	(54)	1331
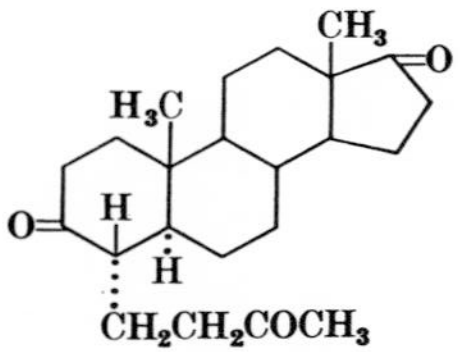 CH_3, O, H_3C, O, H, H, $CH_2CH_2COCH_3$	H_2SO_4	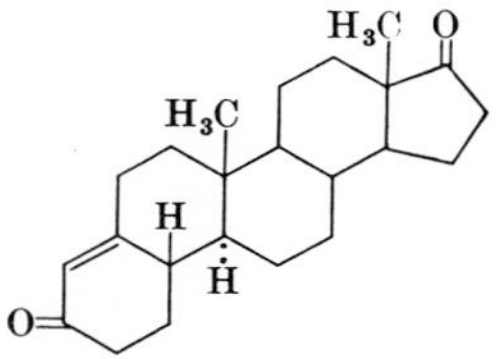 H_3C, O, H_3C, H, H, O	(—)	1332
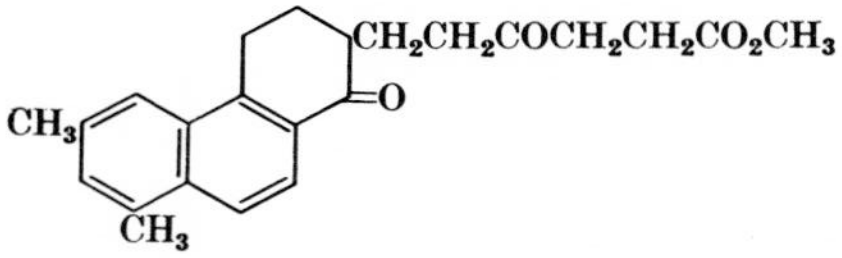 $CH_2CH_2COCH_2CH_2CO_2CH_3$, O, CH_3, CH_3	$NaOCH_3$	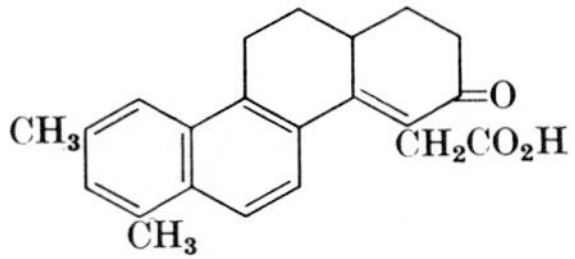 O, CH_2CO_2H, CH_3, CH_3	(96)	387

Note: References 668–2359 are on pp. 403–438.

TABLE VI. SELF-CONDENSATION OF KETONES (*Continued*)

Ketone	Catalyst	Product(s) (Yield, %)	Refs.
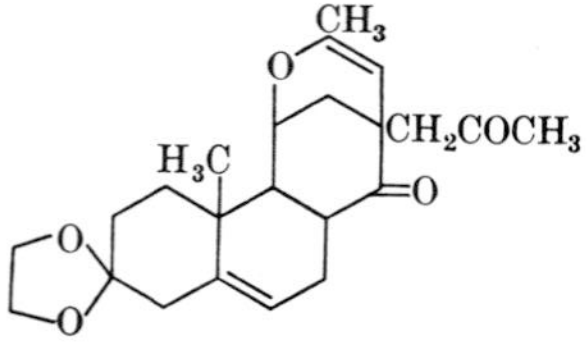 CH_3, H_3C, CH_2COCH_3, O, O, O, O	KOC_4H_9-*t*	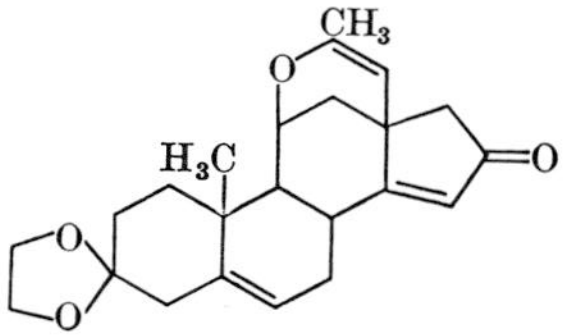 CH_3, H_3C, O, O, O, O (—)	606
$C_6H_5COCH(C_6H_5)CH(C_6H_5)CH_2COCH_3$	HCl	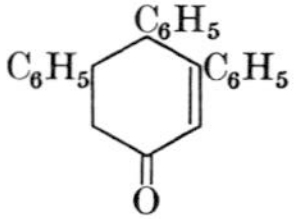 C_6H_5, C_6H_5, C_6H_5, O (50)	1220
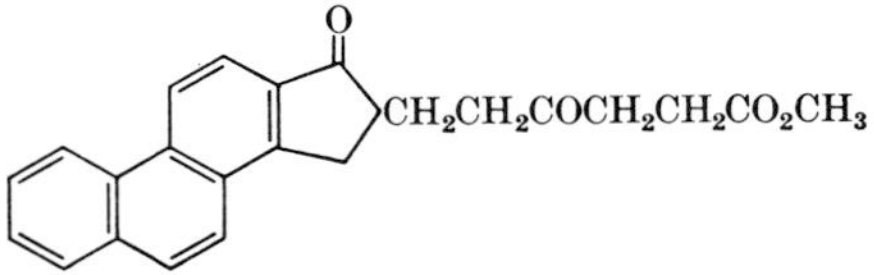 O, $CH_2CH_2COCH_2CH_2CO_2CH_3$	KOH	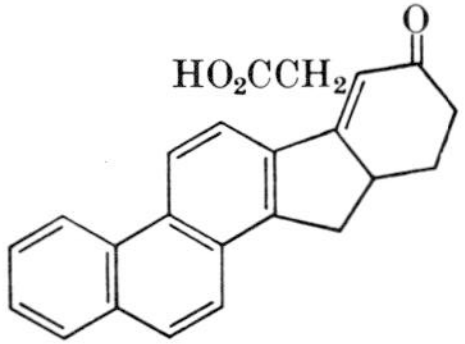 HO_2CCH_2, O (70)	1333
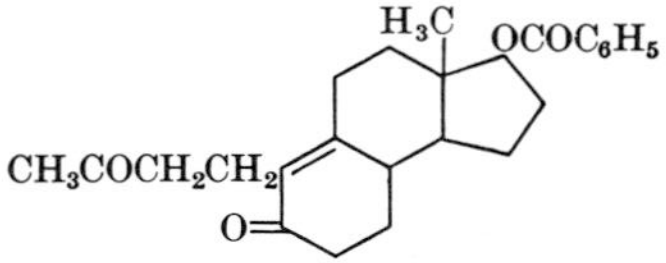 H_3C, $OCOC_6H_5$, $CH_3COCH_2CH_2$, O	$(CH_2)_4NH$	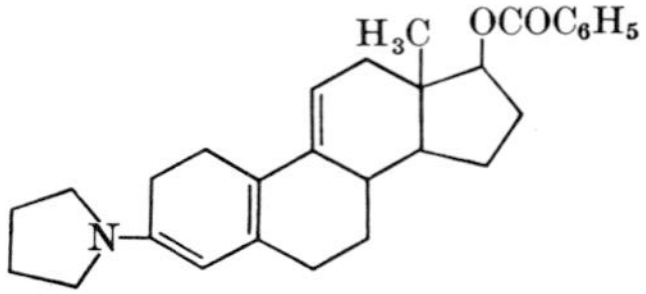 H_3C, $OCOC_6H_5$, N (—)	1334, 1335

Reactants	Catalyst	Product (Yield, %)	Refs.
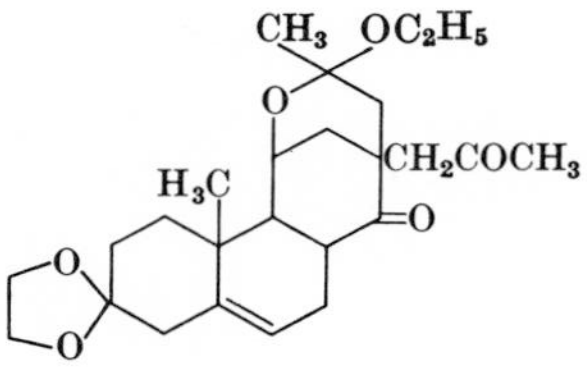 CH_3 OC_2H_5, CH_2COCH_3, H_3C	KOC_4H_9-t	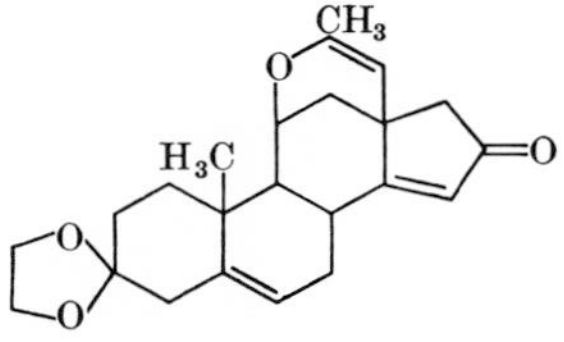 CH_3, H_3C (—)	606
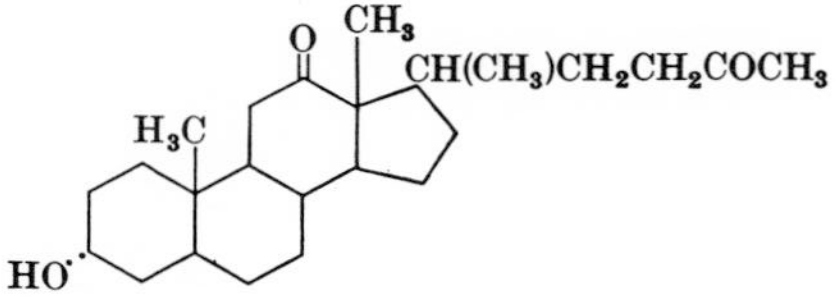 CH_3, $CH(CH_3)CH_2CH_2COCH_3$, H_3C, HO	,,	No condensation	1331
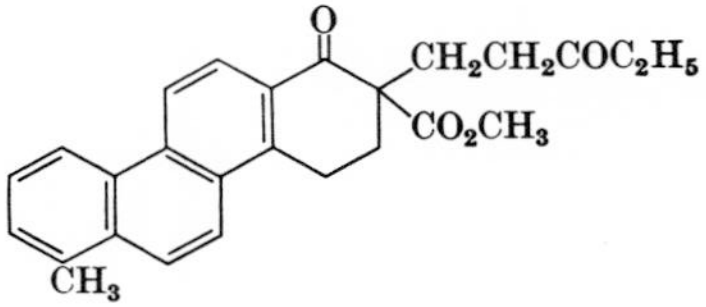 $CH_2CH_2COC_2H_5$, CO_2CH_3, CH_3	HCl, CH_3CO_2H	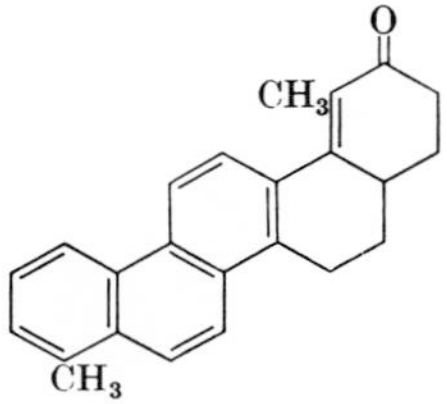 CH_3, CH_3 (—)	1336
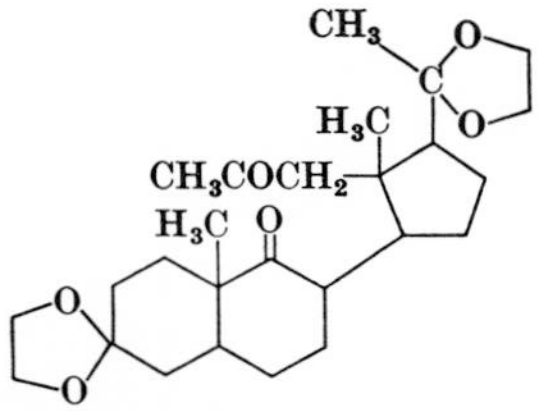 CH_3, H_3C, CH_3COCH_2, H_3C	$NaOCH_3$	No condensation	1330

Note: References 668–2359 are on pp. 403–438.

TABLE VI. SELF-CONDENSATION OF KETONES (*Continued*)

Ketone	Catalyst	Product(s) (Yield, %)	Refs.
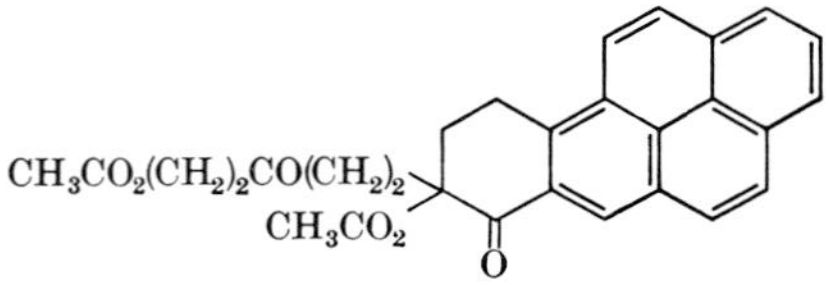 $CH_3CO_2(CH_2)_2CO(CH_2)_2$, CH_3CO_2 (structure)	$NaOCH_3$	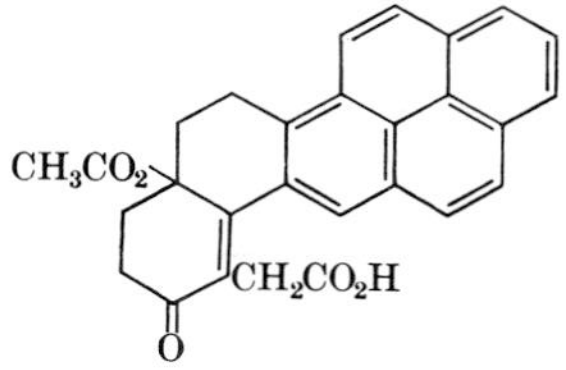 CH_3CO_2, CH_2CO_2H (structure) (—)	391
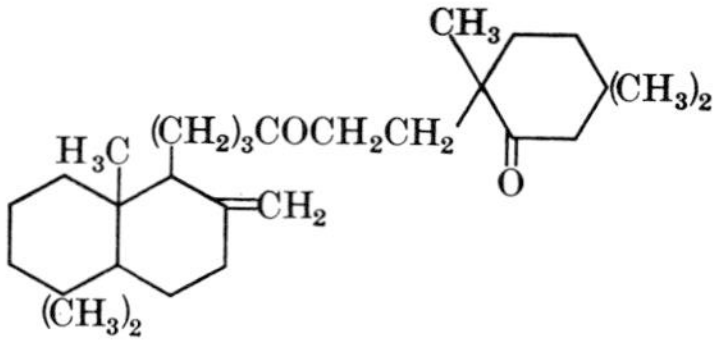 H_3C, $(CH_2)_3COCH_2CH_2$, $=CH_2$, $(CH_3)_2$, CH_3, $(CH_3)_2$ (structure)	KOC_4H_9-*t*	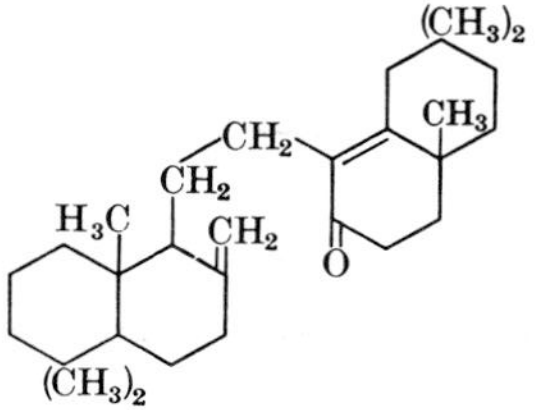 H_3C, CH_2, CH_2, CH_2, $(CH_3)_2$, CH_3, $(CH_3)_2$ (structure) (—) and corresponding β-ketol (—)	1337
$C_6H_5COCH_2CH(C_6H_5)CH(C_6H_5)COCH_2C_6H_5$	$NaOCH_3$	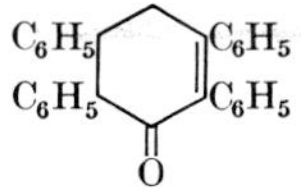 C_6H_5, C_6H_5, C_6H_5, C_6H_5 (structure) (75)	386, 1338
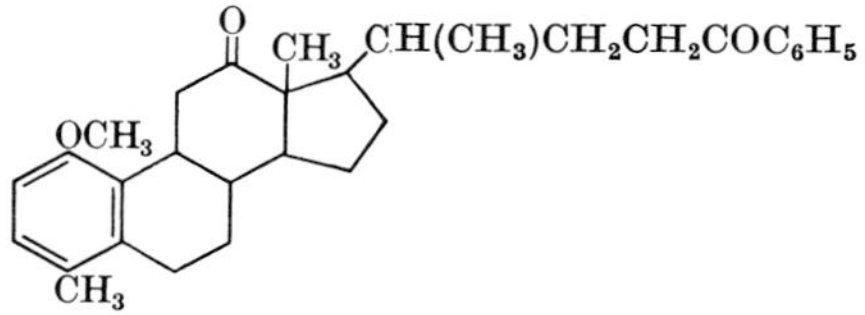 OCH_3, CH_3, CH_3, $CH(CH_3)CH_2CH_2COC_6H_5$ (structure)	$NaOCH_3$, $NaNH_2$, or NaH	No condensation	1331

Reactant	Conditions	Product(s) and Yield(%)	Refs.
$CH(CO_2CH_3)(CH_2)_2COC_3H_7$-*i*	KOC_4H_9-*t*	(—)	1339
	HCl, CH_3CO_2H	(—),	1340
		(—)	

Note: References 668–2359 are on pp. 403–438.

TABLE VI. SELF-CONDENSATION OF KETONES (*Continued*)

Ketone	Catalyst	Product(s) (Yield, %)		Refs.
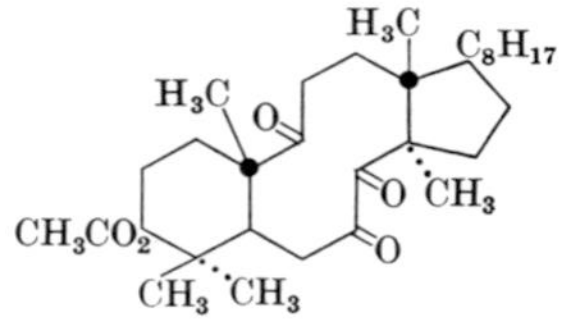 H_3C C_8H_{17} H_3C O O CH_3 CH_3CO_2 O CH_3 CH_3	HCl, CH_3CO_2H	H_3C O H_3C C_8H_{17} CH_3CO_2 CH_3 CH_3 O CH_3 OH	(—)	1340
H_3C C_8H_{17} H_3C O O CH_3 CH_3CO_2 CH_3 CH_3	,,	H_3C C_8H_{17} H_3C CH_3 CH_3CO_2 O CH_3 CH_3	(—)	1340
CH_3O CH_3O $NCH_2CH_2COCH_3$ CH_2COCH_2 CH_3O CH_3O $NCH_2CH_2COCH_3$	$NaOCH_3$	CH_3O CH_3O N $COCH_3$ OH CH_2 CH_3O CH_3O $NCH_2CH_2COCH_3$	(—)	1312
$C_6H_5COC(C_6H_5){=}C(C_6H_5)$-$C(C_6H_5){=}C(C_6H_5)CH_2C_6H_5$	Na	Hexaphenylbenzene (72)		1341

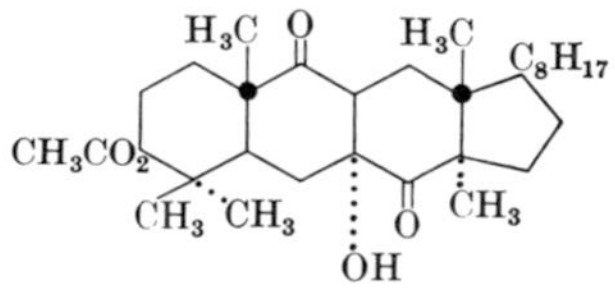

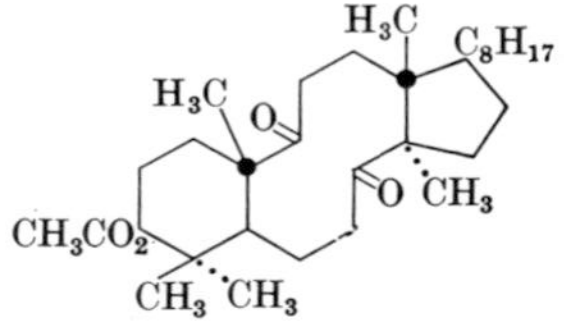

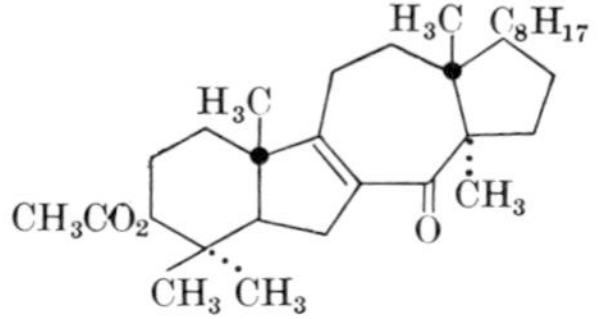

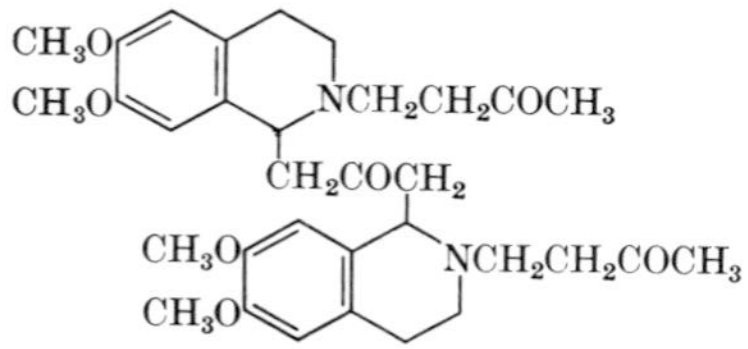

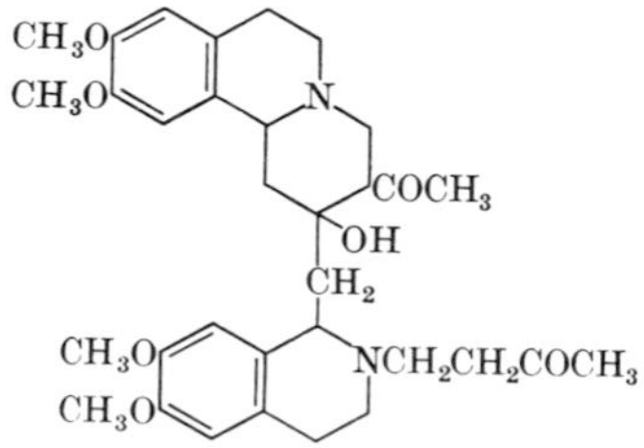

Note: References 668–2359 are on pp. 403–438.

TABLE VII. Mixed Condensation of Ketones

Ketone	Catalyst	Product(s) (Yield, %)	Refs.
CH_3COCO_2H and			
CH_3COCH_3	KOH	$CH_3COCH_2C(OH)(CH_3)CO_2H$ (14–19)	1342
	$(CH_2)_5NH$	$CH_3COCH{=}C(CH_3)CO_2H$ (28)	1343
$CH_3COC_2H_5$	KOH	$CH_3COCH(CH_3)C(OH)(CH_3)CO_2H$ (—)	1342
$C_6H_5COCH_3$	,,	$C_6H_5COCH_2C(OH)(CH_3)CO_2H$ (—)	1344
CH_3COCH_3 and			
 O	NaOH	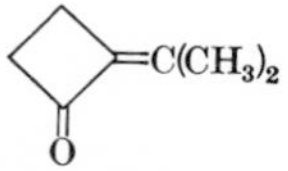 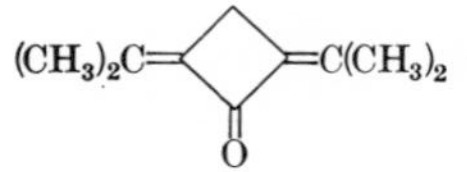 $=C(CH_3)_2$ (20), $(CH_3)_2C=$ $=C(CH_3)_2$ (20) O O	180
$CH_3COC_2H_5$	Metal oxides; also $Ba(OH)_2$, HCl	3-Methyl- and 3-ethyl-2-cyclohexen-1-one (—)	175, 1345–1347
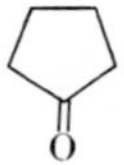 O	$NaOCH_3$	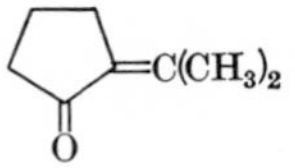 $=C(CH_3)_2$ (39), $(CH_3)_2C=$ $=C(CH_3)_2$ (—) O O 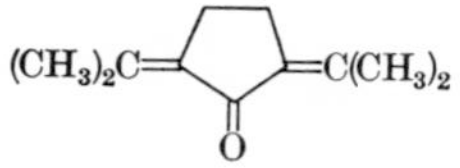	1348, 181, 182, 2344
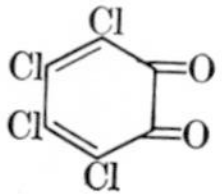 Cl, Cl, Cl, Cl, =O, =O	HCl, −30°	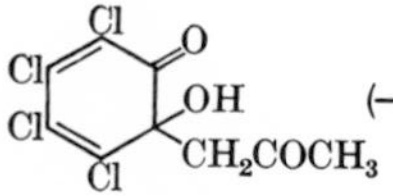 Cl, Cl, Cl, Cl, O, OH, CH_2COCH_3 (—)	1349
	HCl, 25°	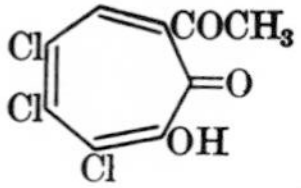 Cl, Cl, Cl, $COCH_3$, =O, OH (—)	1349
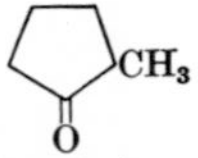 CH_3, O	$NaOC_2H_5$	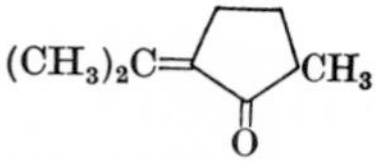 $(CH_3)_2C=$ CH_3 (—) O	182

Note: References 668–2359 are on pp. 403–438.

TABLE VII. Mixed Condensation of Ketones (*Continued*)

Ketone	Catalyst	Product(s) (Yield, %)	Refs.
CH_3COCH_3 (***contd.***) **and**			
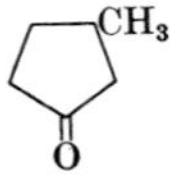 CH_3, O	$NaOC_2H_5$	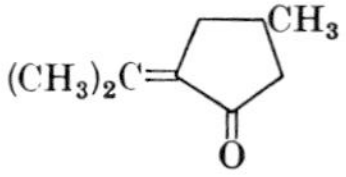 $(CH_3)_2C=$, CH_3, O (—)	182
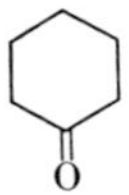 O	,,	 CH_2COCH_3 (23)	182
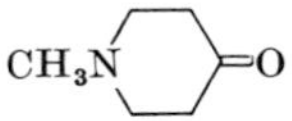 CH_3N, =O	Ion-exchange resin Amberlite IRA 400	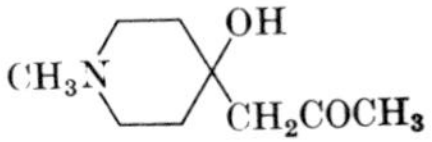 CH_3N, OH, CH_2COCH_3 (—)	1350
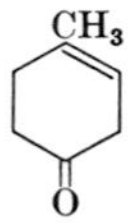 CH_3, O	$NaOC_2H_5$	Solid, m.p. 73° (—)	1351
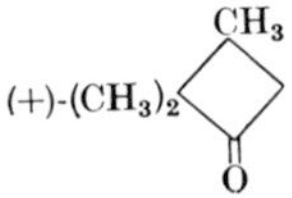 (+)-$(CH_3)_2$, CH_3, O	NaOH	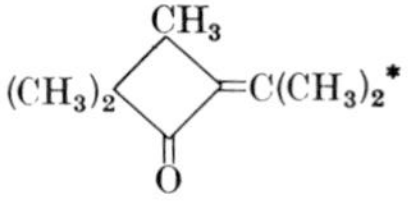 $(CH_3)_2$, CH_3, $=C(CH_3)_2$* , O (—)	1352
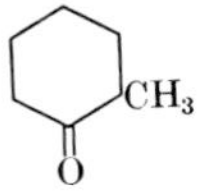 CH_3, O	$NaOC_2H_5$	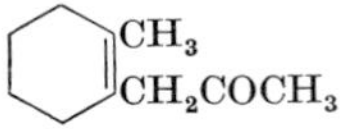 CH_3, CH_2COCH_3 (—)	182

	Catalyst	Product (Yield, %)	Refs.
CH_3; O	"	CH_3; CH_2COCH_3 (—),	1353, 182, 1146, 1354
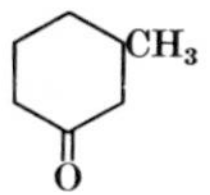		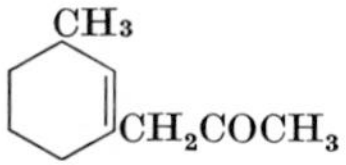 $(CH_3)_2C{=}CHCOCH_3$ (—), $(CH_3)_2C{=}CHCOCH{=}C(CH_3)_2$ (—), $C_{13}H_{20}O$ (—)	
CH_3; O	"	CH_3; CH_2COCH_3 (17)	182
$(CH_3)_2$; O; $(CH_3)_2$; =O	$NaOCH_3$	$=CHCOCH_3$; $(CH_3)_2$; O; $(CH_3)_2$ (—)	1355, 1356
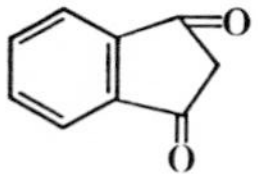 O; O	$(CH_2)_5NH$	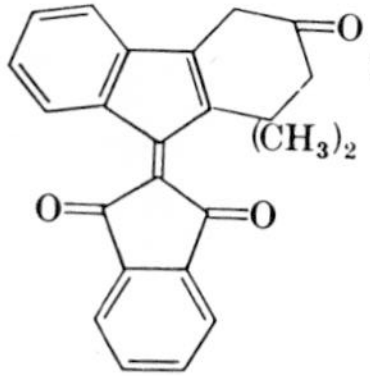 O; $(CH_3)_2$; O; O (—)	1357
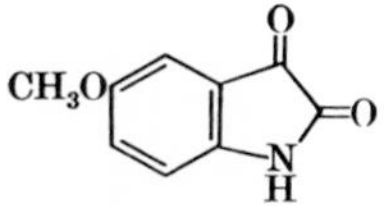 CH_3O; O; O; N; H	"	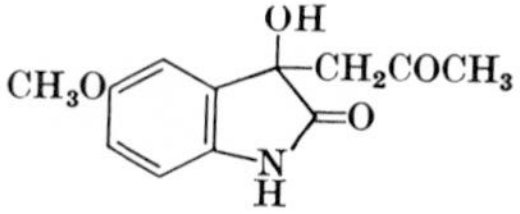 CH_3O; OH; CH_2COCH_3; O; N; H (—)	1371

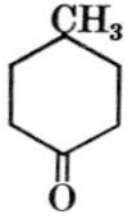

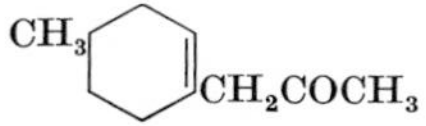

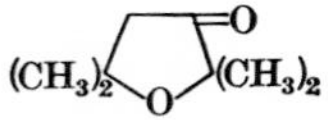

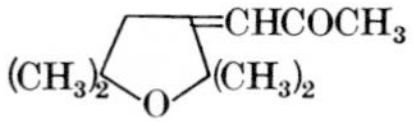

Note: References 668–2359 are on pp. 403–438.

* The optical activity was preserved in the product.

TABLE VII. MIXED CONDENSATION OF KETONES (*Continued*)

Ketone	Catalyst	Product(s) (Yield, %)	Refs.
CH_3COCH_3 (***contd.***) **and**			
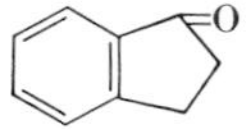	KOH	$C_{12}H_{12}O$ (—)	1235, 2344
$C_6H_5CH_2COCO_2H$	,,	$CH_3COCH_2C(OH)(CH_2C_6H_5)CO_2H$ (—)	1358
i-$C_3H_7CH_2COCH_2C_3H_7$-i	i-C_3H_7MgCl	$(CH_3)_2C(OH)CH(C_3H_7$-$i)COCH_2C_3H_7$-i (—)	176
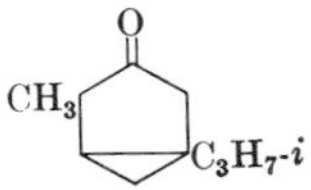 CH_3, C_3H_7-i	$NaOCH_3$	Solid, m.p. 78.5° (8)	1359
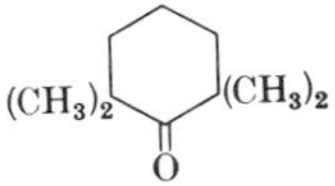 $(CH_3)_2$, $(CH_3)_2$	$C_6H_5N(CH_3)MgCl$	No condensation	1360
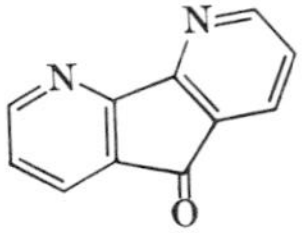	Al_2O_3	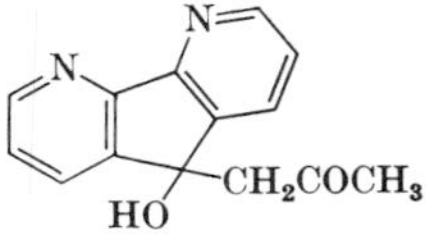 HO, CH_2COCH_3 (75)	1361
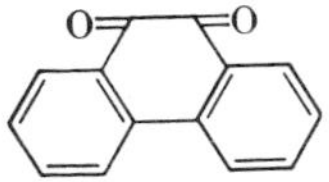	KOH	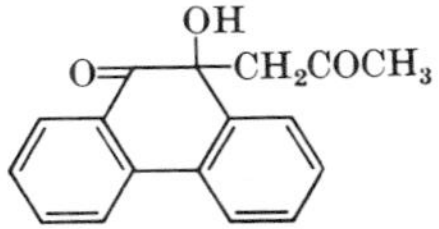 OH, CH_2COCH_3 (—),	1362

Reactants	Catalyst	Product (Yield, %)	Refs.
		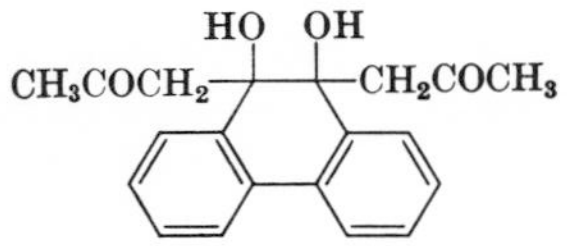 HO OH; CH_3COCH_2—C—C—CH_2COCH_3 (phenanthrene ring drawing) (—)	
$C_6H_5COCOC_6H_5$	,,	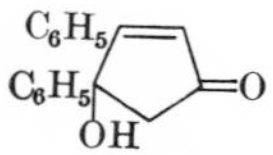 C_6H_5, C_6H_5, OH, =O (cyclopentenone ring drawing) (60), $C_{17}H_{16}O_3$ (—), $C_{31}H_{24}O_4$ (—)	1362–1364
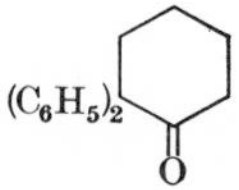 $(C_6H_5)_2$ (cyclohexanone ring drawing) O	C_6H_5MgX (X = Cl, Br, I)	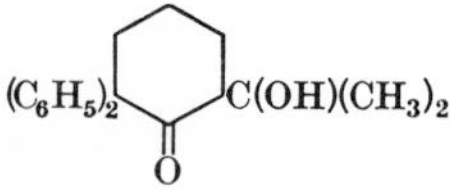 $(C_6H_5)_2$ (cyclohexanone ring drawing) $C(OH)(CH_3)_2$, O (58)	1365
$CH_3COCH{=}CH_2$ **and** $(CH_3)_2C{=}CHCOCH_3$	BaO, MgO, or CaO	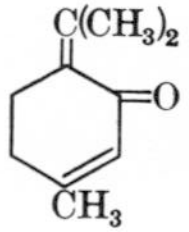 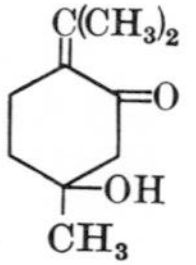 $C(CH_3)_2$, =O, CH_3 (cyclohexenone ring drawing), $C(CH_3)_2$, =O, OH, CH_3 (cyclohexanone ring drawing) (Total, 15)	1366, 1196
CH_3COCH_2CHO **and** $(C_6H_5)_2CO$	KNH_2	$(C_6H_5)_2C(OH)CH_2COCH_2CHO$† (67)	212
$C_2H_5COCO_2H$ **and** $C_6H_5COCH_3$	KOH	$C_6H_5COCH_2C(OH)(C_2H_5)CO_2H$ (—)	1344
$CH_3COC_2H_5$ **and** $CH_3COCH(CH_3)CH_2OH$	,,	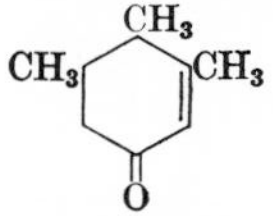 CH_3, CH_3, CH_3, O (cyclohexenone ring drawing) (52)	1367

Note: References 668–2359 are on pp. 403–438.

† This compound was isolated as its copper enolate.

TABLE VII. MIXED CONDENSATION OF KETONES (*Continued*)

Ketone	Catalyst	Product(s) (Yield, %)	Refs.
$CH_3COC_2H_5$ (***contd.***) and			
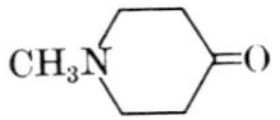 CH_3N (piperidone, =O)	Ion-exchange resin Amberlite IRA 400	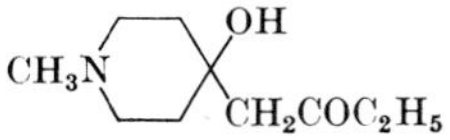 CH_3N (piperidine, OH, $CH_2COC_2H_5$) (—)	1350
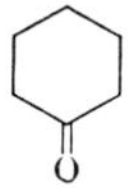 (cyclohexanone, O)	NaOH	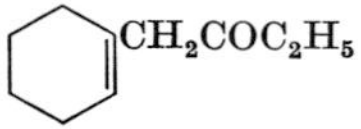 (cyclohexene, $CH_2COC_2H_5$) (—)	178
$C_6H_5COCH_3$	$C_6H_5(R)NMgX$, (R = H, CH_3, C_2H_5, C_6H_5)	$C_2H_5COCH{=}C(C_6H_5)CH_3$ (28), $C_6H_5COCH_2C(OH)(CH_3)C_2H_5$ (34)	198
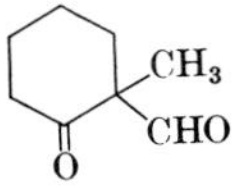 (cyclohexanone, CH_3, CHO, O)	$(CH_2)_5NH$, CH_3CO_2H	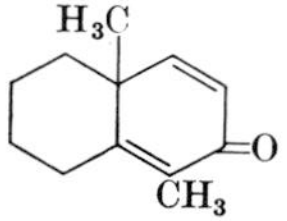 (H_3C, =O, CH_3) (2)	1368
$C_6H_5COCOC_6H_5$	NaOH	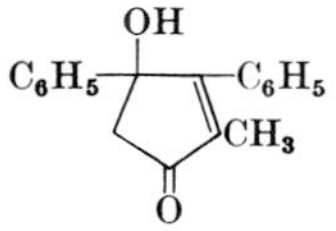 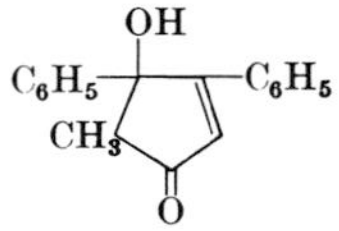 (OH, C_6H_5, C_6H_5, CH_3, O) (—), (OH, C_6H_5, C_6H_5, CH_3, O) (—)	1369, 199, 1370
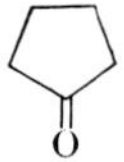 (cyclopentanone, O) and			
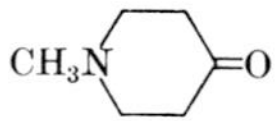 CH_3N (piperidone, =O)	Ion-exchange resin Amberlite IRA 400	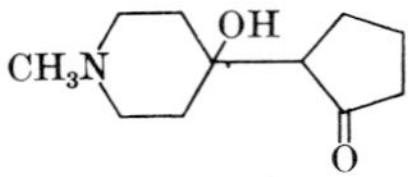 CH_3N (OH, cyclopentanone, O) (—)	1350

 COCH$_3$	NaNH$_2$	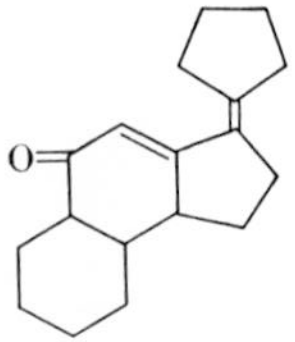 O (—)	1143
$C_6H_5CH_2COCO_2H$	NaOH; KOH	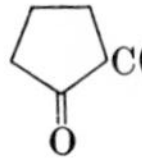 C(OH)(CH$_2$C$_6$H$_5$)CO$_2$H (4–39), O	184
		HO$_2$C(C$_6$H$_5$CH$_2$)(HO)C 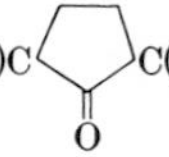 C(OH)(CH$_2$C$_6$H$_5$)CO$_2$H O	
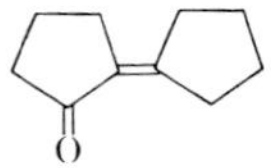 O	HCl	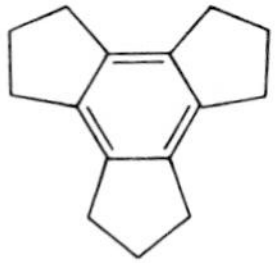 (55)	1157
$CH_3COCH_2CH_2CO_2H$ **and** $C_6H_5COCOC_6H_5$	KOH	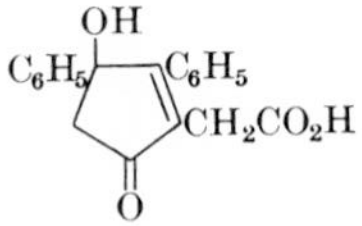 OH C$_6$H$_5$ C$_6$H$_5$ CH$_2$CO$_2$H (40) O	200
$CH_3COC_3H_7$-*n* **and** 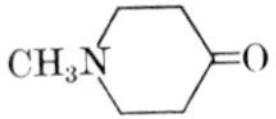 CH$_3$N =O	Ion-exchange resin Amberlite IRA 400	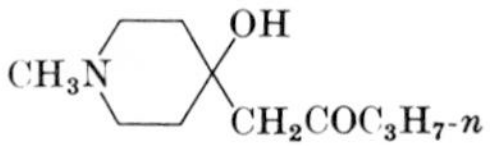 CH$_3$N OH CH$_2$COC$_3$H$_7$-*n* (—)	1350
$C_6H_5COCH_3$	C_6H_5(R)NMgX, (R = H, CH_3, C_2H_5, C_6H_5)	$C_6H_5COCH{=}C(CH_3)C_3H_7$-*n* (32)	198
$C_6H_5COCOC_6H_5$	KOH	$C_{19}H_{18}O_2$ (—)	199

Note: References 668–2359 are on pp. 403–438.

TABLE VII. Mixed Condensation of Ketones (*Continued*)

Ketone	Catalyst	Product(s) (Yield, %)	Refs.
$CH_3COC_3H_7$-*i* **and**			
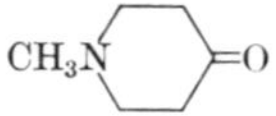 CH_3N =O	Ion-exchange resin Amberlite IRA 400	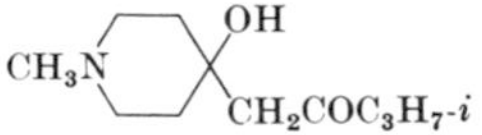 CH_3N OH $CH_2COC_3H_7$-*i* (—)	1350
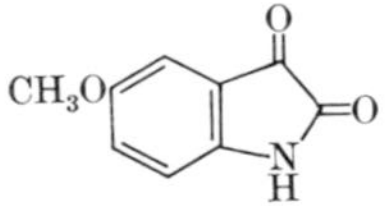 CH_3O O =O N H	$(C_2H_5)_2NH$	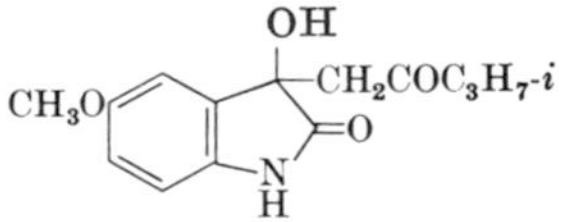 CH_3O OH $CH_2COC_3H_7$-*i* =O N H (—)	1371
$(C_2H_5)_2CO$ **and** $C_6H_5COCOC_6H_5$	KOH	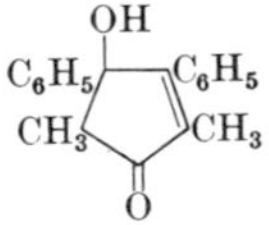 OH C_6H_5 C_6H_5 CH_3 CH_3 O (88)	201, 199
$CH_3COCH{=}C(CH_3)_2$ **and** $C_6H_5COCH_3$	$C_6H_5N(CH_3)MgBr$	$C_6H_5COCH_2C(OH)(CH_3)CH{=}C(CH_3)_2$ (31), $C_6H_5C(OH)(CH_3)CH_2COCH{=}C(CH_3)_2$ (33)	198
 O **and**			
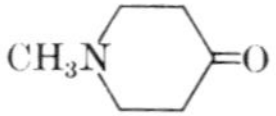 CH_3N =O	Ion-exchange resin Amberlite IRA 400	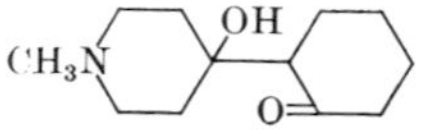 CH_3N OH O= (—)	1350
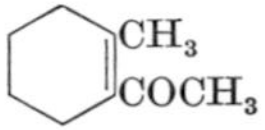 CH_3 $COCH_3$	KOC_4H_9-*t*, C_5H_5N	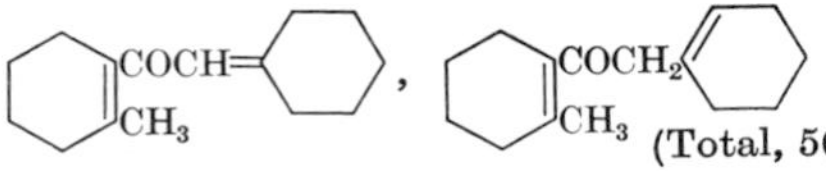 $COCH{=}$ CH_3, $COCH_2$ CH_3 (Total, 56)	179
$C_6H_5CH_2CH_2COCO_2H$	KOH or NaOH	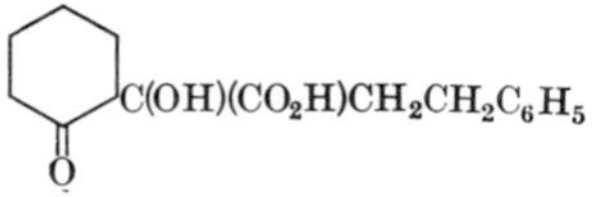 $C(OH)(CO_2H)CH_2CH_2C_6H_5$ O (15–40),	1372

		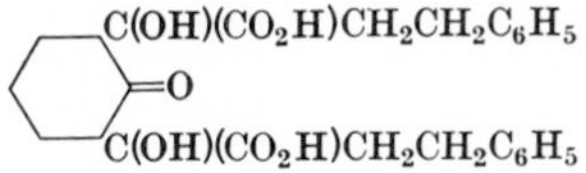 $C(OH)(CO_2H)CH_2CH_2C_6H_5$ / $C(OH)(CO_2H)CH_2CH_2C_6H_5$ (—)	
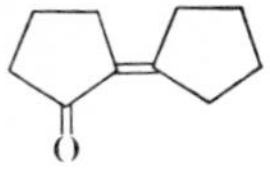	HCl	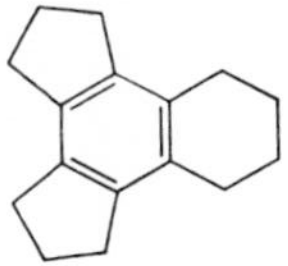 (29)	1157
	Primary aliphatic or cycloaliphatic amines	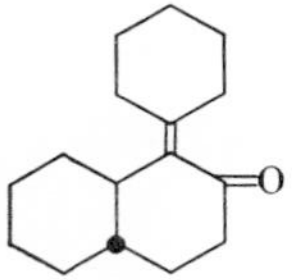 (—)	1181
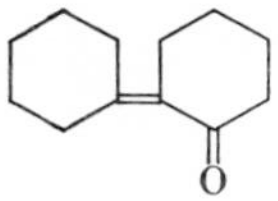	H_2SO_4	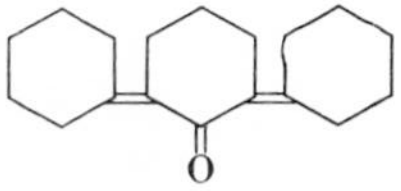 (30),  (30)	1178
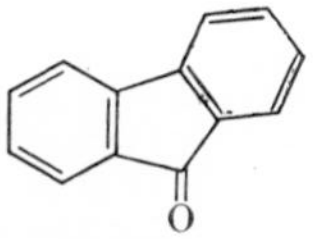	$C_6H_5N(CH_3)MgBr$	No condensation	1373
$C_6H_5COCOC_6H_5$	KOH	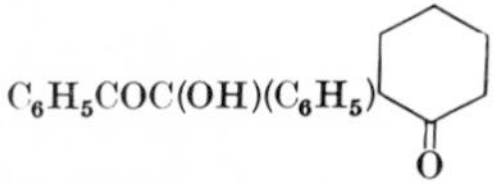 $C_6H_5COC(OH)(C_6H_5)$ (70–90)	207

Note: References 668–2359 are on pp. 403–438.

TABLE VII. Mixed Condensation of Ketones (*Continued*)

Ketone	Catalyst	Product(s) (Yield, %)	Refs.
$CH_3COC_4H_9$-*t* **and**			
$C_6H_5COCH_3$	$C_6H_5N(CH_3)MgI$	$C_6H_5COCH_2C(OH)(CH_3)C_4H_9$-*t* (48)	1374
4-$CH_3C_6H_4COCH_3$	,,	4-$CH_3C_6H_4COCH_2C(OH)(CH_3)C_4H_9$-*t* (31)	1374
4-$C_2H_5C_6H_4COCH_3$	,,	4-$C_2H_5C_6H_4COCH_2C(OH)(CH_3)C_4H_9$-*t* (33)	1374
4-*i*-$C_3H_7C_6H_4COCH_3$	,,	4-*i*-$C_3H_7C_6H_4COCH_2C(OH)(CH_3)C_4H_9$-*t* (21)	1374
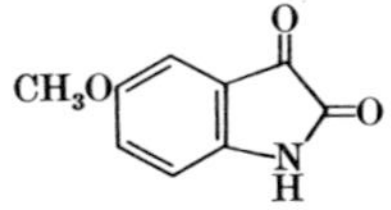 CH_3O, O, O, N, H	$(C_2H_5)_2NH$	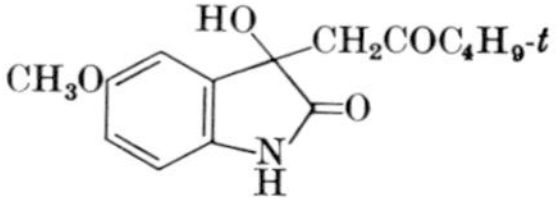 HO, $CH_2COC_4H_9$-*t*, CH_3O, O, N, H (—)	2078
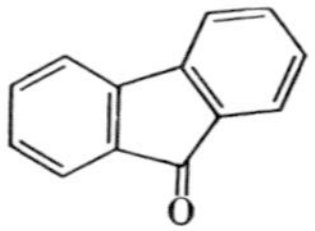 O	$C_6H_5N(CH_3)MgBr$	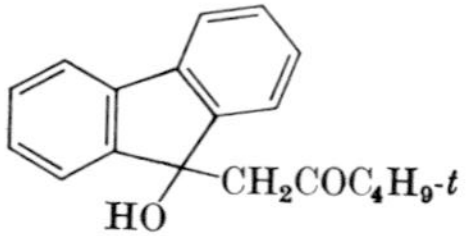 HO, $CH_2COC_4H_9$-*t* (70)	1033
$C_6H_5COC_6H_5$	,,	$(C_6H_5)_2C(OH)CH_2COC_4H_9$-*t* (35)	1033, 159
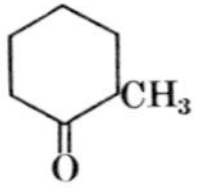 O, CH_3 **and**			
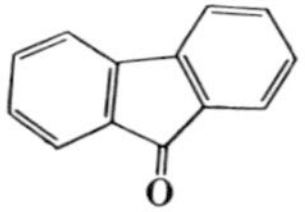 O	,,	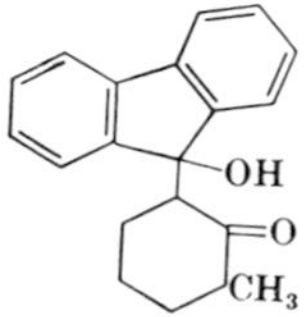 OH, O, CH_3 (90)	1373

$C_6H_5COCOC_6H_5$	KOH	CH_3, O, $C(OH)(C_6H_5)COC_6H_5$	(70–90)	207
$C_2H_5COC_4H_9$-*t* **and** O	$C_6H_5N(CH_3)MgBr$	HO, $CH(CH_3)COC_4H_9$-*t*	(60)	1033
i-$C_3H_7COC_3H_7$-*i* **and** O	,,	HO, $C(CH_3)_2COC_3H_7$-*i*	(60)	1033
O, O, N H **and**				
4-$BrC_6H_4COCH_3$	$(C_2H_5)_2NH$	OH, CH_2COR, O, N H; A, R = C_6H_4Br-4	(87)	1375
4-$ClC_6H_4COCH_3$	,,	A, R = C_6H_4Cl-4	(87)	1375
3-$O_2NC_6H_4COCH_3$	,,	A, R = $C_6H_4NO_2$-3	(66)	1376
$C_6H_5COCH_3$	,,	A, R = C_6H_5	(63)	1375
4-$CH_3C_6H_4COCH_3$	,,	A, R = $C_6H_4CH_3$-4	(86)	1375
4-$CH_3OC_6H_4COCH_3$	,,	A, R = $C_6H_4OCH_3$-4	(85)	1375
F, O	KOH	N, CO_2H, F	(61)	1377

Note: References 668–2359 are on pp. 403–438.

TABLE VII. MIXED CONDENSATION OF KETONES (*Continued*)

Ketone	Catalyst	Product(s) (Yield, %)	Refs.
2-$ClC_6H_4COCH_3$ **and** 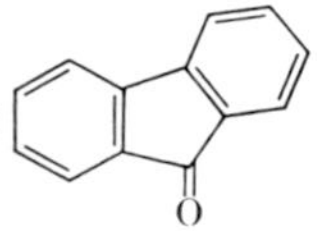 (fluorenone structure)	$C_6H_5N(CH_3)MgBr$	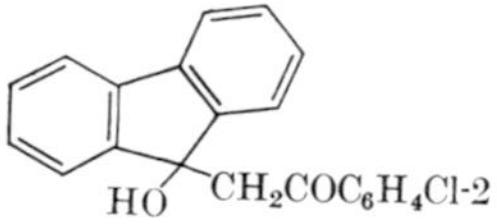 HO, $CH_2COC_6H_4Cl$-2 (45)	1373
4-$O_2NC_6H_4COCH_3$ **and** 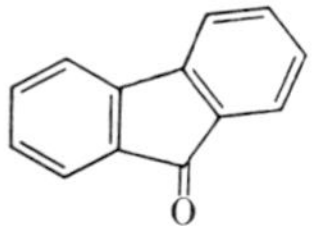 (fluorenone structure)	,,	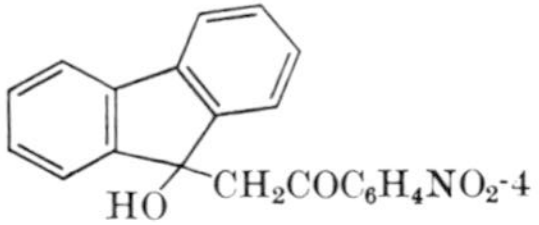 HO, $CH_2COC_6H_4NO_2$-4 (45)	1373
$C_6H_5COCH_3$ **and**			
$C_6H_5CH_2COCO_2H$	KOH	$C_6H_5COCH_2C(OH)(CH_2C_6H_5)CO_2H$ (—)	1378
3,4-$(CH_2O_2)C_6H_3CH_2COCO_2H$	,,	3,4-$(CH_2O_2)C_6H_3CH_2C(OH)$-$(CH_2COC_6H_5)CO_2H$ (62)	1379
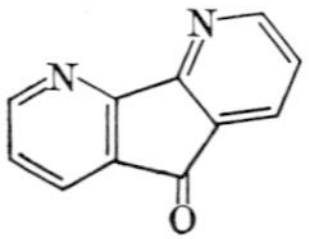 (diazafluorenone structure, N, N)	Al_2O_3	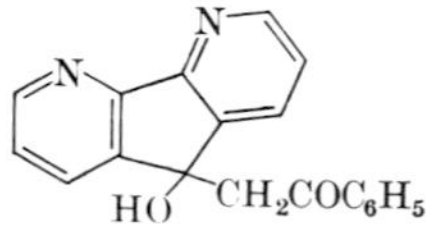 N, N, HO, $CH_2COC_6H_5$ (85)	1361
$C_6H_5COCH_2C_6H_5$	$NaOCH_3$	$C_6H_5COC(C_6H_5){=}C(CH_3)C_6H_5$ (80–85)	1380
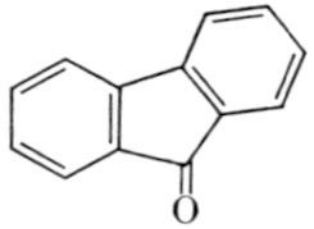 (fluorenone structure)	$C_6H_5N(CH_3)MgBr$	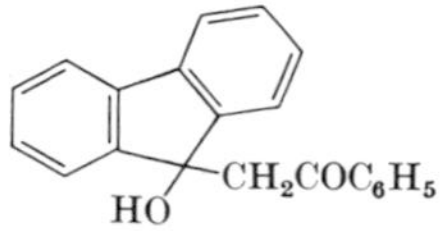 HO, $CH_2COC_6H_5$ (80–85)	1373, 1033
$C_6H_5COCOC_6H_5$	KOH; $NaOCH_3$ or $C_6H_5N(CH_3)$-MgBr	$C_6H_5COC(C_6H_5){=}CHCOC_6H_5$ (80), $C_{22}H_{18}O_3$ (—)	1381, 208, 1362, 1373

Reactants	Catalyst	Product (Yield, %)	Refs.
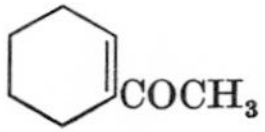 $COCH_3$ **and**			
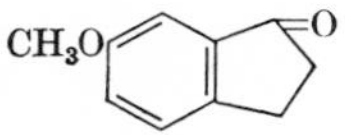 CH_3O	KOC_4H_9-*t*	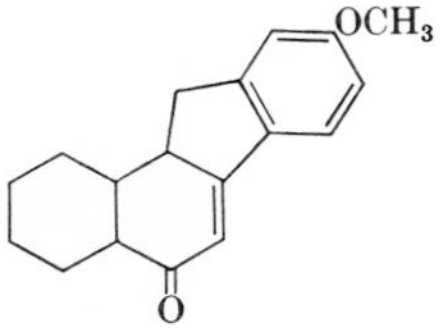 OCH_3 (25)	1382
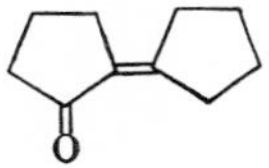	$NaNH_2$	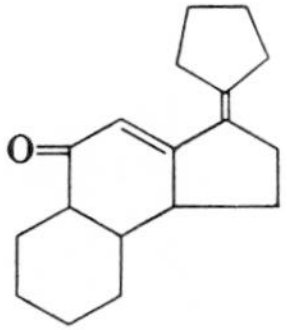 (—)	1143
$CH_3COC_6H_{13}$-*n* **and** $C_6H_5COCOC_6H_5$	KOH	$C_{22}H_{24}O_2$ (—)	199
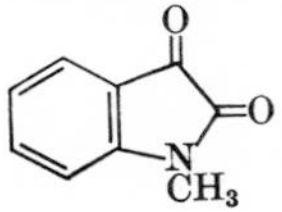 N-CH_3 **and**			
4-$BrC_6H_4COCH_3$	$(C_2H_5)_2NH$	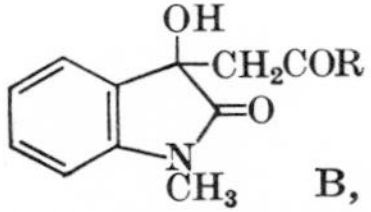 OH, CH_2COR, CH_3 B, R = C_6H_4Br-4 (54)	1376
4-$ClC_6H_4COCH_3$	”	B, R = C_6H_4Cl-4 (67)	1376
3-$O_2NC_6H_4COCH_3$	”	B, R = $C_6H_4NO_2$-3 (75)	1376
$C_6H_5COCH_3$	”	B, R = C_6H_5 (72)	1375
4-$CH_3OC_6H_4COCH_3$	”	B, R = $C_6H_4OCH_3$-4 (60)	1376

Note: References 668–2359 are on pp. 403–438.

TABLE VII. MIXED CONDENSATION OF KETONES (*Continued*)

Ketone	Catalyst	Product(s) (Yield, %)	Refs.
$C_6H_5CH_2COCO_2H$ **and**			
$C_6H_5(CH_2)_2COCH_3$	NaOH	$C_6H_5(CH_2)_2COCH_2C(OH)(CH_2C_6H_5)CO_2H$ (40)	1383, 1384
$4\text{-}CH_3OC_6H_4(CH_2)_2COCH_3$	,,	$4\text{-}CH_3OC_6H_4(CH_2)_2COCH_2\text{-}C(OH)(CH_2C_6H_5)CO_2H$ (—)	1384
$C_6H_5COC_2H_5$ **and** 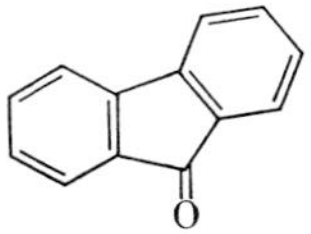 (fluorenone, O)	$C_6H_5N(CH_3)MgBr$	No condensation	1373
(CH_3, $COCH_3$ cyclohexene) **and**			
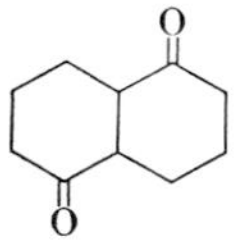 (O, O bicyclic dione)	$KOC_4H_9\text{-}t$ or $Al(OC_4H_9\text{-}t)_3$	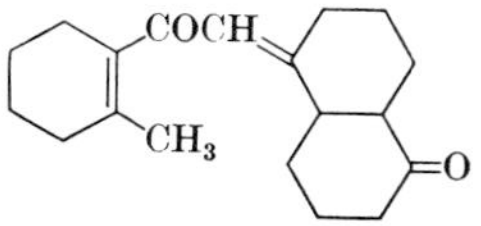 (COCH=, CH_3, =O) (61)	1385
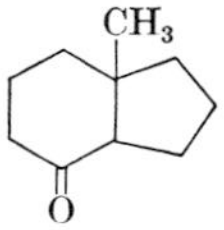 (CH_3, O hydrindanone)	$KOC_4H_9\text{-}t$	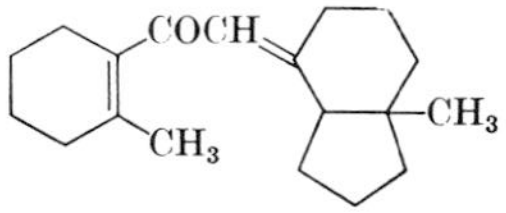 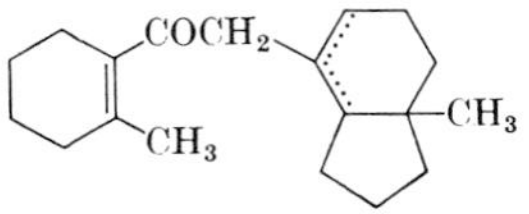 (COCH=, CH_3, CH_3) (—),	1386
		($COCH_2$, CH_3, CH_3) (—)	

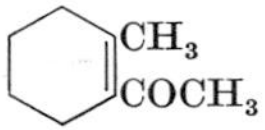

Reactants	Condensing Agent	Product (Yield, %)	Refs.
[structure]	,,	[structure] (37)	1385
[structure] CH_3 $COCH_3$ and [structure]	$Al(OC_4H_9\text{-}t)_3$	[structure] (32)	1385
[structure] and [structure]	$C_6H_5N(CH_3)MgBr$	[structure] (84)	1373
$C_6H_5COC_6H_5$	,,	[structure] $C(OH)(C_6H_5)_2$ (21)	1373
$1\text{-}C_{10}H_7COCH_3$ and [structure]	,,	[structure] $CH_2COC_{10}H_7\text{-}1$ (60)	1373

Note: References 668–2359 are on pp. 403–438.

TABLE VII. Mixed Condensation of Ketones (*Continued*)

Ketone	Catalyst	Product(s) (Yield, %)	Refs.
$2\text{-}C_{10}H_7COCH_3$ and 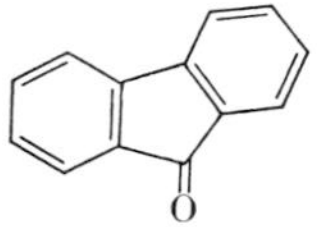 [structure: O]	$C_6H_5N(CH_3)MgBr$	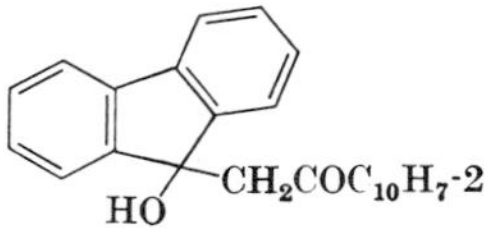 [structure: HO, $CH_2COC_{10}H_7\text{-}2$] (45)	1373
$(C_6H_5)_2CO$	"	$2\text{-}C_{10}H_7COCH_2C(OH)(C_6H_5)_2$ (57)	1373
$C_6H_5(COCH_2)_2COCH_3$ and $(C_6H_5)_2CO$	$NaNH_2$	$C_6H_5(COCH_2)_3C(C_6H_5)_2OH$ (—)	213
$(4\text{-}BrC_6H_4CO)_2$ and $(C_6H_5CH_2)_2CO$	$NaOCH_3$	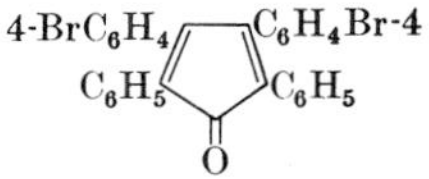 [structure: $4\text{-}BrC_6H_4$, $C_6H_4Br\text{-}4$, C_6H_5, C_6H_5, O] (—)	203
$(C_6H_5CO)_2$ and $(C_6H_5CH_2)_2CO$	KOH	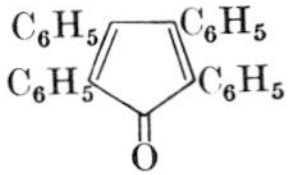 [structure: C_6H_5, C_6H_5, C_6H_5, C_6H_5, O] (91–96)	204, 202, 1387
	KOH, C_2H_5OH, 3–5 min.	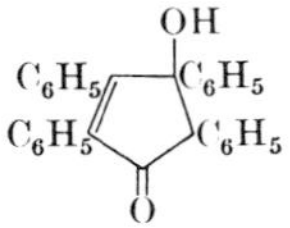 [structure: OH, C_6H_5, C_6H_5, C_6H_5, C_6H_5, O] (—)	1387
$C_6H_5COCH{=}CHC_6H_5$ and $C_6H_5COCH{=}C(CH_3)C_6H_5$	$NaOC_2H_5$	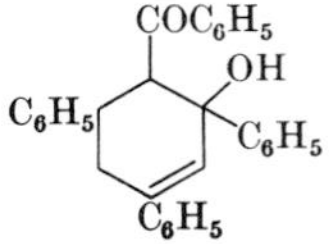 [structure: COC_6H_5, OH, C_6H_5, C_6H_5, C_6H_5] (84)	1388

Reactants	Catalyst	Product (Yield, %)	Refs.
$4\text{-}CH_3OC_6H_4COCOC_6H_5$ and $(C_6H_5CH_2)_2CO$	$NaOCH_3$	2-$(4\text{-}CH_3OC_6H_4)$-3,4,5-$(C_6H_5)_3$-cyclopentadienone (—)	203
$(C_6H_5CH_2)_2CO$ and			
$(4\text{-}CH_3C_6H_4CO)_2$	"	2,5-$(C_6H_5)_2$-3,4-$(4\text{-}CH_3C_6H_4)_2$-cyclopentadienone (—)	203
$(4\text{-}CH_3OC_6H_4CO)_2$	"	2,5-$(C_6H_5)_2$-3,4-$(4\text{-}CH_3OC_6H_4)_2$-cyclopentadienone (—)	203
$4\text{-}(CH_3)_2NC_6H_4COCOC_6H_4Cl\text{-}4$	"	3-$[4\text{-}(CH_3)_2NC_6H_4]$-4-$(4\text{-}ClC_6H_4)$-2,5-$(C_6H_5)_2$-cyclopentadienone (—)	203
$4\text{-}(CH_3)_2NC_6H_4COCOC_6H_5$	"	3-$[4\text{-}(CH_3)_2NC_6H_4]$-2,4,5-$(C_6H_5)_3$-cyclopentadienone (—)	203
$(4\text{-}C_2H_5OC_6H_4CO)_2$	"	2,5-$(C_6H_5)_2$-3,4-$(4\text{-}C_2H_5OC_6H_4)_2$-cyclopentadienone (—)	203
$4\text{-}C_6H_5C_6H_4COCOC_6H_5$	"	3-$(4\text{-}C_6H_5C_6H_4)$-2,4,5-$(C_6H_5)_3$-cyclopentadienone (—)	203

Note: References 668–2359 are on pp. 403–438.

TABLE VII. MIXED CONDENSATION OF KETONES (*Continued*)

Ketone	Catalyst	Product(s) (Yield, %)	Refs.
$(C_6H_5CH_2)_2CO$ (*contd.*) and			
$(4\text{-}i\text{-}C_3H_7C_6H_4CO)_2$	$NaOCH_3$	4-i-$C_3H_7C_6H_4$, $C_6H_4C_3H_7$-i-4, C_6H_5, C_6H_5, O (cyclopentadienone) (—)	203
$(4\text{-}C_6H_5C_6H_4CO)_2$	"	4-$C_6H_5C_6H_4$, $C_6H_4C_6H_5$-4, C_6H_5, C_6H_5, O (cyclopentadienone) (—)	203
$(4\text{-}C_6H_5SC_6H_4CO)_2$	"	4-$C_6H_5SC_6H_4$, $C_6H_4SC_6H_5$-4, C_6H_5, C_6H_5, O (cyclopentadienone) (—)	203
$(4\text{-}C_6H_5OC_6H_4CO)_2$	"	4-$C_6H_5OC_6H_4$, $C_6H_4OC_6H_5$-4, C_6H_5, C_6H_5, O (cyclopentadienone) (—)	203

Note: References 668–2359 are on pp. 403–438.

TABLE VIII. CONDENSATION OF ALIPHATIC ALDEHYDES WITH ACETONE
(R in the product is the group R in the aldehyde RCHO.)

Aldehyde, RCHO	Catalyst	Product(s) (Yield, %)	Refs.
CH_2O	LiOH	$(HOCH_2)_2CHCOCH_3$ (80)	1389–1392
	NaOH	$HOCH_2CH_2COCH_3$ (84)	652, 577, 1123, 1389, 1393–1405
	"	$HOCH_2CH_2COCH_3$ (2.4), O; $HOCH_2$; CH_2OH; O (ring structure) (1.5)	1406
	"	$(C_4H_6O)_n$ (80)	1407, 1408
	KOH or K_2CO_3	$HOCH_2CH_2COCH_3$ (25–29), $CH_2{=}C(CH_2OH)COCH_3$ (5–10), O; O; $COCH_3$ (ring structure) (4), O; O; $COCH{=}C(CH_3)_2$ (ring structure) (11), $(CH_3)_2C(OH)CH_2COCH_3$ (5)	1409–1413
	$Ca(OH)_2$	OH; $(HOCH_2)_2$; $(CH_2OH)_2$; O (ring structure) (—)	911, 910, 1389, 1414, 1415
	$Ca_3(PO_4)_2$, 450°	$CH_2{=}CHCOCH_3$ (97–100)	1416, 125, 126, 1417
CCl_3CHO	CH_3CO_2Na, $(CH_3CO)_2O$	$RCHOHCH_2COCH_3$ (60)	273, 287, 635, 636, 1418, 1419

Note: References 668–2359 are on pp. 403–438.

TABLE VIII. CONDENSATION OF ALIPHATIC ALDEHYDES WITH ACETONE (*Continued*)
(R in the product is the group R in the aldehyde RCHO.)

Aldehyde, RCHO	Catalyst	Product(s) (Yield, %)	Refs.
CH_3CHO	KOH	$RCHOHCH_2COCH_3$ (87)	576, 6, 229, 577, 1393, 1420–1425
	NaOH; $(CO_2H)_2$	$RCH{=}CHCOCH_3$ (45)	613, 577, 1393, 1421–1423, 1425–1429
	$Ca_3(PO_4)_2$	$(CH_3)_2C{=}CHCH{=}CHCOCH_3$ (—)	2326
$C(Cl_2){=}CClCHO$	H_2SO_4	$(RCH{=}CH)_2CO$ (65)	289
$C(Cl_2){=}CHCHO$	,,	$(RCH{=}CH)_2CO$ (40)	638
	NaOH	$RCH{=}CHCOCH_3$ (57)	1430
$O_2NCH(CHO)_2$	,,	$HOC_6H_4NO_2$-4 (65)	1431, 1432
$CH_2{=}CHCHO$	KCN	Polymers (—)	1423
	HCl	$(CH_3)_2C{=}CHCOCH_3$ (—)	1423
$HOCH(CHO)_2$	NaOH	CH_3–(cyclopentenone, C=O)–OH (1.5)	1433, 1434
C_2H_5CHO	,,	$RCHOHCH_2COCH_3$ (60)	1435, 577
	NaOH; $(CO_2H)_2$	$RCH{=}CHCOCH_3$ (50)	1436, 613, 1428, 1435, 1437
$CH_3CH{=}CHCHO$	NaOH	$(RCH{=}CH)_2CO$ (31), $RCH{=}CHCOCH_3$ (24)	1438–1440, 1399, 2359
	KOH	$RCH{=}CHCOCH_3$ (38)	301, 1399, 1425, 1438–1441
n-C_3H_7CHO	NaOH, H_2O, $(C_2H_5)_2O$, 15°	$RCHOHCH_2COCH_3$ (84)	511, 510, 649, 1442
	NaOH; $(CO_2H)_2$	$RCH{=}CHCOCH_3$ (60)	613, 1442, 1443
	Na_2CO_3	$RCH{=}CHCOCH_3$ (8), n-$C_3H_7CH{=}C(C_2H_5)CHO$ (18)	510–512, 763, 1442–1445
i-C_3H_7CHO	NaOH	$RCHOHCH_2COCH_3$ (50)	511, 510, 577, 1446, 1447
	NaOH; I_2	$RCH{=}CHCOCH_3$ (37)	511, 1446
	NaOH	$RCH{=}CHCOCH_3$ (—), $(CH_3)_2C{=}CHCH_2COCH_3$ (—)	511, 510, 577, 1436, 1437, 1448–1450

$C(Cl_2)=CHCH=CHCHO$	HCl	$(RCH=CH)_2CO$ (58)	1451
$CH_3CH=C(CH_3)CHO$	$NaOC_2H_5$	$RCH=CHCOCH_3$ (44)	105, 1452
$(CH_3)_2C=CHCHO$	NaOH	$RCH=CHCOCH_3$ (—)	1453
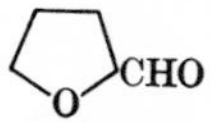 CHO	,,	$RCHOHCH_2COCH_3$ (41)	1454, 1455
n-C_4H_9CHO	,,	$RCHOHCH_2COCH_3$ (—)	1456
i-C_4H_9CHO	,,	$RCHOHCH_2COCH_3$ (76)	511, 1435
	NaOH; I_2	$RCH=CHCOCH_3$ (68)	511, 1422, 1435, 1444, 1450, 1457–1463
t-C_4H_9CHO	$NaOC_2H_5$; I_2	$RCH=CHCOCH_3$ (30), $(CH_3)_2C=CHCOCH_3$ (12)	511
$HO(CH_2)_4CHO$	NaOH	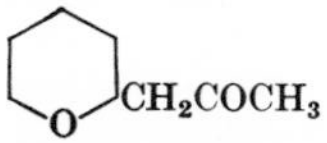 CH_2COCH_3 (54)	1464, 522
$CH_3(CH=CH)_2CHO$	NaOH	$RCH=CHCOCH_3$ (—)	1465
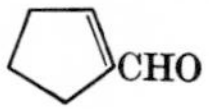 CHO	,,	$RCH=CHCOCH_3$ (51)	1466
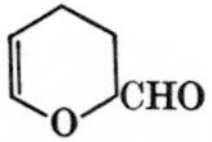 CHO	,,	$RCH=CHCOCH_3$ (65)	1467
n-$C_4H_9O_2CCHO$	Heat	$RCHOHCH_2COCH_3$ (47)	1468
$C_2H_5CH=C(CH_3)CHO$	NaOH	$RCH=CHCOCH_3$ (—)	1437, 1469
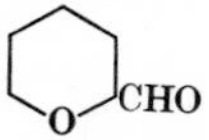 CHO	,,	$RCHOHCH_2COCH_3$ (41)	1467
n-$C_3H_7CH(CH_3)CHO$	$Ba(OH)_2$	$RCH=CHCOCH_3$ (—)	1469
$(C_2H_5)_2CHCHO$	,,	$RCH=CHCOCH_3$ (—)	288, 1470
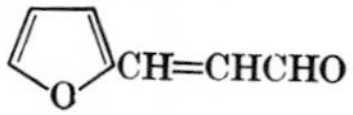 CH=CHCHO	NaOH	$RCH=CHCOCH_3$ (71)	1471, 1039, 1438

Note: References 668–2359 are on pp. 403–438.

TABLE VIII. Condensation of Aliphatic Aldehydes with Acetone (*Continued*)
(R in the product is the group R in the aldehyde RCHO.)

Aldehyde, RCHO	Catalyst	Product(s) (Yield, %)	Refs.
2-Furyl-CH=CHCHO (*contd.*)	NaOH	$(RCH{=}CH)_2CO$ (97)	1471, 1399, 1535
1-Cyclohexenyl-CHO	"	$RCH{=}CHCOCH_3$ (38)	1472
Cyclohexyl-CHO	"	$RCHOHCH_2COCH_3$ (21), $RCH{=}CHCOCH_3$ (33), 1-cyclohexenyl-$CH_2CH_2COCH_3$ (—)	178, 1473
n-$C_6H_{13}CHO$	"	$RCHOHCH_2COCH_3$ (37), $RCH{=}CHCOCH_3$ (41)	90, 1436, 1456, 1474–1477
	NaOH; $(CO_2H)_2$	$RCH{=}CHCOCH_3$ (70)	613, 90, 1422, 1437, 1444, 1474–1478
	NaOH	$RCHOHCH_2COCH_3$ (—), $C_{17}H_{34}O_3$ (5)	1477, 1474–1476
$(CH_3)_2C[CH_2N(CH_3)_2]CHO$	$NaOC_2H_5$	$RCH{=}CHCOCH_3$ (24)	1479
$CH_3(CH{=}CH)_3CHO$	"	$RCH{=}CHCOCH_3$ (17)	1480, 1465
4-Methyl-1-cyclohexenyl-CHO	NaOH	$RCH{=}CHCOCH_3$ (60)	1481
4-Methyl-3-cyclohexenyl-CHO	"	$RCH{=}CHCOCH_3$ (70)	1481, 1484
2-Methyl-4-cyclohexenyl-CHO (stereo)	"	$RCHOHCH_2COCH_3$ (44)	1482, 1483
2-Methyl-4-cyclohexenyl-CHO (stereo, both centers marked)	"	$RCH{=}CHCOCH_3$ (68)	1482–1485
n-$C_7H_{15}CHO$	"	$RCHOHCH_2COCH_3$ (22)	1474

n-$C_4H_9CH(C_2H_5)CHO$	$Ba(OH)_2$	$RCHOHCH_2COCH_3$ (—)	288
O $(CH{=}CH)_2CHO$	NaOH	$(RCH{=}CH)_2CO$ (100)	1486
CHO, CH_3	$NaOC_2H_5$	$RCH{=}CHCOCH_3$ (70)	1487
CH_3, CH_3, CHO	$NaOCH_3$	$RCH{=}CHCOCH_3$ (—)	1484
CH_3, CHO, CH_3	NaOH	$RCH{=}CHCOCH_3$ (60)	1481
CH_3, CHO, CH_3	”	$RCH{=}CHCOCH_3$ (60)	1481
CH_3, CHO, CH_3	”	$RCH{=}CHCOCH_3$ (70)	1481, 1484
CH_3, CHO, CH_3	$NaOCH_3$	$RCH{=}CHCOCH_3$ (—)	1484
$(CH_3)_2C{=}CHCH_2SC(CH_3){=}CHCHO$	NaOH	$RCH{=}CHCOCH_3$ (50)	1488
$CH_3CO_2CH_2CH{=}CH(CH_2)_3CHO$	$(CH_2)_5NH$, CH_3CO_2H	$RCH{=}CHCOCH_3$ (14)	218
$CH_3(CH_2)_5CH{=}CHCHO$	NaOH	$RCH{=}CHCOCH_3$ (55)	1472
CHO, $(CH_3)_2$	KOH	$RCH{=}CHCOCH_3$ (53)	1489

Note: References 668–2359 are on pp. 403–438.

TABLE VIII. CONDENSATION OF ALIPHATIC ALDEHYDES WITH ACETONE (*Continued*)
(R in the product is the group R in the aldehyde RCHO.)

Aldehyde, RCHO	Catalyst	Product(s) (Yield, %)	Refs.
CHO, CH_3, CH_3 (cyclohexane ring)	$NaOC_2H_5$	$RCHOHCH_2COCH_3$ (13), $RCH{=}CHCOCH_3$ (56)	1490
CHO, CH_3, CH_3 (cyclohexane ring)	,,	$RCH{=}CHCOCH_3$ (21)	1490
n-$C_8H_{17}CHO$	NaOH	$RCHOHCH_2COCH_3$ (15)	1474
4-$O_2NC_6H_4CH{=}C(CH_3)CHO$	,,	$RCH{=}CHCOCH_3$ (30)	971
$(CH_3)_2C{=}CHCH{=}CHC(CH_3){=}CHCHO$	$NaOC_2H_5$	$RCH{=}CHCOCH_3$ (62)	1480
CH_3, CHO, $C(CH_3){=}CH_2$ (cyclopentene ring)	NaOH	$RCH{=}CHCOCH_3$ (—)	836
$(CH_3)_2$, CHO, CH_3 (cyclohexadiene ring)	$NaNH_2$	$RCH{=}CHCOCH_3$ (46)	105
CHO, $C(CH_3){=}CH_2$ (cyclohexene ring)	NaOH	$RCH{=}CHCOCH_3$ (—), $(RCH{=}CH)_2CO$ (—)	973
CHO, CH_3 (bicyclooctene)	$NaOCH_3$	$RCH{=}CHCOCH_3$ (—)	1485, 1484
CH_3, CHO, O, CH_3, CH_3 (cyclohexenone ring)	$(CH_2)_5NH$, CH_3CO_2H	$RCH{=}CHCOCH_3$ (—)	1313

$(CH_3)_2C{=}CHCH_2CH_2C(CH_3){=}CHCHO$, citral	$NaOC_6H_5$	$RCH{=}CHCOCH_3$ (81)	611, 612, 654, 1492–1501
$CH_3CH{=}CHCH(CH_3)CH_2$-$C(CH_3){=}CHCHO$	Na_2SO_3	$RCH{=}CHCOCH_3$ (68)	627
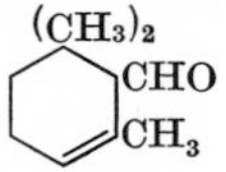 $(CH_3)_2$ CHO CH_3	$NaOC_2H_5$	$RCH{=}CHCOCH_3$ (—)	1502
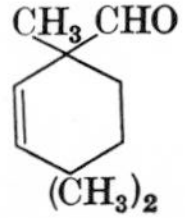 CH_3 CHO $(CH_3)_2$	NaOH	$RCH{=}CHCOCH_3$ (40)	1503
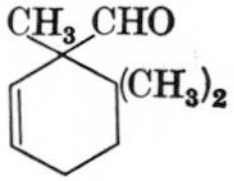 CH_3 CHO $(CH_3)_2$	”	$RCH{=}CHCOCH_3$ (37)	619
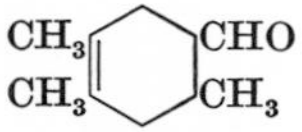 CH_3 CH_3 CHO CH_3	$NaOCH_3$	$RCH{=}CHCOCH_3$ (—)	1484
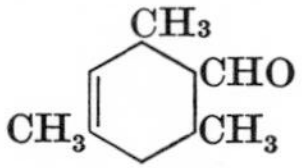 CH_3 CH_3 CHO CH_3	”	$RCH{=}CHCOCH_3$ (—)	1485, 2352
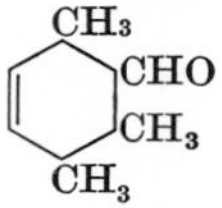 CH_3 CHO CH_3 CH_3	”	$RCH{=}CHCOCH_3$ (—)	1484
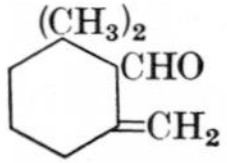 $(CH_3)_2$ CHO $=CH_2$	NaH	$RCH{=}CHCOCH_3$ (11), β-ionone (66)	1504

Note: References 668–2359 are on pp. 403–438.

TABLE VIII. Condensation of Aliphatic Aldehydes with Acetone (*Continued*)
(R in the product is the group R in the aldehyde RCHO.)

Aldehyde, RCHO	Catalyst	Product(s) (Yield, %)	Refs.
$(CH_3)_2C{=}CH(CH_2)_2CH(CH_3)CH_2CHO$	KOH; $(CH_3CO)_2O$	$RCH{=}CHCOCH_3$ (42)	1505, 1496, 1506, 1507
$i\text{-}C_3H_7CH_2CH{=}C(C_3H_7\text{-}i)CHO$	NaOH	$RCH{=}CHCOCH_3$ (—)	1437
$(CH_3)_2$, CHO, CH_3	KOH	$RCH{=}CHCOCH_3$ (55–58)	225
$(CH_3)_2$, CH_3, CHO	$NaOC_2H_5$	$RCH{=}CHCOCH_3$ (9)	1490
$(CH_3)_2$, CHO, CH_3	KOH	$RCH{=}CHCOCH_3$ (—)	225
$i\text{-}C_3H_7$, CH_3, O, CH_3, CH_2CHO	NaOH	$RCH{=}CHCOCH_3$ (—)	1508
$n\text{-}C_9H_{19}CHO$	,,	$RCHOHCH_2COCH_3$ (15)	1474
$(CH_3)_2CH(CH_2)_3CH(CH_3)CH_2CHO$	KOH	$RCH{=}CHCOCH_3$ (71), $(RCH{=}CH)_2CO$ (—)	1496
$C_6H_5(CH{=}CH)_2CHO$	NaOH	$(RCH{=}CH)_2CO$ (56)	1000
$(CH_3)_2C{=}C(CH_3)(CH_2)_2C(CH_3){=}CHCHO$	,,	$RCH{=}CHCOCH_3$ (65)	1509, 1510
CH_3, $CH(CH_3)CH_2CHO$	KOH	$RCH{=}CHCOCH_3$ (—)	144
$(CH_3)_2$, CH_3, CHO, CH_3	$NaNH_2$	$RCH{=}CHCOCH_3$ (28)	621
$(CH_3)_2$, CH_3, CHO, $=CH_2$	NaH	$RCH{=}CHCOCH_3$ (28), 6-Methyl-β-ionone (26)	620

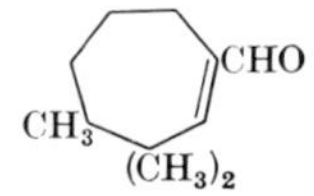 CHO, CH_3, $(CH_3)_2$	$NaNH_2$	RCH=CHCOCH$_3$ (—)	1511
 CH_3 CH(CH_3)CH_2CHO	KOH	RCH=CHCOCH$_3$ (—)	144
2,9-Dimethyl-2,8-decadien-4,6-diyne-1,10-dial	Al(OC_4H_9-t)$_3$	5,12-Dimethyl-3,5,11,13-hexadecatetraene-7,9-diyne-2,15-dione (28)	607
4-$O_2NC_6H_4$CH=C(CH_3)CH=CHCHO	NaOH	RCH=CHCOCH$_3$ (—)	971
2,9-Dimethyl-2,4,6,8-decatetraene-1,10-dial	KOH	5,12-Dimethyl-3,5,7,9,11,13-hexadecahexaene-2,15-dione (34)	607
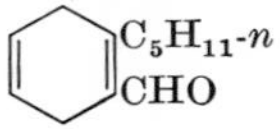 C_5H_{11}-n, CHO	NaOH	RCH=CHCOCH$_3$ (—)	1512
n-$C_{11}H_{23}$CHO	"	RCHOHCH$_2$COCH$_3$ (10)	1474
 CH_3, C_5H_{11}-n, CHO	"	RCH=CHCOCH$_3$ (—)	1512
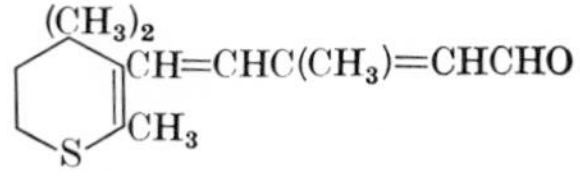 $(CH_3)_2$, CH=CHC(CH_3)=CHCHO, CH_3, S	"	RCH=CHCOCH$_3$ (—)	1513
 CH_3, CH_3, C_5H_{11}-n, CHO	"	RCH=CHCOCH$_3$ (—)	1512
3,7,11-Trimethyl-2,4,6,10-dodecatetraenal	$NaOC_2H_5$	RCH=CHCOCH$_3$ (28)	1480
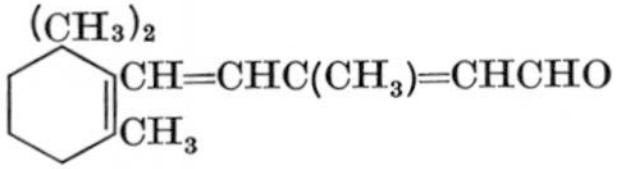 $(CH_3)_2$, CH=CHC(CH_3)=CHCHO, CH_3	NaOH	RCH=CHCOCH$_3$ (98)	1514
Crocetin dialdehyde	KOH	5,9,14,18-Tetramethyl-3,5,7,9,11,13,15,17,19-docosanonaene-2,21-dione (52)	607

Note: References 668–2359 are on pp. 403–438.

TABLE IX. Condensation of Aromatic Aldehydes with Acetone

(R in the product is the group R in the aldehyde RCHO.)

Aldehyde, RCHO	Catalyst	Product(s) (Yield, %)	Refs.
5-Br-thiophene-2-CHO (Br, S, CHO)	NaOH	RCH=CHCOCH$_3$ (—)	1515
5-Br-furan-2-CHO (Br, O, CHO)	KOH	$(RCH=CH)_2CO$ (91)	1516
5-Cl-furan-2-CHO (Cl, O, CHO)	"	$(RCH=CH)_2CO$ (91)	1516
5-I-furan-2-CHO (I, O, CHO)	"	$(RCH=CH)_2CO$ (91)	1516
5-O_2N-furan-2-CHO (O_2N, O, CHO)	H_3PO_4	$(RCH=CH)_2CO$ (52)	1519, 1516–1518
thiophene-2-CHO (S, CHO)	NaOH	RCH=CHCOCH$_3$ (69)	1520, 1035, 1521
	"	$(RCH=CH)_2CO$ (58–90)	1030, 1521, 2334, 2335
selenophene-2-CHO (Se, CHO)	"	RCH=CHCOCH$_3$ (71)	1522
furan-2-CHO (O, CHO)	"	RCH=CHCOCH$_3$ (74)	1471, 3, 1051, 1393, 1523–1538
	"	$(RCH=CH)_2CO$ (90)	1525–1537, 3

	$Ca(OH)_2$	$RCHOHCH_2COCH_3$ (—)	1523
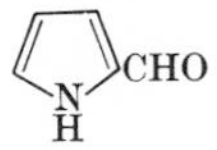 CHO N H	KOH	$RCH{=}CHCOCH_3$ (75), $(RCH{=}CH)_2CO$ (4)	1540, 1541
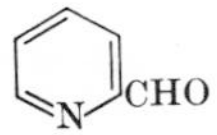 CHO N	NaOH	$RCHOHCH_2COCH_3$ (41)	1542
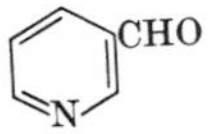 CHO N	,,	$RCHOHCH_2COCH_3$ (14), $RCH{=}CHCOCH_3$ (14), $(RCH{=}CH)_2CO$ (1.4)	1542, 1543
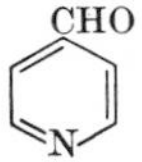 CHO N	,,	$RCH{=}CHCOCH_3$ (21)	1542, 1543
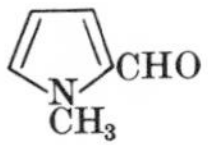 CHO N CH_3	,,	$(RCH{=}CH)_2CO$ (45)	1541
3,5-Br_2-4-HOC_6H_2CHO	,,	$RCH{=}CHCOCH_3$ (48)	1544
4-BrC_6H_4CHO	,,	$(RCH{=}CH)_2CO$ (50)	1545
2-HO-3-BrC_6H_3CHO	,,	$RCH{=}CHCOCH_3$ (56)	1544
2-HO-5-BrC_6H_3CHO	,,	$RCH{=}CHCOCH_3$ (100)	1547, 1546, 1549
2-ClC_6H_4CHO	,,	$RCH{=}CHCOCH_3$ (58)	1548
3-ClC_6H_4CHO	,,	$(RCH{=}CH)_2CO$ (—)	1550
4-ClC_6H_4CHO	,,	$RCH{=}CHCOCH_3$ (—)	1551
2-$O_2NC_6H_4CHO$	,,	$RCHOHCH_2COCH_3$ (100)	476, 1552
	,,	$(RCH{=}CH)_2CO$ (—)	1550
3-$O_2NC_6H_4CHO$	,,	$RCH{=}CHCOCH_3$ (—), $(RCH{=}CH)_2CO$ (—)	1553
	$Ba(OH)_2$	$RCHOHCH_2COCH_3$ (—)	1554, 1548
4-$O_2NC_6H_4CHO$	NaOH	$RCHOHCH_2COCH_3$ (—)	1555, 61, 1556
	,,	$(RCH{=}CH)_2CO$ (—)	1550

Note: References 668–2359 are on pp. 403–438.

TABLE IX. CONDENSATION OF AROMATIC ALDEHYDES WITH ACETONE (*Continued*)
(R in the product is the group R in the aldehyde RCHO.)

Aldehyde, RCHO	Catalyst	Product(s) (Yield, %)	Refs.
2-HO-5-$O_2NC_6H_3CHO$	NaOH	RCH=$CHCOCH_3$ (—)	1547
C_6H_5CHO	,,	*trans*-RCH=$CHCOCH_3$ (78)	655, 6, 83, 89, 236, 613, 1523, 1525, 1539, 1548, 1557–1561
	,,	$(RCH{=}CH)_2CO$ (94)	657, 4, 613, 1525, 1528, 1558–1562
	$Ca(OH)_2$	$RCHOHCH_2COCH_3$ (—)	1523, 61
2-HOC_6H_4CHO	NaOH	RCH=$CHCOCH_3$ (65)	1544, 1546, 1564, 1565
	,,	$(RCH{=}CH)_2CO$ (98)	591, 1565–1568
	$(CH_3)_2NH$	$(CH_3)_2N$, CH_2, O, $N(CH_3)_2$, CH_3 (25)	1569
4-HOC_6H_4CHO	NaOH	RCH=$CHCOCH_3$ (66)	1544, 1570, 1571
	,,	$(RCH{=}CH)_2CO$ (100)	347, 1570, 1572
	HCl	$(RCH{=}CH)_2CO$ (52), RCH=$CHCOCH_3$ (—)	1571, 1570, 1572
2,6-$(HO)_2C_6H_3CHO$	NaOH	RCH=$CHCOCH_3$ (—)	1568
2-$H_2NC_6H_4CHO$	—	N, CH_3 (—)	1573
CH_3, N, CHO	KOH	RCH=$CHCOCH_3$ (—)	1543

$C_6H_4(CHO)_2$-1,2	"	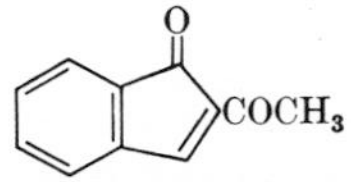 $COCH_3$ (—)	1574
$C_6H_4(CHO)_2$-1,4	NaOH	$C_6H_4(CH{=}CHCOCH_3)_2$-1,4 (75)	1575, 1576
2-$HO_2CC_6H_4CHO$	"	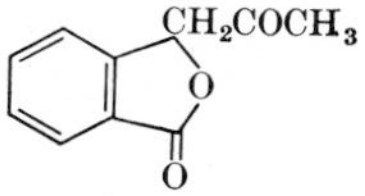 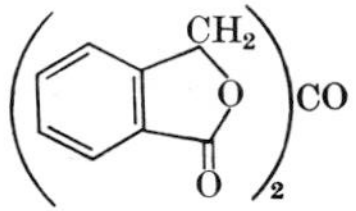 CH_2COCH_3 (—), CH_2 CO (—)	1577
3,4-$(CH_2O_2)C_6H_3CHO$	NaOH	$RCH{=}CHCOCH_3$ (90)	1578–1581, 1546
	"	$(RCH{=}CH)_2CO$ (—)	1581
3-CH_3O-4-HO-5-BrC_6H_2CHO	"	$RCH{=}CHCOCH_3$ (—)	1582
3-O_2N-4-$CH_3C_6H_3CHO$	"	$RCH{=}CHCOCH_3$ (90)	1583
4-$CH_3C_6H_4CHO$	"	$RCH{=}CHCOCH_3$ (—)	1583
2-$CH_3OC_6H_4CHO$	"	$RCH{=}CHCOCH_3$ (—)	1580, 515, 1441, 1546
	"	$(RCH{=}CH)_2CO$ (—)	1584
3-$CH_3OC_6H_4CHO$	"	$(RCH{=}CH)_2CO$ (—)	1584
4-$CH_3OC_6H_4CHO$	"	$RCH{=}CHCOCH_3$ (83), $(RCH{=}CH)_2CO$ (3)	1585, 1580, 1586
	"	$RCHOHCH_2COCH_3$ (—)	61
	Ion-exchange resin Amberlite 400	$(RCH{=}CH)_2CO$ (—)	1563
2-HO-3-$CH_3OC_6H_3CHO$	NaOH	$RCHOHCH_2COCH_3$ (—)	1547
2-HO-4-$CH_3OC_6H_3CHO$	"	$RCH{=}CHCOCH_3$ (—)	1547, 1587
2-HO-5-$CH_3OC_6H_3CHO$	"	$RCH{=}CHCOCH_3$ (—)	1547
3-HO-4-$CH_3OC_6H_3CHO$	HCl	$RCH{=}CHCOCH_3$ (—)	1588
2-CH_3O-4-HOC_6H_3CHO	NaOH	$RCH{=}CHCOCH_3$ (—)	1587
3-CH_3O-4-HOC_6H_3CHO	"	$RCH{=}CHCOCH_3$ (88)	1579, 1546, 1580, 1582, 1589–1591
	HCl	$(RCH{=}CH)_2CO$ (55)	1591, 1593
2-$O_2NC_6H_4CH{=}CHCHO$	NaOH	$RCH{=}CHCOCH_3$ (—)	1594
$C_6H_5CH{=}CHCHO$	"	$RCH{=}CHCOCH_3$ (—), $(RCH{=}CH)_2CO$ (—)	1595, 1561
	$NaOC_2H_5$	$(RCH{=}CH)_2CO$ (—)	1596, 1561, 1563, 1595, 1597

Note: References 668–2359 are on pp. 403–438.

TABLE IX. CONDENSATION OF AROMATIC ALDEHYDES WITH ACETONE (*Continued*)

(R in the product is the group R in the aldehyde RCHO.)

Aldehyde, RCHO	Catalyst	Product(s) (Yield, %)	Refs.
$2-C_2H_5OC_6H_4CHO$	NaOH	$(RCH{=}CH)_2CO$ (—)	1598
$2,4-(CH_3O)_2C_6H_3CHO$	"	$RCH{=}CHCOCH_3$ (—)	1587, 1008
$2,6-(CH_3O)_2C_6H_3CHO$	"	$RCH{=}CHCOCH_3$ (—)	1568
$3,4-(CH_3O)_2C_6H_3CHO$	"	$RCH{=}CHCOCH_3$ (97)	1578, 411, 1580, 1589, 1598, 1599
	"	5-R-3-(CH=CHR)-2-cyclohexen-1-one (structure) (20)	411, 1578, 1589
$2-HO-3-C_2H_5OC_6H_3CHO$	"	$RCH{=}CHCOCH_3$ (—)	1590
$3,5-(OCH_3)_2-4-HOC_6H_2CHO$	"	$RCH{=}CHCOCH_3$ (50)	1600
$4-(CH_3)_2NC_6H_4CHO$	"	$RCH{=}CHCOCH_3$ (—)	1601, 1602
2-Quinolinecarboxaldehyde (structure)	KOH	$RCH{=}CHCOCH_3$ (—)	1543
$4-CH_3OC_6H_4CH{=}CHCHO$	NaOH	$RCH{=}CHCOCH_3$ (—)	1603
$4-i-C_3H_7C_6H_4CHO$	"	$RCH{=}CHCOCH_3$ (—)	1525
	"	$(RCH{=}CH)_2CO$ (90)	1525
$2,3,6-(CH_3)_3C_6H_2CHO$	"	$RCH{=}CHCOCH_3$ (60), $(RCH{=}CH)_2CO$ (5)	1604, 1605
$2,4,5-(CH_3)_3C_6H_2CHO$	$NaOC_2H_5$	$RCH{=}CHCOCH_3$ (59)	1606, 1604
	NaOH	$RCHOHCH_2COCH_3$ (—), $RCH{=}CHCOCH_3$ (—)	1606
$2,4,6-(CH_3)_3C_6H_2CHO$	—	$RCH{=}CHCOCH_3$ (—)	1607
$3-CH_3O-4-C_2H_5OC_6H_3CHO$	NaOH	$RCH{=}CHCOCH_3$ (83), $(RCH{=}CH)_2CO$ (50)	411
$2,4,5-(CH_3O)_3C_6H_2CHO$	"	$RCH{=}CHCOCH_3$ (—)	1608
$2,4,6-(CH_3O)_3C_6H_2CHO$	"	$RCH{=}CHCOCH_3$ (Very good)	1008
$3,4,5-(CH_3O)_3C_6H_2CHO$	"	$RCH{=}CHCOCH_3$ (40)	1600
$2-(CH_3)_2N-5-CH_3C_6H_3CHO$	Various conditions	No condensation	1573

2-Naphthaldehyde (structure)	NaOH	$RCH{=}CHCOCH_3$ (—)	1610
	$NaOCH_3$	$(RCH{=}CH)_2CO$ (—)	1610
2-Hydroxy-1-naphthaldehyde (structure)	KOH	4-(CH_2COCH_3)-2-CH_3-benzochromene (structure) (Poor)	1611
$2,4\text{-}(CH_3O)_2C_6H_3CH{=}CHCHO$	NaOH	$RCH{=}CHCOCH_3$ (80)	1008
$3\text{-}(CH_3)_2NC_6H_4CH{=}CHCHO$	”	$RCH{=}CHCOCH_3$ (—)	1602
$4\text{-}t\text{-}C_4H_9C_6H_4CHO$	”	$RCH{=}CHCOCH_3$ (80)	1612
$2\text{-}n\text{-}C_3H_7O\text{-}3\text{-}CH_3OC_6H_3CHO$	”	$RCH{=}CHCOCH_3$ (—)	1590
$3\text{-}CH_3O\text{-}4\text{-}n\text{-}C_3H_7OC_6H_3CHO$	”	$RCH{=}CHCOCH_3$ (—)	411
$3\text{-}CH_3O\text{-}4\text{-}i\text{-}C_3H_7OC_6H_3CHO$	”	$RCH{=}CHCOCH_3$ (—)	411
$2,4,6\text{-}(CH_3O)_3C_6H_2CH{=}CHCHO$	”	$RCH{=}CHCOCH_3$ (—)	1008
$2\text{-}n\text{-}C_4H_9O\text{-}3\text{-}CH_3OC_6H_3CHO$	”	$RCH{=}CHCOCH_3$ (—)	1590
$2\text{-}n\text{-}C_3H_7O\text{-}3\text{-}C_2H_5OC_6H_3CHO$	”	$RCH{=}CHCOCH_3$ (—)	1590
$2\text{-}n\text{-}C_4H_9O\text{-}3\text{-}C_2H_5OC_6H_3CHO$	”	$RCH{=}CHCOCH_3$ (—)	1590
$4\text{-}C_6H_5CH_2C_6H_4CHO$	”	$RCH{=}CHCOCH_3$ (25), $(RCH{=}CH)_2CO$ (48)	1613
$2\text{-}HO\text{-}4\text{-}C_6H_5CH_2OC_6H_3CHO$	”	$RCH{=}CHCOCH_3$ (50)	1587
$2,6\text{-}(CH_3O)_2\text{-}4\text{-}n\text{-}C_5H_{11}C_6H_2CHO$	”	$RCH{=}CHCOCH_3$ (85)	1614
$3\text{-}CH_3O\text{-}4\text{-}C_6H_5CH_2OC_6H_3CHO$	”	$RCH{=}CHCOCH_3$ (80)	411
1-Pyrenecarboxaldehyde (structure)	”	$RCH{=}CHCOCH_3$ (100)	1615

Note: References 668–2359 are on pp. 403–438.

TABLE X. CONDENSATION OF ALDEHYDES WITH 2-BUTANONE

(R in the product is the group R in the aldehyde RCHO.)

Aldehyde, RCHO	Catalyst	Product(s) (Yield, %)	Refs.
CH_2O	LiOH or NaOH	$CH_3C(CH_2OH)_2COCH_3$ (85)	1389, 551, 577, 927, 1392, 1406, 1616–1619
	KOH	$CH_3CH(CH_2OH)COCH_3$ (97)	1620, 119, 126, 337, 577, 652, 1389, 1394, 1396, 1400, 1405, 1406, 1621–1629
	KOH, or ion-exchange resin Dowex-50-X8	$CH_2{=}C(CH_3)COCH_3$ (82–90)	1630, 125, 634
	$KOCH_3$	CH_3, =O, CH_3, CH_3 (49)	1367, 1624
	$Ca(OH)_2$, H_2O, 50–55°, 2 hr.	CH_3, OH, $(CH_2OH)_2$, $HOCH_2$, O (—)	1415, 1389
	$Ca(OH)_2$, H_2O, 5–50°, 10 hr.	$CH_3C(CH_2OH)_2CHOHCH(CH_2OH)_2$ (70)	1631
	PbO, $MgSiO_3$	$CH_2{=}CHCOC_2H_5$ (—), $CH_2{=}C(CH_3)COCH_3$ (—)	1632, 1620, 1626, 1633
CCl_3CHO	CH_3CO_2K, $(CH_3CO)_2O$	$RCHOHCH(CH_3)COCH_3$ (55)	287, 273, 635
CH_3CHO	KOH	$RCHOHCH(CH_3)COCH_3$ (87)	269, 271, 576, 577, 615, 1634–1637
	NaOH; HBr	$RCH{=}C(CH_3)COCH_3$ (77)	1636, 1638, 1639
$C(Cl_2){=}CHCHO$	HCl	$RCH{=}C(CH_3)COCH_3$ (30)	637

$O_2NCH(CHO)_2$	NaOH	$2\text{-}CH_3\text{-}4\text{-}NO_2C_6H_3OH$ (100)	1640
C_2H_5CHO	KOH	$RCHOHCH(CH_3)COCH_3$ (61)	1641
$CH_3CH{=}CHCHO$	NaOH	$RCH{=}C(CH_3)COCH_3$ (24), $RCH{=}CHCOC(CH_3){=}CHR$ (16)	1438, 583, 1399 1439
$n\text{-}C_3H_7CHO$	"	$RCHOHCH(CH_3)COCH_3$ (82)	1636, 1642
	"	$RCH{=}C(CH_3)COCH_3$ (30), $n\text{-}C_3H_7CH{=}C(C_2H_5)CHO$ (31)	1445, 1444
$i\text{-}C_3H_7CHO$	"	$RCH{=}CHCOC_2H_5$ (21), $RCH{=}C(CH_3)COCH_3$ (17)	231, 1643
	$NaOC_2H_5$	$RCH{=}CHCOC_2H_5$ (85%), $RCH{=}C(CH_3)COCH_3$ (15%) (—)	231
	HCl	$RCH{=}C(CH_3)COCH_3$ (28)	231
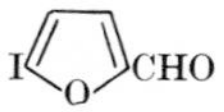 5-I-furan-2-CHO (I, O, CHO ring structure)	NaOH	$RCH{=}CHCOC_2H_5$, $RCH{=}C(CH_3)COCH_3$ (—)	1644
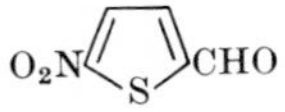 5-O_2N-thiophene-2-CHO (O_2N, S, CHO ring structure)	H_2SO_4, CH_3CO_2H	$RCH{=}CHCOC(CH_3){=}CHR$ (—)	226
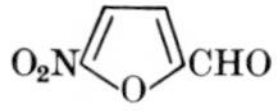 5-O_2N-furan-2-CHO (O_2N, O, CHO ring structure)	HCl	$RCH{=}C(CH_3)COCH_3$ (90)	1645, 226
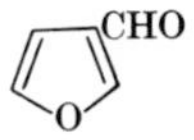 furan-3-CHO (O, CHO ring structure)	NaOH	$RCH{=}CHCOC_2H_5$, $RCH{=}C(CH_3)COCH_3$, $RCH{=}CHCOC(CH_3){=}CHR$ (—)	1646, 590, 1523, 1526, 1530, 1535, 1647–1649
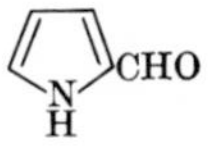 pyrrole-2-CHO (N-H, CHO ring structure)	KOH	$RCH{=}CHCOC_2H_5$ (90)	1541, 1540
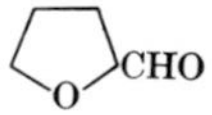 tetrahydrofuran-2-CHO (O, CHO ring structure)	NaOH	$RCHOHCH(CH_3)COCH_3$ (67)	1454

Note: References 668–2359 are on pp. 403–438.

TABLE X. CONDENSATION OF ALDEHYDES WITH 2-BUTANONE (*Continued*)
(R in the product is the group R in the aldehyde RCHO.)

Aldehyde, RCHO	Catalyst	Product(s) (Yield, %)	Refs.
n-C_4H_9CHO	KOH	$RCHOHCH(CH_3)COCH_3$ (35)	1641
sec-C_4H_9CHO	,,	$RCHOHCH_2COC_2H_5$ (—)	1650
i-C_4H_9CHO	,,	$RCHOHCH(CH_3)COCH_3$ (34)	1651, 141
$HOCH_2(CH_2)_3CHO$	NaOH	O, $CH(CH_3)COCH_3$ (37)	1464, 522
CHO, N, CH_3	KOH	$RCH{=}CHCOC_2H_5$ (43)	1541
$(C_2H_5)_2CHCHO$	,,	$RCHOHCH_2COC_2H_5$ (—)	1650, 1652
2-$O_2NC_6H_4CHO$	H_2SO_4	$RCH{=}C(CH_3)COCH_3$ (—)	1653
3-$O_2NC_6H_4CHO$	,,	$RCH{=}C(CH_3)COCH_3$ (—)	1653, 1654
4-$O_2NC_6H_4CHO$	NaOH	$RCHOHCH_2COC_2H_5$, $RCHOHCH(CH_3)COCH_3$ (—)	40
	H_2SO_4	$RCH{=}C(CH_3)COCH_3$ (—)	1653
C_6H_5CHO	NaOH, H_2O, 3–5°, 6 hr.	$RCHOHCH_2COC_2H_5$, $RCHOHCH(CH_3)COCH_3$ (Total, 83)	1656
	NaOH, H_2O, 8 days	$RCH{=}CHCOC_2H_5$ (95), $C_{18}H_{18}O_2$ (—)	232, 240, 523, 582, 1524, 1656–1658, 1663, 1664
	HCl	$RCH{=}C(CH_3)COCH_3$ (42–85), $RCH{=}CHCOC(CH_3){=}CHR$ (22)	233, 232, 234–236, 240, 1444, 1656–1661
	HBr	$RCH{=}CHCOC(CH_3){=}CHR$ (—), $C_{25}H_{20}O$ (—)	1659, 1664, 1665
2-HOC_6H_4CHO	NaOH	$RCH{=}CHCOC_2H_5$ (—)	1524, 569, 1666, 1667
	HCl	CH_3, $CH{=}CHC_6H_4OH$-2, $\overset{\oplus}{O}$, $Cl^{\ominus}$ (—)	1668

	HCl, $FeCl_3$	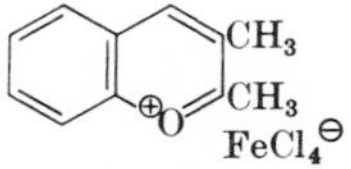 [structure] CH_3, CH_3, $FeCl_4^{\ominus}$ (15), [structure] CH_2CH_3, $FeCl_4^{\ominus}$ (—) 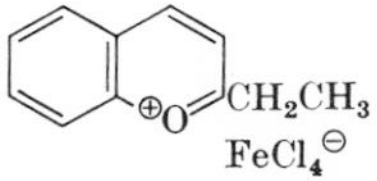	569, 568
3-HOC_6H_4CHO	HCl	RCH=CHCOC(CH_3)=CHR (—)	1669
4-HOC_6H_4CHO	,,	RCH=C(CH_3)$COCH_3$ (—)	1669, 1666
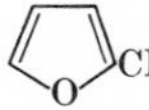 [structure] CH=CHCHO	NaOH	RCH=C(CH_3)$COCH_3$ (24), RCH=CHCOC(CH_3)=CHR (—)	1438, 1399
2,4-$(HO)_2C_6H_3CHO$	HCl	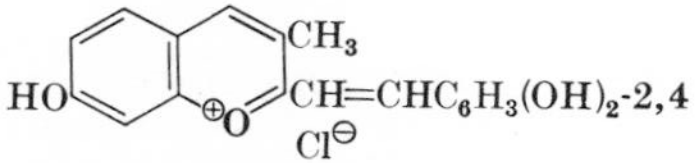 [structure] HO, CH_3, CH=CH$C_6H_3(OH)_2$-2,4, $Cl^{\ominus}$ (—)	1668
3,4-$(HO)_2C_6H_3CHO$	,,	RCH=CHCOC(CH_3)=CHR (—)	1669
2-$H_2NC_6H_4CHO$	NaOH	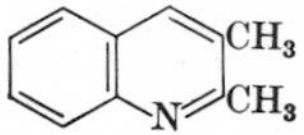 [structure] CH_3, CH_3 (—)	1670
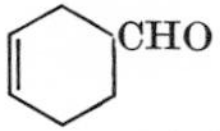 [structure] CHO	,,	RCH=CHCOC_2H_5 (—)	1671
	$NaOC_2H_5$	RCH=CHCOC_2H_5 (—), RCH=C(CH_3)$COCH_3$ (—)	1671
n-$C_6H_{13}CHO$	KOH	RCHOHCH(CH_3)$COCH_3$ (50)	1641
$C_6H_4(CHO)_2$-1,2	NaOH	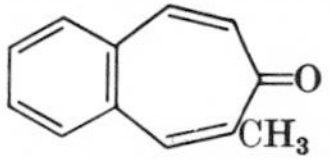 [structure] =O, CH_3 (80)	1672, 223
3,4-$(CH_2O_2)C_6H_3CHO$	,,	RCH=CHCOC_2H_5 (—)	1666, 1669, 1673
	HCl, −3°	RCH=C(CH_3)$COCH_3$ (—)	1666
	HCl, C_2H_5OH	$C_{28}H_{20}O_7$ (—)	1664
2-$CH_3C_6H_4CHO$	NaOH	RCH=CHCOC_2H_5 (21)	1674

Note: References 668–2359 are on pp. 403–438.

TABLE X. CONDENSATION OF ALDEHYDES WITH 2-BUTANONE (*Continued*)
(R in the product is the group R in the aldehyde RCHO.)

Aldehyde, RCHO	Catalyst	Product(s) (Yield, %)	Refs.
CHO	$NaOC_2H_5$	$RCH{=}C(CH_3)COCH_3$ or $RCH{=}CHCOC_2H_5$* (60)	1487
$2\text{-}CH_3OC_6H_4CHO$	NaOH	$RCH{=}CHCOC_2H_5$ (—)	1666
	HCl	$RCH{=}C(CH_3)COCH_3$ (81)	1661, 1666
$3\text{-}CH_3OC_6H_4CHO$	,,	$RCH{=}CHCOC(CH_3){=}CHR$ (—)	1669
	,,	$RCH{=}C(CH_3)COCH_3$ (61)	1661
$4\text{-}CH_3OC_6H_4CHO$	NaOH	$RCH{=}CHCOC_2H_5$ (—)	1673, 240, 1669
	HCl	$RCH{=}C(CH_3)COCH_3$ (84)	1661, 240, 1669
$2\text{-}HO\text{-}3\text{-}CH_3OC_6H_3CHO$	KOH	$RCH{=}CHCOC_2H_5$ (—)	1570
$2\text{-}HO\text{-}4\text{-}CH_3OC_6H_3CHO$	NaOH	$RCH{=}CHCOC_2H_5$ (—)	1570
$3\text{-}CH_3O\text{-}4\text{-}HOC_6H_3CHO$	,,	$RCH{=}CHCOC_2H_5$ (—)	1675, 1667
	HCl	$RCH{=}C(CH_3)COCH_3$ (—)	1675
$n\text{-}C_4H_9CH(C_2H_5)CHO$	KOH	$RCHOHCH_2COC_2H_5$ (—)	1650
$C_6H_5CH{=}CHCHO$	NaOH	$RCH{=}C(CH_3)COCH_3$ (44)	583, 1662, 1676, 1677
	HCl	$RCH{=}C(CH_3)COCH_3$* (63)	583
$C_6H_5CH(CH_3)CHO$	NaOH	$RCH{=}C(CH_3)COCH_3$ (62)	231
$3,4\text{-}(CH_3O)_2C_6H_3CHO$	HCl	$RCH{=}CHCOC(CH_3){=}CHR$ (—)	1669
CHO	$NaOCH_3$	$RCH{=}CHCOC_2H_5$ or $RCH{=}C(CH_3)COCH_3$* (—)	1678, 1474
$4\text{-}CH_3OC_6H_4CH{=}CHCHO$	NaOH	$RCH{=}C(CH_3)COCH_3$ (25)	583
	HCl	$RCH{=}C(CH_3)COCH_3$ (67)	583
$4\text{-}i\text{-}C_3H_7C_6H_4CHO$	NaOH	$RCH{=}CHCOC_2H_5$ (68)	1666, 1679, 1787
	HCl	$RCH{=}C(CH_3)COCH_3$ (45)	1679, 1680, 1787
CH_3 CHO	$NaOCH_3$	$RCH{=}CHCOC_2H_5$ (—)	1484

$(CH_3)_2C{=}CHCH_2CH_2$- $C(CH_3){=}CHCHO$	$LiOC_2H_5$	$RCH{=}CHCOC_2H_5$ (53), $RCH{=}C(CH_3)COCH_3$ (12)	582, 1491, 1506, 1592, 1681–1688
	Triton B	$RCH{=}CHCOC_2H_5$ (25), $RCH{=}C(CH_3)COCH_3$ (35)	582
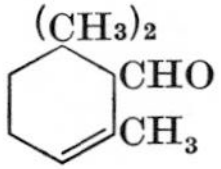 $(CH_3)_2$, CHO, CH_3	$NaOC_2H_5$	$RCH{=}CHCOC_2H_5$ (52)	1685, 1689
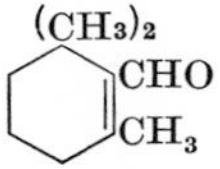 $(CH_3)_2$, CHO, CH_3	"	$RCH{=}CHCOC_2H_5$ (63)	1685, 1689
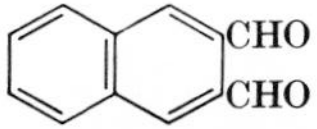 CHO, CHO	NaOH	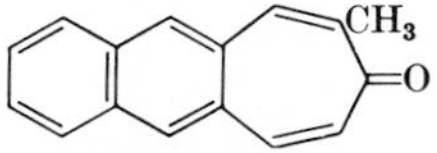 CH_3, =O (70)	1690

Note: References 668–2359 are on pp. 403–438.

* The structure of the product was not proved.

TABLE XI. CONDENSATION OF ALDEHYDES WITH ALIPHATIC METHYL KETONES OTHER THAN ACETONE AND 2-BUTANONE

(R in the product is the group R in the aldehyde RCHO.)

Ketone and Ald^hyde, RCHO	Catalyst	Product(s) (Yield, %)	Refs.
$Cl_2CHCOCH_3$ **and**			
CH_2O	Na_2CO_3	$Cl_2CHCOCH_2CH_2OH$ (75)	1691
2-$O_2NC_6H_4CHO$	NaOH	RCH=$CHCOCHCl_2$ (—)	1653
3-$O_2NC_6H_4CHO$	,,	RCH=$CHCOCHCl_2$ (—)	1653
4-$O_2NC_6H_4CHO$	,,	RCH=$CHCOCHCl_2$ (30)	1653
CH_3COCO_2H **and**			
CH_2O	H_2SO_4, CH_3CO_2H	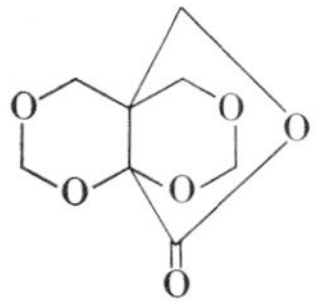 [structure] (40)	1692
$C(Cl_2)$=CHCHO	HCl	RCH=$CHCOCO_2H$ (66)	1693
n-C_3H_7CHO	$(C_2H_5)_2NH$	RCH=$CHCOCO_2H$ (4)	219
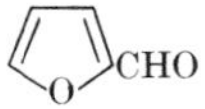 [structure] CHO	NaOH	RCH=$CHCOCO_2H$* (Excellent)	1694, 1695
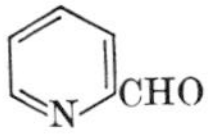 [structure] CHO	KOH	RCH=$CHCOCO_2H$ (16)	1696
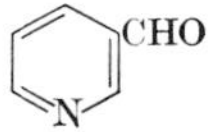 [structure] CHO	,,	RCH=$CHCOCO_2H$ (51)	1696
2-$O_2NC_6H_4CHO$	,,	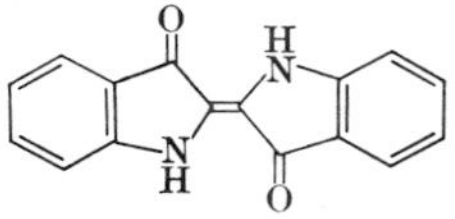 [structure] (—)	1695

3-$O_2NC_6H_4CHO$	"	$RCH{=}CHCOCO_2H$ (—)	1695
4-$O_2NC_6H_4CHO$	"	$RCH{=}CHCOCO_2H$ (—)	1695
C_6H_5CHO	"	$RCH{=}CHCOCO_2H$* (80)	1694, 1695, 1697, 1698
	HCl	$C_6H_5CH{=}$[ring: O, =O, C_6H_5, O] (—)	1700
2-HOC_6H_4CHO	"	$RCH{=}CHCOCO_2H$ (—)	1695
4-HOC_6H_4CHO	"	$RCH{=}CHCOCO_2H$ (—)	1695
2,4-$(HO)_2C_6H_3CHO$	"	$RCH{=}CHCOCO_2H$ (—)	1695
n-$C_6H_{13}CHO$	H_2SO_4	n-C_6H_{13}[ring: O, =O, O] (25)	219
3,4-$(CH_2O_2)C_6H_3CHO$	NaOH	$RCH{=}CHCOCO_2H$ (67)	1699, 1695
4-$CH_3OC_6H_4CHO$	KOH	$RCH{=}CHCOCO_2H$ (—)	1695, 1697
$C_6H_5CH{=}CHCHO$	"	$RCH{=}CHCOCO_2H$ (—)	1695
2,5-$(CH_3)_2$-4-$CH_3OC_6H_2CHO$	"	$RCH{=}CHCOCO_2H$ (90)	1701
$(CH_3)_2C{=}CHCH_2CH_2$-$C(CH_3){=}CHCHO$	"	$RCH{=}CHCOCO_2H$ (—)	1695
2,5-$(CH_3)_2$-4-$C_6H_5CH_2OC_6H_2CHO$	"	$RCH{=}CHCOCO_2H$ (—)	1701
CH_3COCH_2Br **and** O_2N-[furan]-CHO	H_2SO_4, CH_3CO_2H	$RCH{=}C(Br)COCH{=}CHR$ (—)	1702

Note: References 668–2359 are on pp. 403–438.

* This compound was isolated as the sodium or potassium salt.

TABLE XI. Condensation of Aldehydes with Aliphatic Methyl Ketones Other than Acetone and 2-Butanone (*Continued*)

(R in the product is the group R in the aldehyde RCHO.)

Ketone and Aldehyde, RCHO	Catalyst	Product(s) (Yield, %)	Refs.
CH_3COCH_2Cl **and** CH_2O	NaOH	$CH_3COCCl(CH_2OH)_2$ (33),	1691
		$CH_3COC(Cl)$ [1,3-dioxane ring: C–CH_2–O–CH_2–O–CH_2] (—)	
	,,	$CH_3COC(OH)(CH_2OH)_2$ (36)	1691
O_2N-[furan]-CHO	H_2SO_4, CH_3CO_2H	RCH=CClCOCH=CHR (—)	1702
2-HO-4-ClC_6H_3CHO	KOH	[Cl-benzofuran]-$COCH_3$ (60)	1703
2-HOC_6H_4CHO	,,	[benzofuran]-$COCH_3$ (80)	1704, 1703
2-HO-3-$CH_3OC_6H_3CHO$	,,	[OCH_3-benzofuran]-$COCH_3$ (38)	1703
4-$(CH_3)_2NC_6H_4CHO$	HCl, CH_3CO_2H	RCH=CClCOCH$_3$ (—)	1705
CH_3COCH_2I **and** O_2N-[furan]-CHO	H_2SO_4, CH_3CO_2H	RCH=CICOCH=CHR (—)	1702
CH_3COCH=NOH **and** 3-$O_2NC_6H_4CHO$	NaOH	RCH=CHCOCH=NOH (100)	1706

Reactant	Catalyst	Product (yield %)	Refs.
C_6H_5CHO	"	$RCH{=}CHCOCH{=}NOH$ (90)	1706
4-i-$C_3H_7C_6H_4CHO$	"	$RCH{=}CHCOCH{=}NOH$ (—)	1706
$CH_3COCH_2NO_2$ and			
(2-furyl)CHO	"	$RCH{=}CHCOCH_2NO_2$ (90)	1707
4-ClC_6H_4CHO	"	$RCH{=}CHCOCH_2NO_2$ (49)	1707
C_6H_5CHO	"	$RCH{=}CHCOCH_2NO_2$ (90)	1707
2-$H_2NC_6H_4CHO$	None	2-CH_3-3-NO_2-quinoline (69)	1707
2-H_2N-4,5-$(CH_2O_2)C_6H_2CHO$	CH_3CO_2H	6,7-methylenedioxy-2-CH_3-3-NO_2-quinoline (—)	1707
4-$CH_3C_6H_4CHO$	NaOH	$RCH{=}CHCOCH_2NO_2$ (66)	1707
4-$CH_3OC_6H_4CHO$	"	$RCH{=}CHCOCH_2NO_2$ (81)	1707
$CH_3COCOCH_3$ and			
Cl_3CCHO	$(CH_2)_5NH$, CH_3CO_2H	bicyclic bis(tetrahydrofuran) with two OH and two CCl_3 groups (41)	1708, 593
(2-furyl)CHO	"	$RCH{=}CHCOCOCH{=}CHR$ (22)	592, 1486
	KOH	$C_{23}H_{22}O_9$ (—)	1709
4-BrC_6H_4CHO	$(CH_2)_5NH$, CH_3CO_2H	$RCH{=}CHCOCOCH{=}CHR$ (5–21)	593
4-ClC_6H_4CHO	"	$RCH{=}CHCOCOCH{=}CHR$ (5–21)	593
C_6H_5CHO	"	$RCH{=}CHCOCOCH{=}CHR$ (10)	594, 592, 1710
	KOH	$C_{29}H_{28}O_6$ (45)	1709
(2-furyl)$CH{=}CHCHO$	$(CH_2)_5NH$	$RCH{=}CHCOCOCH{=}CHR$ (—)	1486

Note: References 668–2359 are on pp. 403–438.

TABLE XI. Condensation of Aldehydes with Aliphatic Methyl Ketones Other than Acetone and 2-Butanone (*Continued*)
(R in the product is the group R in the aldehyde RCHO.)

Ketone and Aldehyde, RCHO	Catalyst	Product(s) (Yield, %)	Refs.
$CH_3COCOCH_3$ (***contd.***) **and**			
$3,4\text{-}(CH_2O_2)C_6H_3CHO$	$(CH_2)_5NH$, CH_3CO_2H	$RCH{=}CHCOCOCH{=}CHR$ (5–21)	593
$2\text{-}CH_3OC_6H_4CHO$	,,	$RCH{=}CHCOCOCH{=}CHR$ (5–21)	593
$4\text{-}CH_3OC_6H_4CHO$	,,	$RCH{=}CHCOCOCH{=}CHR$ (5–21)	593
	KOH	$C_{32}H_{34}O_9$ (—)	1709
$CH_3(CH{=}CH)_3CHO$	$(CH_2)_5NH$	$RCH{=}CHCOCOCH_3$ (4), $RCH{=}CHCOCOCH{=}CHR$ (—)	1711
$C_6H_5CH{=}CHCHO$	$(CH_2)_5NH$, CH_3CO_2H	$RCH{=}CHCOCOCH{=}CHR$ (57)	594, 592, 1712
2-Furyl$(CH{=}CH)_2CHO$	$(CH_2)_5NH$	$RCH{=}CHCOCOCH{=}CHR$ (—)	1486
$2\text{-}C_2H_5OC_6H_4CHO$	$(CH_2)_5NH$, CH_3CO_2H	$RCH{=}CHCOCOCH{=}CHR$ (5–21)	593
$C_6H_5(CH{=}CH)_2CHO$	,,	$RCH{=}CHCOCOCH{=}CHR$ (32)	592, 594, 1712
$C_6H_5(CH{=}CH)_3CHO$	,,	$RCH{=}CHCOCOCH_3$ (9), $RCH{=}CHCOCOCH{=}CHR$ (18)	592, 1712
$C_6H_5(CH{=}CH)_5CHO$	,,	$RCH{=}CHCOCOCH_3$ (—)	592
$CH_3COCH_2OCH_3$ **and**			
$3,4\text{-}Cl_2C_6H_2(CHO)_2\text{-}1,2$	NaOH	(40)	1713
$3\text{-}BrC_6H_3(CHO)_2\text{-}1,2$	,,	(40)	1713
$3\text{-}ClC_6H_3(CHO)_2\text{-}1,2$	,,	(40)	1713
$C_6H_4(CHO)_2\text{-}1,2$	,,	(40)	1713

$CH_3COCHOHCH_3$ and			
CH_2O	KOH	$CH_3COC(OH)(CH_2OH)CH_3$ (60)	230
CH_3CHO	,,	$CH_3COC(OH)(CH_3)CHOHCH_3$ (60)	230
$CH_3COC(CH_3){=}CH_2$ and CH_2O	,,	$HOCH_2CH_2COC(CH_3){=}CH_2$ (20)	1714
	K_2CO_3	$CH_3COCH(CH_3)CH_2OCH_3$ (20), $CH_3COC(CH_3)(CH_2OH)CH_2OCH_3$ (47)	1714
▷$COCH_3$ and			
CH_2O	KOH	$HOCH_2CH_2CO$◁ (—), $CH_2{=}CHCO$◁ (—)	1717
CCl_3CHO	,,	$RCHOHCH_2CO$◁ (Low)	1717
CH_3CHO	,,	$RCHOHCH_2CO$◁ (—), $RCH{=}CHCO$◁ (—)	1717
n-C_3H_7CHO	,,	$RCHOHCH_2CO$◁ (28), n-$C_3H_7CH{=}C(C_2H_5)CHO$ (18)	1717
i-C_3H_7CHO	,,	$RCHOHCH_2CO$◁ (26), $RCH{=}CHCO$◁ (29)	1715, 1520, 1717
2-Furyl-CHO	,,	$RCH{=}CHCO$◁ (95)	1716
4-$O_2NC_6H_4CHO$	NaOH	$RCH{=}CHCO$◁ (33)	1717
C_6H_5CHO	,,	$RCH{=}CHCO$◁ (72)	1717
2-$CH_3OC_6H_4CHO$	,,	$RCH{=}CHCO$◁ (62)	1717
$C_6H_5CH{=}CHCHO$	,,	$RCH{=}CHCO$◁ (66)	1717, 1718

Note: References 668–2359 are on pp. 403–438.

TABLE XI. CONDENSATION OF ALDEHYDES WITH ALIPHATIC METHYL KETONES OTHER THAN ACETONE AND 2-BUTANONE (*Continued*)
(R in the product is the group R in the aldehyde RCHO.)

Ketone and Aldehyde, RCHO	Catalyst	Product(s) (Yield, %)	Refs.
[cyclopropyl]$COCH_3$ (***contd.***) **and**			
2-$C_2H_5OC_6H_4CHO$	NaOH	RCH=CHCO[cyclopropyl] (74)	1717
[isoquinoline structure bearing CHO]	"	RCH=CHCO[cyclopropyl] (65)	1717
1-$C_{10}H_7CHO$	"	RCH=CHCO[cyclopropyl] (48)	1717
3,4-$(C_2H_5O)_2C_6H_3CHO$	"	RCH=CHCO[cyclopropyl] (78)	1717
$CH_3COCH_2COCH_3$ **and**			
CH_2O	$(C_2H_5)_3N$	[cyclohexanone ring with CH_3CO, $COCH_3$, CH_3, OH, =O] (70)	220
CH_3CHO	"	[cyclohexanone ring with CH_3, CH_3CO, $COCH_3$, CH_3, OH, =O] (89)	220, 1719
$Cl_2C{=}CClCHO$	HCl	$RCH{=}C(COCH_3)_2$ (44)	216, 1719
	H_2SO_4	$RCH{=}C(COCH_3)_2$ (50)	289
C_2H_5CHO	$(C_2H_5)_2NH$	$RCH{=}C(COCH_3)_2$ (43)	220, 216
n-C_3H_7CHO	$(CH_2)_5NH$	$RCH{=}C(COCH_3)_2$ (85)	220, 217
[thiophene]CHO	"	$RCH{=}C(COCH_3)_2$ (—)	1720

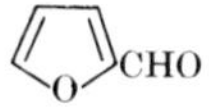 [structure] CHO	"	RCH=C$(COCH_3)_2$ (80)	590, 1721
C_6H_5CHO	"	RCH=C$(COCH_3)_2$ (66)	216, 1722
2-HOC_6H_4CHO	NaOH	RCH=C$(COCH_3)_2$ (—)	591, 1720
	"	RCH=CHCOCH=CHR (—)	591
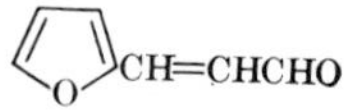 [structure] CH=CHCHO	Ion-exchange resin, Amberlite	RCH=C$(COCH_3)_2$ (—)	1720
$C_6H_4(CHO)_2$-1,2	$(C_2H_5)_2NH$	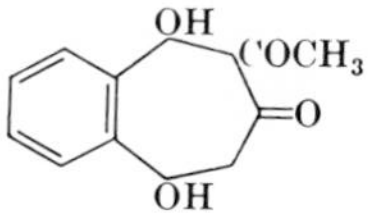 [structure: OH, $COCH_3$, =O, OH] (85)	601
4-$CH_3C_6H_4CHO$	$(CH_2)_5NH$	RCH=C$(COCH_3)_2$ (48)	589
3-$CH_3OC_6H_4CHO$	Ion-exchange resin, Amberlite	RCH(OH)CH$(COCH_3)_2$ (—)	1720
C_6H_5CH=CHCHO	$(CH_2)_5NH$	RCH=C$(COCH_3)_2$ (—)	1723, 1720
3,4-$(CH_3O)_2C_6H_3CHO$	Ion-exchange resin, Amberlite	RCH(OH)CH$(COCH_3)_2$ (—)	1720
	$NaNH_2$	RCH=C$(COCH_3)_2$ (—)	1720
3-HO-5-$C_2H_5OC_6H_3CHO$	$(CH_2)_5NH$	RCH=C$(COCH_3)_2$ (—)	1720
4-i-$C_3H_7C_6H_4CHO$	"	RCH=C$(COCH_3)_2$ (—)	1720
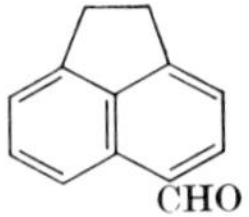 [structure] CHO	"	RCH=C$(COCH_3)_2$ (—)	1720
4-$C_6H_5C_6H_4CH$=CHCHO	"	RCH=C$(COCH_3)_2$ (—)	1720
$CH_3COCO_2C_2H_5$ and			
C_2H_5CHO	$(C_2H_5)_2NH$	$CH_3CH(CHO)C(OH)(CH_3)CO_2C_2H_5$ (45)	215
n-C_3H_7CHO	"	$C_2H_5CH(CHO)C(OH)(CH_3)CO_2C_2H_5$ (45–50)	214, 215
n-C_4H_9CHO	"	n-$C_3H_7CH(CHO)C(OH)(CH_3)CO_2C_2H_5$ (37–48)	214, 215
n-$C_5H_{11}CHO$	"	n-$C_4H_9CH(CHO)C(OH)(CH_3)CO_2C_2H_5$ (37–57)	214, 215
n-$C_6H_{13}CHO$	"	n-$C_5H_{11}CH(CHO)C(OH)(CH_3)CO_2C_2H_5$ (57)	214, 215
n-$C_7H_{15}CHO$	"	n-$C_6H_{13}CH(CHO)C(OH)(CH_3)CO_2C_2H_5$ (56)	214

Note: References 668–2359 are on pp. 403–438.

TABLE XI. Condensation of Aldehydes with Aliphatic Methyl Ketones Other than Acetone and 2-Butanone (*Continued*)

(R in the product is the group R in the aldehyde RCHO.)

Ketone and Aldehyde, RCHO	Catalyst	Product(s) (Yield, %)	Refs.
$CH_3COCH_2CH_2CO_2H$ **and**			
CH_2O	H_2SO_4, CH_3CO_2H	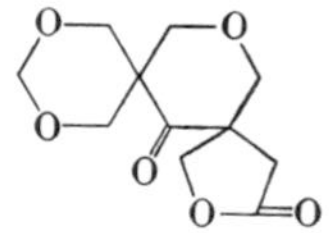 (—)	1724–1727
i-C_3H_7CHO	NaOH	$RCH{=}CHCOCH_2CH_2CO_2H$ (—)	1728
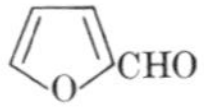 2-Furyl-CHO	"	$RCH{=}CHCOCH_2CH_2CO_2H$ (—)	1729
	CH_3CO_2Na	$RCH{=}C(COCH_3)CH_2CO_2H$ (—)	1730
	HCl	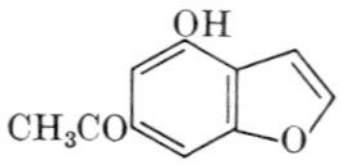 (70)	1731
3,4-$Cl_2C_6H_3CHO$	$(CH_2)_5NH$, CH_3CO_2H	$RCH{=}CHCOCH_2CH_2CO_2H$ (40)	1732
2-ClC_6H_4CHO	"	$RCH{=}CHCOCH_2CH_2CO_2H$ (—)	1732
3-ClC_6H_4CHO	NaOH	$RCH{=}CHCOCH_2CH_2CO_2H$ (—)	1733
4-ClC_6H_4CHO	$(CH_2)_5NH$, CH_3CO_2H	$RCH{=}CHCOCH_2CH_2CO_2H$ (50)	1732
2-$O_2NC_6H_4CHO$	HCl	$RCH{=}C(COCH_3)CH_2CO_2H$ (60)	1731
3-$O_2NC_6H_4CHO$	NaOH	$RCH{=}CHCOCH_2CH_2CO_2H$ (40)	1731
	HCl	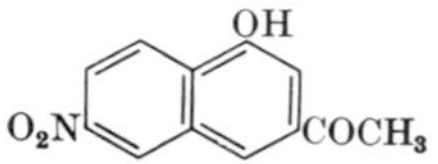 (95)	1731
4-$O_2NC_6H_4CHO$	NaOH	$RCH{=}CHCOCH_2CH_2CO_2H$ (30)	1731
	HCl	$RCH{=}C(COCH_3)CH_2CO_2H$ (50)	1731, 2331

C_6H_5CHO	NaOH or $(CH_2)_5NH$, CH_3CO_2H	$RCH{=}CHCOCH_2CH_2CO_2H$	(50–60)	1731, 1732, 1734, 2331
	CH_3CO_2Na or ion-exchange resin, Amberlite IR-120	$RCH{=}C(COCH_3)CH_2CO_2H$	(12–48)	1735–1737
2-HOC_6H_4CHO	NaOH	$RCH{=}CHCOCH_2CH_2CO_2H$	(55)	1732, 1731
3-HOC_6H_4CHO	$(CH_2)_5NH$, CH_3CO_2H	$RCH{=}CHCOCH_2CH_2CO_2H$	(50)	1732, 1731
	HCl	OH, HO, $COCH_3$	(50)	1731
4-HOC_6H_4CHO	NaOH	$RCH{=}CHCOCH_2CH_2CO_2H$	(25)	1731, 1738
	$(CH_2)_5NH$, CH_3CO_2H	$RCH{=}CHCOCH_2CH_2CO_2H$	(55)	1732
	HCl	$RCH{=}C(COCH_3)CH_2CO_2H$	(60)	1731
2,4-$(HO)_2C_6H_3CHO$	NaOH	$RCH{=}CHCOCH_2CH_2CO_2H$	(40)	1731
	HCl	$RCH{=}C(COCH_3)CH_2CO_2H$	(80)	1731
3,4-$(CH_2O_2)C_6H_3CHO$	NaOH	$RCH{=}CHCOCH_2CH_2CO_2H$	(43)	1732
	HCl	OH, $COCH_3$, O, H_2C–O	(50)	1731
2-$CH_3OC_6H_4CHO$	NaOH	$RCH{=}CHCOCH_2CH_2CO_2H$	(55)	1732
3-$CH_3OC_6H_4CHO$	"	$RCH{=}CHCOCH_2CH_2CO_2H$	(50)	1732
4-$CH_3OC_6H_4CHO$	$(CH_2)_5NH$, CH_3CO_2H	$RCH{=}CHCOCH_2CH_2CO_2H$	(67)	1732, 1739
	HCl	$RCH{=}C(COCH_3)CH_2CO_2H$	(60)	1731
3-CH_3O-4-HOC_6H_3CHO	NaOH	$RCH{=}CHCOCH_2CH_2CO_2H$	(30)	1731, 1732
	HCl	OH, HO, CH_3O, $COCH_3$	(60)	1731
$C_6H_5CH{=}CHCHO$	NaOH	$RCH{=}CHCOCH_2CH_2CO_2H$	(60)	1731
	HCl	$RCH{=}C(COCH_3)CH_2CO_2H$	(80)	1731

Note: References 668–2359 are on pp. 403–438.

TABLE XI. CONDENSATION OF ALDEHYDES WITH ALIPHATIC METHYL KETONES OTHER THAN ACETONE AND 2-BUTANONE (*Continued*)

(R in the product is the group R in the aldehyde RCHO.)

Ketone and Aldehyde, RCHO	Catalyst	Product(s) (Yield, %)	Refs.
$CH_3COCH_2CH_2CO_2H$ (***contd.***) **and**			
2,4-$(CH_3O)_2C_6H_3CHO$	NaOH	$RCH{=}CHCOCH_2CH_2CO_2H$ (63)	1731
3,4-$(CH_3O)_2C_6H_3CHO$	,,	$RCH{=}CHCOCH_2CH_2CO_2H$ (40)	1731
2-$C_2H_5OC_6H_4CHO$	,,	$RCH{=}CHCOCH_2CH_2CO_2H$ (50)	1731
4-$C_2H_5OC_6H_4CHO$	,,	$RCH{=}CHCOCH_2CH_2CO_2H$ (54)	1731
4-$(CH_3)_2NC_6H_4CHO$	,,	$RCH{=}CHCOCH_2CH_2CO_2H$ (66)	1731
$(CH_3)_2C{=}CHCH_2CH_2$-$C(CH_3){=}CHCHO$	HCl	$RCH{=}C(COCH_3)CH_2CO_2H$ (25)	1731
CHO, OH (structure)	,,	$COCH_3$, O, O (structure) (40)	1731
$CH_3COC_3H_7$-*n* **and**			
CH_2O	H_2SO_4; CH_3CO_2H	O, O, O, O, CH_2OCOCH_3, C_2H_5 (structure) (—)	1740
	$KOCH_3$	C_2H_5, O, C_2H_5, CH_3 (structure) (20)	1624
	NaOH	$RCHOHCH(C_2H_5)COCH_3$ (54)	1622, 1624
	NaOH, SiO_2, 260°	$RCH{=}C(C_2H_5)COCH_3$ (23)	125
CCl_3CHO	CH_3CO_2H	$RCHOHCH_2COC_3H_7$-*n* (14)	635
CH_3CHO	$KOCH_3$	$RCHOHCH(C_2H_5)COCH_3$ (55)	270
	—	$RCH{=}CHCOC_3H_7$-*n* (60%), $RCH{=}C(C_2H_5)COCH_3$ (40%) (—)	271

C_2H_5CHO	NaOH	$RCH{=}CHCOC_3H_7$-*n* and/or $RCH{=}C(C_2H_5)COCH_3$ (—)	1436
n-C_3H_7CHO	—	$RCH{=}CHCOC_3H_7$-*n* (—), $RCH{=}C(C_2H_5)COCH_3$ (—)	1444
i-C_3H_7CHO	KOH	$RCHOHCH_2COC_3H_7$-*n* (—)	274
	NaOH	$RCH{=}CHCOC_3H_7$-*n* (30)	1643
I, O, CHO (furan ring)	,,	$RCH{=}CHCOC_3H_7$-*n* and/or $RCH{=}C(C_2H_5)COCH_3$ (—)	1644
O_2N, O, CHO (furan ring)	HCl	$RCH{=}C(C_2H_5)COCH_3$ (—)	1645
O, CHO (furan ring)	NaOH	$RCH{=}CHCOC_3H_7$-*n* (60), $RCH{=}C(C_2H_5)COCH_3$ (—)	1655, 1649
$HOCH_2(CH_2)_3CHO$	,,	O, $CH(C_2H_5)COCH_3$ (six-membered ring) (21)	522
CH_3, O, CHO (furan ring)	,,	$RCH{=}CHCOC_3H_7$-*n* (75)	1741
3-$O_2NC_6H_4CHO$	$POCl_3$	$RCH{=}CHCOC_3H_7$-*n* and/or $RCH{=}C(C_2H_5)COCH_3$ (—)	1654
C_6H_5CHO	NaOH	$RCH{=}CHCOC_3H_7$-*n* (100)	1742, 1743
	HCl	$RCH{=}C(C_2H_5)COCH_3$ (90)	236, 1742
2-HOC_6H_4CHO	NaOH	$RCH{=}CHCOC_3H_7$-*n* (52)	1667
3-HOC_6H_4CHO	HCl	$RCH{=}CHCOC(C_2H_5){=}CHR$ (—)	237
4-HOC_6H_4CHO	,,	$RCH{=}C(C_2H_5)COCH_3$ (76)	237
	NaOH	$RCH{=}CHCOC_3H_7$-*n* (—)	1570
$C_6H_4(CHO)_2$-1,2	,,	O, C_2H_5 (ring structure) (78), O, COC_3H_7-*n* (ring structure) (—)	1672, 223

Note: References 668–2359 are on pp. 403–438.

TABLE XI. CONDENSATION OF ALDEHYDES WITH ALIPHATIC METHYL KETONES OTHER THAN ACETONE AND 2-BUTANONE (*Continued*)

(R in the product is the group R in the aldehyde RCHO.)

Ketone and Aldehyde, RCHO	Catalyst	Product(s) (Yield, %)	Refs.
$CH_3COC_3H_7$-*n* (***contd.***) **and**			
3,4-$(CH_2O_2)C_6H_3CHO$	NaOH	$RCH{=}CHCOC_3H_7$-*n* (53)	1666
3-$CH_3OC_6H_4CHO$	HCl	$RCH{=}CHCOC(C_2H_5){=}CHR$ (—)	237
4-$CH_3OC_6H_4CHO$	"	$RCH{=}C(C_2H_5)COCH_3$ (48)	1666, 237
	NaOH	$RCH{=}CHCOC_3H_7$-*n* (46)	1666
2-HO-3-$CH_3OC_6H_3CHO$	"	$RCH{=}CHCOC_3H_7$-*n* (—)	1570
2-HO-4-$CH_3OC_6H_3CHO$	KOH	$RCH{=}CHCOC_3H_7$-*n* (—)	1570
3-CH_3O-4-HOC_6H_3CHO	NaOH	$RCH{=}CHCOC_3H_7$-*n* (—)	1667, 1593
$C_6H_5CH{=}CHCHO$	"	$RCH{=}CHCOC_3H_7$-*n* (60)	1718
4-*i*-$C_3H_7C_6H_4CHO$	"	$RCH{=}CHCOC_3H_7$-*n* (96)	1679, 1666
	HCl	$RCH{=}C(C_2H_5)COCH_3$ (69)	1679
$(CH_3)_2C{=}CHCH_2CH_2$-$C(CH_3){=}CHCHO$	$NaOC_2H_5$	$RCH{=}CHCOC_3H_7$-*n* and/or $RCH{=}C(C_2H_5)COCH_3$ (Total, 70)	1492
	NaOH	$RCH{=}CHCOC_3H_7$-*n* (28), $RCH{=}C(C_2H_5)COCH_3$ (8)	1687, 1682
1-$C_{10}H_7CHO$	"	$RCH{=}CHCOC_3H_7$-*n* (—)	1744
$CH_3COC_3H_7$-*i* **and**			
CH_2O	K_2CO_3	$CH_3COC(CH_3)_2CH_2OH$ (40), $HOCH_2CH_2COC(CH_3)_2CH_2OH$ (—)	577, 1618
	H_2SO_4	(16)	1740
CCl_3CHO	CH_3CO_2Na, $(CH_3CO)_2O$	$RCHOHCH_2COC_3H_7$-*i* (—)	273
CH_3CHO	KOH	$RCHOHC(CH_3)_2COCH_3$ (70)	269, 272, 577, 615, 1745
C_2H_5CHO	"	$C_2H_5CH{=}C(CH_3)CHO$ (34), $C_2H_5CHOHCH(CH_3)CHO$ (11)	272

n-C_3H_7CHO	"	$RCHOHC(CH_3)_2COCH_3$ and $RCHOHCH_2COC_3H_7$-i (Total 28), $RCH{=}CHCOC_3H_7$-i (12)	272
i-C_3H_7CHO	NaOH	$RCH{=}CHCOC_3H_7$-i (35)	1643
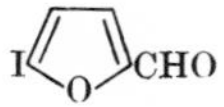 I, O, CHO	"	$RCH{=}CHCOC_3H_7$-i† (—)	1644
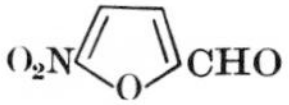 O_2N, O, CHO	KOH	$RCH{=}CHCOC_3H_7$-i (33)	273
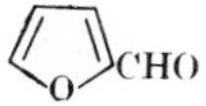 O, CHO	NaOH	$RCH{=}CHCOC_3H_7$-i (60)	1655
$(C_2H_5)_2CHCHO$	KOH	$RCHOHCH_2COC_3H_7$-i (6), $RCH{=}CHCOC_3H_7$-i (47)	272
C_6H_5CHO	NaOH	$RCH{=}CHCOC_3H_7$-i (—)	1663, 1524, 1666, 1743, 1746
2-HOC_6H_4CHO	"	$RCH{=}CHCOC_3H_7$-i (—)	1570
4-HOC_6H_4CHO	"	$RCH{=}CHCOC_3H_7$-i (57)	1747
$C_6H_4(CHO)_2$-1,2	$KOCH_3$	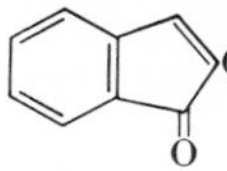 COC_3H_7-i, O (—)	1672
3,4-$(CH_2O_2)C_6H_3CHO$	NaOH	$RCH{=}CHCOC_3H_7$-i (—)	1666
2-$CH_3OC_6H_4CHO$	"	$RCH{=}CHCOC_3H_7$-i (65)	1666
4-$CH_3OC_6H_4CHO$	"	$RCH{=}CHCOC_3H_7$-i (72)	1548
$(CH_3)_2C{=}CHCH_2CH_2$-$C(CH_3){=}CHCHO$	"	$RCH{=}CHCOC_3H_7$-i (—)	1683
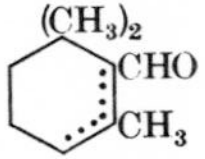 $(CH_3)_2$, CHO, CH_3	$NaOC_2H_5$	$RCH{=}CHCOC_3H_7$-i (—)	1689
$C_{20}H_{24}O_2$, crocetin dialdehyde	KOH	$R(CH{=}CHCOC_3H_7$-$i)_2$ (75)	607

Note: References 668–2359 are on pp. 403–438.

† The structure of this compound was not established.

TABLE XI. Condensation of Aldehydes with Aliphatic Methyl Ketones Other than Acetone and 2-Butanone (*Continued*)
(R in the product is the group R in the aldehyde RCHO.)

Ketone and Aldehyde, RCHO	Catalyst	Product(s) (Yield, %)	Refs.
$CH_3COC(OH)(CH_3)_2$ and			
[2-furyl]CHO	NaOH	$RCH{=}CHCOC(OH)(CH_3)_2$ (91)	1748
C_6H_5CHO	$NaOCH_3$	$RCH{=}CHCOC(OH)(CH_3)_2$ (89)	1749, 1167, 1748
$2\text{-}H_2NC_6H_4CHO$	NaOH	[2-quinolyl]$C(OH)(CH_3)_2$ (52)	1748
$3,4\text{-}(CH_2O_2)C_6H_3CHO$	"	$RCH{=}CHCOC(OH)(CH_3)_2$ (89)	1748
$C_6H_5CH{=}CHCHO$	"	$RCH{=}CHCOC(OH)(CH_3)_2$ (95)	1748
$CH_3CO(CH_2)_2CH{=}CH_2$ and $(CH_3)_2C{=}CHCH_2CH_2\text{-}C(CH_3){=}CHCHO$	$NaOCH_3$	$RCH{=}CHCO(CH_2)_2CH{=}CH_2$ (64)	1750, 1751
$CH_3COCH{=}C(CH_3)_2$ and			
CH_2O	MgO	$HOCH_2CH_2COCH{=}C(CH_3)_2$ (25), $CH_3COCH(CH_2OH)C(CH_3){=}CH_2$ (20)	1752, 1753
CCl_3CHO	CH_3CO_2H	$RCHOHCH_2COCH{=}C(CH_3)_2$ (10)	635
$CH_2{=}CHCHO$	Mg, 200°	$RCHOHCH_2COCH{=}C(CH_3)_2$ (5)	1754
C_2H_5CHO	Mg, 300°	$RCH{=}CHCOCH{=}C(CH_3)_2$ (2)	1754
$CH_3CH{=}CHCHO$	BaO	$RCH{=}CHCOCH{=}C(CH_3)_2$ (5), [6-methyl-1,3-cyclohexadiene-1-carboxaldehyde] (CHO, CH_3) (12)	559, 1755
$n\text{-}C_3H_7CHO$	KOH	$RCH{=}CHCOCH{=}C(CH_3)_2$ (22), $n\text{-}C_3H_7CH{=}C(C_2H_5)CHO$ (14)	513, 559, 1755
[2-furyl]CHO	NaOH	$RCH{=}CHCOCH{=}C(CH_3)_2$ (17)	1530, 1529

C_6H_5CHO	$(CH_2)_5NH$, CH_3CO_2H	$RCH{=}CHCOCH{=}C(CH_3)_2$ (—)	1560, 4
2-HOC_6H_4CHO	NaOH	$RCH{=}CHCOCH{=}C(CH_3)_2$ (Small)	1570
$(CH_3)_2C{=}CHCH_2CH_2$-$C(CH_3){=}CHCHO$	"	$RCH{=}CHCOCH{=}C(CH_3)_2$ (—)	1756
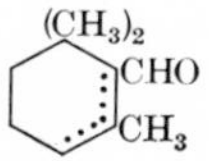 $(CH_3)_2$ CHO CH_3	$NaOC_2H_5$	$RCH{=}CHCOCH{=}C(CH_3)_2$ (—)	1689
$CH_3CO(CH_2)_3CO_2H$ and 4-HOC_6H_4CHO	NaOH	$RCH{=}CHCO(CH_2)_3CO_2H$ (40)	1738
$CH_3COC_4H_9$-*n* and			
CH_2O	—	$RCH{=}C(C_3H_7$-*n*$)COCH_3$ (—)	125
Cl_3CCHO	CH_3CO_2H	$RCHOHCH_2COC_4H_9$-*n* (15)	635
CH_3CHO	*i*-C_3H_7MgCl	$RCHOHCH_2COC_4H_9$-*n* (25)	1757
i-C_3H_7CHO	KOH	$RCHOHCH_2COC_4H_9$-*n* (—)	274
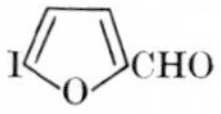 I O CHO	NaOH	$RCH{=}CHCOC_4H_9$-*n*† (—)	1644
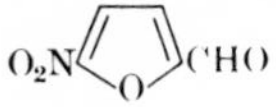 O_2N O CHO	HCl	$RCH{=}C(C_3H_7$-*n*$)COCH_3$ (—)	1645
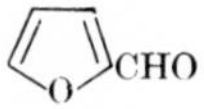 O CHO	NaOH	$RCH{=}CHCOC_4H_9$-*n* (—)	1649
C_6H_5CHO	"	$RCH{=}CHCOC_4H_9$-*n* (—)	1663, 1524
	HCl	$RCH{=}C(C_3H_7$-*n*$)COCH_3$ (90)	236, 1663
3-HOC_6H_4CHO	"	$RCH{=}C(C_3H_7$-*n*$)COCH_3$ (—)	237
4-HOC_6H_4CHO	"	$RCH{=}C(C_3H_7$-*n*$)COCH_3$ (—)	237

Note: References 668–2359 are on pp. 403–438.

† The structure of this compound was not established.

TABLE XI. CONDENSATION OF ALDEHYDES WITH ALIPHATIC METHYL KETONES OTHER THAN ACETONE AND 2-BUTANONE (*Continued*)

(R in the product is the group R in the aldehyde RCHO.)

Ketone and Aldehyde, RCHO	Catalyst	Product(s) (Yield, %)	Refs.
$CH_3COC_4H_9$-*n* (***contd.***) **and**			
$C_6H_4(CHO)_2$-1,2	NaOH	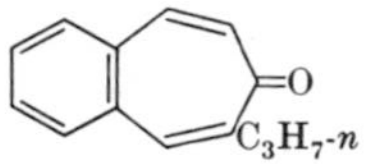 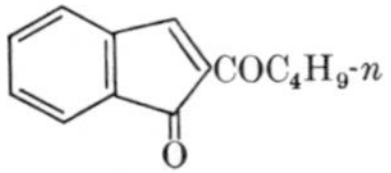 =O, C_3H_7-*n* (60),	1672, 223
		COC_4H_9-*n*, O (30)	
3-$CH_3OC_6H_4CHO$	HCl	$RCH{=}C(C_3H_7$-*n*$)COCH{=}CHR$ (—)	237
4-$CH_3OC_6H_4CHO$	”	$RCH{=}C(C_3H_7$-*n*$)COCH_3$ (—)	237
3-CH_3O-4-HOC_6H_3CHO	NaOH	$RCH{=}CHCOC_4H_9$-*n* (—)	1593
$CH_3COC_4H_9$-*i* **and**			
CH_2O	”	$CH_3COCH(CH_2OH)C_3H_7$-*i* (54)	1622, 1740
Cl_3CCHO	CH_3CO_2H	$RCHOHCH_2COC_4H_9$-*i* (32)	635, 1287
CH_3CHO	KOH	$RCHOHCH_2COC_4H_9$-*i* (—)	1758, 1745
	i-C_3H_7MgCl	$RCHOHCH_2COC_4H_9$-*i* (38)	141
C_2H_5CHO	KOH	$RCHOHCH_2COC_4H_9$-*i* (24)	579
n-C_3H_7CHO	”	$RCHOHCH_2COC_4H_9$-*i* (38)	579, 1652
i-C_3H_7CHO	”	$RCHOHCH_2COC_4H_9$-*i* (—)	274
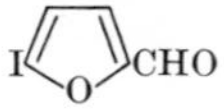 I, O, CHO	NaOH	$RCH{=}CHCOC_4H_9$-*i*† (—)	1644
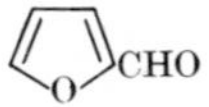 O, CHO	”	$RCH{=}CHCOC_4H_9$-*i* (80)	1530, 590, 1648, 1759, 1760
$HOCH_2(CH_2)_3CHO$	”	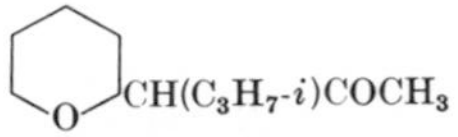 O, $CH(C_3H_7$-*i*$)COCH_3$ (3)	522
$(C_2H_5)_2CHCHO$	”	$RCH{=}CHCOC_4H_9$-*i* (—)	1470
C_6H_5CHO	”	$RCH{=}CHCOC_4H_9$-*i* (64)	243, 238, 239, 1743

Reactant	Catalyst	Product (Yield, %)	References
2-HOC_6H_4CHO	HCl	$RCH{=}CHCOC_4H_9$-*i* (38)	238, 239
	NaOH	$RCH{=}CHCOC_4H_9$-*i* (—)	239
	HCl	$RCH{=}CHCOC_4H_9$-*i* (—)	239
$C_6H_4(CHO)_2$-1,2	$KOCH_3$	2-(COC_4H_9-*i*)-inden-1-one (55)	1672
3,4-$(CH_2O_2)C_6H_3CHO$	NaOH	$RCH{=}CHCOC_4H_9$-*i* (—)	1743
4-$CH_3OC_6H_4CHO$	"	$RCH{=}CHCOC_4H_9$-*i* (89)	1762, 240
	HCl	$RCH{=}CHCOC_4H_9$-*i* (—)	240
3-CH_3O-4-HOC_6H_3CHO	NaOH	$RCH{=}CHCOC_4H_9$-*i* (—)	1763
n-$C_3H_7CH{=}C(C_2H_5)CHO$	(—)	$RCHOHCH_2COC_4H_9$-*i* (—)	1764
n-$C_4H_9CH(C_2H_5)CHO$	(—)	$RCHOHCH_2COC_4H_9$-*i* (—)	1764
$C_6H_5CH{=}CHCHO$	$NaOC_2H_5$	$RCH{=}CHCOC_4H_9$-*i* (—)	1765
2-Furyl-$(CH{=}CH)_2CHO$	NaOH	$RCH{=}CHCOC_4H_9$-*i* (71)	1766
4-*i*-$C_3H_7C_6H_4CHO$	"	$RCH{=}CHCOC_4H_9$-*i* (50)	1666, 1743
1-$C_{10}H_7CHO$	"	$RCH{=}CHCOC_4H_9$-*i* (—)	1744
CH_3-cyclohexenyl-$CH(CH_3)CH_2CHO$	KOH	$RCH{=}CHCOC_4H_9$-*i* (—)	144
CH_3-cyclohexyl-$CH(CH_3)CH_2CHO$	"	$RCH{=}CHCOC_4H_9$-*i* (—)	144
$CH_3COC_4H_9$-*t* and CH_2O	NaOH	$HOCH_2CH_2COC_4H_9$-*t* (—), $(HOCH_2)_2CHCOC_4H_9$-*t* (10)	577, 2339
CCl_3CHO	CH_3CO_2Na, $(CH_3CO)_2O$	$RCHOHCH_2COC_4H_9$-*t* (—)	273
CH_3CHO	*i*-C_3H_7MgCl	$RCHOHCH_2COC_4H_9$-*t* (—)	1757, 513
5-O_2N-2-furyl-CHO	KOH	$RCH{=}CHCOC_4H_9$-*t* (45)	273

Note: References 668–2359 are on pp. 403–438.

† The structure of this compound was not established.

TABLE XI. CONDENSATION OF ALDEHYDES WITH ALIPHATIC METHYL KETONES OTHER THAN ACETONE AND 2-BUTANONE (*Continued*)
(R in the product is the group R in the aldehyde RCHO.)

Ketone and Aldehyde, RCHO	Catalyst	Product(s) (Yield, %)	Refs.
$CH_3COC_4H_9$-*t* (***contd.***) **and**			
2-Furaldehyde (furan-CHO)	$NaOCH_3$	$RCH{=}CHCOC_4H_9$-*t* (68)	1532, 1526, 1655
2-Pyridinecarboxaldehyde (pyridine-CHO)	NaOH	$RCH{=}CHCOC_4H_9$-*t* (43)	1542
C_6H_5CHO	"	$RCH{=}CHCOC_4H_9$-*t* (93)	1768, 524, 1524, 1663
$2\text{-}HOC_6H_4CHO$	"	$RCH{=}CHCOC_4H_9$-*t* (—)	1570
$3,4\text{-}(CH_2O_2)C_6H_3CHO$	$NaOC_2H_5$	$RCH{=}CHCOC_4H_9$-*t* (95)	513, 1769
$4\text{-}CH_3OC_6H_4CHO$	NaOH	$RCH{=}CHCOC_4H_9$-*t* (65)	1771
$3\text{-}CH_3O\text{-}4\text{-}HOC_6H_3CHO$	KOH	$RCH{=}CHCOC_4H_9$-*t* (—)	1579, 1593
$C_6H_5CH{=}CHCHO$	$NaOC_2H_5$	$RCH{=}CHCOC_4H_9$-*t* (38)	513
2-Furyl-$(CH{=}CH)_2CHO$	NaOH	$RCH{=}CHCOC_4H_9$-*t* (51)	1766
$3,5\text{-}(CH_3)_2\text{-}4\text{-}HOC_6H_2CHO$	"	$RCH{=}CHCOC_4H_9$-*t* (43)	2337
$3,4\text{-}(CH_2O_2)C_6H_3CH{=}CHCHO$	$NaOC_2H_5$	$RCH{=}CHCOC_4H_9$-*t* (73)	513
2,7-Dimethyl-2,6-octadien-4-yne-1,8-dial	$Al(OC_4H_9\text{-}t)_3$	$R(CH{=}CHCOC_4H_9\text{-}t)_2$ (30)	1772
2,7-Dimethyl-2,4,6-octatriene-1,8-dial	"	$R(CH{=}CHCOC_4H_9\text{-}t)_2$ (35)	1772
$(CH_3)_2C{=}CHCH_2CH_2C(CH_3){=}CHCHO$, citral	NaOH	$RCH{=}CHCOC_4H_9$-*t* (68)	1773
2,9-Dimethyl-2,8-decadiene-4,6-diyne-1,10-dial	$Al(OC_4H_9\text{-}t)_3$	$R(CH{=}CHCOC_4H_9\text{-}t)_2$ (46)	607
2,9-Dimethyl-2,4,6,8-decatetraene-1,10-dial	"	$R(CH{=}CHCOC_4H_9\text{-}t)_2$ (79)	607
Pyrene-3-carboxaldehyde	$NaOC_2H_5$	$RCH{=}CHCOC_4H_9$-*t* (80)	1615
$C_{20}H_{42}O_2$, crocetin dialdehyde	KOH	$R(CH{=}CHCOC_4H_9\text{-}t)_2$ (49)	607
$CH_3COCH{=}CHC_3H_7$-*n* **and** $n\text{-}C_4H_9CH(C_2H_5)CHO$	NaOH	$RCHOHCH_2COCH{=}CHC_3H_7$-*n* (—)	1652

$CH_3COCOC_4H_9$-*t* **and** C_6H_5CHO	$(CH_2)_5NH$	$RCH{=}CHCOCOC_4H_9$-*t* (34)	325
$CH_3CO(CH_2)_4CO_2H$ **and** 4-HOC_6H_4CHO	NaOH	$RCH{=}CHCO(CH_2)_4CO_2H$ (24)	1738
$CH_3COC_5H_{11}$-*n* **and**			
CH_2O	K_2CO_3	$CH_3COCH(CH_2OH)C_4H_9$-*n*† (42)	577
CCl_3CHO	CH_3CO_2H	$RCHOHCH_2COC_5H_{11}$-*n* (20)	635
n-C_3H_7CHO	NaOH	$RCHOHCH_2COC_5H_{11}$-*n*† (—)	1652
i-C_3H_7CHO	KOH	$RCHOHCH_2COC_5H_{11}$-*n* (—), $RCHOHCH(C_4H_9$-*n*$)COCH_3$ (Trace)	274
I–(furan ring)–CHO	NaOH	$RCH{=}CHCOC_5H_{11}$-*n*† (—)	1644
O_2N–(furan ring)–CHO	HCl	$RCH{=}C(C_4H_9$-*n*$)COCH_3$ (—)	1645
(furan ring)–CHO	NaOH	$RCH{=}CHCOC_5H_{11}$-*n* (—)	1649
$(C_2H_5)_2CHCHO$	"	$RCH{=}CHCOC_5H_{11}$-*n*† (—)	1470
C_6H_5CHO	"	$RCH{=}CHCOC_5H_{11}$-*n* (45)	243
	HCl	$RCH{=}C(C_4H_9$-*n*$)COCH_3$ (23), $RCH{=}CHCOC_5H_{11}$-*n* (2)	243, 236
3,4-$(CH_2O_2)C_6H_3CHO$	NaOH	$RCH{=}CHCOC_5H_{11}$-*n*† (—)	1761
3-CH_3O-4-HOC_6H_3CHO	"	$RCH{=}CHCOC_5H_{11}$-*n* (40)	1774
n-$C_4H_9CH(C_2H_5)CHO$	"	$RCHOHCH_2COC_5H_{11}$-*n*† (—)	1652
$CH_3COCH_2C_4H_9$-*i* **and** CCl_3CHO	CH_3CO_2H	$RCHOHCH_2COCH_2C_4H_9$-*i* (15)	635
$CH_3COCH_2C_4H_9$-*t* **and** CH_3CHO	KOH, CH_3OH	$RCHOHCH_2COCH_2C_4H_9$-*t* (40), $RCH{=}CHCOCH_2C_4H_9$-*t* (10), $CH_3CH(OCH_3)CH_2COCH_2C_4H_9$-*t* (10)	1775, 1757, 1758
	KOH, C_2H_5OH	$RCH{=}CHCOCH_2C_4H_9$-*t* (35), $CH_3CH(OC_2H_5)CH_2COCH_2C_4H_9$-*t* (30)	1775–1777
$CH_3COCH(CH_3)CH(OCH_3)_2$ **and** 9-(2,6,6-Trimethylcyclohexen-1-yl)-3,7-dimethyl-2,4,6,8-nonatetraen-1-al	$NaOCH_3$	$RCH{=}CHCOCH(CH_3)CH(OCH_3)_2$ (65)	1778

Note: References 668–2359 are on pp. 403–438.

† The structure of this compound was not established.

TABLE XI. CONDENSATION OF ALDEHYDES WITH ALIPHATIC METHYL KETONES OTHER THAN ACETONE AND 2-BUTANONE (*Continued*)

(R in the product is the group R in the aldehyde RCHO.)

Ketone and Aldehyde, RCHO	Catalyst	Product(s) (Yield, %)	Refs.
$CH_3COCH(CH_3)CH(OCH_3)_2$ (***contd.***) **and**			
13-(2,6,6-Trimethylcyclohexen-1-yl)-2,7,11-trimethyl-2,4,6-8,10,12-tridecahexaen-1-al	$NaOCH_3$	$RCH{=}CHCOCH(CH_3)CH(OCH_3)_2$ (—)	1778
$O_2N(C_4H_2O)CH{=}CHCOCH_3$ **and**			
$4\text{-}CH_3OC_6H_4CHO$	H_2SO_4, CH_3CO_2H	$2\text{-}RCH{=}CHCOCH{=}CH(C_4H_2O)NO_2\text{-}5$ (35)	646
$2,4\text{-}(CH_3O)_2C_6H_3CHO$	"	$2\text{-}RCH{=}CHCOCH{=}CH(C_4H_2O)NO_2\text{-}5$ (40)	646
$(C_4H_3S)CH{=}CHCOCH_3$ **and**			
$(C_4H_3S)CHO$	NaOH	$2\text{-}RCH{=}CHCOCH{=}CH(C_4H_3S)$ (52)	1030
$(C_4H_3O)CHO$	"	$2\text{-}RCH{=}CHCOCH{=}CH(C_4H_3S)$ (87)	1779
$4\text{-}O_2NC_6H_4CHO$	"	$2\text{-}RCH{=}CHCOCH{=}CH(C_4H_3S)$ (52)	647
$(C_4H_3O)CH{=}CHCHO$	"	$2\text{-}RCH{=}CHCOCH{=}CH(C_4H_3S)$ (79)	1779
$2,4\text{-}(CH_3O)_2C_6H_3CHO$	"	$2\text{-}RCH{=}CHCOCH{=}CH(C_4H_3S)$ (69)	1008
$2,4,6\text{-}(CH_3O)_3C_6H_2CHO$	"	$2\text{-}RCH{=}CHCOCH{=}CH(C_4H_3S)$ (85)	1008
$(C_4H_3Se)CH{=}CHCOCH_3$ **and**			
$(C_4H_3Se)CHO$	"	$2\text{-}RCH{=}CHCOCH{=}CH(C_4H_3Se)$ (52)	1522

2-C_4H_3OCH=CHCOCH$_3$ **and**				
5-Br-2-C_4H_2OCHO	KOH	2-RCH=CHCOCH=CH(C_4H_3O)	(88)	1516
5-Cl-2-C_4H_2OCHO	,,	2-RCH=CHCOCH=CH(C_4H_3O)	(91)	1516
5-I-2-C_4H_2OCHO	,,	2-RCH=CHCOCH=CH(C_4H_3O)	(84)	1516
2-C_4H_3SCHO	NaOH	2-RCH=CHCOCH=CH(C_4H_3O)	(87)	1779
2-C_4H_3SeCHO	,,	2-RCH=CHCOCH=CH(C_4H_3O)	(64)	1522
2-C_4H_3OCHO	,,	2-RCH=CHCOCH=CH(C_4H_3O)	(95)	1780
2-ClC$_6$H$_4$CHO	,,	2-RCH=CHCOCH=CH(C_4H_3O)	(—)	1781
2-O$_2$NC$_6$H$_4$CHO	,,	2-RCH=CHCOCH=CH(C_4H_3O)	(29)	1781
3-O$_2$NC$_6$H$_4$CHO	,,	2-RCH=CHCOCH=CH(C_4H_3O)	(33)	1781
4-O$_2$NC$_6$H$_4$CHO	H_2SO_4, CH_3CO_2H	2-RCH=CHCOCH=CH(C_4H_3O)	(86)	646, 1781
C$_6$H$_5$CHO	NaOH	2-RCH=CHCOCH=CH(C_4H_3O)	(—)	1525
2-C_4H_3SCH=CHCHO	,,	2-RCH=CHCOCH=CH(C_4H_3O)	(49)	1779
3,4-(CH$_2$O$_2$)C$_6$H$_3$CHO	,,	2-RCH=CHCOCH=CH(C_4H_3O)	(—)	1780
4-CH$_3$C$_6$H$_4$CHO	,,	2-RCH=CHCOCH=CH(C_4H_3O)	(50)	1781
4-CH$_3$OC$_6$H$_4$CHO	,,	2-RCH=CHCOCH=CH(C_4H_3O)	(—)	1780
4-(CH$_3$)$_2$NC$_6$H$_4$CHO	,,	2-RCH=CHCOCH=CH(C_4H_3O)	(33)	1780

Note: References 668–2359 are on pp. 403–438.

TABLE XI. Condensation of Aldehydes with Aliphatic Methyl Ketones Other than Acetone and 2-Butanone (*Continued*)

(R in the product is the group R in the aldehyde RCHO.)

Ketone and Aldehyde, RCHO	Catalyst	Product(s) (Yield, %)	Refs.
2-$C_4H_3O(CH_2)_2COCH_3$ **and**			
2-C_4H_3OCHO	NaOH	2-RCH=CHCO$(CH_2)_2(C_4H_3O)$ (54)	1533
CH_3, O, $COCH_3$, =O (pyranone ring) **and**			
2-C_4H_3SCHO	$(CH_2)_5NH$	RCH=CHCO, O, =O, CH_3 (pyranone ring) (—)	595
		A, R = 2-C_4H_3S	
2-C_4H_3OCHO	"	A, R = 2-C_4H_3O (85)	595
3,4-$Cl_2C_6H_3CHO$	"	A, R = 3,4-$Cl_2C_6H_3$ (46)	595
2-$O_2NC_6H_4CHO$	"	A, R = 2-$O_2NC_6H_4$ (65)	595
3-$O_2NC_6H_4CHO$	"	A, R = 3-$O_2NC_6H_4$ (60)	595
C_6H_5CHO	"	A, R = C_6H_5 (55)	595, 1782
$C_6H_5CH{=}CHCHO$	"	A, R = $C_6H_5CH{=}CH$ (57)	595
2,3-$(CH_3O)_2C_6H_3CHO$	"	A, R = 2,3-$(CH_3O)_2C_6H_3$ (47)	595
3,4-$(CH_3O)_2C_6H_3CHO$	"	A, R = 3,4$(CH_3O)_2C_6H_3$ (46)	595
4-$(CH_3)_2NC_6H_4CHO$	"	A, R = 4-$(CH_3)_2NC_6H_4$ (63)	595
4-*i*-$C_3H_7C_6H_4CHO$	"	A, R = *i*-$C_3H_7C_6H_4$ (65)	595
1-$C_{10}H_7CHO$	"	A, R = 1-$C_{10}H_7$ (62)	595
3,4-$(C_2H_5O)_2C_6H_3CHO$	"	A, R = $(C_2H_5O)_2C_6H_3$ (43)	595
4-$(C_2H_5)_2NC_6H_4CHO$	"	A, R = $(C_2H_5)_2NC_6H_4$ (58)	595

(1-cyclohexenyl)$COCH_3$ **and** C_6H_5CHO	$NaOC_2H_5$	$RCH{=}CHCO$(1-cyclohexenyl) (40)	316
$CH_3COC(CH_3){=}CHC_3H_7\text{-}n$ **and** $n\text{-}C_3H_7CH{=}C(C_2H_5)CHO$	NaOH	$RCHOHCH_2COC(CH_3){=}CHC_3H_7\text{-}n$ (—)	1652
$CH_3COCH_2CH_2CH{=}C(CH_3)_2$ **and** CH_2O	$NaNH_2$	$CH_2{=}C(COCH_3)CH_2CH{=}C(CH_3)_2$ (15)	1783
	CH_3CO_2Na	$HOCH_2CH(COCH_3)CH_2CH{=}C(CH_3)_2$ (10)	1783
(cyclohexyl)$COCH_3$ **and** $i\text{-}C_4H_9CHO$	NaOH	$RCH{=}CHCO$(cyclohexyl) (13)	1459
$CH_3CO(CH_2)_5CO_2H$ **and** $4\text{-}HOC_6H_4CHO$	”	$RCH{=}CHCO(CH_2)_5CO_2H$ (42)	1738
$CH_3COC_6H_{13}\text{-}n$ **and** CH_2O	K_2CO_3	$CH_3COCH(CH_2OH)C_5H_{11}\text{-}n$† (46)	577
	$KOCH_3$	2-C_5H_{11}-n-4-n-C_5H_{11}-3-CH_3-2-cyclohexen-1-one (ring structure: $C_5H_{11}\text{-}n$, =O, $n\text{-}C_5H_{11}$, CH_3) (32)	1624
CCl_3CHO	CH_3CO_2H	$RCHOHCH_2COC_6H_{13}\text{-}n$ (15)	635
$i\text{-}C_3H_7CHO$	KOH	$RCHOHCH_2COC_6H_{13}\text{-}n$ (—)	274
I-(2-furyl)-CHO (5-iodofurfural)	NaOH	$RCH{=}CHCOC_6H_{13}\text{-}n$† (—)	1644
O_2N-(2-furyl)-CHO (5-nitrofurfural)	HCl	$RCH{=}C(C_5H_{11}\text{-}n)COCH_3$ (—)	1645
C_6H_5CHO	KOH	$RCH{=}CHCOC_6H_{13}\text{-}n$ (—)	1743
	HCl	$RCH{=}C(C_5H_{11}\text{-}n)COCH_3$ (90)	236
$2\text{-}HOC_6H_4CHO$	NaOH	$RCH{=}CHCOC_6H_{13}\text{-}n$ (—)	1667

Note: References 668–2359 are on pp. 403–438.

† The structure of this compound was not established.

TABLE XI. CONDENSATION OF ALDEHYDES WITH ALIPHATIC METHYL KETONES OTHER THAN ACETONE AND 2-BUTANONE (*Continued*)

(R in the product is the group R in the aldehyde RCHO.)

Ketone and Aldehyde, RCHO	Catalyst	Product(s) (Yield, %)	Refs.
$CH_3COC_6H_{13}$-*n* (***contd.***) **and**			
CH=CHCHO	NaOH	RCH=CHCOC$_6$H$_{13}$-*n* (44)	1766
3,4-(CH$_2$O$_2$)C$_6$H$_3$CHO	"	RCH=CHCOC$_6$H$_{13}$-*n*† (87)	1761
4-CH$_3$OC$_6$H$_4$CHO	"	RCH=CHCOC$_6$H$_{13}$-*n* (—)	1743
C$_6$H$_5$CH=CHCHO	NaOC$_2$H$_5$	RCH=CHCOC$_6$H$_{13}$-*n* (—)	1784
(CH=CH)$_2$CHO	NaOH	RCH=CHCOC$_6$H$_{13}$-*n* (92)	1766
CH$_3$ CHO	NaOC$_2$H$_5$	RCH=CHCOC$_6$H$_{13}$-*n*† (35)	1487
CH$_3$COCH(CH$_3$)C$_4$H$_9$-*n* **and**			
n-C$_3$H$_7$CH=C(C$_2$H$_5$)CHO	NaOH	RCHOHCH$_2$COCH(CH$_3$)C$_4$H$_9$-*n* (—)	1652
CH$_3$CO(CH$_2$)$_2$CH(CH$_3$)C$_2$H$_5$-(*d,l*) **and**			
C$_6$H$_5$CHO	"	RCH=CHCO(CH$_2$)$_2$CH(CH$_3$)C$_2$H$_5$ (—)	1785
4-CH$_3$OC$_6$H$_4$CHO	"	RCH=CHCO(CH$_2$)$_2$CH(CH$_3$)C$_2$H$_5$ (—)	1785
CH$_3$CO(CH$_2$)$_2$CH(CH$_3$)C$_2$H$_5$-(+) **and**			
C$_6$H$_5$CHO	"	RCH=CHCO(CH$_2$)$_2$CH(CH$_3$)C$_2$H$_5$ (—)	1785
4-CH$_3$OC$_6$H$_4$CHO	"	RCH=CHCO(CH$_2$)$_2$CH(CH$_3$)C$_2$H$_5$ (—)	1785
CH$_3$CO(CH$_2$)$_3$C$_3$H$_7$-*i* **and**			
CHO	"	RCH=CHCO(CH$_2$)$_3$C$_3$H$_7$-*i* (—)	1649
(CH$_3$)$_2$ CHO CH$_3$	NaOC$_2$H$_5$	RCH=CHCO(CH$_2$)$_3$C$_3$H$_7$-*i*† (—)	1689

$C_6H_5CH_2COCH_3$ and			
[2-furyl]CHO	NaOH	$RCH{=}CHCOCH_2C_6H_5$ (73)	590, 1759
[2-pyridyl]CHO	HBr, CH_3CO_2H	$RCH{=}C(C_6H_5)COCH_3$ (72)	1786
[3-pyridyl]CHO	"	$RCH{=}C(C_6H_5)COCH_3$ (72)	1786
[4-pyridyl]CHO	"	$RCH{=}C(C_6H_5)COCH_3$ (70)	1786
2-$O_2NC_6H_4CHO$	$(CH_2)_5NH$, n-$C_6H_{13}CO_2H$	$RCH{=}C(C_6H_5)COCH_3$ (22)	84
3-$O_2NC_6H_4CHO$	"	$RCH{=}C(C_6H_5)COCH_3$ (57)	84
C_6H_5CHO	KOH	$RCH{=}CHCOCH_2C_6H_5$ (48)	264, 2327
	$(CH_2)_5NH$	$RCH{=}C(C_6H_5)COCH_3$ (80)	265, 84, 597
	HCl	$RCH{=}C(C_6H_5)COCH_3$ (—)	264, 1788
	H_2SO_4	*trans*-$C_6H_5CH{=}CHC_6H_5$ (—)	264, 1788
2-HOC_6H_4CHO	$(CH_2)_5NH$	$RCH{=}CHCOCH_2C_6H_5$ (50)	266, 239
$C_6H_4(CHO)_2$-1,2	"	[benzo-fused seven-membered ring: OH, C_6H_5, =O, OH] (73)	601
3,4-$(CH_2O_2)C_6H_3CHO$	KOH	$RCH{=}CHCOCH_2C_6H_5$ (64)	1789
4-$CH_3C_6H_4CHO$	"	$RCH{=}CHCOCH_2C_6H_5$ (—)	1789
2-$CH_3OC_6H_4CHO$	HCl	$RCH{=}C(C_6H_5)COCH_3$ (37), $RCH{=}C(C_6H_5)COCH{=}CHR$ (—)	266, 239
4-$CH_3OC_6H_4CHO$	KOH	$RCH{=}CHCOCH_2C_6H_5$ (—)	1789
	HCl	$RCH{=}C(C_6H_5)COCH_3$ (27)	1790
$C_6H_5CH{=}CHCHO$	NaOH	$RCH{=}CHCOCH_2C_6H_5$ (—)	1791

Note: References 668–2359 are on pp. 403–438.

† The structure of this compound was not established.

TABLE XI. CONDENSATION OF ALDEHYDES WITH ALIPHATIC METHYL KETONES OTHER THAN ACETONE AND 2-BUTANONE (*Continued*)

(R in the product is the group R in the aldehyde RCHO.)

Ketone and Aldehyde, RCHO	Catalyst	Product(s) (Yield, %)	Refs.
$C_6H_5OCH_2COCH_3$ **and**			
C_6H_5CHO	NaOH, H_2O, 25°	$RCH{=}C(OC_6H_5)COCH_3$ (—)	268
	NaOH, C_2H_5OH, 25°	$RCH{=}C(OC_6H_5)COCH{=}CHR$ (—)	268
	HCl, $(C_2H_5)_2O$, 25°, 48 hr.	$RCH{=}C(OC_6H_5)COCH_3$ (—)	268
2-HOC_6H_4CHO	NaOH	$RCH{=}C(OC_6H_5)COCH_3$ (40)	268
4-$CH_3OC_6H_4CHO$	"	$RCH{=}C(OC_6H_5)COCH_3$ (—)	268
$CH_3COC_7H_{15}$-*n* **and**			
Cl_3CCHO	CH_3CO_2H	$RCHOHCH_2COC_7H_{15}$-*n* (20)	635
5-nitro-2-furaldehyde (O_2N–furan–CHO)	HCl	$RCH{=}C(COCH_3)C_6H_{13}$-*n* (—)	1645
C_6H_5CHO	KOH	$RCH{=}CHCOC_7H_{15}$-*n* (70)	1743
2-HO-5-$BrC_6H_3CH{=}CHCOCH_3$ **and** 4-$(CH_3)_2NC_6H_4CHO$	NaOH	2-HO-5-$BrC_6H_3CH{=}CHCOCH{=}CHR$ (80)	1547
4-$O_2NC_6H_4CH{=}CHCOCH_3$ **and**			
5-nitro-2-thiophenecarboxaldehyde (O_2N–thiophene–CHO)	H_2SO_4, CH_3CO_2H	4-$O_2NC_6H_4CH{=}CHCOCH{=}CHR$ (40)	647
5-nitro-2-furaldehyde (O_2N–furan–CHO)	"	4-$O_2NC_6H_4CH{=}CHCOCH{=}CHR$ (30)	646
2-HO-5-$O_2NC_6H_3CH{=}CHCOCH_3$ **and** 4-$(CH_3)_2NC_6H_4CHO$	NaOH	2-HO-5-$O_2NC_6H_3CH{=}CHCOCH{=}CHR$ (—)	1547
$C_6H_5CH{=}CHCOCH_3$ **and**			
CCl_3CHO	CH_3CO_2K, $(CH_3CO)_2O$	$RCHOHCH_2COCH{=}CHC_6H_5$ (65)	287
5-nitro-2-thiophenecarboxaldehyde (O_2N–thiophene–CHO)	CH_3CO_2H	$RCH{=}CHCOCH{=}CHC_6H_5$ (20)	647

O_2N-(furan)-CHO	H_2SO_4	$RCH{=}CHCOCH{=}CHC_6H_5$ (41)	646, 1517
(thiophene)-CHO	NaOH	$RCH{=}CHCOCH{=}CHC_6H_5$ (—)	1521, 1793
(selenophene)-CHO	..	$RCH{=}CHCOCH{=}CHC_6H_5$ (52)	1522
(furan)-CHO	..	$RCH{=}CHCOCH{=}CHC_6H_5$ (87)	1794, 1525, 1539
$3\text{-}O_2NC_6H_4CHO$	..	$RCH{=}CHCOCH{=}CHC_6H_5$ (68)	1795
C_6H_5CHO	..	$RCH{=}CHCOCH{=}CHC_6H_5$ (100)	1558, 1525, 1539, 1796, 1797
$3,4\text{-}(CH_2O_2)C_6H_3CHO$	..	$RCH{=}CHCOCH{=}CHC_6H_5$ (—)	1796
$4\text{-}CH_3OC_6H_4CHO$	..	$RCH{=}CHCOCH{=}CHC_6H_5$ (—)	1584
(furan)-$CH{=}CHCHO$	Na_2CO_3	$RCH{=}CHCOCH{=}CHC_6H_5$ (57)	1766
$2\text{-}O_2NC_6H_4CH{=}CHCHO$	NaOH	$RCH{=}CHCOCH{=}CHC_6H_5$ (19)	1798, 1594
$3\text{-}O_2NC_6H_4CH{=}CHCHO$	..	$RCH{=}CHCOCH{=}CHC_6H_5$ (24)	1798
$C_6H_5CH{=}CHCHO$	..	$RCH{=}CHCOCH{=}CHC_6H_5$ (—)	1589
$4\text{-}(CH_3)_2NC_6H_4CHO$	..	$RCH{=}CHCOCH{=}CHC_6H_5$ (—)	1799, 1800
$3,4\text{-}(CH_2O_2)C_6H_3CH{=}CHCHO$	..	$RCH{=}CHCOCH{=}CHC_6H_5$ (40)	1603
$3\text{-}CH_3OC_6H_4CH{=}CHCHO$	..	$RCH{=}CHCOCH{=}CHC_6H_5$ (—)	1603
$C_6H_5(CH{=}CH)_2CHO$	..	$RCH{=}CHCOCH{=}CHC_6H_5$ (—)	1801
Anthracene-9-carboxaldehyde	Na_2CO_3	$RCH{=}CHCOCH{=}CHC_6H_5$ (—)	1801
$2\text{-}HOC_6H_4CH{=}CHCOCH_3$ and			
(furan)-CHO	NaOH	$RCH{=}CHCOCH{=}CHC_6H_4OH\text{-}2$ (—)	347
C_6H_5CHO	..	$RCH{=}CHCOCH{=}CHC_6H_4OH\text{-}2$ (—)	1796
$3\text{-}HOC_6H_4CHO$	..	$RCH{=}CHCOCH{=}CHC_6H_4OH\text{-}2$ (—)	347
$3,4\text{-}(CH_2O_2)C_6H_3CHO$	..	$RCH{=}CHCOCH{=}CHC_6H_4OH\text{-}2$ (85)	347, 1796
$2\text{-}CH_3OC_6H_4CHO$	..	$RCH{=}CHCOCH{=}CHC_6H_4OH\text{-}2$ (90)	347
$3\text{-}CH_3OC_6H_4CHO$	..	$RCH{=}CHCOCH{=}CHC_6H_4OH\text{-}2$ (9)	347
$4\text{-}CH_3OC_6H_4CHO$	..	$RCH{=}CHCOCH{=}CHC_6H_4OH\text{-}2$ (—), $(RCH{=}CH)_2CO$ (20)	347, 515

Note: **References 668–2359 are on pp. 403–438.**

TABLE XI. Condensation of Aldehydes with Aliphatic Methyl Ketones Other than Acetone and 2-Butanone (*Continued*)

(R in the product is the group R in the aldehyde RCHO.)

Ketone and Aldehyde, RCHO	Catalyst	Product(s) (Yield, %)	Refs.
$2\text{-}HOC_6H_4CH{=}CHCOCH_3$ (***contd.***) **and**			
$3,4\text{-}(CH_3O)_2C_6H_3CHO$	NaOH	$RCH{=}CHCOCH{=}CHC_6H_4OH\text{-}2$ (—)	347
$4\text{-}(CH_3)_2NC_6H_4CHO$	,,	$RCH{=}CHCOCH{=}CHC_6H_4OH\text{-}2$ (75)	515
$C_6H_5COCH_2COCH_3$ **and**			
C_6H_5CHO	KNH_2, NH_3, then $4\text{-}CH_3C_6H_4SO_3H$	$RCHOHCH_2COCH_2COC_6H_5$ (—), $RCH{=}CHCOCH_2COC_6H_5$ (28)	1802, 211
$4\text{-}CH_3C_6H_4CHO$	$(CH_2)_5NH$	$RCH{=}CHCOCH_2COC_6H_5$ (23)†	589
$C_6H_5CH_2CH_2COCH_3$ **and**			
(furan ring)CHO	NaOH	$RCH{=}CHCOCH_2CH_2C_6H_5$ (46)	1803
C_6H_5CHO	,,	$RCH{=}CHCOCH_2CH_2C_6H_5$ (80)	1803, 1787
$3,4\text{-}(CH_2O_2)C_6H_3CHO$	,,	$RCH{=}CHCOCH_2CH_2C_6H_5$ (73)	1803
$2\text{-}CH_3OC_6H_4CHO$	,,	$RCH{=}CHCOCH_2CH_2C_6H_5$ (52)	1803
$4\text{-}CH_3OC_6H_4CHO$	,,	$RCH{=}CHCOCH_2CH_2C_6H_5$ (80)	1803
$3,4\text{-}(CH_3O)_2C_6H_3CHO$	,,	$RCH{=}CHCOCH_2CH_2C_6H_5$ (73)	1803
$4\text{-}(CH_3)_2NC_6H_4CHO$	,,	$RCH{=}CHCOCH_2CH_2C_6H_5$ (34)	1803
$C_6H_5CH_2OCH_2COCH_3$ **and**			
C_6H_5CHO	NaOH, H_2O, dioxane, 100°	$RCH{=}CHCOCH_2OCH_2C_6H_5$ (30)	1792
	HCl, 25°, 24 hr.	O, =O; C_6H_5, C_6H_5 (—)	1792
	HCl, 25°, 48 hr.	$C_6H_5CH{=}$; O, =O; C_6H_5, C_6H_5 (83)	1792

Reactants	Catalyst	Product (Yield, %)	Refs.
$CH_3COC_8H_{17}$-*n* **and**			
O_2N–(furan)–CHO	HCl	$RCH{=}C(COCH_3)C_7H_{15}$-*n* (—)	1645
C_6H_5CHO	KOH	$RCH{=}CHCOC_8H_{17}$-*n* (—)	1743
3-CH_3O-4-HOC_6H_3CHO	"	$RCH{=}CHCOC_8H_{17}$-*n* (—)	1579
$(CH_3)_2C(OH)(CH_2)_3CH(CH_3)COCH_3$ **and**			
$C_{20}H_{24}O_2$, crocetin dialdehyde	"	$R[CH{=}CHCOCH(CH_3)(CH_2)_3C(OH)(CH_3)_2]_2$ (8)	607
$C_6H_5CH{=}C(CH_3)COCH_3$ **and**			
C_6H_5CHO	HCl	$RCH{=}CHCOC(CH_3){=}CHC_6H_5$ (—)	1660
$C_6H_5CH{=}CHCHO$	NaOH	$RCH{=}CHCOC(CH_3){=}CHC_6H_5$ (—)	1801
4-$CH_3OC_6H_4CH{=}CHCOCH_3$ **and**			
(thiophene)–CHO	"	$RCH{=}CHCOCH{=}CHC_6H_4OCH_3$-4 (—)	1793
(furan)–CHO	"	$RCH{=}CHCOCH{=}CHC_6H_4OCH_3$-4 (89)	1794
4-$O_2NC_6H_4CHO$	"	$RCH{=}CHCOCH{=}CHC_6H_4OCH_3$-4 (25)	1798
4-$CH_3OC_6H_4CHO$	"	$RCH{=}CHCOCH{=}CHC_6H_4OCH_3$-4 (—)	1585
2-$O_2NC_6H_4CHO$	"	$RCH{=}CHCOCH{=}CHC_6H_4OCH_3$-4 (73)	1798
4-$O_2NC_6H_4CHO$	"	$RCH{=}CHCOCH{=}CHC_6H_4OCH_3$-4 (13)	1798
4-$(CH_3)_2NC_6H_4CHO$	"	$RCH{=}CHCOCH{=}CHC_6H_4OCH_3$-4 (42)	1798
4-$CH_3OC_6H_4CH{=}CHCHO$	"	$RCH{=}CHCOCH{=}CHC_6H_4OCH_3$-4 (—)	1603
2-HO-4-$CH_3OC_6H_3CH{=}CHCOCH_3$ **and** 4-$(CH_3)_2NC_6H_4CHO$	"	$RCH{=}CHCOCH{=}CHC_6H_3OH$-2-$OCH_3$-4 (—)	1547
2-HO-5-$CH_3OC_6H_3CH{=}CHCOCH_3$ **and** 4-$(CH_3)_2NC_6H_4CHO$	"	$RCH{=}CHCOCH{=}CHC_6H_3OH$-2-$OCH_3$-5 (—)	1547
3-CH_3O-4-$HOC_6H_3CH{=}CHCOCH_3$ **and** 2-HOC_6H_4CHO	"	$RCH{=}CHCOCH{=}CHC_6H_3OCH_3$-3-OH-4 (60)	1582

Note: References 668–2359 are on pp. 403–438.

† The structure of this compound was not established.

TABLE XI. CONDENSATION OF ALDEHYDES WITH ALIPHATIC METHYL KETONES OTHER THAN ACETONE AND 2-BUTANONE (*Continued*)

(R in the product is the group R in the aldehyde RCHO.)

Ketone and Aldehyde, RCHO	Catalyst	Product(s) (Yield, %)	Refs.
$C_6H_5(CH_2)_3COCH_3$ **and**			
2-Furaldehyde (furan ring with CHO)	NaOH	$RCH{=}CHCO(CH_2)_3C_6H_5$ (48)	1803
C_6H_5CHO	”	$RCH{=}CHCO(CH_2)_3C_6H_5$ (64)	1803
3,4-$(CH_2O_2)C_6H_3CHO$	”	$RCH{=}CHCO(CH_2)_3C_6H_5$ (68)	1803
2-$CH_3OC_6H_4CHO$	”	$RCH{=}CHCO(CH_2)_3C_6H_5$ (28)	1803
4-$CH_3OC_6H_4CHO$	”	$RCH{=}CHCO(CH_2)_3C_6H_5$ (85)	1803
3,4-$(CH_3O)_2C_6H_3CHO$	”	$RCH{=}CHCO(CH_2)_3C_6H_5$ (61)	1803
4-$(CH_3)_2NC_6H_4CHO$	”	$RCH{=}CHCO(CH_2)_3C_6H_5$ (31)	1803
2-HO-3-$CH_3OC_6H_3CHOHCH_2$-$COCH_3$ **and** 4-$(CH_3)_2NC_6H_4CHO$	”	$RCH{=}CHCOCH{=}CHC_6H_3OH$-2-$OCH_3$-3 (—)	1547
Cyclohexene ring with CH_3, $COCH_3$, $(CH_3)_2$ **and** 3-$CH_3OC_6H_4CHO$	KOH	Cyclohexene ring with CH_3, $COCH{=}CHR$, $(CH_3)_2$ (60)	1804
n-$C_4H_9CH(C_2H_5)CH{=}CHCOCH_3$ **and** $(C_2H_5)_2CHCHO$	NaOH	$RCH{=}CHCOCH{=}CHCH(C_2H_5)C_4H_9$-$n$ (—)	1470
$CH_3COC_9H_{19}$-n **and**			
5-Bromo-2-furaldehyde (Br–furan–CHO)	”	$RCH{=}CHCOC_9H_{19}$-n (—)	1644
5-Nitro-2-furaldehyde (O_2N–furan–CHO)	HCl	$RCH{=}CH(COCH_3)C_8H_{17}$-$n$ (—)	1645
2-ClC_6H_4CHO	NaOH	$RCH{=}CHCOC_9H_{19}$-n (—)	1805
3-ClC_6H_4CHO	”	$RCH{=}CHCOC_9H_{19}$-n (—)	1805
4-ClC_6H_4CHO	”	$RCH{=}CHCOC_9H_{19}$-n (60)	1805
C_6H_5CHO	KOH	$RCH{=}CHCOC_9H_{19}$-n (—)	1806, 1524, 1791, 1805, 1806

	HCl	RCH=C($COCH_3$)C_8H_{17}-*n* (—)	1791
2-HOC_6H_4CHO	NaOH	RCH=CHCOC_9H_{19}-*n* (48)	1805
(2-furyl)CH=CHCHO	″	RCH=CHCOC_9H_{19}-*n* (45)	1766
3,4-(CH_2O_2)C_6H_3CHO	″	RCH=CHCOC_9H_{19}-*n* (—)	1791
4-$CH_3OC_6H_4CHO$	″	RCH=CHCOC_9H_{19}-*n* (—)	1791
C_6H_5CH=CHCHO	″	RCH=CHCOC_9H_{19}-*n* (—)	1791
(2-furyl)(CH=CH)$_2$CHO	″	RCH=CHCOC_9H_{19}-*n* (90)	1766
3,4-(CH_3O)$_2C_6H_3CHO$	″	RCH=CHCOC_9H_{19}-*n* (72)	1805
4-*i*-$C_3H_7C_6H_4CHO$	″	RCH=CHCOC_9H_{19}-*n* (—)	1805, 1791
2,4,5-(CH_3O)$_3C_6H_2CHO$	″	RCH=CHCOC_9H_{19}-*n* (—)	1608
n-$C_4H_9CH(C_2H_5)(CH_2)_2COCH_3$ and (C_2H_5)$_2$CHCHO	″	RCH=CHCO(CH_2)$_2$CH(C_2H_5)C_4H_9-*n* (—)	1470
C_6H_5(CH=CH)$_2COCH_3$ and CCl_3CHO	CH_3CO_2K, CH_3CO_2H	RCHOHCH$_2$CO(CH=CH)$_2C_6H_5$ (36)	1808
(2-thienyl)CHO	NaOH	RCH=CHCO(CH=CH)$_2C_6H_5$ (—)	1793
(2-furyl)CHO	″	RCH=CHCO(CH=CH)$_2C_6H_5$ (—)	1809
C_6H_5CHO	″	RCH=CHCO(CH=CH)$_2C_6H_5$ (—)	1676
2-HOC_6H_4CHO	″	RCH=CHCO(CH=CH)$_2C_6H_5$ (—)	1589
3,4-(CH_3O)$_2C_6H_3CHO$	″	RCH=CHCO(CH=CH)$_2C_6H_5$ (—)	1589
4-$CH_3OC_6H_4CHO$	″	RCH=CHCO(CH=CH)$_2C_6H_5$ (—)	1589
3-CH_3O-4-HOC_6H_3CHO	″	RCH=CHCO(CH=CH)$_2C_6H_5$ (—)	1589
C_6H_5CH=CHCHO	″	RCH=CHCO(CH=CH)$_2C_6H_5$ (—)	1595
3,4-(CH_3O)$_2C_6H_3CHO$	″	RCH=CHCO(CH=CH)$_2C_6H_5$ (—)	1589
C_6H_5(CH=CH)$_2$CHO	″	RCH=CHCO(CH=CH)$_2C_6H_5$ (—)	1801
Anthracene-9-carboxaldehyde	Na_2CO_3	RCH=CHCO(CH=CH)$_2C_6H_5$ (—)	1801
$C_6H_5(COCH_2)_2COCH_3$ and C_6H_5CHO	$NaNH_2$	$C_6H_5(COCH_2)_3CHOHC_6H_5$ (45)	213

Note: References 668–2359 are on pp. 403–438.

TABLE XI. CONDENSATION OF ALDEHYDES WITH ALIPHATIC METHYL KETONES OTHER THAN ACETONE AND 2-BUTANONE (*Continued*)

(R in the product is the group R in the aldehyde RCHO.)

Ketone and Aldehyde, RCHO	Catalyst	Product(s) (Yield, %)	Refs.
2,4-$(CH_3O)_2C_6H_3CH{=}CHCOCH_3$			
and 2-thiophenecarboxaldehyde (S-ring)CHO	NaOH	$RCH{=}CHCOCH{=}CHC_6H_3(OCH_3)_2$-2,4 (80)	1008
4-$(CH_3)_2NC_6H_4CH{=}CClCOCH_3$ and			
C_6H_5CHO	"	$RCH{=}CHCOC(Cl){=}CHC_6H_4N(CH_3)_2$-4 (—)	1705
3,4-$(CH_2O_2)C_6H_3CHO$	"	$RCH{=}CHCOC(Cl){=}CHC_6H_4N(CH_3)_2$-4 (—)	1705
$C_6H_5CH{=}CHCHO$	"	$RCH{=}CHCOC(Cl){=}CHC_6H_4N(CH_3)_2$-4 (—)	1705
4-$(CH_3)_2NC_6H_4CH{=}CHCOCH_3$ and			
2-HO-5-$O_2NC_6H_3CHO$	"	$RCH{=}CHCOCH{=}CHC_6H_4N(CH_3)_2$-4 (—)	1547
C_6H_5CHO	"	$RCH{=}CHCOCH{=}CHC_6H_4N(CH_3)_2$-4 (75)	1798
4-$CH_3OC_6H_4CHO$	"	$RCH{=}CHCOCH{=}CHC_6H_4N(CH_3)_2$-4 (—)	1798
4-$(CH_3)_2NC_6H_4CHO$	"	$RCH{=}CHCOCH{=}CHC_6H_4N(CH_3)_2$-4 (65)	1798
2,4,6-$(CH_3O)_3C_6H_2CHO$	Na_2CO_3	$RCH{=}CHCOCH{=}CHC_6H_4N(CH_3)_2$-4 (66)	1810
2-C_3H_7-n-2-[$CH_2C(CH_3)_2COCH_3$]-1,3-dioxolane and			
2,9-Dimethyl-2,8-decadiene-4,6-diyne-1,10-dial	$Al(OC_4H_9\text{-}t)_3$	R[CH=CHCOC$(CH_3)_2CH_2$-(2-n-C_3H_7-1,3-dioxolan-2-yl)]$_2$ (19)	607
2,9-Dimethyl-2,4,6,8-decatetraene-1,10-dial	"	R[CH=CHCOC$(CH_3)_2CH_2$-(2-n-C_3H_7-1,3-dioxolan-2-yl)]$_2$ (11)	607
$C_{20}H_{24}O_2$, crocetin dialdehyde	KOH	R[CH=CHCOC$(CH_3)_2CH_2$-(2-n-C_3H_7-1,3-dioxolan-2-yl)]$_2$ (48)	607

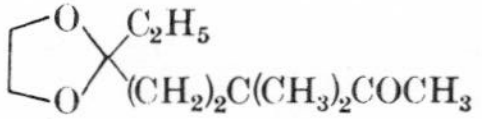 and			
2,9-Dimethyl-2,8-decadiene-4,6-diyne-1,10-dial	$Al(OC_4H_9\text{-}t)_3$	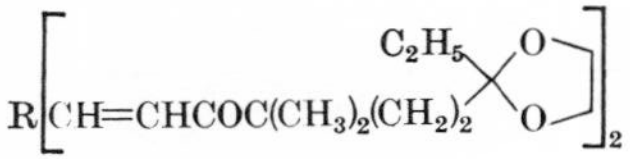 (21)	608
2,9-Dimethyl-2,4,6,8-decatetraene-1,10-dial	"	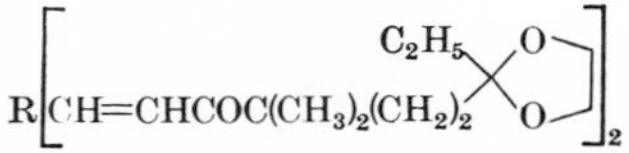 (30)	608
$C_{20}H_{24}O_2$, crocetin dialdehyde	KOH	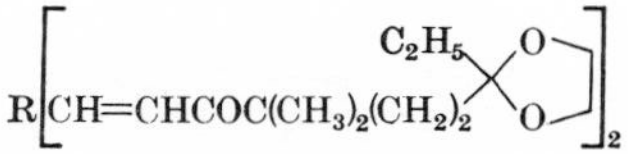 (55)	608
β-*apo*-2-Carotenaldehyde	"	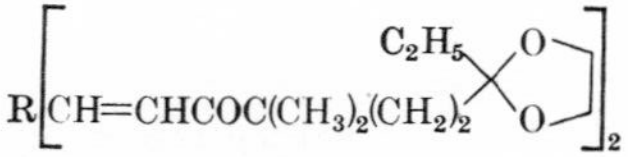 (32)	608
$CH_3COC_{10}H_{21}\text{-}n$ and $3\text{-}CH_3O\text{-}4\text{-}HOC_6H_3CHO$	"	$RCH{=}CHCOC_{10}H_{21}\text{-}n$ (—)	1579
$4\text{-}CH_3OC_6H_4(CH{=}CH)_2COCH_3$ and $4\text{-}CH_3OC_6H_4CHO$	NaOH	$RCH{=}CHCO(CH{=}CH)_2C_6H_4OCH_3\text{-}4$ (—)	1603
$2,4,6\text{-}(CH_3O)_3C_6H_2CH{=}CHCOCH_3$ and 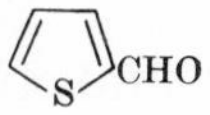	"	$RCH{=}CHCOCH{=}CHC_6H_2(OCH_3)_3\text{-}2,4,6$ (83)	1008
$C_6H_5CH{=}C(OC_6H_5)COCH_3$ and C_6H_5CHO	"	$RCH{=}CHCOC(OC_6H_5){=}CHC_6H_5$ (—)	268
$4\text{-}CH_3OC_6H_4CHO$	"	$RCH{=}CHCOC(OC_6H_5){=}CHC_6H_5$ (—)	268

Note: References 668–2359 are on pp. 403–438.

TABLE XI. CONDENSATION OF ALDEHYDES WITH ALIPHATIC METHYL KETONES OTHER THAN ACETONE AND 2-BUTANONE (*Continued*)
(R in the product is the group R in the aldehyde RCHO.)

Ketone and Aldehyde, RCHO	Catalyst	Product(s) (Yield, %)	Refs.
$C_6H_5COC(OH)(C_6H_5)CH_2COCH_3$ **and**			
C_6H_5CHO	KOH	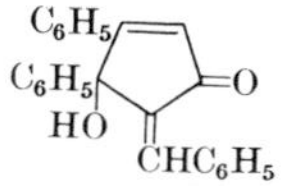 [structure: cyclopentenone bearing C_6H_5, C_6H_5, HO, =O, =CHC_6H_5] (—)	324
$C_6H_5(CH{=}CH)_4COCOCH_3$ **and**			
$C_6H_5(CH{=}CH)_3CHO$	$(CH_2)_5NH$, CH_3CO_2H	$RCH{=}CHCOCO(CH{=}CH)_4C_6H_5$ (100)	592
$4\text{-}CH_3OC_6H_4CH{=}C(OC_6H_5)COCH_3$ **and**			
C_6H_5CHO	NaOH	$RCH{=}CHCOC(OC_6H_5){=}CHC_6H_4OCH_3\text{-}4$ (—)	268
$4\text{-}CH_3OC_6H_4CHO$	"	$RCH{=}CHCOC(OC_6H_5){=}CHC_6H_4OCH_3\text{-}4$ (—)	268
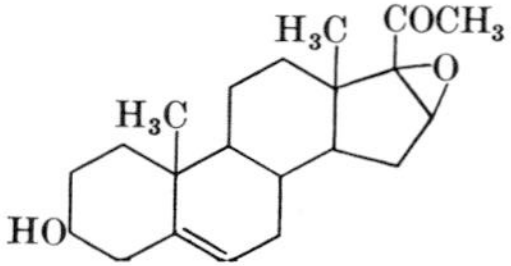 [structure: steroid with HO, H_3C, H_3C, $COCH_3$, epoxide O] **and**			
C_6H_5CHO	$NaOCH_3$	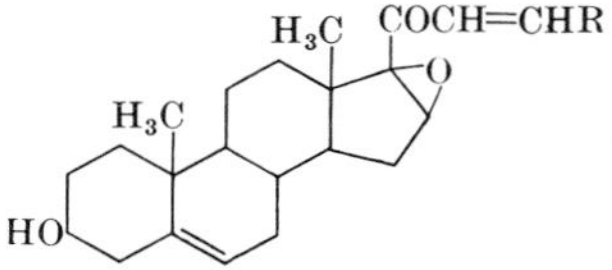 [structure: steroid with HO, H_3C, H_3C, $COCH{=}CHR$, epoxide O] (90)	1811

Note: References 668–2359 are on pp. 403–438.

TABLE XII. CONDENSATION OF ALDEHYDES WITH ACYCLIC NON-METHYL KETONES
(R in the product is the group R in the aldehyde RCHO.)

Ketone and Aldehyde, RCHO	Catalyst	Product(s) (Yield, %)	Refs.
$HOCH_2COCH_2OH$ **and**			
CH_2O	HCl	Furfural (—)	975
$HOCH_2CHO$	Glycine	Threopentulose (2), other pentoses (—)	229, 975
$HOCH_2CHOHCHO$	$Ba(OH)_2$	Fructose, sorbose (—)	1812
D-$HOCH_2(CHOH)_2CHO$	$Ca(OH)_2$	D-*gluco*-Heptulose (22), D-*altro*-heptulose (12), D-*allo*-heptulose (3)	228
D- 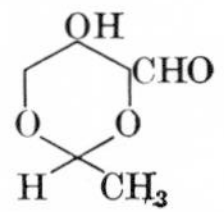 [ring structure: OH, CHO, O, O, H, CH_3]	$Ba(OH)_2$	D-*gluco*-Heptulose (20), 5,7-O-ethylidene-D-*altro*-heptulose (17), 5,7-O-ethylidene-D-*manno*-heptulose (7)	227
$(C_2H_5)_2CO$ **and**			
CH_2O	NaOH	$CH_3CH(CH_2OH)COC_2H_5$ (75), $CH_3C(CH_2OH)_2COC_2H_5$ (—), $CH_3C(CH_2OH)_2COCH(CH_2OH)CH_3$ (—)	652, 1389, 1400, 1406, 1618, 1622, 1813
	$Ca(OH)_2$	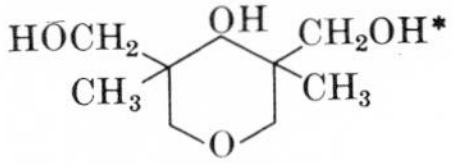 [ring structure: $HOCH_2$, OH, CH_2OH*, CH_3, CH_3, O] (—)	1415
	H_2SO_4, CH_3CO_2H	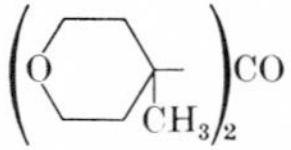 [structure: (O-ring, CH_3)$_2$CO] (37)	1740
CCl_3CHO	CH_3CO_2K, $(CH_3CO)_2O$	$RCHOHCH(CH_3)COC_2H_5$ (70)	287
CH_3CHO	KOH	$RCHOHCH(CH_3)COC_2H_5$ (71)	580, 269, 581, 1124
C_2H_5CHO	"	$RCHOHCH(CH_3)COC_2H_5$ (50)	580, 581

Note: References 668–2359 are on pp. 403–438.

* This product was isolated as the diallyl ether.

TABLE XII. CONDENSATION OF ALDEHYDES WITH ACYCLIC NON-METHYL KETONES (*Continued*)
(R in the product is the group R in the aldehyde RCHO.)

Ketone and Aldehyde, RCHO	Catalyst	Product(s) (Yield, %)	Refs.
$(C_2H_5)_2CO$ **(*contd.*) and**			
n-C_3H_7CHO	KOH	$RCHOHCH(CH_3)COC_2H_5$ (83)	580, 581
i-C_3H_7CHO	NaOH	$RCHOHCH(CH_3)COC_2H_5$ (25)	1124, 274, 1643
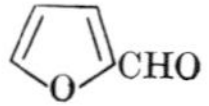	"	$RCH{=}C(CH_3)COC_2H_5$ (—)	1526
n-$C_4H_9O_2CCHO$	None, heat	$RCHOHCH(CH_3)COC_2H_5$ (81)	1468
n-$C_5H_{11}CHO$	KOH	$RCHOHCH(CH_3)COC_2H_5$ (83)	580, 581
C_6H_5CHO	"	$RCH{=}C(CH_3)COC_2H_5$ (40), 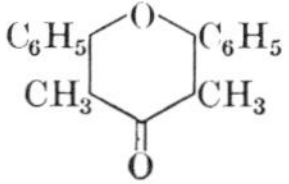 (20)	1548, 523, 1663, 1814, 2332
n-$C_6H_{13}CHO$	"	$RCHOHCH(CH_3)COC_2H_5$ (80)	580, 581
	"	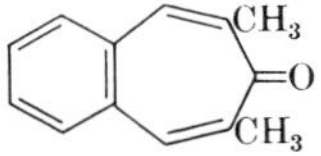 (65)	1672, 223
	$NaOC_2H_5$	$RCH{=}C(CH_3)COC_2H_5$ (45)	1487
$(CH_3)_2C{=}CHCH_2CH_2C(CH_3){=}CH$-$CHO$, citral	NaOH	$RCH{=}C(CH_3)COC_2H_5$ (33)	1687, 1492
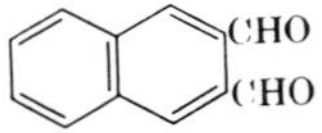	"	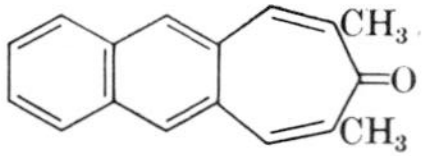 (70)	1690
$C_2H_5COCHOHCH_3$ **and**			
CH_2O	KOH	$C_2H_5COC(OH)(CH_2OH)CH_3$ (61)	230, 1767
CH_3CHO	"	$C_2H_5COC(OH)(CH_3)CHOHCH_3$ (47)	230, 1767
$(CH_3OCH_2)_2CO$ **and** $O_2NCH(CHO)_2$	NaOH	2,6-$(CH_3O)_2$-4-$O_2NC_6H_2OH$ (62)	1815

$C_2H_5COC_3H_7$-*n* **and** *n*-C_3H_7CHO	KOH	*n*-$C_3H_7CHOHCH(CH_3)COC_3H_7$-*n* (1), *n*-$C_3H_7CH{=}C(CH_3)COC_3H_7$-*n* (9), *n*-$C_3H_7CH{=}C(C_2H_5)CHO$ (53)	222
$C_2H_5COC_4H_9$-*n* **and**  CH_3, CHO	$NaOC_2H_5$	$RCH{=}C(CH_3)OC_4H_9$-*n*† (40)	1487
(*n*-$C_3H_7)_2CO$ **and**			
CH_2O	H_2SO_4, CH_3CO_2H	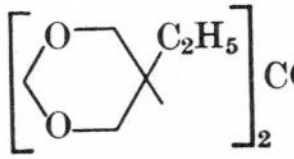 O, O, C_2H_5, 2, CO (32)	1740
n-C_3H_7CHO	KOH	$RCHOHCH(C_2H_5)COC_3H_7$-*n* (3.6), *n*-$C_3H_7CHOHCH(C_2H_5)CHO$ (26), *n*-$C_3H_7CH{=}C(C_2H_5)CHO$ (47)	222
n-$C_4H_9O_2CCHO$	None, 125°	$RCHOHCH(C_2H_5)COC_3H_7$-*n* (84)	1468
C_6H_5CHO	NaOH	$RCH{=}C(C_2H_5)COC_3H_7$-*n* (19.4), $C_{21}H_{24}O_2$ (15)	524, 2332
 CHO, CHO	$NaOC_2H_5$	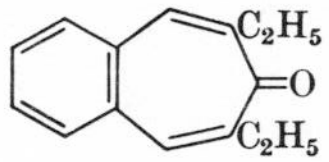 C_2H_5, O, C_2H_5 (73)	223
n-$C_3H_7CH{=}C(C_2H_5)CHO$	KOH	*n*-$C_3H_7CHOHCH(C_2H_5)COC_3H_7$-*n* (1.6), *n*-$C_3H_7CH{=}C(C_2H_5)COC_3H_7$-*n* (2.5)	222
n-$C_4H_9CH(C_2H_5)CHO$	NaOH	$RCHOHCH(C_2H_5)COC_3H_7$-*n* (—)	1652
(*i*-$C_3H_7)_2CO$ **and**			
CH_2O	"	$CH_2OHC(CH_3)_2COC_3H_7$-*i* (—)	1618
n-C_3H_7CHO	$C_6H_5N(CH_3)MgBr$	$RCHOHC(CH_3)_2COC_3H_7$-*i* (82)	177, 141
$(C_2H_5)_2CHCHO$	"	$RCHOHC(CH_3)_2COC_3H_7$-*i* (67)	177
C_6H_5CHO	"	$RCHOHC(CH_3)_2COC_3H_7$-*i* (80)	177
$C_6H_5CH_2CHO$	"	$RCHOHC(CH_3)_2COC_3H_7$-*i* (62)	177
t-$C_4H_9COCH{=}CHCH_3$ **and** C_6H_5CHO	$NaOC_2H_5$	$R(CH{=}CH)_2COC_4H_9$-*t* (28)	513
$C_2H_5COC(CH_3)_2C_2H_5$ **and** 3,4-$(CH_2O_2)C_6H_4CHO$	Na_2CO_3	$RCH{=}C(CH_3)COC(CH_3)_2C_2H_5$ (—)	1816

Note: References 668–2359 are on pp. 403–438.

† The structure of this compound was not established.

TABLE XII. Condensation of Aldehydes with Acyclic Non-Methyl Ketones (*Continued*)
(R in the product is the group R in the aldehyde RCHO.)

Ketone and Aldehyde, RCHO	Catalyst	Product(s) (Yield, %)	Refs.
$(n\text{-}C_4H_9)_2CO$ and 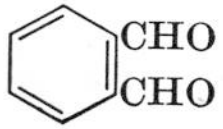 [structure] CHO, CHO	$NaOC_2H_5$	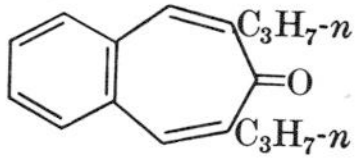 [structure] $C_3H_7\text{-}n$, =O, $C_3H_7\text{-}n$ (97)	223
$(i\text{-}C_4H_9)_2CO$ and			
CH_3CHO	$C_6H_5N(CH_3)MgBr$	$RCHOHCH(C_3H_7\text{-}i)COC_4H_9\text{-}i$ (73)	177, 141
$n\text{-}C_3H_7CHO$	$i\text{-}C_3H_7MgCl$	$RCHOHCH(C_3H_7\text{-}i)COC_4H_9\text{-}i$ (50)	141
$i\text{-}C_4H_9CHO$	"	$RCHOHCH(C_3H_7\text{-}i)COC_4H_9\text{-}i$ (50)	141
$n\text{-}C_6H_{13}CHO$	$C_6H_5N(CH_3)MgBr$	$RCHOHCH(C_3H_7\text{-}i)COC_4H_9\text{-}i$ (88)	177
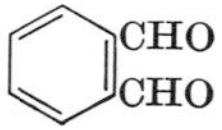 [structure] CHO, CHO	$NaOC_2H_5$	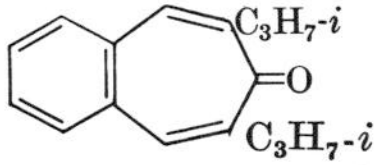 [structure] $C_3H_7\text{-}i$, =O, $C_3H_7\text{-}i$ (27)	223
$C_6H_5CH_2COC_2H_5$ and [structure] CHO, CHO	KOH	[structure] CH_3, =O, C_6H_5 (70)	1817
$C_6H_5CH_2COCH_2OCH_3$ and [structure] CHO, CHO	"	[structure] OCH_3, =O, C_6H_5 (83)	1818
$n\text{-}C_3H_7COC(CH_3)_2C_3H_7\text{-}n$ and $3,4\text{-}(CH_2O_2)C_6H_3CHO$	Na_2CO_3	$RCH{=}C(C_2H_5)COC(CH_3)_2C_3H_7\text{-}n$ (—)	1816
$(n\text{-}C_5H_{11})_2CO$ and $n\text{-}C_3H_7CHO$	$C_6H_5N(CH_3)MgBr$	$RCHOHCH(C_4H_9\text{-}n)COC_5H_{11}\text{-}n$ (15)	177
$(t\text{-}C_4H_9CH_2)_2CO$ and 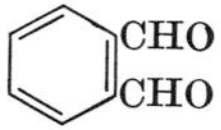 [structure] CHO, CHO	$NaOC_2H_5$	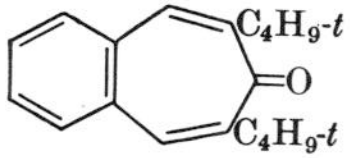 [structure] $C_4H_9\text{-}t$, =O, $C_4H_9\text{-}t$ (4.5)	223
$4\text{-}CH_3OC_6H_4CH{=}CHCOC_2H_5$ and $3,4\text{-}(CH_2O_2)C_6H_3CHO$	HCl	$C_{28}H_{26}O_4$ (—)	1673

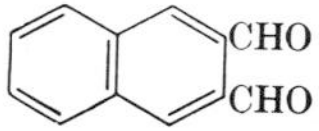

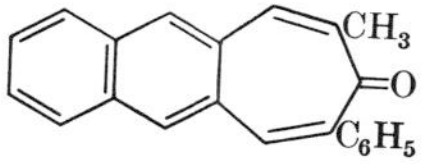

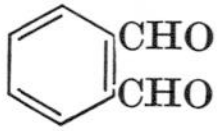

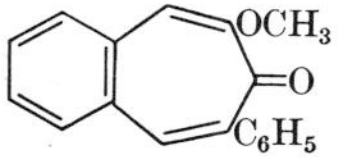

4-$CH_3OC_6H_4CHO$	NaOH	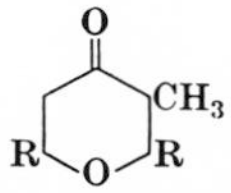 [structure: O, CH_3, R, O, R] (—)	1673
($C_6H_5CH_2)_2CO$ and			
$C(Cl_2)=CClCHO$	H_2SO_4	$[RCH=C(C_6H_5)]_2CO$ (31)	289
$C(Cl_2)=CHCHO$	,,	$[RCH=C(C_6H_5)]_2CO$ (39)	639
$O_2NCH(CHO)_2$	NaOH	2,6-$(C_6H_5)_2$-4-$O_2NC_6H_2OH$ (100)	1640
C_6H_5CHO	HCl	$C_6H_5CHClCH(C_6H_5)COCH_2C_6H_5$ (—)	1788, 1819
	KOH	$RCH=C(C_6H_5)COCH_2C_6H_5$ (—), $[RCH=C(C_6H_5)]_2CO$ (—)	1819
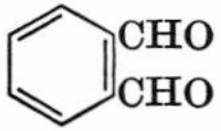 [structure: CHO, CHO]	,,	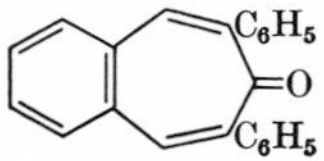 [structure: C_6H_5, =O, C_6H_5] (95)	1672
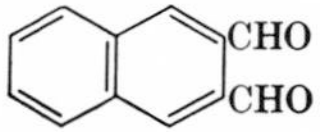 [structure: CHO, CHO]	NaOH	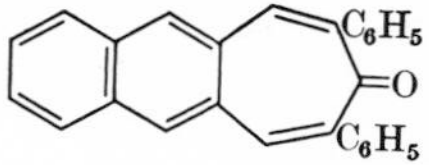 [structure: C_6H_5, =O, C_6H_5] (92)	1690
$(n\text{-}C_7H_{15})_2CO$ and 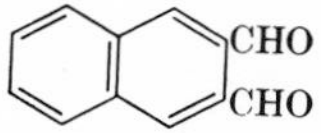 [structure: CHO, CHO]	$NaOC_2H_5$	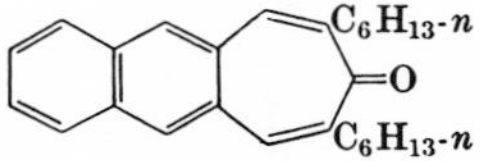 [structure: C_6H_{13}-*n*, =O, C_6H_{13}-*n*] (76)	223
$(C_6H_5COCH_2)_2CH_2$ and CH_2O	NaOH	Polymer (—)	1820
$(C_6H_5COCH_2)_2CHCH_3$ and CH_2O	,,	$C_6H_5COCH_2CH(CH_3)CH(CH_2OH)COC_6H_5$ (91)	1820
$[C_6H_5COCH(CH_3)]_2CH_2$ and CH_2O	,,	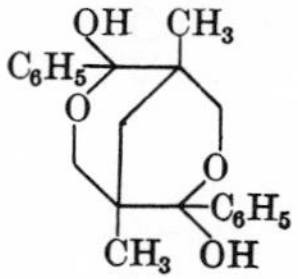 [structure: OH, CH_3, C_6H_5, O, O, C_6H_5, CH_3, OH] (54)	1820
$(C_6H_5COCH_2)_2CHC_6H_5$ and CH_2O	,,	$C_6H_5COCH_2CH(C_6H_5)CH(CH_2OH)COC_6H_5$ (89)	1820

Note: References 668–2359 are on pp. 403–438.

TABLE XIII. CONDENSATION OF ALDEHYDES WITH CYCLOPENTANONES

(R in the product is the group R in the aldehyde RCHO.)

Substituent(s) in Cyclopentanone	Aldehyde, RCHO	Catalyst	Substituent(s) in Cyclopentanone Product(s) (Yield, %)	Refs.
2-Oxo-3-Cl	C_6H_5CHO	HCl	2-Oxo-3-Cl-5-RCH= (84)*	1821
2,2,3-Cl_3	"	"	2,2,3-Cl_3-5-RCH= (16)*	1821
None	CH_2O	K_2CO_3	2-CH_2OH (18), 2,2-$(CH_2OH)_2$ (32), 2,2,5,5-$(CH_2OH)_4$ (39)	276, 278, 1822, 2340
	"	$KOCH_3$	(2-oxocyclopentyl)$_2CH_2$† (30)	550
	CH_3CHO	K_2CO_3	2-RCHOH (31), 2,2-$(RCHOH)_2$ (16), 2,2,5,5-$(RCHOH)_4$ (32)	279
	"	KOH	2-RCH= (8), 2,5-(RCH=) (18), CH_3CH=CHCHO† (25)	290
	$C(Cl_2)$=CClCHO	H_2SO_4	2,5-$(RCH=)_2$ (86)	289
	$C(Cl_2)$=CHCHO	"	2,5-$(RCH=)_2$ (88)	638
	C_2H_5CHO	—	2-RCHOH (—)	1206
	CH_3CH=CHCHO	KOH	2-RCH= (30)	301
	n-C_3H_7CHO	NaOH	2-RCH= (46), 2,5-$(RCH=)_2$ (—)	286, 1206
	2-Thiophenecarboxaldehyde (thiophene ring, CHO)	$NaOC_2H_5$	2,5-$(RCH=)_2$ (48)	1823, 303, 1824, 1825
	Furfural (furan ring, CHO)	NaOH	2-RCH= (60)	293, 1826
	"	KOH	2,5-$(RCH=)_2$ (100)	1826, 1535, 1827, 1828
	OHC–(thiophene-2,5-diyl)–CHO	K_2CO_3	2,5-$(RCH=)_2$ (—)	365

N CHO	NaOH	2,5-$(RCH{=})_2$ (Excellent)	1825
N CHO	”	2,5-$(RCH{=})_2$ (Excellent)	1825
N CHO	”	2,5-$(RCH{=})_2$ (Excellent)	1825
CH_3 O CHO	”	2,5-$(RCH{=})_2$ (Excellent)	1825
n-$C_4H_9O_2CCHO$	None, 100°	2-RCHOH (82)	1468
2-ClC_6H_4CHO	NaOH	2-RCH= (24), 2,5-$(RCH{=})_2$ (—)	293, 292
3-ClC_6H_4CHO	”	2,5-$(RCH{=})_2$ (—)	1829
4-ClC_6H_4CHO	”	2-RCH= (—), 2,5-$(RCH{=})_2$ (—)	293, 292
2-$O_2NC_6H_4CHO$	”	2,5-$(RCH{=})_2$ (—)	292
3-$O_2NC_6H_4CHO$	”	2,5-$(RCH{=})_2$ (66)	1830, 292
4-$O_2NC_6H_4CHO$	”	2,5-$(RCH{=})_2$ (74)	1830, 292
C_6H_5CHO	—	2,5-$(RCHOH)_2$ (10)	291
”	NaOH	2-RCH= (74)	299, 290, 295, 1831
”	”	2,5-$(RCH{=})_2$ (95)	294, 290, 292, 1828, 1831
O CH=CHCHO	”	2,5-$(RCH{=})_2$ (90)	1471
2-HOC_6H_4CHO	”	2,5-$(RCH{=})_2$ (—)	1830, 1572, 1832
4-HOC_6H_4CHO	”	2,5-$(RCH{=})_2$ (63)	1830, 1572
n-$C_6H_{13}CHO$	—	2-RCHOH (—)	1206

Note: References 668–2359 are on pp. 403–438.

* This product may contain some of the isomers containing an α,β-endocyclic double bond.[304]
† This is a complete structural formula.

TABLE XIII. Condensation of Aldehydes with Cyclopentanones (*Continued*)

(R in the product is the group R in the aldehyde RCHO.)

Substituent(s) in Cyclohexanone	Aldehyde, RCHO	Catalyst	Substituent(s) in Cyclopentanone Product(s) (Yield, %)	Refs.
None (*contd.*)	3,4-$(CH_2O_2)C_6H_3CHO$	KOH	2,5-(RCH=)$_2$ (—)	1830, 292
	″	$NaOC_5H_{11}$-*t*	2-RCH= (44)	295
	3-$CH_3OC_6H_4CHO$	NaOH	2,5-(RCH=)$_2$ (—)	1829
	4-$CH_3OC_6H_4CHO$	″	2-RCH= (72)	293
	″	″	2,5-(RCH=)$_2$ (—)	1828, 292, 1563
	3-CH_3O-4-HOC_6H_3CHO	″	2,5-(RCH=)$_2$ (—)	1830
	[bicyclo[2.2.1]heptene-CHO structure]	$NaOC_2H_5$	2-RCH= (60)	1487
	[benzofuran-2-CHO structure]	NaOH	2,5-(RCH=)$_2$ (—)	1829
	$C_6H_5CH{=}CHCHO$	″	2-RCH= (71), 2,5-(RCH=)$_2$ (—)	290
	″	″	2,5-(RCH=)$_2$ (—)	1828, 292
	4-$CH_3CONHC_6H_4CHO$	″	2,5-(RCH=)$_2$ (—)	1834
	2-$C_2H_5OC_6H_4CHO$	″	2,5-(RCH=)$_2$ (—)	1598
	3,4-$(CH_3O)_2C_6H_3CHO$	″	2,5-(RCH=)$_2$ (—)	1598
	4-$(CH_3)_2NC_6H_4CHO$	″	2,5-(RCH=)$_2$ (—)	292
	4-$(CH_3)_2CHC_6H_4CHO$	KOH	2,5-(RCH=)$_2$ (—)	1830, 292
2-CH_3	C_6H_5CHO	$NaOCH_3$	2-CH_3-5-RCH= (—)	304, 1835, 1836
	″	HCl	[$C_6H_5CH_2$-cyclopentenone-CH_3 structure]† (—)	304, 1835–1837
3-CH_3	″	NaOH	2,5-(RCH=)$_2$-3-CH_3 (42)	1838, 318
2-CH_3CH=	″	KOH	2-(CH_3CH=)-5-RCH= (—)	290
2-C_2H_5	″	″	2-C_2H_5-5-RCH= (100)	290
	$C_6H_5CH{=}CHCHO$	NaOH	2-C_2H_5-5-RCH= (—)	290
2,4-$(CH_3)_2$	3-$O_2NC_6H_4CHO$	″	2,4-$(CH_3)_2$-5-RCH= (—)	292

	$4\text{-}O_2NC_6H_4CHO$	,,	$2,4\text{-}(CH_3)_2\text{-}5\text{-}RCH=$ (—)	292
	C_6H_5CHO	,,	$2,4\text{-}(CH_3)_2\text{-}5\text{-}RCH=$ (—)	292
	$3,4\text{-}(CH_2O_2)C_6H_3CHO$	,,	$2,4\text{-}(CH_3)_2\text{-}5\text{-}RCH=$ (—)	292
	$4\text{-}CH_3OC_6H_4CHO$	,,	$2,4\text{-}(CH_3)_2\text{-}5\text{-}RCH=$ (—)	292
	$C_6H_5CH{=}CHCHO$	,,	$2,4\text{-}(CH_3)_2\text{-}5\text{-}RCH=$ (—)	292
	$4\text{-}(CH_3)_2NC_6H_4CHO$	,,	$2,4\text{-}(CH_3)_2\text{-}5\text{-}RCH=$ (—)	292
	$4\text{-}i\text{-}C_3H_7C_6H_4CHO$	,,	$2,4\text{-}(CH_3)_2\text{-}5\text{-}RCH=$ (—)	292
$2,5\text{-}(CH_3)_2$	C_6H_5CHO	HCl	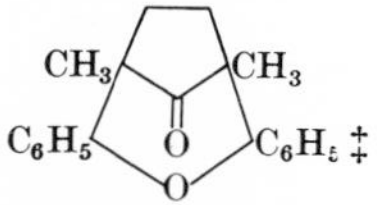 (bicyclic structure: CH_3, CH_3, C_6H_5, C_6H_5, O, O) ‡ (86)	530
	$4\text{-}CH_3C_6H_4CHO$	,,	$C_{23}H_{26}O_2$ (25)	530
	$4\text{-}i\text{-}C_3H_7C_6H_4CHO$	,,	$C_{27}H_{34}O_2$ (37)	530
$3\text{-}n\text{-}C_3H_7$	$4\text{-}O_2NC_6H_4CHO$	NaOH	$2,5\text{-}(RCH{=})_2\text{-}3\text{-}n\text{-}C_3H_7$ (—)	1839
$3\text{-}C_2H_5\text{-}3\text{-}CH_3$	,,	,,	$2,5\text{-}(RCH{=})_2\text{-}3\text{-}CH_3\text{-}3\text{-}C_2H_5$ (—)	1839
$3\text{-}C_2H_5\text{-}4\text{-}CH_3$	,,	,,	$2,5\text{-}(RCH{=})_2\text{-}3\text{-}C_2H_5\text{-}4\text{-}CH_3$ (—)	1839
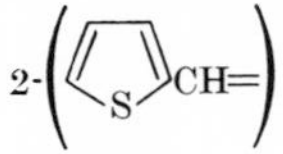 2-(2-thienyl-CH=)	C_6H_5CHO	KOH	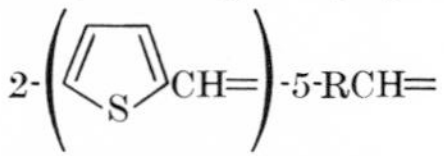 2-(2-thienyl-CH=)-5-RCH= (82)	303
$3\text{-}C_6H_5$	,,	NaOH	$2,5\text{-}(RCH{=})_2\text{-}3\text{-}C_6H_5$ (—)	1259
$2\text{-}\overline{(CH_2)_5CH}$	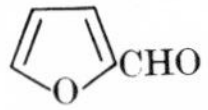 2-furyl-CHO	KOH	$2\text{-}\overline{(CH_2)_5CH}\text{-}5\text{-}RCH=$ (—)	1840
	C_6H_5CHO	,,	$2\text{-}\overline{(CH_2)_5CH}\text{-}5\text{-}RCH=$ (—)	1840
	$4\text{-}CH_3OC_6H_4CHO$	,,	$2\text{-}\overline{(CH_2)_5CH}\text{-}5\text{-}RCH=$ (—)	1840
	$C_6H_5CH{=}CHCHO$	,,	$2\text{-}\overline{(CH_2)_5CH}\text{-}5\text{-}RCH=$ (—)	1840
	$3,4\text{-}(CH_3O)_2C_6H_3CHO$	,,	$2\text{-}\overline{(CH_2)_5CH}\text{-}5\text{-}RCH=$ (—)	1840
	$4\text{-}i\text{-}C_3H_7C_6H_4CHO$	,,	$2\text{-}\overline{(CH_2)_5CH}\text{-}5\text{-}RCH=$ (—)	1840
$2\text{-}C_6H_5CH=$	C_6H_5CHO	,,	$2\text{-}(C_6H_5CH{=})\text{-}5\text{-}RCH=$ (100)	290, 1828
$2\text{-}C_6H_5CH_2$	,,	NaOH	$2\text{-}C_6H_5CH_2\text{-}5\text{-}RCH=$ (95)	290, 1835

Note: References 668–2359 are on pp. 403–438.

† This is a complete structural formula.
‡ This is the structure proposed for the product.

TABLE XIV. CONDENSATION OF ALDEHYDES WITH CYCLOHEXANONES

(R in the product is the group R in the aldehyde RCHO.)

Substituent(s) in Cyclohexanone	Aldehyde, RCHO	Catalyst	Substituent(s) in Cyclohexanone Product(s) (Yield, %)	Refs.
2-Oxo	$C_6H_5CH{=}CHCHO$	—	2-Oxo-3,6-$(RCH{=})_2$ (—)	327
	$C_6H_5CH_2CH_2CHO$	—	2-Oxo-3,6-$(RCH{=})_2$ (—)	327
4-Oxo	 $C_6H_4(CHO)_2$ (o-)	KOH	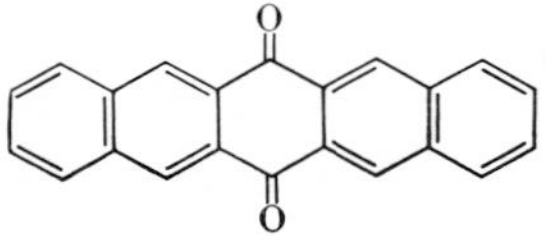 * (90), * (—)	334
None	CH_2O	KOH	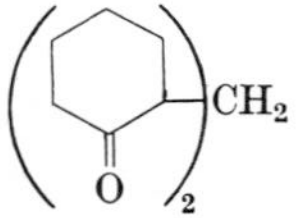 $(C_6H_9O)_2CH_2$ (68)	1841, 277, 550, 552, 1842
	"	$Ca(OH)_2$	2,2,6,6-$(CH_2OH)_4$ (40)	278
	"	$KOCH_3$	2-CH_2OH (60)	550, 277, 278, 1822, 1841
	"	K_2CO_3	2,2-$(CH_2OH)_2$ (—)	277
	" (4 mole equiv.)	H_2SO_4, CH_3CO_2H	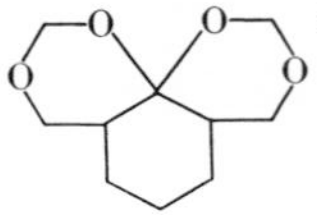 * (8), 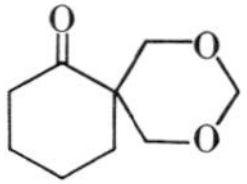 * (—)	460

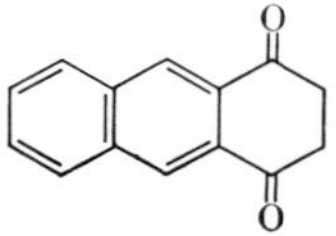

,, (10 mole equiv.)	,,	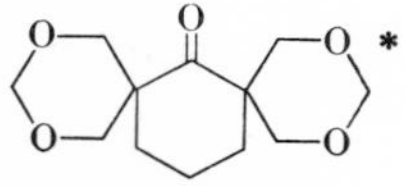 (30)	460
CCl_3CHO	CH_3CO_2K, $(CH_3CO)_2O$	2-RCHOH (70)	287
CH_3CHO	$Ca(OH)_2$	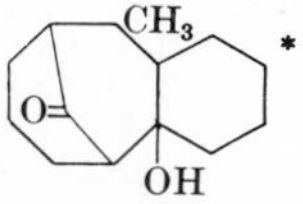 (24)	552–555
,,	NaOH	2-RCHOH (30)	285, 82, 277, 1843
,,	Ion-exchange resin Amberlite IR 120	2-RCH= (Small)	283, 82, 277, 285, 1842
$C(Cl_2)$=CClCHO	H_2SO_4	2,6-$(RCH=)_2$ (26)	289
$C(Cl_2)$=CHCHO	HCl	2,6-$(RCH=)_2$ (76)	639, 638
C_2H_5CHO	$NaOC_2H_5$	2-RCHOH (—)	1843
,,	NaOH, 15–20°	2-RCH= (15)	285, 282, 552, 1843
CH_3CH=CHCHO	KOH	2-RCH= (—)	301
n-C_3H_7CHO	Ion-exchange resin Amberlite IR 120	2-RCH= (—)	283, 282
,,	NaOH	2-RCHOH (56)	1844, 552, 1843
$C(Cl_2)$=CHCH=CHCHO	HCl	2,6-$(RCH=)_2$ (70)	1451
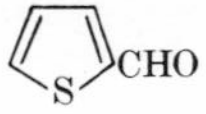	$NaOC_2H_5$	2,6-$(RCH=)_2$ (51)	1823, 303, 1824, 1825, 1845
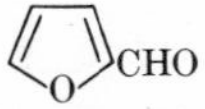	NaOH	2-RCH= (85)	1846–1848, 293, 618

Note: References 668–2359 are on pp. 403–438.

* This is a complete structural formula.

TABLE XIV. Condensation of Aldehydes with Cyclohexanones (*Continued*)

(R in the product is the group R in the aldehyde RCHO.)

Substituent(s) in Cyclohexanone	Aldehyde, RCHO	Catalyst	Substituent(s) in Cyclohexanone Product(s) (Yield, %)	Refs.
None (*contd*).	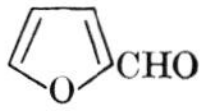 (O)CHO (*contd*).	NaOH	2,6-(RCH=)$_2$ (100)	1824, 618, 1535, 1827, 1828, 1847, 1848
	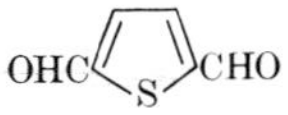 OHC(S)CHO	K_2CO_3	2,6-(RCH=)$_2$ (—)	365
	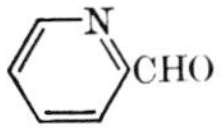 (N)CHO	NaOH	2,6-(RCH=)$_2$ (Excellent)	1825
	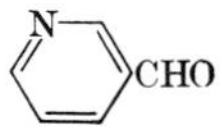 (N)CHO	,,	2,6-(RCH=)$_2$ (Excellent)	1825
	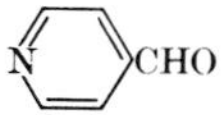 (N)CHO	,,	2,6-(RCH=)$_2$ (Excellent)	1825
	 CH_3(O)CHO	,,	2,6-(RCH=)$_2$ (Excellent)	1825
	n-$C_4H_9O_2CCHO$	None, 100°	2-RCHOH (84)	1468
	$(C_2H_5)_2CHCHO$	$Ba(OH)_2$	2-RCH= (—)	288
	4-BrC_6H_4CHO	NaOH	2,6-RCH=)$_2$ (89)	1849, 282
	,,	,,	2-RCHOH (epimers) (66)	1850
	2-ClC_6H_4CHO	,,	2-RCH= (70), 2,6-(RCH=)$_2$ (—)	302, 292, 293
	4-ClC_6H_4CHO	NaOH, H_2O, C_2H_5OH, rfx., 1.5 hr.	2,6-(RCH=) (91)	1849, 292
	,,	NaOH, H_2O, rfx., 6 hr.	2-RCH= (—)	293
	,,	NaOH, H_2O, 25°, 10 hr.	2-RCHOH (88)	1850

4-IC_6H_4CHO	NaOH, H_2O, C_2H_5OH, rfx., 1.5 hr.	2,6-$(RCH{=})_2$ (84)	1849
"	NaOH, H_2O, 25°, 11 hr.	2-RCHOH (epimers) (84)	1850
2-$O_2NC_6H_4CHO$	HCO_2H	2,6-$(RCH{=})_2$ (—)	292
3-$O_2NC_6H_4CHO$	NaOH	2,6-$(RCH{=})_2$ (34)	1798, 292
4-$O_2NC_6H_4CHO$	NaOH, C_2H_5OH, warm	2,6-$(RCH{=})_2$ (39)	1798, 292
"	NaOH, H_2O, 0°	2-RCHOH (*threo*) (54), 2-RCHOH (*erythro*) (30)	64, 291
$C(Cl_2){=}CH(CH{=}CH)_2$-CHO	HCl	2,6-$(RCH{=})_2$ (66)	1451
C_6H_5CHO	NaOH, H_2O, 25°, 4 hr.	2-RCHOH (100)	291, 292, 294
"	NaOH, H_2O, 25°, 10 days	2-RCH= (76)	294, 293, 295, 298, 299, 1097, 1846, 1851, 1852
"	NaOH, C_2H_5OH	2,6-$(RCH{=})_2$ (98)	294, 292, 1097, 1828, 1851, 1852
2-Furyl-CH=CHCHO	NaOH	2,6-$(RCH{=})_2$ (93)	1471
2-HOC_6H_4CHO	"	2,6-$(RCH{=})_2$ (—)	1832
4-HOC_6H_4CHO	"	2,6-$(RCH{=})_2$ (—)	1572
n-$C_6H_{13}CHO$	Ion-exchange resin Amberlite IR 120	2-RCH= (—)	283, 282
3,4-$(CH_2O_2)C_6H_3CHO$	$NaOC_5H_{11}$-t	2-RCH= (61)	295, 299
"	NaOH	2,6-$(RCH{=})_2$ (—)	292
2-$CH_3C_6H_4CHO$	KOH	2-RCH= (71), 2,6-$(RCH{=})_2$ (—)	301
4-$CH_3C_6H_4CHO$	"	2-RCH= (44)	298
"	NaOH	2,6-$(RCH{=})_2$ (70)	282

Note: References 668–2359 are on pp. 403–438.

TABLE XIV. CONDENSATION OF ALDEHYDES WITH CYCLOHEXANONES (*Continued*)

(R in the product is the group R in the aldehyde RCHO.)

Substituent(s) in Cyclohexanone	Aldehyde, RCHO	Catalyst	Substituent(s) in Cyclohexanone Product(s) (Yield, %)	Refs.
None (*contd*).	2-$CH_3OC_6H_4CHO$	NaOH	2-RCHOH (98)	291
	3-$CH_3OC_6H_4CHO$	KOH	2-RCH= (70)	302
	4-$CH_3OC_6H_4CHO$	NaOH, H_2O, 25°	2-RCHOH (83)	291
	”	KOH, H_2O, rfx.	2-RCH= (50)	298, 1852
	”	NaOH, C_2H_5OH	2,6-$(RCH=)_2$ (—)	1572, 292, 1563
	=CHCHO	NaOH	2-RCH= (—)	1853
	CHO	”	2,6-$(RCH=)_2$ (—)	1854
	$C_6H_5CH=CHCHO$	”	2,6-$(RCH=)_2$ (—)	292, 1563
	3-$CH_3CONHC_6H_4$-CHO	”	2,6-$(RCH=)_2$ (82)	1798
	4-$CH_3CONHC_6H_4$-CHO	”	2,6-$(RCH=)_2$ (75)	1798, 1834
	2,4-$(CH_3)_2C_6H_3CHO$	”	2,6-$(RCH=)_2$ (70)	282
	3,4-$(CH_3O)_2C_6H_3CHO$	”	2-RCH= (52)	299
	4-$(CH_3)_2NC_6H_4CHO$	$NaOC_2H_5$	2-RCH= (62)	1798
	”	NaOH	2,6-$(RCH=)_2$ (64)	1849, 292, 1798
	4-*i*-$C_3H_7C_6H_4CHO$	”	2,6-$(RCH=)_2$ (—)	292, 1855
	i-$C_3H_7(CH_2)_3CH$-$(CH_3)CH_2CHO$	”	2-RCH= (70)	1496
	”	KOH, NaOH	2-RCH= (29), 2,6-$(RCH=)_2$ (16)	1496
	$C_6H_5(CH=CH)_2CHO$	NaOH	2,6-$(RCH=)_2$ (71)	1000
	CHO, OCH_3	”	2,6-$(RCH=)_2$ (—)	1854

	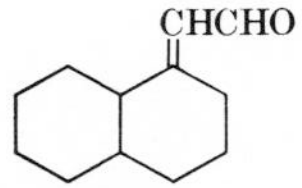 CHCHO	,,	2-RCH= (—), 2,6-(RCH=)$_2$ (—)	1856
2-HO	CH_2O	KOH	2-CH_2OH (52)	230, 1767
4-HO	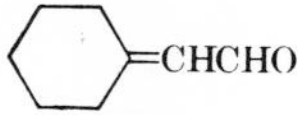 =CHCHO	NaOH	2-(RCH=)-4-HO (68)	1857, 1858
	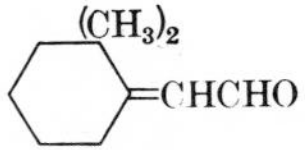 $(CH_3)_2$ =CHCHO	,,	2-(RCH=)-4-HO (—)	1857
2-CH_3	CH_2O	$Ca(OH)_2$	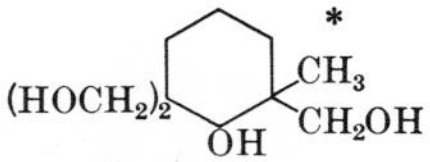 $(HOCH_2)_2$ OH CH_3 CH_2OH * (—)	278
	,,	$KOCH_3$, CH_3OH, < 40°	2-CH_3-2-CH_2OH (18)	550, 1822
	,,	$KOCH_3$, CH_3OH, rfx., few min.	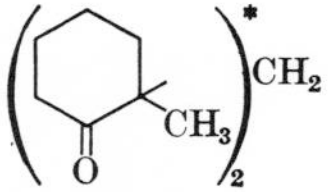 (O CH_3)$_2$ CH_2 * (30)	550, 1842
	$C(Cl_2)$=CHCHO	HCl	2-CH_3-2-[$C(Cl_2)$=CHCHOH]-6-[$C(Cl_2)$=CHCH=] (53)	638
	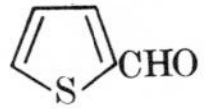 S CHO	NaOH	2-CH_3-6-RCH= (60)	1845
	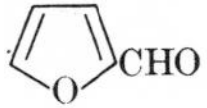 O CHO	$NaOCH_3$	2-CH_3-6-RCH= (84)	1272, 618, 1847, 1859, 1860

Note: References 668–2359 are on pp. 403–438.

* This is a complete structural formula.

TABLE XIV. CONDENSATION OF ALDEHYDES WITH CYCLOHEXANONES (*Continued*)

(R in the product is the group R in the aldehyde RCHO.)

Substituent(s) in Cyclohexanone	Aldehyde, RCHO	Catalyst	Substituent(s) in Cyclohexanone Product(s) (Yield, %)	Refs.
2-CH_3 (*contd.*)	4-ClC_6H_4CHO	$NaOC_2H_5$	2-CH_3-6-RCH= (32)	304
	C_6H_5CHO	NaOH	*trans*-2-CH_3-6-RCH= (69)	1861, 80, 304, 530, 1835–1837, 1851
	,,	HCl	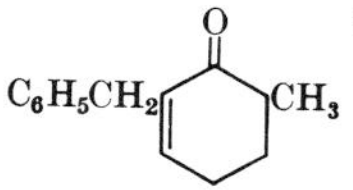 2-$C_6H_5CH_2$-6-CH_3-cyclohex-2-enone (structure) * (83)	304, 530, 1835, 1836
3-CH_3	CH_2O	$KOCH_3$, CH_3OH, <40°	2-CH_2OH-5-CH_3 (—)	550
	,,	$KOCH_3$, CH_3OH, rfx., 2–3 min.	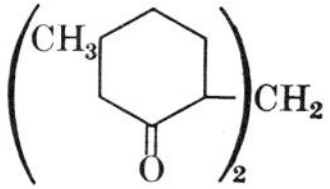 (3-CH_3-cyclohexanon-6-yl)$_2CH_2$ (structure) *, † (—)	550
	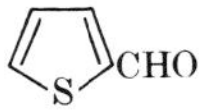 2-thiophenecarboxaldehyde (structure)	NaOH	2,6-$(RCH=)_2$ (—)	1829
	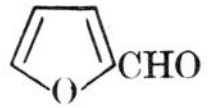 2-furaldehyde (structure)	KOH, 6 hr.	2- or 6-RCH= (70)	1847, 618
	,,	$NaOCH_3$	2,6-$(RCH=)_2$ (55)	618
	2-ClC_6H_4CHO	NaOH	6-RCH= (—)	1862
	4-ClC_6H_4CHO	,,	6-RCH= (—)	1862
	2-HO-5-ClC_6H_3CHO	,,	6-RCH= (—)	1863
	2-$O_2NC_6H_4CHO$	HCO_2H	2,6-$(RCH=)_2$ (—)	292
	C_6H_5CHO	KOH	6-RCH= (45)	1851, 318, 1862, 1864, 1865
	,,	$NaOC_2H_5$	6- and 2,6-$(RCH=)_2$ (—)	318
	,,	,,	6-RCH= (—), $C_{14}H_{20}O$ (—)	1865

	2-HOC_6H_4CHO	NaOH	6-RCH= (35)	1832
	4-HOC_6H_4CHO	HCl	2,6-$(RCH{=})_2$‡ (80)	1572
	3,4-$(CH_2O_2)C_6H_3CHO$	NaOH	6-RCH= (—)	1862
	4-$CH_3C_6H_4CHO$	,,	6-RCH= (—)	1862
	4-$CH_3OC_6H_4CHO$	$NaOCH_3$	6-RCH= (—)	1864, 1572, 1862
	,,	HCl	2,6-$(RCH{=})_2$‡ (—)	1572
	2-HO-5-$CH_3C_6H_3CHO$	NaOH	6-RCH= (—)	1863
	2,3-$(CH_3O)_2C_6H_3CHO$	,,	6-RCH= (—)	1862
	4-$(CH_3)_2NC_6H_4CHO$	,,	6-RCH= (—)	1862
	4-*i*-$C_3H_7C_6H_4CHO$	$NaOCH_3$	6-RCH= (—), 2,6-$(RCH{=})_2$ (—)	1864, 1862
4-CH_3	CH_2O	$Ca(OH)_2$	[cyclohexanol ring: 1-CH_3, $(HOCH_2)_2$, $(CH_2OH)_2$, OH]* (—)	278
	,,	$KOCH_3$	2-CH_2OH-4-CH_3 (34)	550, 1822
	,,	,,	[bis(4-methyl-2-oxocyclohexyl)methane: $(CH_3$-cyclohexanone$)_2CH_2$]* (50)	550
	$C(Cl_2){=}CHCHO$	HCl	2,6-$(RCH{=})_2$-4-CH_3 (70)	638
	2-thiophenecarboxaldehyde (thienyl-CHO)	NaOH	2,6-$(RCH{=})_2$-4-CH_3 (100)	1829, 1845
	2-furaldehyde (furyl-CHO)	KOH	2-$(RCH{=})_2$-4-CH_3 (75)	1847, 618, 1272, 1860
	,,	$NaOCH_3$	2,6-$(RCH{=})_2$-4-CH_3 (55)	618, 1847, 1860

Note: References 668–2359 are on pp. 403–438.

* This is a complete structural formula.
† This structure was suggested but not established.
‡ This product might contain an α,β-unsaturated endocyclic double bond.[304]

TABLE XIV. Condensation of Aldehydes with Cyclohexanones (*Continued*)
(R in the product is the group R in the aldehyde RCHO.)

Substituent(s) in Cyclohexanone	Aldehyde, RCHO	Catalyst	Substituent(s) in Cyclohexanone Product(s) (Yield, %)	Refs.
4-CH_3 (*contd.*)	N CHO	NaOH	2,6-(RCH=)$_2$-4-CH_3 (Excellent)	1825
	N CHO	NaOH	2,6-(RCH=)$_2$-4-CH_3 (Excellent)	1825
	2-$O_2NC_6H_4CHO$	HCO_2H	2,6-(RCH=)$_2$-4-CH_3 (—)	292
	C_6H_5CHO	KOH	2-(RCH=)-4-CH_3 (60), 2,6-(RCH=)$_2$-4-CH_3 (35)	1851, 1866
	2-HOC_6H_4CHO	NaOH	2,6-(RCH=)$_2$-4-CH_3 (50)	1832
	n-$C_6H_{13}CHO$	"	2-(RCH=)-4-CH_3 (36)	284
	4-$CH_3OC_6H_4CHO$	KOH	2-(RCH=)-4-CH_3 (65), 2,6-(RCH=)$_2$-4-CH_3 (—)	300
	4-$CH_3CONHC_6H_4$-CHO	NaOH	2,6-(RCH=)$_2$-4-CH_3 (—)	1834
	4-i-$C_3H_7C_6H_4CHO$	KOH	2-(RCH=)-4-CH_3 (—)	1867
	$C_2H_5O_2C(CH_2)_7CHO$	NaOH	2-(RCH=)-4-CH_3 (—)	1868
4-CH_3O	=CHCHO	"	2-(RCH=)-4-CH_3O (25)	1869, 1856
	2-Cholestanylidene-ethanal	"	*trans*-2-(RCH=)-4-(CH_3O) (10), *cis*-2-(RCH=)-4-(CH_3O) (10), 2,6-(RCH=)$_2$-4-CH_3O (67)	1870, 1088
4-CH_3CO_2	H CHCHO CH_3 C_9H_{17}	$NaOC_2H_5$	2-(RCH=)-4-CH_3CO_2 (17)	1871

2,5-$(OH)_2$-2,5-$(CH_3)_2$-4-oxo	C_6H_5CHO	KOH	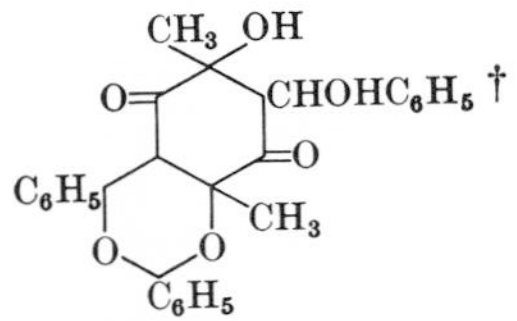	(—)	1709
2,2-$(CH_3)_2$	,,	$NaOC_2H_5$	2,2-$(CH_3)_2$-6-RCH=	(30)	531, 1833
	,,	HCl	2,2-$(CH_3)_2$-6-RCH=	(46)	531
2,4-$(CH_3)_2$	2-ClC_6H_4CHO	NaOH	2,4-$(CH_3)_2$-6-RCH=	(—)	292
	4-ClC_6H_4CHO	,,	2,4-$(CH_3)_2$-6-RCH=	(—)	292
	3-$O_2NC_6H_4CHO$	,,	2,4-$(CH_3)_2$-6-RCH=	(—)	292
	4-$O_2NC_6H_4CHO$	,,	2,4-$(CH_3)_2$-6-RCH=	(—)	292
	C_6H_5CHO	,,	2,4-$(CH_3)_2$-6-RCHOH	(—)	292
	,,	,,	2,4-$(CH_3)_2$-6-RCH=	(—)	292
	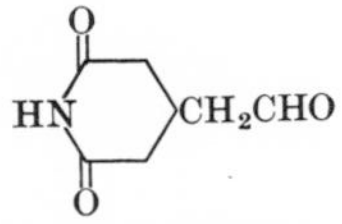	$C_6H_5N(CH_3)$-MgBr	2,4-$(CH_3)_2$-6-RCHOH	(12)	1872
	3,4-$(CH_2O_2)C_6H_3CHO$	NaOH	2,4-$(CH_3)_2$-6-RCH=	(—)	292
	4-$CH_3OC_6H_4CHO$	,,	2,4-$(CH_3)_2$-6-RCH=	(—)	292
	C_6H_5CH=$CHCHO$	,,	2,4-$(CH_3)_2$-6-RCH=	(—)	292
	4-$(CH_3)_2NC_6H_4CHO$	,,	2,4-$(CH_3)_2$-6-RCH=	(—)	292
	4-*i*-$C_3H_7C_6H_4CHO$	,,	2,4-$(CH_3)_2$-6-RCH=	(—)	292
2,5-$(CH_3)_2$	2-ClC_6H_4CHO	,,	2,5-$(CH_3)_2$-6-RCH=	(—)	292
	4-ClC_6H_4CHO	,,	2,5-$(CH_3)_2$-6-RCH=	(—)	292
	3-$O_2NC_6H_4CHO$	,,	2,5-$(CH_3)_2$-6-RCH=	(—)	292
	4-$O_2NC_6H_4CHO$	,,	2,5-$(CH_3)_2$-6-RCH=	(—)	292
	C_6H_5CHO	,,	2,5-$(CH_3)_2$-6-RCHOH (—), 2,5-$(CH_3)_2$-6-RCH=	(—)	292
	3,4-$(CH_2O_2)C_6H_3CHO$	,,	2,5-$(CH_3)_2$-6-RCH=	(—)	292
	4-$CH_3OC_6H_4CHO$	,,	2,5-$(CH_3)_2$-6-RCH=	(—)	292

Note: References 668–2359 are on pp. 403–438.

† This structure was suggested but not established.

TABLE XIV. CONDENSATION OF ALDEHYDES WITH CYCLOHEXANONES (*Continued*)
(R in the product is the group R in the aldehyde RCHO.)

Substituent(s) in Cyclohexanone	Aldehyde, RCHO	Catalyst	Substituent(s) in Cyclohexanone Product(s) (Yield, %)	Refs.
2,5-$(CH_3)_2$ (*contd.*)	$C_6H_5CH{=}CHCHO$	NaOH	2,5-$(CH_3)_2$-6-RCH= (—)	292
	4-$(CH_3)_2NC_6H_4CHO$	,,	2,5-$(CH_3)_2$-6-RCH= (—)	292
	4-*i*-$C_3H_7C_6H_4CHO$	,,	2,5-$(CH_3)_2$-6-RCH= (—)	292
2,6-$(CH_3)_2$	C_6H_5CHO	HCl	CH_3 CH_3 O R O R (88)	530
4-C_3H_7-*i*	,,	$NaOC_2H_5$	2,6-$(RCH{=})_2$-4-C_3H_7-*i* (—)	320
2-CH_3-2-C_2H_5 §	,,	HCl	2-CH_3-2-C_2H_5-6-RCH=‡ (32)	531
4-CH_3-4-C_2H_5	,,	$(CH_2)_5NH$	2,6-$(RCH{=})_2$-4-CH_3-4-C_2H_5 (80)	1873
	3,4-$(CH_2O_2)C_6H_3CHO$	,,	2,6-$(RCH{=})_2$-4-CH_3-4-C_2H_5 (80)	1873
3,3,4-$(CH_3)_3$	4-$O_2NC_6H_4CHO$	NaOH	2,6-$(RCH{=})_2$-3,3,4-$(CH_3)_3$ (—)	1839
2-CH_3-5-$CH_3CH{=}CH_2$	C_6H_5CHO	$NaOC_2H_5$	2-CH_3-5-$CH_3CH{=}CH_2$-6-RCH= (20)	317, 319
2-[$(CH_3)_2C{=}$]-5-CH_3	,,	,,	2-[$(CH_3)_2C{=}$]-5-CH_3-6-RCH= (—)	318, 319
	2-HOC_6H_4CHO	NaOH	$(2\text{-}HOC_6H_4CH{=}CH)_2CO$ (—), CH_3 O (—)	1832
2-$(CH_2{=}CHCH_2)$-4-CH_3	C_6H_5CHO	$NaOCH_3$	2-$(CH_2{=}CHCH_2)$-4-CH_3-6-RCH= (40), CH_3† $CH_2{=}CHCH_2$ $CHOHC_6H_5$ OH (2.5)	473
4-*t*-C_4H_9	*n*-$C_6H_{13}CHO$	NaOH	2-(RCH=)-4-C_4H_9-*t* (35)	284
	4-$CH_3CONHC_6H_4$-CHO	,,	2,6-$(RCH{=})_2$-4-C_4H_9-*t* (—)	1834
	4-$(CH_3)_2NC_6H_4CHO$	,,	2,6-$(RCH{=})_2$-4-C_4H_9-*t* (—)	1834

2-CH_3-2-n-C_3H_7	C_6H_5CHO	$NaOCH_3$	2-CH_3-2-n-C_3H_7-6-RCH= (—)	1835
	,,	HCl	2-CH_3-2-n-C_3H_7-6-RCH= (—)	473, 1835
2-n-C_3H_7-4-CH_3	,,	$NaOCH_3$	2-n-C_3H_7-4-CH_3-6-RCH= (30), $C_{17}H_{26}O_2$ (8)	473
	,,	HCl	2-n-C_3H_7-4-CH_3-6-RCH=‡ (38)	473
2-i-C_3H_7-5-CH_3	,,	,,	2-i-C_3H_7-5-CH_3-6-RCH=‡ (88)	317–319
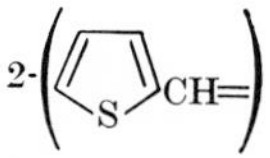	,,	KOH	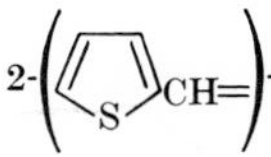-6-RCH= (80)	303
2,3,4,5-$(CH_3)_4$	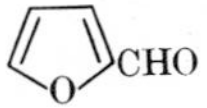	,,	2,3,4,5-$(CH_3)_4$-6-RCH= (86)	1874
4,4-$(CH_2)_5$	C_6H_5CHO	$(CH_2)_5NH$	2,6-$(RCH{=})_2$-4,4-$(CH_2)_5$ (80)	1873
	3,4-$(CH_2O_2)C_6H_3CHO$	,,	2,6-$(RCH{=})_2$-4,4-$(CH_2)_5$ (80)	1873
2-C_6H_5	C_6H_5CHO	NaOH	2-C_6H_5-6-RCH= (—)	1875
2-[$(CH_2)_5C$=]	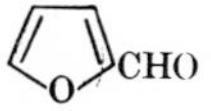	,,	2-[$(CH_2)_5C$=]-6-RCH= (—)	282
	C_6H_5CHO	,,	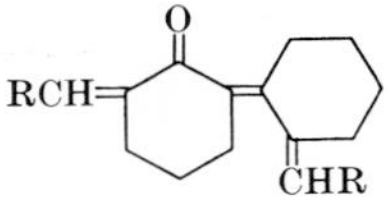 (40)	1178

Note: References 668–2359 are on pp. 403–438.

† This structure was suggested but not established.
‡ This product might contain an α,β-unsaturated endocyclic double bond.[304]
§ The sample of this compound used contained 15% of 2-ethyl-6-methylcyclohexanone which formed 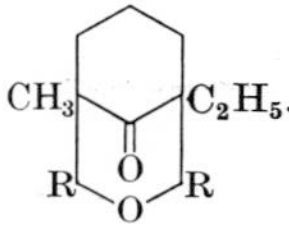.

TABLE XIV. CONDENSATION OF ALDEHYDES WITH CYCLOHEXANONES (*Continued*)
(R in the product is the group R in the aldehyde RCHO.)

Substituent(s) in Cyclohexanone	Aldehyde, RCHO	Catalyst	Substituent(s) in Cyclohexanone Product(s) (Yield, %)	Refs.
2-[$(CH_2)_5C$=] (*contd.*)	4-$CH_3OC_6H_5CHO$	NaOH	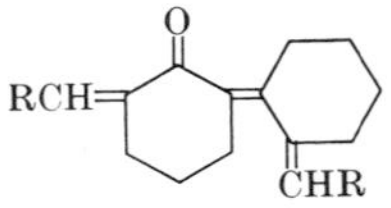 (—)	1178
4-$(CH_2)_5CH$	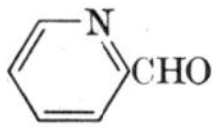	,,	2,6-(RCH=)$_2$-4-[$(CH_2)_5CH$] (Excellent)	1825
	4-ClC_6H_4CHO	,,	2,6-(RCH=)$_2$-4-[$(CH_2)_5CH$] (—)	1876
	4-FC_6H_4CHO	,,	2,6-(RCH=)$_2$-4-[$(CH_2)_5CH$] (—)	1876
	C_6H_5CHO	,,	2,6-(RCH=)$_2$-4-[$(CH_2)_5CH$] (80)	1876
	3,4-$(CH_2O_2)C_6H_3CHO$	,,	2,6-(RCH=)$_2$-4-[$(CH_2)_5CH$] (80)	1876
	4-$(CH_3)_2NC_6H_4CHO$	,,	2,6-(RCH=)$_2$-4-[$(CH_2)_5CH$] (—)	1876
2-C_6H_5CH=	C_6H_5CHO	,,	2-(C_6H_5CH=)-6-RCHOH (—)	1834
	,,	KOH	2-(C_6H_5CH=)-6-RCH= (40)	1852, 294
	4-$CH_3OC_6H_4CHO$	,,	2-(C_6H_5CH=)-6-RCH= (35)	1852
	4-*i*-$C_3H_7C_6H_4CHO$	,,	2-(C_6H_5CH=)-6-RCH= (50)	1867
2-$C_6H_5CH_2$	C_6H_5CHO	,,	2-$C_6H_5CH_2$-6-RCH= (—)	1851
2-C_6H_5CHOH	,,	NaOH	2-C_6H_5CHOH-6-RCH= (48)	294
2-(C_6H_5CH=)-4-CH_3	,,	KOH	2-(C_6H_5CH=)-4-CH_3-6-RCH= (40)	1851
3-CH_3-3-(4-$CH_3C_6H_4$)	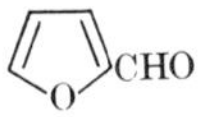	NaOH	3-CH_3-3-(4-$CH_3C_6H_4$)-6-RCH= (—)	1877

2-(4-$CH_3C_6H_4CH_2$)	C_6H_5CHO	KOH	2-(4-$CH_3C_6H_4CH_2$)-6-RCH= (—)	1851
2-[2,3-$(CH_3O)_2C_6H_3$]	"	"	2-[2,3-$(CH_3O)_2C_6H_3$]-6-RCH= (—)	1878
2-($NCCH_2CH_2$)-2-C_6H_5	"	NaOH	2-($NCCH_2CH_2$)-2-C_6H_5-6-RCH= (95)	1875
2-(4-$CH_3OC_6H_4CH$=)-4-CH_3	"	KOH	2-(4-$CH_3OC_6H_4CH$=)-4-CH_3-6-RCH= (100)	300
	4-$CH_3OC_6H_4CHO$	"	2-(4-$CH_3OC_6H_4CH$=)-4-CH_3-6-RCH= (—)	300
2-(4-$CH_3OC_6H_4CH$=)-5-CH_3	4-HOC_6H_4CHO	HCl	2-(4-$CH_3OC_6H_4CH$=)-5-CH_3-6-RCH=‡ (82)	1572
2-[4-$(CH_3)_2NC_6H_4CH$=]	C_6H_5CHO	$NaOC_2H_5$	2-[4-$(CH_3)_2NC_6H_4CH$=]-6-RCH= (90)	1798
	4-$CH_3OC_6H_4CHO$	"	2-[4-$(CH_3)_2NC_6H_4CH$=]-6-RCH= (—)	1798
2-[2-CH_3O-3-$C_2H_5OC_6H_3$]	C_6H_5CHO	KOH	2-[2-CH_3O-3-$C_2H_5OC_6H_3$]-6-RCH= (—)	1878
2,2-$(C_6H_5)_2$	(furan ring, O) CHO	$NaOCH_3$	2,2-$(C_6H_5)_2$-6-RCH= (66)	1272
	2-ClC_6H_4CHO	KOH	2,2-$(C_6H_5)_2$-6-RCH= (84)	1365
	C_6H_5CHO	"	2,2-$(C_6H_5)_2$-6-RCH= (62)	1365, 304

Note: References 668–2359 are on pp. 403–438.

‡ This product might contain an α,β-unsaturated endocyclic double bond.[304]

TABLE XV. Condensation of Aldehydes with Alicyclic Ketones Other than Cyclopentanones and Cyclohexanones

(R in the product is the group R in the aldehyde RCHO.)

Ketone	Aldehyde, RCHO	Catalyst	Product(s) (Yield, %)	Refs.
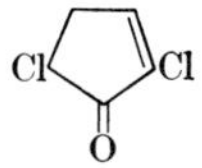 Cl, Cl, O	C_6H_5CHO	HCl	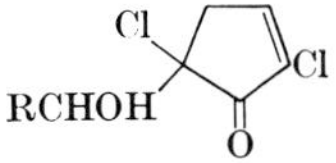 Cl, RCHOH, Cl, O (40)	1821
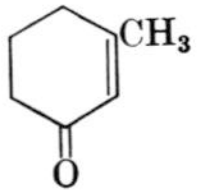 CH_3, O	,,	$NaOC_2H_5$	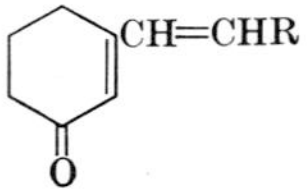 CH=CHR, O (—)	321
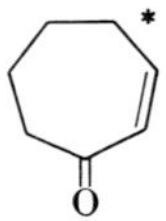 *, O	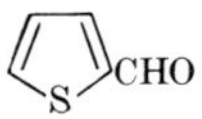 CHO, S	HCl	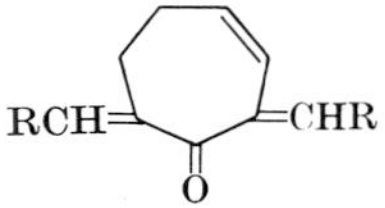 RCH=, =CHR, O (—) A, R = 2-(C_4H_3S)	1879
	C_6H_5CHO	,,	A, R = C_6H_5 (21)	1879
	3,4-$(CH_2O_2)C_6H_3CHO$	,,	A, R = 3,4-$(CH_2O_2)C_6H_3$ (8)	1879
	2-$CH_3OC_6H_4CHO$	,,	A, R = 2-$CH_3OC_6H_4$ (22)	1879
	2-$C_{10}H_7CHO$	,,	A, R = 2-$C_{10}H_7$ (31)	1879
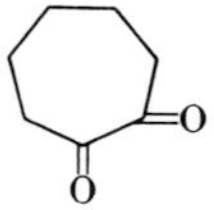 O, O	C_6H_5CHO	$(CH_2)_5NH$	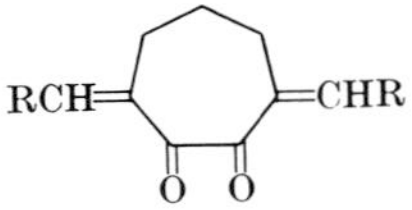 RCH=, =CHR, O, O (34)	1880, 325, 326
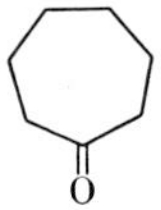 O	CH_2O	$KOCH_3$	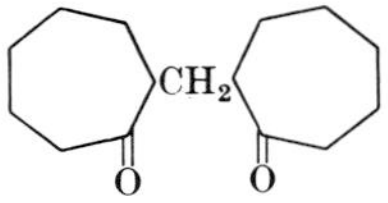 CH_2, O, O (40)	550

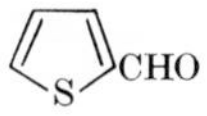 S CHO	$NaOC_2H_5$	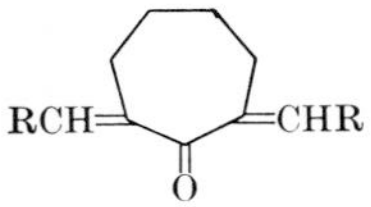 RCH= =CHR, O (48)	1823, 1845
		B, R = 2-(C_4H_3S)	
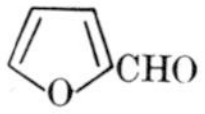 O CHO	KOH	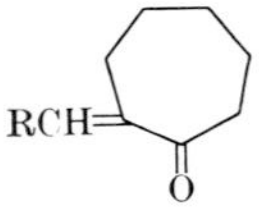 RCH=, O (28),	1848
		B, R = 2-(C_4H_3O) (6)	
,,	$NaOCH_3$	B, R = 2-(C_4H_3O) (80)	1272, 303, 1881
 OHC S CHO	K_2CO_3	B, R = 2-(C_4H_2S)CHO-5 (—)	365
2-ClC_6H_4CHO	$NaOC_2H_5$	B, R = 2-ClC_6H_4 (53)	1881
3-$O_2NC_6H_4CHO$	NaOH	B, R = 3-$O_2NC_6H_4$ (—)	292
4-$O_2NC_6H_4CHO$	,,	B, R = 4-$O_2NC_6H_4$ (—)	292
C_6H_5CHO	KOH	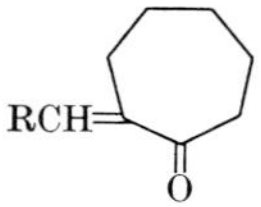 RCH=, O (89)	302
,,	$NaOCH_3$	B, R = C_6H_5 (78)	1882, 292, 318, 1881, 2332
2-HOC_6H_4CHO	NaOH	B, R = 2-HOC_6H_4 (11)	1832

Note: References 668–2359 are on pp. 403–438.

* The cycloheptenone was formed *in situ* from its ethylene ketal.

TABLE XV. CONDENSATION OF ALDEHYDES WITH ALICYCLIC KETONES OTHER THAN CYCLOPENTANONES AND CYCLOHEXANONES (*Continued*)

(R in the product is the group R in the aldehyde RCHO.)

Ketone	Aldehyde, RCHO	Catalyst	Product(s) (Yield, %)	Refs.
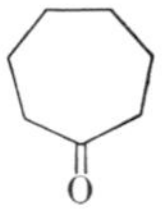 Cycloheptanone (*contd.*)	 o-Phthalaldehyde ($C_6H_4(CHO)_2$)	NaOH	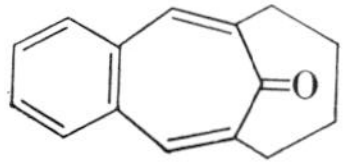 [structure] (31)	309
	3,4-$(CH_2O_2)C_6H_3CHO$	$NaOC_2H_5$	B, R = 3,4-$(CH_2O_2)C_6H_3$ (47)	1881, 292
	4-$CH_3C_6H_4CHO$	$NaOCH_3$	B, R = 4-$CH_3C_6H_4$ (60)	1883, 1881
	2-$CH_3OC_6H_4CHO$	KOH; $NaOCH_3$	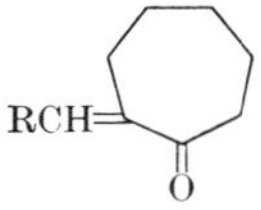 2-(RCH=)cycloheptanone (—), B, R = 2-$CH_3OC_6H_4$ (—)	305
	3-$CH_3OC_6H_4CHO$	"	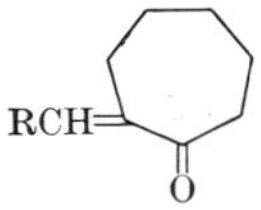 2-(RCH=)cycloheptanone (—), B, R = 3-$CH_3OC_6H_4$ (—)	305
	4-$CH_3OC_6H_4CHO$	$NaOCH_3$	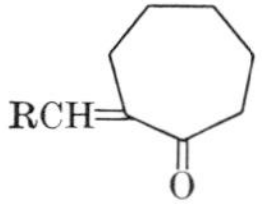 2-(RCH=)cycloheptanone (50), B, R = 4-$CH_3OC_6H_4$ (35–94)	1884, 292, 305, 1881
	$C_6H_5CH{=}CHCHO$	$NaOC_2H_5$	B, R = $C_6H_5CH{=}CH$ (26)	1881, 292
	2,3-$(CH_3O)_2C_6H_3CHO$	"	B, R = 2,3-$(CH_3O)_2C_6H_3$ (53)	1881

	$3,4\text{-}(CH_3O)_2C_6H_3CHO$	$NaOCH_3$	RCH= O (64),	1884, 305
			B, R = $3,4\text{-}(CH_3O)_2C_6H_3$ (—)	
	,,	,,	B, R = $3,4\text{-}(CH_3O)_2C_6H_3$ (90)	1884, 1881
	$4\text{-}(CH_3)_2NC_6H_4CHO$	$NaOC_2H_5$	B, R = $4\text{-}(CH_3)_2NC_6H_4$ (25)	1881
	$4\text{-}i\text{-}C_3H_7C_6H_4CHO$	,,	B, R = $4\text{-}i\text{-}C_3H_7C_6H_4$ (43)	1881
	$1\text{-}C_{10}H_7CHO$	,,	B, R = $1\text{-}C_{10}H_7$ (40)	1881
CH_3 CH_3 O	C_6H_5CHO	,,	RCH=CH CH_3 O (—)	321, 46, 1837
CH_3 O	CHO O	$NaOCH_3$	RCH= CH_3 O (75)	1272
O	$O_2NCH(CHO)_2$	NaOH	$(CH_2)_5$ O= =$NO_2^{\ominus}Na^{\oplus}$ (15)	310
	C_6H_5CHO	KOH	RCHOH O (25)	307
	,,	,,	RCH= O (35)	307

Note: References 668–2359 are on pp. 403–438.

TABLE XV. CONDENSATION OF ALDEHYDES WITH ALICYCLIC KETONES OTHER THAN CYCLOPENTANONES AND CYCLOHEXANONES (*Continued*)

(R in the product is the group R in the aldehyde RCHO.)

Ketone	Aldehyde, RCHO	Catalyst	Product(s) (Yield, %)	Refs.
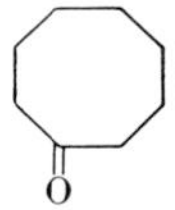 Cyclooctanone (*contd.*)	 o-Phthalaldehyde ($C_6H_4(CHO)_2$)	NaOH	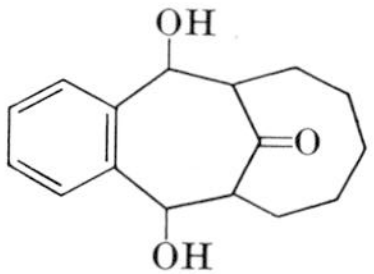 Benzo-fused dihydroxy bicyclic ketone (OH, OH, =O) (56)	309
	2-$CH_3OC_6H_4CHO$	$NaOC_2H_5$	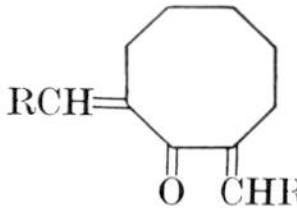 RCH= cyclooctanone =CHR (O, CHR) (—) C, R=2-$CH_3OC_6H_4$	306, 305
	3-$CH_3OC_6H_4CHO$	,,	C, R=3-$CH_3OC_6H_4$ (—)	306, 305
	4-$CH_3OC_6H_4CHO$	,,	C, R=4-$CH_3OC_6H_4$ (—)	306, 305
	2,4-$(CH_3O)_2C_6H_3CHO$	,,	C, R=2,4-$(CH_3O)_2C_6H_3$ (—)	306, 305
	2,5-$(CH_3O)_2C_6H_3CHO$	,,	C, R=2,5-$(CH_3O)_2C_6H_3$ (—)	306, 305
	3,4-$(CH_3O)_2C_6H_3CHO$	,,	C, R=3,4-$(CH_3O)_2C_6H_3$ (—)	306, 305
	3,4,5-$(CH_3O)_3C_6H_2CHO$	,,	C, R=3,4,5-$(CH_3O)_3C_6H_2$ (—)	306, 305
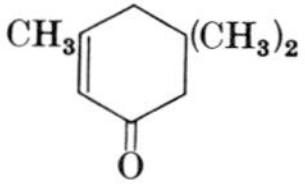 CH_3, $(CH_3)_2$ cyclohexenone	C_6H_5CHO	,,	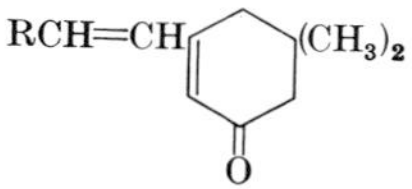 RCH=CH–, $(CH_3)_2$ cyclohexenone (—)	321, 322, 1837
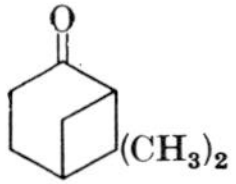 Bicyclic ketone, $(CH_3)_2$	,,	,,	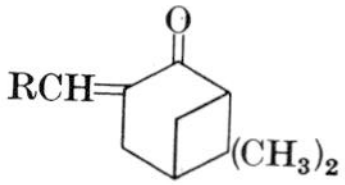 RCH= bicyclic ketone, $(CH_3)_2$ (—)	1885, 315
	,,	HCl	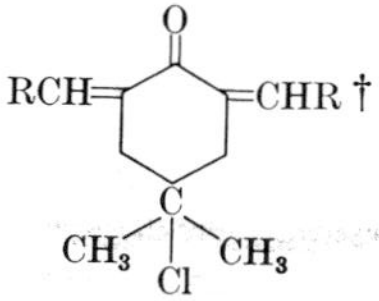 RCH= cyclohexanone =CHR †, C(CH_3)(CH_3)Cl (—)	315

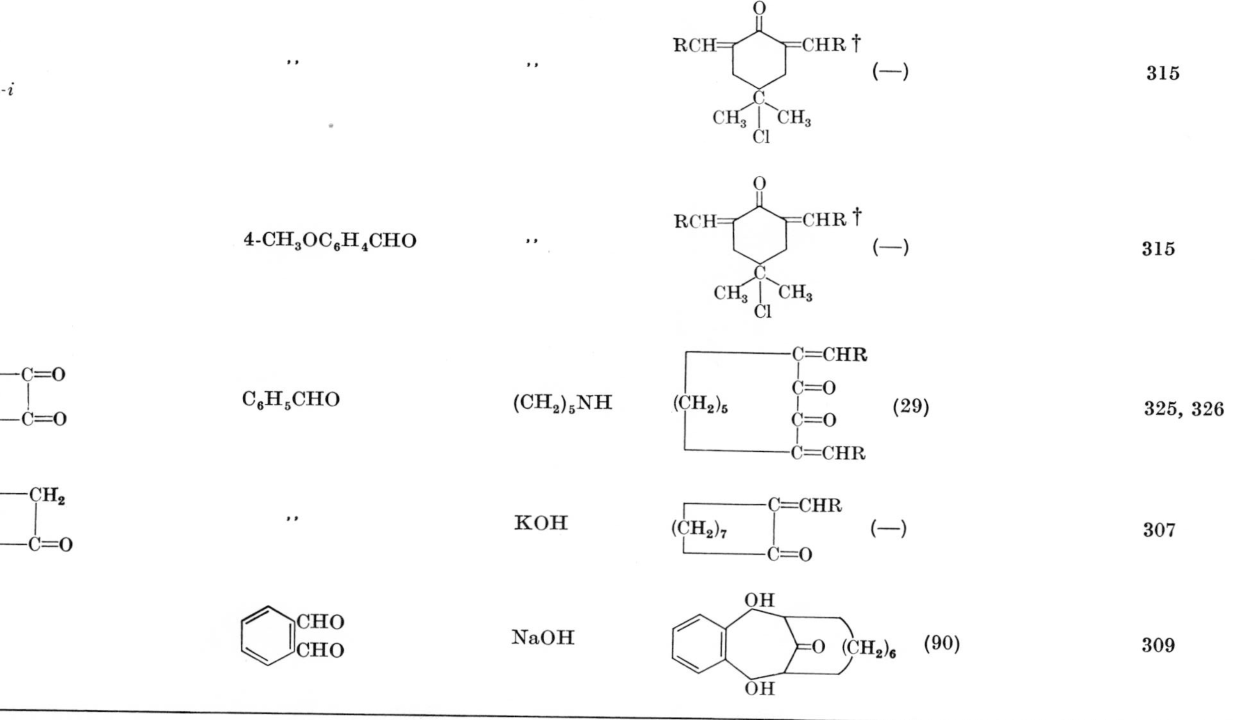

Ketone	Aldehyde	Catalyst	Product	(Yield, %)	Refs.
	,,	,,		(—)	315
	4-$CH_3OC_6H_4CHO$	,,		(—)	315
	C_6H_5CHO	$(CH_2)_5NH$		(29)	325, 326
	,,	KOH		(—)	307
		NaOH		(90)	309

Note: References 668–2359 are on pp. 403–438.

† This product might contain an α,β-unsaturated endocyclic double bond.[304]

TABLE XV. CONDENSATION OF ALDEHYDES WITH ALICYCLIC KETONES OTHER THAN CYCLOPENTANONES AND CYCLOHEXANONES (*Continued*)

(R in the product is the group R in the aldehyde RCHO.)

Ketone	Aldehyde, RCHO	Catalyst	Product(s) (Yield, %)	Refs.
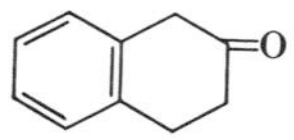	C_6H_5CHO	HCl	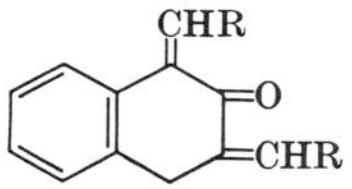 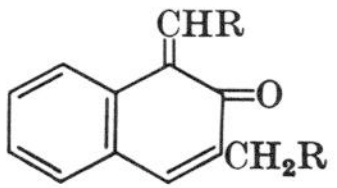 CHR, =O, =CHR or CHR, =O, CH_2R (75)	1886
	2-HOC_6H_4CHO	,,	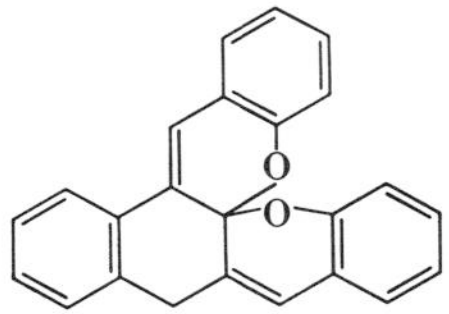 O, O (35)	1886
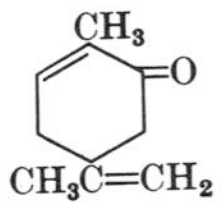 CH_3, =O, $CH_3C{=}CH_2$	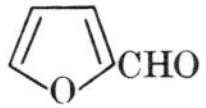 CHO, O	NaOH	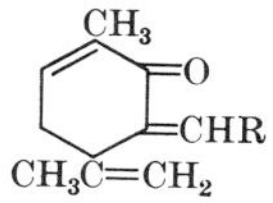 CH_3, =O, =CHR, $CH_3C{=}CH_2$ (15)	1887
	C_6H_5CHO	,,	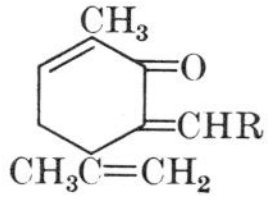 CH_3, =O, =CHR, $CH_3C{=}CH_2$ (23)	1887, 317
	,,	$NaOC_2H_5$	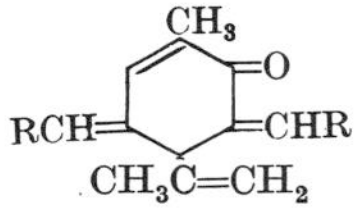 CH_3, =O, RCH=, =CHR, $CH_3C{=}CH_2$ (—)	316–319
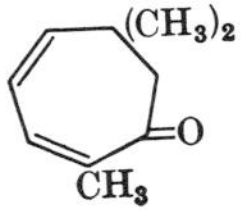 $(CH_3)_2$, =O, CH_3	,,	$NaOC_2H_5$	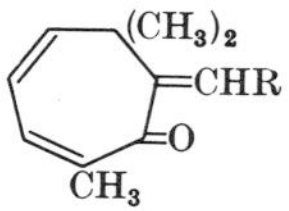 $(CH_3)_2$, =CHR, =O, CH_3 (—)	318

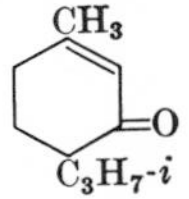 CH_3, =O, C_3H_7-*i*	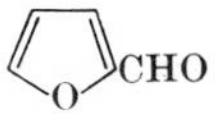 2-Furaldehyde (CHO)	$NaOC_2H_5$	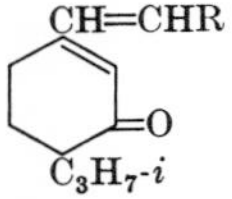 CH=CHR, =O, C_3H_7-*i* (—) D, R = 2-(C_4H_3O)	323
	C_6H_5CHO	,,	D, R = C_6H_5 (90)	323, 317, 319, 1888–1891
	2-HOC_6H_4CHO	NaOH or HCl	D, R = 2-HOC_6H_4 (25)	1891
	3,4-$(CH_2O_2)C_6H_3CHO$	$NaOC_2H_5$	D, R = 3,4-$(CH_2O_2)C_6H_3$ (85)	1891
	4-$CH_3OC_6H_4CHO$	$NaOC_2H_5$ or HCl	D, R = 4-$CH_3OC_6H_4$ (72)	1891
	4-$(CH_3)_2NC_6H_4CHO$	$NaOC_2H_5$	D, R = 4-$(CH_3)_2NC_6H_4$ (—)	323
	2-HO_2C-3,4-$(CH_3O)_2$-C_6H_2CHO	,,	D, R = 2-HO_2C-3,4-$(CH_3O)_2C_6H_2$ (35)	1891
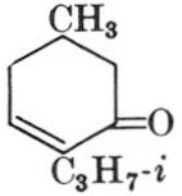 CH_3, =O, C_3H_7-*i*	C_6H_5CHO	,,	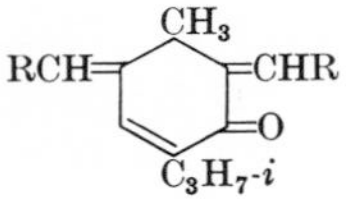 RCH=, CH_3, =CHR, =O, C_3H_7-*i* (—)	319, 320
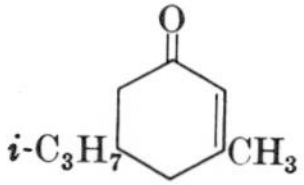 O, *i*-C_3H_7, CH_3	,,	,,	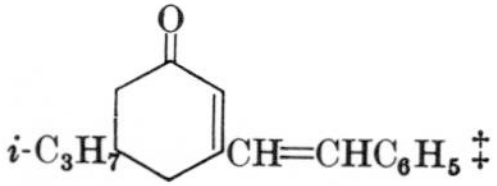 O, *i*-C_3H_7, CH=CHC$_6H_5$ ‡ (50)	1837
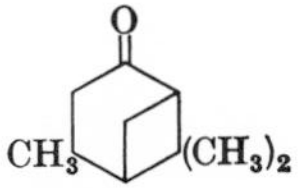 O, CH_3, $(CH_3)_2$	,,	HCl	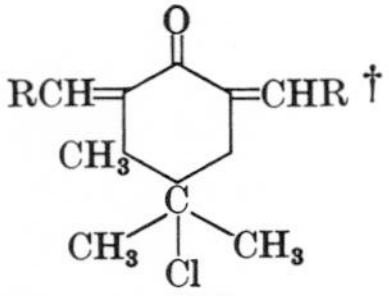 O, RCH=, =CHR †, CH_3, C, CH_3, CH_3, Cl (—)	315

Note: References 668–2359 are on pp. 403–438.

† The product might contain an α,β-unsaturated endocyclic double bond.[304]
‡ This structure has been suggested but not established.

TABLE XV. CONDENSATION OF ALDEHYDES WITH ALICYCLIC KETONES OTHER THAN CYCLOPENTANONES AND CYCLOHEXANONES (*Continued*)

(R in the product is the group R in the aldehyde RCHO.)

Ketone	Aldehyde, RCHO	Catalyst	Product(s) (Yield, %)	Refs.
O, CH_3, C_3H_7-*i*	C_6H_5CHO	HCl	O, CH_3, =CHR, C_3H_7-*i* (47)	1837
CH_3, $(CH_3)_2$, =O	CCl_3CHO	CH_3CO_2K, $(CH_3CO)_2O$	CH_3, $(CH_3)_2$, =O, CHOHR (4)	287
	CHO, O (2-furaldehyde)	$NaNH_2$	CH_3, $(CH_3)_2$, =O, =CHR (20) E, R = 2-(C_4H_3O)	618, 1526, 1892
	C_6H_5CHO	Na	E, R = C_6H_5 (—)	313
	3,4-$(CH_2O_2)C_6H_3CHO$	,,	E, R = 3,4-$(CH_2O_2)_2C_6H_3$ (—)	1893
	2-$CH_3C_6H_4CHO$	$NaNH_2$	E, R = 2-$CH_3C_6H_4$ (—)	314
	3-$CH_3C_6H_4CHO$	,,	E, R = 3-$CH_3C_6H_4$ (—)	314
	4-$CH_3C_6H_4CHO$	,,	E, R = 4-$CH_3C_6H_4$ (—)	314
	3-$CH_3OC_6H_4CHO$	Na	E, R = 3-$CH_3OC_6H_4$ (—)	1893
	4-$CH_3OC_6H_4CHO$	,,	E, R = 4-$CH_3OC_6H_4$ (—)	313
	$C_6H_5CH{=}CHCHO$	,,	E, R = $C_6H_5CH{=}CH$ (—)	313
	2-$CH_3CO_2C_6H_4CHO$	$NaNH_2$	E, R = 2-$CH_3CO_2C_6H_4$ (Good)	314
	3-$CH_3CO_2C_6H_4CHO$	,,	E, R = 3-$CH_3CO_2C_6H_4$ (Poor)	314
	4-$CH_3CO_2C_6H_4CHO$	,,	E, R = 4-$CH_3CO_2C_6H_4$ (Good)	314
	4-$C_2H_5OC_6H_4CHO$	Na	E, R = 4-$C_2H_5OC_6H_4$ (—)	313
	4-$(CH_3)_2NC_6H_4CHO$	$NaNH_2$	E, R = $(CH_3)_2NC_6H_4$ (—)	314
	4-*i*-$C_3H_7C_6H_4CHO$	Na	E, R = 4-*i*-$C_3H_7C_6H_4$ (—)	313
	4-$(C_2H_5)_2NC_6H_4CHO$	$NaNH_2$	E, R = 4-$(C_2H_5)_2NC_6H_4$ (—)	314
=O (2-decalone)	CHO, S (2-thiophenecarboxaldehyde)	NaOH	CHR, =O, =CHR (—) F, R = 2-(C_4H_3S)	1829

	$4\text{-}FC_6H_4CHO$	,,	F, R = $4\text{-}FC_6H_4$ (—)	1829
	C_6H_5CHO	—	[bicyclic ketone with =O and CHOHR] ‡ (25)	291
	,,	—	F, R = C_6H_5 (—)	1894
	$1\text{-}C_{10}H_7CHO$	NaOH	F, R = $1\text{-}C_{10}H_7$ (—)	1829
$(CH_2)_8$ ring with C=O, C=O	C_6H_5CHO	$(CH_2)_5NH$	$(CH_2)_7$ ring with C=CHR, C=O, C=O (40),	325
			$(CH_2)_6$ ring with C=CHR, C=O, C=O, C=CHR (15)	
			G, R = C_6H_5	
	,,	,,	G, R = C_6H_5 (80)	325, 326
	$3\text{-}CH_3C_6H_4CHO$	$(CH_2)_5NH$, CH_3CO_2H	G, R = $3\text{-}CH_3C_6H_4$ (47)	325
	$4\text{-}CH_3C_6H_4CHO$	,,	G, R = $4\text{-}CH_3C_6H_4$ (77)	325
	$4\text{-}CH_3OC_6H_4CHO$	,,	$(CH_2)_7$ ring with C=CHR, C=O, C=O (—),	325
			G, R = $4\text{-}CH_3OC_6H_4$ (—)	
	$4\text{-}(CH_3)_2NC_6H_4CHO$	,,	G, R = $4\text{-}(CH_3)_2NC_6H_4$ (37)	325
	$4\text{-}i\text{-}C_3H_7C_6H_4CHO$	,,	G, R = $4\text{-}i\text{-}C_3H_7C_6H_4$ (47)	325

Note: References 668–2359 are on pp. 403–438.

‡ This structure has been suggested but not established.

TABLE XV. Condensation of Aldehydes with Alicyclic Ketones Other than Cyclopentanones and Cyclohexanones (*Continued*)

(R in the product is the group R in the aldehyde RCHO.)

Ketone	Aldehyde, RCHO	Catalyst	Product(s) (Yield, %)	Refs.
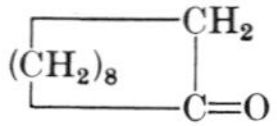 $(CH_2)_8$, CH_2, $C{=}O$ (ring)	C_6H_5CHO	KOH	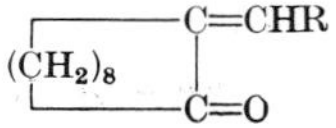 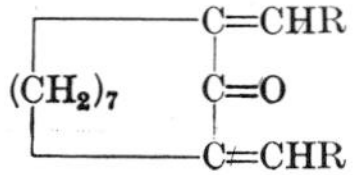 $(CH_2)_8$, $C{=}CHR$, $C{=}O$ (ring) (21),	308
			$(CH_2)_7$, $C{=}CHR$, $C{=}O$, $C{=}CHR$ (ring) (16)	
	$C_6H_4(CHO)_2$ (o-phthalaldehyde)	NaOH	benzo ring, $C{=}O$, $(CH_2)_7$ (64)	309
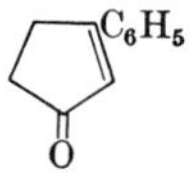 3-C_6H_5-cyclopent-2-enone (C_6H_5, O)	C_6H_5CHO	"	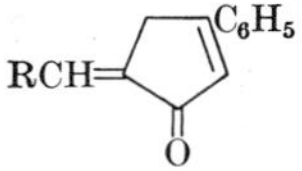 $RCH{=}$, C_6H_5, O (—)	1259
	2-HOC_6H_4CHO	"	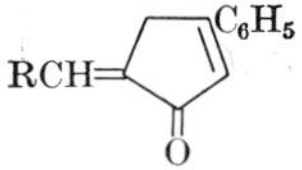 $RCH{=}$, C_6H_5, O (—)	1259
	"	$(CH_2)_5NH$, CH_3CO_2H	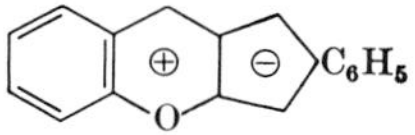 ⊕, ⊖, O, C_6H_5 (41)	1895
	4-$(CH_3)_2NC_6H_4CHO$	NaOH	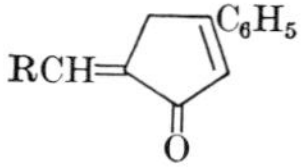 $RCH{=}$, C_6H_5, O (—)	1259
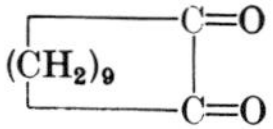 $(CH_2)_9$, $C{=}O$, $C{=}O$ (ring)	C_6H_5CHO	$(CH_2)_5NH$	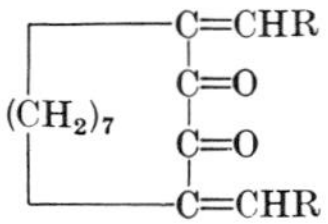 $(CH_2)_7$, $C{=}CHR$, $C{=}O$, $C{=}O$, $C{=}CHR$ (ring) (60)	325

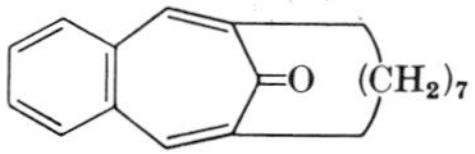

Ketone	Aldehyde	Catalyst	Product (Yield, %)	Refs.
$(CH_2)_9$ [C=O, CH_2 ring]	$C_6H_4(CHO)_2$ (o-)	NaOH	benzo-fused ring C=O, $(CH_2)_8$ (16)	309
benzocyclooctanone	$4\text{-}O_2NC_6H_4CHO$	KOH	RCH= benzocyclooctanone, O (—)	1896
$(CH_2)_{10}$ [C=O, C=O ring]	C_6H_5CHO	$(CH_2)_5NH$	$(CH_2)_8$ [C=CHR, C=O, C=O, C=CHR ring] (—)	325
$(CH_2)_{10}$ [C=O, CH_2 ring]	”	”	$(CH_2)_9$ [C=CHR, C=O, C=CHR ring] (51)	326
	$C_6H_4(CHO)_2$ (o-)	NaOH	benzo-fused ring C=O, $(CH_2)_9$ (79)	309
O, CH_2CH_2CN decalone	$3,4\text{-}(CH_2O_2)C_6H_3CHO$	”	RCH=, O, CH_2CH_2CN decalone (—)	1897
$(CH_2)_{11}$ [C=O, CH_2 ring]	C_6H_5CHO	$(CH_2)_5NH$	$(CH_2)_{10}$ [C=CHR, C=O, C=CHR ring] (32)	325

Note: References 668–2359 are on pp. 403–438.

TABLE XV. CONDENSATION OF ALDEHYDES WITH ALICYCLIC KETONES OTHER THAN CYCLOPENTANONES AND CYCLOHEXANONES (*Continued*)

(R in the product is the group R in the aldehyde RCHO.)

Ketone	Aldehyde, RCHO	Catalyst	Product(s) (Yield, %)	Refs.
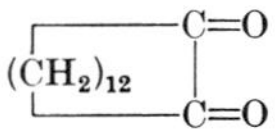 $(CH_2)_{12}$, C=O, C=O	C_6H_5CHO	$(CH_2)_5NH$	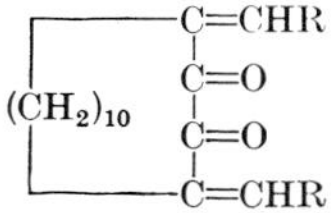 $(CH_2)_{10}$, C=CHR, C=O, C=O, C=CHR (44)	325
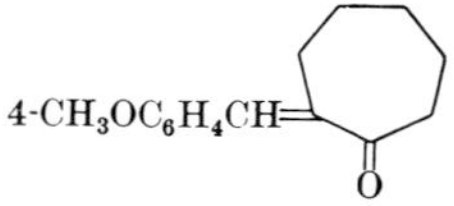 $4\text{-}CH_3OC_6H_4CH=$, O	$3,4\text{-}(CH_3O)_2C_6H_3CHO$	$NaOCH_3$	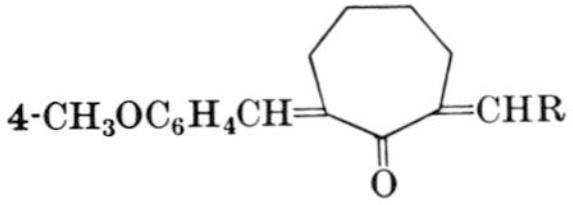 $4\text{-}CH_3OC_6H_4CH=$, =CHR, O (—)	1884
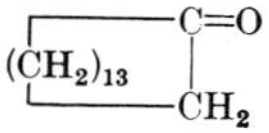 $(CH_2)_{13}$, C=O, CH_2	 CHO, CHO	NaOH	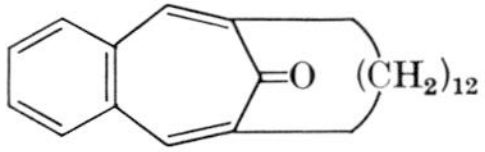 =O, $(CH_2)_{12}$ (52)	309
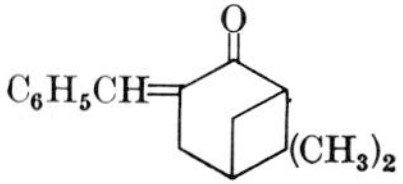 $C_6H_5CH=$, O, $(CH_3)_2$	C_6H_5CHO	HCl	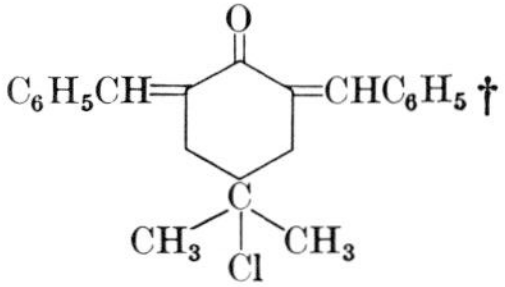 $C_6H_5CH=$, O, $=CHC_6H_5$ †, C, CH_3, CH_3, Cl (—)	315
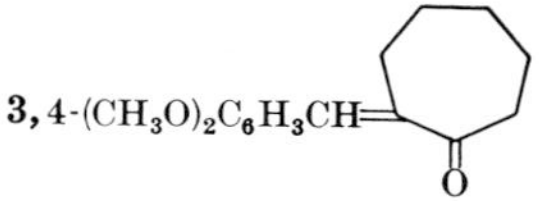 $3,4\text{-}(CH_3O)_2C_6H_3CH=$, O	$4\text{-}CH_3OC_6H_4CHO$	$NaOCH_3$	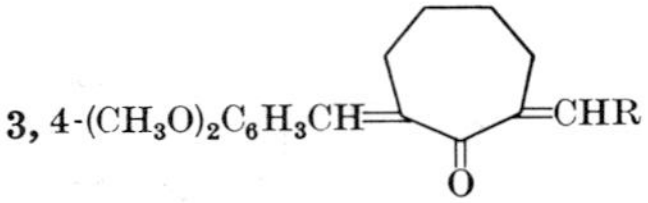 $3,4\text{-}(CH_3O)_2C_6H_3CH=$, =CHR, O (—)	1884
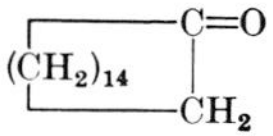 $(CH_2)_{14}$, C=O, CH_2	 CHO, CHO	NaOH	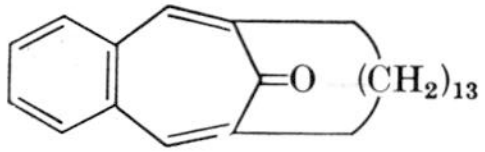 =O, $(CH_2)_{13}$ (76)	309

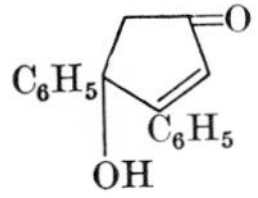 C_6H_5, C_6H_5, OH, =O	C_6H_5CHO	KOH	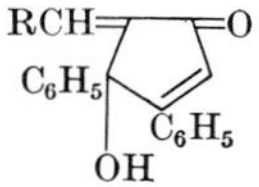 RCH=, =O, C_6H_5, C_6H_5, OH (70)	324
	$C_6H_5CH{=}CHCHO$	,,	RCH=, =O, C_6H_5, C_6H_5, OH (—)	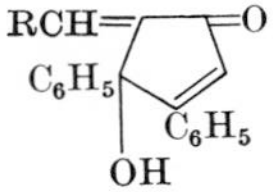 324
	4-*i*-$C_3H_7C_6H_4CHO$ 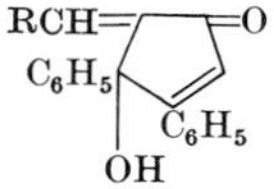	,,	RCH=, =O, C_6H_5, C_6H_5, OH (—)	324
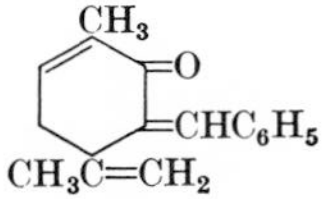 CH_3, =O, =CHC_6H_5, $CH_3C{=}CH_2$	C_6H_5CHO	$NaOC_2H_5$ 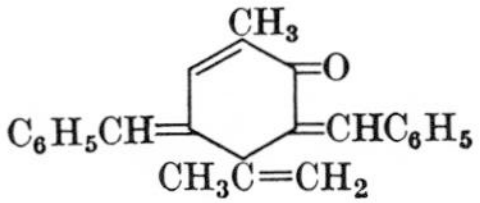	CH_3, =O, C_6H_5CH=, =CHC_6H_5, $CH_3C{=}CH_2$ (19)	316
=O, C_6H_5, CH_3, HO, C_6H_5	,,	KOH	RCH=, =O, C_6H_5, CH_3, HO, C_6H_5 (50)	324
O		NaOH	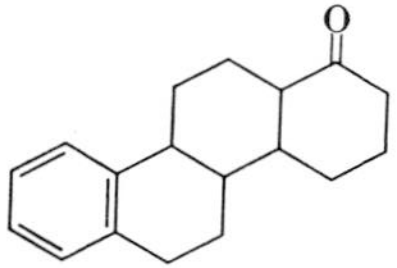 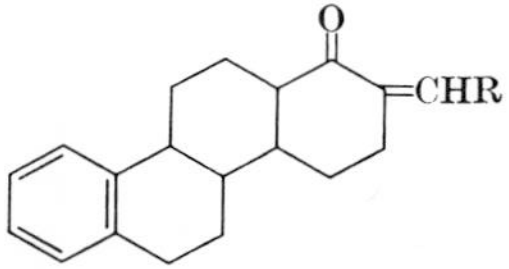 O, =CHR (72)	1898

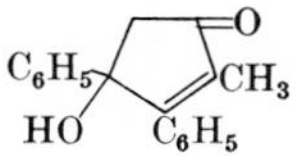

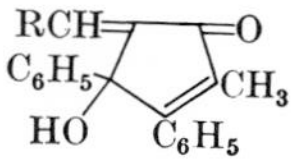

Note: References 668–2359 are on pp. 403–438.

† This product might contain an α,β-unsaturated endocyclic double bond.[304]

TABLE XV. CONDENSATION OF ALDEHYDES WITH ALICYCLIC KETONES OTHER THAN CYCLOPENTANONES AND CYCLOHEXANONES (*Continued*)
(R in the product is the group R in the aldehyde RCHO.)

Ketone	Aldehyde, RCHO	Catalyst	Product(s) (Yield, %)	Refs.
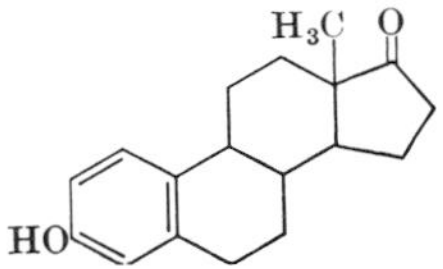 HO, H_3C, O	HO_2CCHO	NaOH	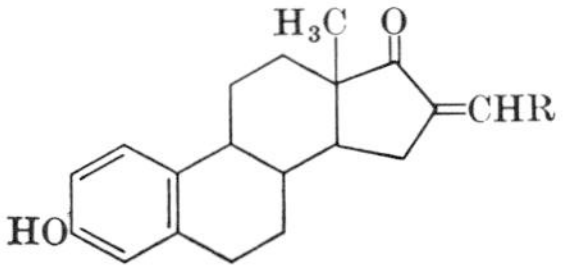 HO, H_3C, O, =CHR (72)	1899
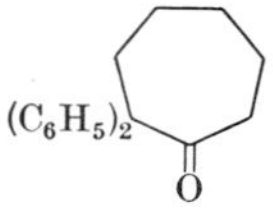 $(C_6H_5)_2$, O	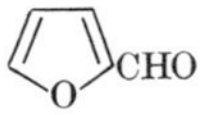 O, CHO	$NaOCH_3$	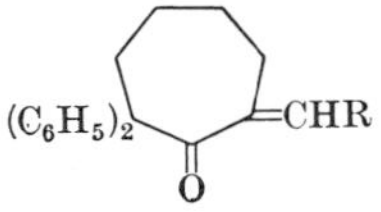 $(C_6H_5)_2$, O, =CHR (50)	1272
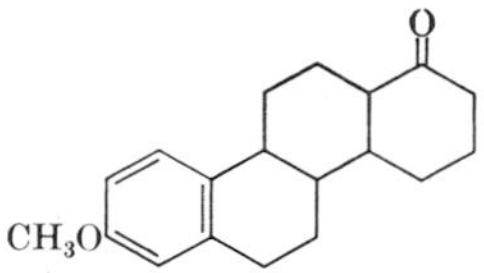 CH_3O, O	,,	NaOH	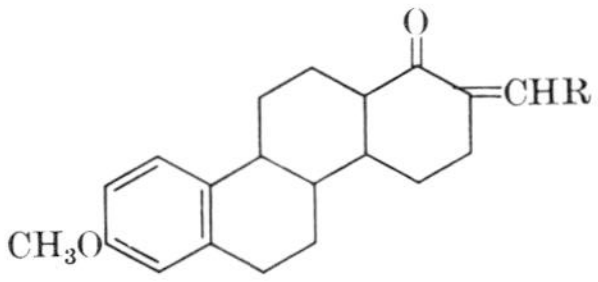 CH_3O, O, =CHR (93)	1901
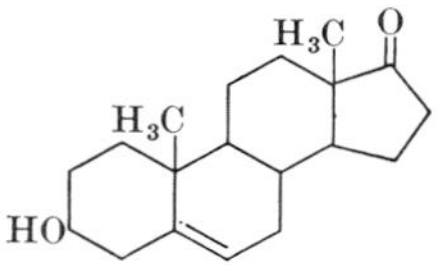 HO, H_3C, H_3C, O	,,	,,	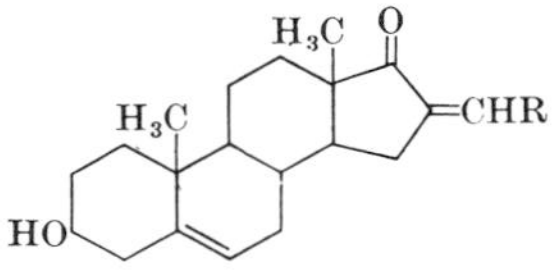 HO, H_3C, H_3C, O, =CHR (100)	2118
			H, R = 2-(C_4H_3O)	

	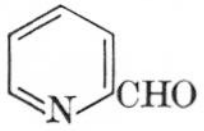 (pyridine ring, N) CHO	KOH	H, R = 2-(C_5H_4N) (—)	
	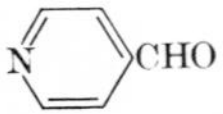 (pyridine ring, N) CHO	"	H, R = 4-(C_5H_4N) (—)	1900
	2,4-$Cl_2C_6H_3CHO$	"	H, R = 2,4-$Cl_2C_6H_3$ (—)	1900
	4-ClC_6H_4CHO	"	H, R = ClC_6H_4 (—)	1900
	2-$CH_3OC_6H_4CHO$	"	H, R = 2-$CH_3OC_6H_4$ (—)	1900
	4-$CH_3OC_6H_4CHO$	"	H, R = 4-$CH_3OC_6H_4$ (—)	1900
	$C_6H_5CH{=}CHCHO$	"	H, R = $C_6H_5CH{=}CH$ (—)	1900
	4-$(CH_3)_2NC_6H_4CHO$	"	H, R = 4-$(CH_3)_2NC_6H_4$ (—)	1900
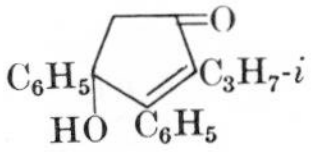 (cyclopentenone: =O, C_6H_5, HO, C_6H_5, C_3H_7-*i*)	C_6H_5CHO	"	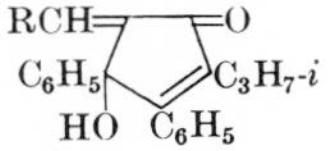 RCH= (cyclopentenone: =O, C_6H_5, HO, C_6H_5, C_3H_7-*i*) (75)	324
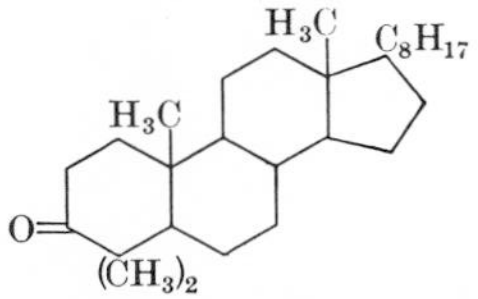 (steroid: O=, $(CH_3)_2$, H_3C, H_3C, C_8H_{17})	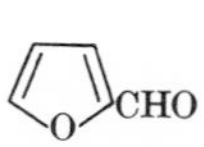 (furan ring, O) CHO	NaOH	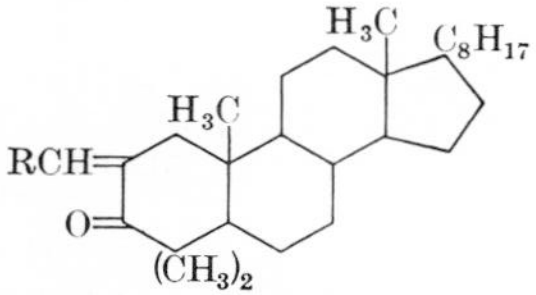 RCH= (steroid: O=, $(CH_3)_2$, H_3C, H_3C, C_8H_{17}) (80)	1902, 1903

Note: References 668–2359 are on pp. 403–438.

TABLE XVI. CONDENSATION OF ALDEHYDES WITH ACETOPHENONE
(R in the product is the group R in the aldehyde RCHO.)

Aldehyde, RCHO	Catalyst	Product(s) (Yield, %)	Refs.
CH_2O	$NaOCH_3$	$CH_2(OCH_2)_2CHCOC_6H_5$ (cyclic: CH_2 bonded to two $O—CH_2$ groups that close the ring at $CHCOC_6H_5$) (20)	1904, 336
	K_2CO_3, CH_3OH	$(CH_3OCH_2)_2CHCOC_6H_5$ (19), $CH_2{=}C(COC_6H_5)CH_2OCH_3$ (7)	1904, 335, 336
	K_2CO_3, C_2H_5OH	$(C_2H_5OCH_2)_2CHCOC_6H_5$ (5), $C_2H_5OCH_2CH_2COC_6H_5$ (5), $HOCH_2CH_2COC_6H_5$ (3)	1904
	$(NH_4)_2CO_3$	$HOCH_2CH_2COC_6H_5$ (—), $C_{19}H_{22}O_5$ (—)	1905, 1906
	$4\text{-}CH_3C_6H_4SO_3H$	$CH_2{=}CHCOC_6H_5$ (1)	1907
	H_2SO_4	$HOCH_2CH_2COC_6H_5$ (25)	335
CCl_3CHO	CH_3CO_2H	$RCHOHCH_2COC_6H_5$ (44)	636, 1419
$C(Cl_2){=}CClCHO$	H_2SO_4	$RCH{=}CHCOC_6H_5$ (43)	289
$C(Cl_2){=}CHCHO$	H_2SO_4; also BaO	$RCH{=}CHCOC_6H_5$ (91)	639, 637, 1908
$CH_3CHClCCl_2CHO$	CH_3CO_2H	$RCHOHCH_2COC_6H_5$ (11)	1909
$i\text{-}C_3H_7CHO$	NaOH	$RCH{=}CHCOC_6H_5$ (90)	1643, 1910, 1911
thiophene ring (S) with O_2N and CHO on the two carbons next to S	H_2SO_4	$RCH{=}CHCOC_6H_5$ (48)	647
thiophene ring (S) with CHO on the carbon next to S and NO_2 on the adjacent carbon	NaOH	$RCHOHCH_2COC_6H_5$ (30), $RCH{=}CHCOC_6H_5$ (36)	1912
furan ring (O) with O_2N and CHO on the two carbons next to O	H_2SO_4	$RCH{=}CHCOC_6H_5$ (55)	646, 1517
thiophene ring (S) with CHO on a carbon next to S	NaOH	$RCH{=}CHCOC_6H_5$ (96)	1913–1915, 1521

[selenophene-2-yl]CHO	$NaOCH_3$	$RCH{=}CHCOC_6H_5$ (50)	1522
[furan-2-yl]CHO	NaOH	$RCH{=}CHCOC_6H_5$ (93)	1916–1920, 1794
[pyrrol-2-yl]CHO	KOH	$RCH{=}CHCOC_6H_5$ (—)	1540, 1541
[tetrahydrofuran-2-yl]CHO	NaOH	$RCH{=}CHCOC_6H_5$ (35)	1454
i-C_4H_9CHO	,,	$RCH{=}CHCOC_6H_5$ (21)	1459, 1436, 1910
OHC[thiophene-2,5-diyl]CHO	K_2CO_3	$C_6H_5COCH{=}CH$[thiophene-2,5-diyl]$CH{=}CHCOC_6H_5$* (—)	365
OHC[furan-2,5-diyl]CHO	Na_2CO_3	$C_6H_5COCH{=}CH$[furan-2,5-diyl]$CH{=}CHCOC_6H_5$* (74)	1921
[2-pyridyl]CHO	NaOH	$RCH{=}CHCOC_6H_5$ (42),	545
		[2-pyridyl]$CH(CH_2COC_6H_5)_2$* (11)	
	,,	$RCHOHCH_2COC_6H_5$ (27)	545
[3-pyridyl]CHO	,,	$RCH{=}CHCOC_6H_5$ (40)	545, 1543

Note: References 668–2359 are on pp. 403–438.

* This is a complete structural formula.

TABLE XVI. CONDENSATION OF ALDEHYDES WITH ACETOPHENONE (*Continued*)
(R in the product is the group R in the aldehyde RCHO.)

Aldehyde, RCHO	Catalyst	Product(s) (Yield, %)	Refs.
4-Pyridine-CHO (structure)	NaOH	$RCH{=}CHCOC_6H_5$ (20), 4-pyridyl-$CH(CH_2COC_6H_5)_2$* (6)	545, 1543
1-Methylpyrrole-2-CHO (structure, N–CH_3)	KOH	$RCH{=}CHCOC_6H_5$ (—)	1541
3,4-Dihydro-2H-pyran-2-CHO (structure)	NaOH	$RCH{=}CHCOC_6H_5$ (—), $RCHOHCH_2COC_6H_5$ (—)	1467
Tetrahydropyran-2-CHO (structure)	,,	$RCH{=}CHCOC_6H_5$ (40)	1467
n-$C_4H_9O_2CCHO$	None, 100°	$RCHOHCH_2COC_6H_5$ (54)	1468
2-HO-3,5-$Br_2C_6H_2CHO$	NaOH	$RCH{=}CHCOC_6H_5$ (55)	549
3,5-Br_2-4-HOC_6H_2CHO	,,	$RCH{=}CHCOC_6H_5$ (54)	549
3,4-$Cl_2C_6H_3CHO$	,,	$RCH{=}CHCOC_6H_5$ (—)	1923
2-BrC_6H_4CHO	,,	$RCH{=}CHCOC_6H_5$ (—)	1924
3-BrC_6H_4CHO	,,	$RCH{=}CHCOC_6H_5$ (93)	1924, 1925
4-BrC_6H_4CHO	,,	$RCH{=}CHCOC_6H_5$ (99)	1924, 543, 1926
2-HO-5-BrC_6H_3CHO	,,	$RCH{=}CHCOC_6H_5$ (—)	1927
2-ClC_6H_4CHO	,,	$RCH{=}CHCOC_6H_5$ (83)	1924, 1928–1932
3-ClC_6H_4CHO	,,	$RCH{=}CHCOC_6H_5$ (94)	1924, 1930, 1933
4-ClC_6H_4CHO	,,	$RCH{=}CHCOC_6H_5$ (89)	1924, 1551, 1930, 1934
2-FC_6H_4CHO	BF_3, CH_3CO_2H	$RCH{=}CHCOC_6H_5$ (—)	1935
4-FC_6H_4CHO	,,	$RCH{=}CHCOC_6H_5$ (—)	1935
2-IC_6H_4CHO	NaOH	$RCH{=}CHCOC_6H_5$ (76)	1924
3-IC_6H_4CHO	,,	$RCH{=}CHCOC_6H_5$ (92)	1924

Aldehyde	Catalyst	Product (Yield, %)	References
4-IC_6H_4CHO	,,	$RCH{=}CHCOC_6H_5$ (92)	1924, 1545
2-$O_2NC_6H_4CHO$	,,	$RCH{=}CHCOC_6H_5$ (48)	1924, 360, 1930, 1936–1939
3-$O_2NC_6H_4CHO$	Na_3PO_4	$RCHOHCH_2COC_6H_5$ (—)	361
	NaOH	$RCH{=}CHCOC_6H_5$ (96)	1924, 360, 1654, 1930, 1938–1943
4-$O_2NC_6H_4CHO$	,,	$RCH{=}CHCOC_6H_5$ (94)	1924, 353, 360, 1930, 1938–1943
$O_2N(C_4H_2S)CH{=}CHCHO$	H_2SO_4	$RCH{=}CHCOC_6H_5$ (48)	647
$O_2N(C_4H_2O)CH{=}CHCHO$	,,	$RCH{=}CHCOC_6H_5$ (56)	646
$C(Cl_2){=}CH(CH{=}CH)_2CHO$	HCl	$RCH{=}CHCOC_6H_5$ (71)	1451
C_6H_5CHO	$NaOC_2H_5$	$RCH{=}CHCOC_6H_5$ (90)	5, 353, 640, 1525, 1558, 1940, 1944–1949
2-Thienyl-CH=CHCHO	NaOH	$RCH{=}CHCOC_6H_5$ (—)	1793
2-HOC_6H_4CHO	HCl	$RCH{=}CHCOC_6H_5$ (53)	640, 1940, 1950–1953
3-HOC_6H_4CHO	,,	$RCH{=}CHCOC_6H_5$ (53)	640, 1951, 1953
4-HOC_6H_4CHO	KOH	$RCH{=}CHCOC_6H_5$ (98)	1948, 353, 1951, 1953
2-Furyl-CH=CHCHO	NaOH	$RCH{=}CHCOC_6H_5$ (87)	1794, 1039
3,4-$(HO)_2C_6H_3CHO$	KOH	$RCH{=}CHCOC_6H_5$ (89)	1948, 348
2-$H_2NC_6H_4CHO$	NaOH	$RCH{=}CHCOC_6H_5$ (16)	1924

Note: References 668–2359 are on pp. 403–438.

* This is a complete structural formula.

TABLE XVI. Condensation of Aldehydes with Acetophenone (*Continued*)

(R in the product is the group R in the aldehyde RCHO.)

Aldehyde, RCHO	Catalyst	Product(s) (Yield, %)	Refs.
3-$H_2NC_6H_4CHO$	NaOH	$RCH{=}CHCOC_6H_5$ (44)	1924
4-$H_2NC_6H_4CHO$	,,	$RCH{=}CHCOC_6H_5$ (11)	1924
[structure: CH_3, N, CHO]	KOH	$RCH{=}CHCOC_6H_5$ (—)	1543
2,5,6-Br_3-3-CH_3O-4-HOC_6CHO	NaOH	$RCH{=}CHCOC_6H_5$ (47)	1954
2-O_2N-3-CH_3O-4-HO-5-BrC_6HCHO	,,	$RCH{=}CHCOC_6H_5$ (42)	1955
2,5-Br_2-3-CH_3O-4-HOC_6HCHO	,,	$RCH{=}CHCOC_6H_5$ (15)	1954
2,6-Br_2-3-CH_3O-4-HOC_6HCHO	,,	$RCH{=}CHCOC_6H_5$ (10)	1954
3-CH_3O-4-HO-5,6-Br_2C_6HCHO	,,	$RCH{=}CHCOC_6H_5$ (20)	1954
3,5-Cl_2-2-CH_3O-4-HOC_6HCHO	,,	$RCH{=}CHCOC_6H_5$ (Low)	1955
$C_6H_4(CHO)_2$-1,2	KOH	[structure: COC_6H_5, O] (6)	1956, 1574
$C_6H_4(CHO)_2$-1,3	NaOH	$C_6H_4(CH{=}CHCOC_6H_5)_2$-1,3* (87)	1957, 1956, 2338
$C_6H_4(CHO)_2$-1,4	,,	$C_6H_4(CH{=}CHCOC_6H_5)_2$-1,4* (96)	1956, 544, 1958, 2338
3,4-$(CH_2O_2)C_6H_3CHO$	,,	$RCH{=}CHCOC_6H_5$ (—)	1929, 1549, 1940, 1949, 1952, 1959
2-$HO_2CC_6H_4CHO$	,,	[structure: $CH_2COC_6H_5$, O, O] (—)	1577
2-Br-3-CH_3O-4-HOC_6H_2CHO	,,	$RCH{=}CHCOC_6H_5$ (18)	1954
3-CH_3O-4-HO-5-BrC_6H_2CHO	,,	$RCH{=}CHCOC_6H_5$ (33)	1954
3-CH_3O-4-HO-6-BrC_6H_2CHO	,,	$RCH{=}CHCOC_6H_5$ (52)	1954
3-CH_3O-4-HO-5-ClC_6H_2CHO	,,	$RCH{=}CHCOC_6H_5$ (10)	1954
3-CH_3O-4-HO-6-ClC_6H_2CHO	,,	$RCH{=}CHCOC_6H_5$ (10)	1954

3-O_2N-4-$CH_3C_6H_3CHO$	,,	$RCH{=}CHCOC_6H_5$ (90)	1583
2-O_2N-3-CH_3O-4-HOC_6H_2CHO	,,	$RCH{=}CHCOC_6H_5$ (52)	1955
3-CH_3O-4-HO-5-$O_2NC_6H_2CHO$	,,	$RCH{=}CHCOC_6H_5$ (Low)	1955
2-$CH_3C_6H_4CHO$	,,	$RCH{=}CHCOC_6H_5$ (92)	829, 1939
3-$CH_3C_6H_4CHO$	$NaOCH_3$	$RCH{=}CHCOC_6H_5$ (60)	1939, 543, 1960
4-$CH_3C_6H_4CHO$	,,	$RCH{=}CHCOC_6H_5$ (45)	353, 1583
2-$CH_3OC_6H_4CHO$	NaOH	$RCH{=}CHCOC_6H_5$ (80–90)	1961, 1930, 1962
3-$CH_3OC_6H_4CHO$	,,	$RCH{=}CHCOC_6H_5$ (—)	1603, 1953, 1940
4-$CH_3OC_6H_4CHO$	$NaOC_2H_5$	$RCH{=}CHCOC_6H_5$ (95)	1930, 353, 1549, 1563, 1654, 1949, 1952, 1953, 1959, 1963, 1964
2-HO-5-$CH_3C_6H_3CHO$	NaOH	$RCH{=}CHCOC_6H_5$ (—)	1965
2-HO-3-$CH_3OC_6H_3CHO$	,,	$RCH{=}CHCOC_6H_5$ (36)	1966
3-CH_3O-4-HOC_6H_4CHO	KOH	$RCH{=}CHCOC_6H_5$ (—)	1953, 1590
(structure: CHO, O)	NaOH	$RCH{=}CHCOC_6H_5$ (—)	1703
(structure: CHO, N, H)	$(CH_2)_5NH$	$RCH{=}CHCOC_6H_5$ (59)	600
2-$O_2NC_6H_4CH{=}CHCHO$	NaOH	$RCH{=}CHCOC_6H_5$ (57)	1798
$C_6H_5CH{=}CHCHO$	,,	$RCH{=}CHCOC_6H_5$ (84)	1924, 1563, 1596, 1967, 1968
(structure: O)$(CH{=}CH)_2CHO$	Na_2CO_3	$RCH{=}CHCOC_6H_5$ (90)	1766
2-C_2H_5O-5-BrC_6H_3CHO	NaOH	$RCH{=}CHCOC_6H_5$ (—)	1927
3-$CH_3CONHC_6H_4CHO$	,,	$RCH{=}CHCOC_6H_5$ (76)	1924
4-$CH_3CONHC_6H_4CHO$	,,	$RCH{=}CHCOC_6H_5$ (23)	1924
2-$C_2H_5OC_6H_4CHO$	,,	$RCH{=}CHCOC_6H_5$ (—)	1969, 1549
2,3-$(CH_3O)_2C_6H_3CHO$	,,	$RCH{=}CHCOC_6H_5$ (—)	1923, 1953
2,4-$(CH_3O)_2C_6H_3CHO$	,,	$RCH{=}CHCOC_6H_5$ (49)	1970
3,4-$(CH_3O)_2C_6H_3CHO$	,,	$RCH{=}CHCOC_6H_5$ (70)	411, 1929, 1953, 1971

Note: References 668–2359 are on pp. 403–438.

* This is a complete structural formula.

TABLE XVI. CONDENSATION OF ALDEHYDES WITH ACETOPHENONE (*Continued*)

(R in the product is the group R in the aldehyde RCHO.)

Aldehyde, RCHO	Catalyst	Product(s) (Yield, %)	Refs.
2-HO-3-$C_2H_5OC_6H_3CHO$	NaOH	RCH=CHCOC_6H_5 (—)	1590
3-O_2N-4-$(CH_3)_2NC_6H_3CHO$	"	RCH=CHCOC_6H_5 (—)	1601
4-$(CH_3)_2NC_6H_4CHO$	"	RCH=CHCOC_6H_5 (60)	1972, 1601, 1602, 1798
2-HO-4-$(CH_3)_2NC_6H_3CHO$	KOH	RCH=CHCOC_6H_5 (50)	1973
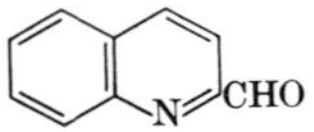 [quinoline-2-carboxaldehyde structure] CHO	"	RCH=CHCOC_6H_5 (—)	1543
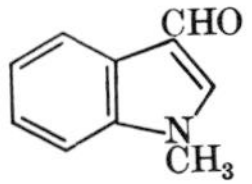 [1-methylindole-3-carboxaldehyde structure] CHO, N, CH_3	$(CH_2)_5NH$	RCH=CHCOC_6H_5 (70)	600
1,3,5-$(CH_3)_3C_6H_2CHO$	—	RCH=CHCOC_6H_5 (—)	1607
4-*i*-$C_3H_7C_6H_4CHO$	$NaOCH_3$	RCH=CHCOC_6H_5 (—)	1974
2,4,6-$(CH_3O)_3C_6H_2CHO$	NaOH	RCH=CHCOC_6H_5 (98)	1970
$(CH_3)_2C{=}CHCH_2CH_2C(CH_3){=}CHCHO$	$NaOC_2H_5$	RCH=CHCOC_6H_5 (35)	1492
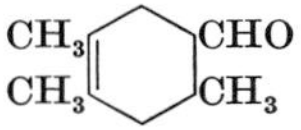 [cyclohexene structure] CH_3, CH_3, CHO, CH_3	$NaOCH_3$	RCH=CHCOC_6H_5 (—)	1678
1-$C_{10}H_7CHO$	NaOH	RCH=CHCOC_6H_5 (—)	1744
Ferrocenecarboxaldehyde	"	RCH=CHCOC_6H_5 (92)	663, 1975
4-$(CH_3)_2NC_6H_4CH{=}CHCHO$	"	RCH=CHCOC_6H_5 (—)	1602
2-*n*-C_3H_7O-3-$CH_3OC_6H_3CHO$	"	RCH=CHCOC_6H_5 (—)	1590
3,4-$(C_2H_5O)_2C_6H_3CHO$	"	RCH=CHCOC_6H_5 (—)	1923
2-*n*-C_4H_9O-3-$CH_3OC_6H_3CHO$	"	RCH=CHCOC_6H_5 (—)	1590
2-*n*-C_3H_7O-3-$C_2H_5OC_6H_3CHO$	"	RCH=CHCOC_6H_5 (—)	1590
3-*n*-C_4H_9O-3-$C_2H_5OC_6H_3CHO$	"	RCH=CHCOC_6H_5 (—)	1590
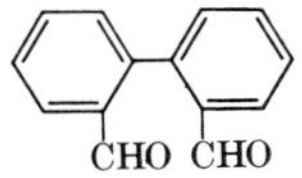 [biphenyl-2,2′-dicarboxaldehyde structure] CHO CHO	"	 [biphenyl structure] $CHOHCH_2COC_6H_5$* / $CHOHCH_2COC_6H_5$ (55)	1976

4-$C_6H_5CH_2C_6H_4CHO$	,,	$RCH{=}CHCOC_6H_5$	(52)	1613
2-$C_6H_5CH_2O$-3,5-$Br_2C_6H_2CHO$	,,	$RCH{=}CHCOC_6H_5$	(8)	549
2-$C_6H_5CH_2O$-5-BrC_6H_3CHO	,,	$RCH{=}CHCOC_6H_5$	(81)	549
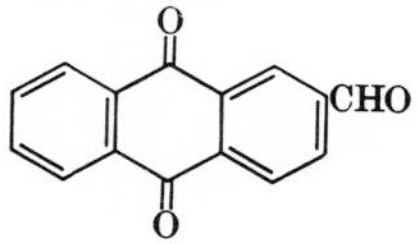 O, CHO, O	KOH	$RCH{=}CHCOC_6H_5$	(—)	1977
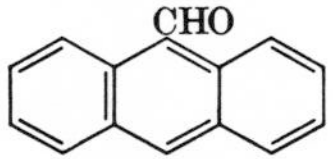 CHO	HCl	$RCH{=}CHCOC_6H_5$	(68)	1978
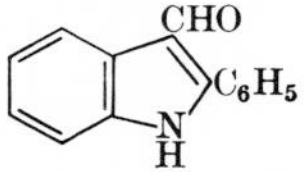 CHO, C_6H_5, N, H	$(CH_2)_5NH$	$RCH{=}CHCOC_6H_5$	(67)	1979
Pyrene-3-carboxaldehyde	$NaOC_2H_5$	$RCH{=}CHCOC_6H_5$	(100)	1615
9-Formyl-3,4-benzacridine	NaOH	$RCH{=}CHCOC_6H_5$	(75–95)	372
10-Formyl-1,2-benzanthracene	,,	$RCH{=}CHCOC_6H_5$	(93)	372

Note: References 668–2359 are on pp. 403–438.

* This is a complete structural formula.

TABLE XVII. Condensation of Aldehydes with Substituted Acetophenones, CH_3COAr

(R in the product is the group R in the aldehyde RCHO.)

Substituent(s) in Acetophenone	Aldehyde, RCHO	Catalyst	Product(s) (Yield, %)	Refs.
2-HO-3-Br-5-F	C_6H_5CHO	KOH	RCH=CHCOAr (—)	1980
	2-HOC_6H_4CHO	,,	RCH=CHCOAr (—)	1980
	3,4-$(CH_2O_2)C_6H_3CHO$	,,	RCH=CHCOAr (—)	1980
	4-$CH_3OC_6H_4CHO$	,,	RCH=CHCOAr (—)	1980
	3-CH_3O-4-HOC_6H_3CHO	,,	RCH=CHCOAr (—)	1980
	4-$(CH_3)_2NC_6H_4CHO$	,,	RCH=CHCOAr (—)	1980
4-NO_2-3-Br	2-$O_2NC_6H_4CHO$	Na_3PO_4	$RCHOHCH_2COAr$ (—)	361
3-NO_2-4-Br	2-$O_2NC_6H_4CHO$	HCl	RCH=CHCOAr (—)	1936
2,4-$(HO)_2$-3-Br-5-NO_2	2-HO-3,5-$Br_2C_6H_2CHO$	KOH	RCH=CHCOAr (—)	1981
	C_6H_5CHO	,,	RCH=CHCOAr (—)	1981
	2-HOC_6H_4CHO	,,	RCH=CHCOAr (—)	1981
	3-Br-4-$CH_3OC_6H_3CHO$	,,	RCH=CHCOAr (—)	1981
	2-CH_3O-5-BrC_6H_3CHO	,,	RCH=CHCOAr (—)	1981
	3-CH_3O-4-HO-5-BrC_6H_2CHO	,,	RCH=CHCOAr (—)	1981
	2-$CH_3OC_6H_4CHO$	,,	RCH=CHCOAr (—)	1981
	4-$CH_3OC_6H_4CHO$	,,	RCH=CHCOAr (—)	1981
	3-CH_3O-4-HOC_6H_3CHO	,,	RCH=CHCOAr (—)	1981
3,5-Br_2-4-HO	C_6H_5CHO	,,	RCH=CHCOAr (—)	1982
	2-HOC_6H_4CHO	,,	RCH=CHCOAr (—)	1982
	3-HOC_6H_4CHO	,,	RCH=CHCOAr (—)	1982
	4-HOC_6H_4CHO	,,	RCH=CHCOAr (—)	1982
	4-$CH_3OC_6H_4CHO$	,,	RCH=CHCOAr (—)	1982
	3-CH_3O-4-HOC_6H_3CHO	,,	RCH=CHCOAr (—)	1982
2,4-$(HO)_2$-3,5-Br_2	(2-pyridyl)CHO	NaOH	RCH=CHCOAr (—)	1983
	3-$O_2NC_6H_4CHO$	H_2SO_4	RCH=CHCOAr (—)	1984
	4-$O_2NC_6H_4CHO$	,,	RCH=CHCOAr (—)	1984
	C_6H_5CHO	,,	RCH=CHCOAr (—)	1984
	2-Br-5-$CH_3OC_6H_3CHO$	KOH	RCH=CHCOAr (—)	1985
	2-$CH_3OC_6H_4CHO$	,,	RCH=CHCOAr (—)	1985
2-HO-3-Cl-4-F	3,4-$(CH_2O_2)C_6H_3CHO$	,,	RCH=CHCOAr (—)	1980
	4-$CH_3OC_6H_4CHO$	,,	RCH=CHCOAr (—)	1980
	4-$(CH_3)_2NC_6H_4CHO$	,,	RCH=CHCOAr (—)	1980
2-HO-3-Cl-5-F	C_6H_5CHO	,,	RCH=CHCOAr (—)	1986

	$2\text{-}HOC_6H_4CHO$	"	$RCH{=}CHCOAr$ (—)	1986
	$4\text{-}HOC_6H_4CHO$	"	$RCH{=}CHCOAr$ (—)	1986
	$3\text{-}CH_3O\text{-}4\text{-}HOC_6H_3CHO$	"	$RCH{=}CHCOAr$ (—)	1986
$2,5\text{-}Cl_2$	C_6H_5CHO	NaOH	$RCH{=}CHCOAr$ (40–50)	1987
	$2\text{-}HOC_6H_4CHO$	"	$RCH{=}CHCOAr$ (40–50)	1987
	$4\text{-}HOC_6H_4CHO$	"	$RCH{=}CHCOAr$ (40–50)	1987
	$3,4\text{-}(CH_2O_2)C_6H_4CHO$	"	$RCH{=}CHCOAr$ (40–50)	1987
	$4\text{-}CH_3OC_6H_4CHO$	"	$RCH{=}CHCOAr$ (40–50)	1987
	$3\text{-}CH_3O\text{-}4\text{-}HOC_6H_3CHO$	"	$RCH{=}CHCOAr$ (40–50)	1987
$3,4\text{-}Cl_2$	[pyridine-2-CHO]	"	$RCH{=}CHCOAr$ (—)	1988
	[pyridine-3-CHO]	"	$RCH{=}CHCOAr$ (—)	1988
	[pyridine-4-CHO]	"	$RCH{=}CHCOAr$ (—)	1988
	[5-CH_3-thiophene-2-CHO]	"	$RCH{=}CHCOAr$ (—)	1988
	[5-CH_3-furan-2-CHO]	"	$RCH{=}CHCOAr$ (—)	1988
$2\text{-}HO\text{-}3,5\text{-}Cl_2$	[furan-2-CHO]	KOH	$RCH{=}CHCOAr$ (75)	1989
	[pyridine-2-CHO]	"	$RCH{=}CHCOAr$ (—)	1990
	[pyridine-3-CHO]	"	$RCH{=}CHCOAr$ (97)	371
	[pyridine-4-CHO]	"	$RCH{=}CHCOAr$ (97)	370

Note: References 668–2359 are on pp. 403–438.

TABLE XVII. CONDENSATION OF ALDEHYDES WITH SUBSTITUTED ACETOPHENONES, CH_3COAr (*Continued*)

(R in the product is the group R in the aldehyde RCHO.)

Substituent(s) in Acetophenone	Aldehyde, RCHO	Catalyst	Product(s) (Yield, %)	Refs.
2-HO-3,5-Cl_2 (*contd.*)	2-ClC_6H_4CHO	KOH	RCH=CHCOAr (65)	1989, 1991
	4-$O_2NC_6H_4CHO$	”	RCH=CHCOAr (82)	1989
	C_6H_5CHO	”	RCH=CHCOAr (82)	1989, 1991
	2-HOC_6H_4CHO	”	RCH=CHCOAr (73)	1989, 1991
	3-HOC_6H_4CHO	”	RCH=CHCOAr (73)	1989, 1991
	4-HOC_6H_4CHO	”	RCH=CHCOAr (74)	1989, 1991
	3,4-$(CH_2O_2)C_6H_3CHO$	”	RCH=CHCOAr (—)	1991
	4-$CH_3C_6H_4CHO$	”	RCH=CHCOAr (69)	1989
	2-$CH_3OC_6H_4CHO$	”	RCH=CHCOAr (—)	1991
	4-$CH_3OC_6H_4CHO$	”	RCH=CHCOAr (66)	1989, 1991
	3-CH_3O-4-HOC_6H_3CHO	”	RCH=CHCOAr (60)	1989
	3,4-$(CH_3O)_2C_6H_3CHO$	”	RCH=CHCOAr (—)	1991
	[naphthalene structure: CHO, OH]	”	RCH=CHCOAr (76)	1989
	[naphthalene structure: OH, CHO]	”	RCH=CHCOAr (87)	1989
2-Br	C_6H_5CHO	NaOH	RCH=CHCOAr (62)	1924, 1992
3-Br	3-BrC_6H_5CHO	”	RCH=CHCOAr (83)	1957
	C_6H_5CHO	NaOH	RCH=CHCOAr (94)	1924, 349
	$C_6H_4(CHO)_2$-1,3	”	$C_6H_4(CH{=}CHCOAr)_2$-1,3 (87)	1957
	[biphenyl structure: CHO, CHO]	HCl	[structure: CH=CHCOAr]$_2$ (98)	1957

4-Br	$C(Cl_2)$=CHCHO	BaO	RCH=CHCOAr (69)	1908
	2-C_4H_3OCHO (furan-2-carboxaldehyde)	NaOH	RCH=CHCOAr (—)	1916
	2-HO-3,5-$Br_2C_6H_2CHO$	,,	RCH=CHCOAr (67)	549
	2-HO-5-BrC_6H_3CHO	,,	RCH=CHCOAr (50)	549
	2-ClC_6H_4CHO	,,	RCH=CHCOAr (99)	1924
	2-$O_2NC_6H_4CHO$	Na_3PO_4	$RCHOHCH_2COAr$ (—)	361
	,,	KOH	RCH=CHCOAr (—)	1936
	3-$O_2NC_6H_4CHO$	NaOH	RCH=CHCOAr (93)	1924
	4-$O_2NC_6H_4CHO$	,,	RCH=CHCOAr (65)	1993
	C_6H_5CHO	,,	RCH=CHCOAr (94)	1924, 349, 540, 1994–1996
	3-NCC_6H_4CHO	,,	RCH=CHCOAr (40–70)	1997
	4-NCC_6H_4CHO	,,	RCH=CHCOAr (40–70)	1997
	3-CH_3O-4-HO-5-BrC_6H_2CHO	,,	RCH=CHCOAr (15)	1955
	3-CH_3O-4-HO-6-BrC_6H_2CHO	,,	RCH=CHCOAr (32)	1955
	4-$(CH_3)_2NC_6H_4CHO$	,,	RCH=CHCOAr (—)	1995
3-Br-4-HO	2-C_5H_4NCHO (pyridine-2-carboxaldehyde)	,,	RCH=CHCOAr (—)	1988
	4-C_5H_4NCHO (pyridine-4-carboxaldehyde)	,,	RCH=CHCOAr (—)	1988
2-HO-4-Br	2-C_5H_4NCHO (pyridine-2-carboxaldehyde)	KOH	RCH=CHCOAr (—)	1990
	3-C_5H_4NCHO (pyridine-3-carboxaldehyde)	,,	RCH=CHCOAr (—)	371
	4-C_5H_4NCHO (pyridine-4-carboxaldehyde)	,,	RCH=CHCOAr (75)	370
	4-$O_2NC_6H_4CHO$	NaOH	RCH=CHCOAr (85)	1993
	C_6H_5CHO	KOH	RCH=CHCOAr (—)	1998

Note: References 668–2359 are on pp. 403–438.

TABLE XVII. Condensation of Aldehydes with Substituted Acetophenones, CH_3COAr (*Continued*)

(R in the product is the group R in the aldehyde RCHO.)

Substituent(s) in Acetophenone	Aldehyde, RCHO	Catalyst	Product(s) (Yield, %)	Refs.
2-HO-5-Br	[2-furyl structure]CHO	NaOH	RCH=CHCOAr (100)	340, 1993
	[2-pyridyl structure]CHO	"	RCH=CHCOAr (—)	1983, 1990
	[3-pyridyl structure]CHO	KOH	RCH=CHCOAr (—)	371
	[4-pyridyl structure]CHO	"	RCH=CHCOAr (97)	370
	2-HO-3,5-$Br_2C_6H_2CHO$	NaOH	RCH=CHCOAr (60)	1993
	3,5-Br_2-4-HOC_6H_2CHO	"	RCH=CHCOAr (80)	1993
	3-BrC_6H_4CHO	"	RCH=CHCOAr (70)	1993
	3-Br-4-HOC_6H_3CHO	"	RCH=CHCOAr (65)	1993
	2-HO-5-BrC_6H_3CHO	"	RCH=CHCOAr (70)	1993
	2-HO-5-$O_2NC_6H_3CHO$	"	RCH=CHCOAr (—)	1983
	C_6H_5CHO	"	RCH=CHCOAr (95)	340, 1965
	2-HOC_6H_4CHO	"	RCH=CHCOAr (55)	1993, 340
	3-HOC_6H_4CHO	"	RCH=CHCOAr (50)	1993
	4-HOC_6H_4CHO	"	RCH=CHCOAr (80)	1993, 340
	3-Cl-4-$CH_3C_6H_3CHO$	"	RCH=CHCOAr (—)	341
	4-$CH_3OC_6H_4CHO$	"	RCH=CHCOAr (90)	340
	3-CH_3O-4-HOC_6H_3CHO	"	RCH=CHCOAr (60)	340
2,4-$(HO)_2$-5-Br	C_6H_5CHO	"	RCH=CHCOAr (40)	2000
	3-$O_2NC_6H_4CHO$	"	RCH=CHCOAr (40)	2000
	3,4-$(CH_2O_2)C_6H_3CHO$	"	RCH=CHCOAr (45)	2000
	4-$CH_3OC_6H_4CHO$	"	RCH=CHCOAr (45)	2000

3,5-Br_2-4-NH_2	2-$O_2NC_6H_4CHO$	HCl	RCH=CHCOAr (—)	1936
2-Cl	2-HO-3,5-$Br_2C_6H_2CHO$	NaOH	RCH=CHCOAr (51)	549
	3,5-Br_2-4-HOC_6H_2CHO	″	RCH=CHCOAr (7)	549
	2,4-$Cl_2C_6H_3CHO$	″	RCH=CHCOAr (—)	1923
	C_6H_5CHO	″	RCH=CHCOAr (76)	1924
	3,4-$(CH_2O_2)C_6H_3CHO$	″	RCH=CHCOAr (—)	1923
	3-CH_3O-4-HO-5-BrC_6H_2CHO	HCl	RCH=CHCOAr (38)	1955
	4-$CH_3OC_6H_4CHO$	NaOH	RCH=CHCOAr (—)	1923
3-Cl	C_6H_5CHO	″	RCH=CHCOAr (96)	1924, 349
4-Cl	$C(Cl_2)$=CHCHO	HCl	RCH=CHCOAr (75)	637
	$C(Cl_2)$=CHCH=CHCHO	″	RCH=CHCOAr (72)	1451
	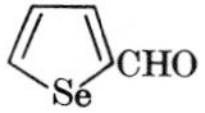 (Se ring) CHO	NaOH	RCH=CHCOAr (78)	366
	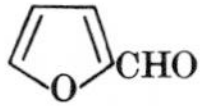 (O ring) CHO	″	RCH=CHCOAr (92)	2001, 1989
	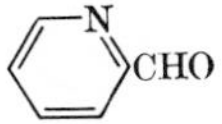 (N ring) CHO	″	RCH=CHCOAr (40)	545
	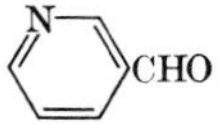 (N ring) CHO	″	RCH=CHCOAr (—)	1988
	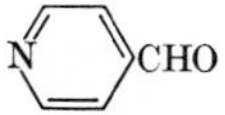 (N ring) CHO	″	RCH=CHCOAr (—)	1988
	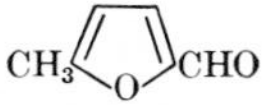 CH_3 (O ring) CHO	″	RCH=CHCOAr (—)	1988
	2-HO-3,5-$Br_2C_6H_2CHO$	″	RCH=CHCOAr (64)	549
	2-HO-5-BrC_6H_3CHO	″	RCH=CHCOAr (21)	549

Note: References 668–2359 are on pp. 403–438.

TABLE XVII. CONDENSATION OF ALDEHYDES WITH SUBSTITUTED ACETOPHENONES, CH_3COAr (*Continued*)
(R in the product is the group R in the aldehyde RCHO.)

Substituent(s) in Acetophenone	Aldehyde, RCHO	Catalyst	Product(s) (Yield, %)	Refs.
4-Cl (*contd.*)	2-ClC_6H_4CHO	NaOH	RCH=CHCOAr (—)	1923
	3-ClC_6H_4CHO	HCl	RCH=CHCOAr (53)	640
	4-ClC_6H_4CHO	NaOH	RCH=CHCOAr (—)	2002
	2-$O_2NC_6H_4CHO$	Na_3PO_4	$RCHOHCH_2COAr$ (—)	361
	". ."	HCl	RCH=CHCOAr (—)	1936
	3-$O_2NC_6H_4CHO$	NaOH	RCH=CHCOAr (90)	1924
	4-$O_2NC_6H_4CHO$	". ."	RCH=CHCOAr (92)	1924
	$C(Cl_2)=CH(CH=CH)_2CHO$	HCl	RCH=CHCOAr (70)	1451
	C_6H_5CHO	$NaOCH_3$	RCH=CHCOAr (90)	540, 349, 1924, 1994, 1996
	3-HOC_6H_4CHO	HCl	RCH=CHCOAr (42)	640
	4-HOC_6H_4CHO	". ."	RCH=CHCOAr (35)	640
	3,4-$(CH_2O_2)C_6H_3CHO$	NaOH	RCH=CHCOAr (—)	2003
	3-CH_3O-4-HO-5-BrC_6H_2CHO	HCl	RCH=CHCOAr (31)	1955
	3-CH_3O-4-HO-6-BrC_6H_2CHO	". ."	RCH=CHCOAr (76)	1955
	3-CH_3O-4-HOC_6H_3CHO	NaOH	RCH=CHCOAr (—)	1923
	3,4-$(CH_3O)_2C_6H_3CHO$	". ."	RCH=CHCOAr (—)	1995
	4-$(CH_3)_2NC_6H_4CHO$	". ."	RCH=CHCOAr (—)	1800
	(structure) N CHO	$NaOCH_3$	RCH=CHCOAr (30)	2004
	2-$C_6H_5CH_2O$-5-BrC_6H_3CHO	NaOH	RCH=CHCOAr (24)	549
2-HO-3-Cl	(structure) O CHO	KOH	RCH=CHCOAr (80)	1989
	2-HOC_6H_4CHO	". ."	RCH=CHCOAr (—)	2005
3-Cl-4-HO	C_6H_5CHO	". ."	RCH=CHCOAr (—)	2005
	2-HOC_6H_4CHO	". ."	RCH=CHCOAr (—)	2005
	3,4-$(CH_2O_2)C_6H_3CHO$	". ."	RCH=CHAOAr (—)	2005
	4-$CH_3OC_6H_4CHO$	". ."	RCH=CHCOAr (—)	2005
	3-CH_3O-4-HOC_6H_3CHO	". ."	RCH=CHCOAr (—)	2005

2-HO-4-Cl	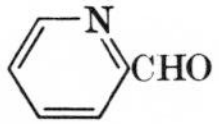 N CHO	,,	RCH=CHCOAr	(—)	1990
	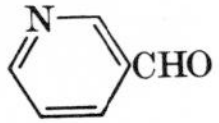 N CHO	,,	RCH=CHCOAr	(—)	371
	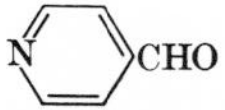 N CHO	,,	RCH=CHCOAr	(76)	370
	C_6H_5CHO	,,	RCH=CHCOAr	(—)	2005, 1992
2-HO-5-Cl	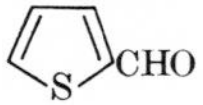 S CHO	NaOH	RCH=CHCOAr	(60)	2006
	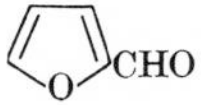 O CHO	KOH	RCH=CHCOAr	(73)	1989, 2006
	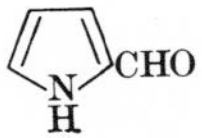 N H CHO	$(CH_2)_5NH$	RCH=CHCOAr	(32)	2006
	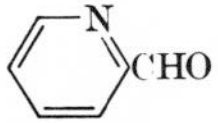 N CHO	NaOH	RCH=CHCOAr	(—)	1983, 1990
	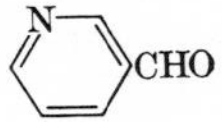 N CHO	KOH	RCH=CHCOAr	(98)	371, 2006
	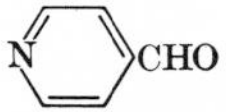 N CHO	,,	RCH=CHCOAr	(95)	370
	2-HO-5-ClC_6H_3CHO	NaOH	RCH=CHCOAr	(—)	1983
	2-HO-5-$O_2NC_6H_3CHO$	,,	RCH=CHCOAr	(—)	1983
	C_6H_5CHO	,,	RCH=CHCOAr	(30)	2006, 2005, 2007
	2-HOC_6H_4CHO	KOH	RCH=CHCOAr	(—)	2005

Note: References 668–2359 are on pp. 403–438.

TABLE XVII. Condensation of Aldehydes with Substituted Acetophenones, CH_3COAr (*Continued*)

(R in the product is the group R in the aldehyde RCHO.)

Substituent(s) in Acetophenone	Aldehyde, RCHO	Catalyst	Product(s) (Yield, %)	Refs.
2-HO-5-Cl (*contd.*)	CH_3 N CHO	NaOH	RCH=CHCOAr (—)	1983
	3,4-$(CH_2O_2)C_6H_3CHO$	..	RCH=CHCOAr (—)	1923
	2-Br-4-$CH_3C_6H_3CHO$	KOH	RCH=CHCOAr (—)	2008
	2-Cl-4-$CH_3C_6H_3CHO$	NaOH	RCH=CHCOAr (—)	341
	2-CH_3-4-BrC_6H_3CHO	KOH	RCH=CHCOAr (—)	2008
	2-HO-5-$CH_3C_6H_3CHO$	NaOH	RCH=CHCOAr (—)	1983
	N CHO	..	RCH=CHCOAr (—)	1983
2,3,4-$(HO)_3$-4-Cl	C_6H_5CHO	..	RCH=CHCOAr (—)	2009
ω-F	..	NaOH	RCH=CFCOAr (42)	2010
3-F	..	HCl	RCH=CHCOAr (28)	349
4-F	O CHO	—	RCH=CHCOAr (—)	2011
	2-FC_6H_4CHO	BF_3, CH_3CO_2H	RCH=CHCOAr (—)	1935
	4-FC_6H_4CHO	..	RCH=CHCOAr (—)	1935, 2354
	2-$O_2NC_6H_4CHO$	..	RCH=CHCOAr (—)	1935
	4-$O_2NC_6H_4CHO$	..	RCH=CHCOAr (—)	1935
	C_6H_5CHO	HCl	RCH=CHCOAr (55)	349, 2003
	3-HOC_6H_4CHO	..	RCH=CHCOAr (45)	640
	4-$CH_3OC_6H_4CHO$	BF_3, CH_3CO_2H	RCH=CHCOAr (—)	1935
	O CHO	NaOH	RCH=CHCOAr (75–95)	1854
3-F-4-HO	N CHO	..	RCH=CHCOAr (70)	1988

	NC_5H_4CHO (pyridine ring, CHO opposite N)	,,	RCH=CHCOAr	(—)	1988
2-HO-5-F	C_6H_5CHO	KOH	RCH=CHCOAr	(—)	1980, 2012
	$2\text{-}HOC_6H_4CHO$	,,	RCH=CHCOAr	(—)	1980, 2012
	$3\text{-}HOC_6H_4CHO$	,,	RCH=CHCOAr	(—)	2012
	$4\text{-}HOC_6H_4CHO$	,,	RCH=CHCOAr	(—)	1980
	$3,4\text{-}(CH_2O_2)C_6H_3CHO$	,,	RCH=CHCOAr	(—)	1980
	$4\text{-}CH_3OC_6H_4CHO$	,,	RCH=CHCOAr	(—)	1980, 2012
	$3\text{-}CH_3O\text{-}4\text{-}HOC_6H_3CHO$	,,	RCH=CHCOAr	(—)	1980
	$4\text{-}(CH_3)_2NC_6H_4CHO$	,,	RCH=CHCOAr	(—)	1980
2-I	C_6H_5CHO	NaOH	RCH=CHCOAr	(53)	1924
3-I	,,	,,	RCH=CHCOAr	(89)	1924, 349
4-I	$4\text{-}FC_6H_4CHO$	,,	RCH=CHCOAr	(—)	1995
	$4\text{-}IC_6H_4CHO$	,,	RCH=CHCOAr	(Good)	1545
	C_6H_5CHO	,,	RCH=CHCOAr	(94)	1924, 349, 1995
	$4\text{-}CH_3OC_6H_4CHO$	,,	RCH=CHCOAr	(—)	1995
	$3,4\text{-}(CH_3O)_2C_6H_3CHO$	,,	RCH=CHCOAr	(—)	1995
	$4\text{-}C_6H_5C_6H_4CHO$	,,	RCH=CHCOAr	(91)	1924
2-HO-5-I	$2\text{-}ClC_6H_4CHO$	KOH	RCH=CHCOAr	(—)	2013
	C_6H_5CHO	,,	RCH=CHCOAr	(—)	2013
	$3,4\text{-}(CH_2O_2)C_6H_3CHO$	,,	RCH=CHCOAr	(—)	2013
	$2\text{-}CH_3OC_6H_4CHO$	,,	RCH=CHCOAr	(—)	2013
$2\text{-}NO_2$	$2\text{-}ClC_6H_4CHO$	,,	RCH=CHCOAr	(—)	2014
	$4\text{-}ClC_6H_4CHO$	,,	RCH=CHCOAr	(—)	2014
	$2\text{-}O_2NC_6H_4CHO$	Na_3PO_4	$RCHOHCH_2COAr$	(—)	361
	,,	$NaOC_2H_5$; also HCl	RCH=CHCOAr	(50)	1948, 360, 1936
	$3\text{-}O_2NC_6H_4CHO$	$NaOC_2H_5$	RCH=CHCOAr	(61)	1948, 360, 1934
	$4\text{-}O_2NC_6H_4CHO$	,,	RCH=CHCOAr	(49)	1948, 360
	C_6H_5CHO	HCl	RCH=CHCOAr	(98)	349, 360, 1937, 1939, 2014, 2015

Note: References 668–2359 are on pp. 403–438.

TABLE XVII. Condensation of Aldehydes with Substituted Acetophenones, CH_3COAr (*Continued*)

(R in the product is the group R in the aldehyde RCHO.)

Substituent(s) in Acetophenone	Aldehyde, RCHO	Catalyst	Product(s) (Yield, %)	Refs.
$2\text{-}NO_2$ (*contd.*)	$4\text{-}HOC_6H_4CHO$	HCl	RCH=CHCOAr (—)	2016
	$4\text{-}CH_3C_6H_4CHO$	KOH	RCH=CHCOAr (—)	2014
	$4\text{-}CH_3OC_6H_4CHO$	$NaOCH_3$	RCH=CHCOAr (—)	1937
	$2,4,6\text{-}(CH_3)_3C_6H_2CHO$	KOH	RCH=CHCOAr (—)	2014
$3\text{-}NO_2$	S CHO (thiophene ring)	NaOH	RCH=CHCOAr (—)	2017
	CH_3 S CHO (thiophene ring)	,,	RCH=CHCOAr (—)	1988
	$2\text{-}ClC_6H_4CHO$	,,	RCH=CHCOAr (74)	1924
	$2\text{-}O_2NC_6H_4CHO$	$NaOC_2H_5$	RCH=CHCOAr (35)	1948, 360, 1934, 2018
	$3\text{-}O_2NC_6H_4CHO$	NaOH	RCH=CHCOAr (80)	1553, 360, 1940, 1948, 2019
	$4\text{-}O_2NC_6H_4CHO$	$NaOCH_3$	RCH=CHCOAr (48)	1941, 360, 1934, 1948
	C_6H_5CHO	NaOH	RCH=CHCOAr (75)	353, 360, 1937, 1939–1941, 2020, 2021
	,,	HCl	RCH=CHCOAr (100)	349
	$4\text{-}HOC_6H_4CHO$	,,	RCH=CHCOAr (—)	2016
	$3,4\text{-}(HO)_2C_6H_3CHO$	,,	RCH=CHCOAr (—)	2019
	$3\text{-}CH_3O\text{-}4\text{-}HO\text{-}5\text{-}BrC_6H_2CHO$	$CH_3CO_2NH_4$	RCH=CHCOAr (92)	1955
	$3\text{-}CH_3O\text{-}4\text{-}HO\text{-}6\text{-}BrC_6H_2CHO$	$CH_3CO_2NH_4$, CH_3CO_2H	RCH=CHCOAr (49)	1955
	$4\text{-}CH_3OC_6H_4CHO$	NaOH	RCH=CHCOAr (100)	356, 1937
	$C_6H_5CH{=}CHCHO$	,,	RCH=CHCOAr (—)	2022
	$4\text{-}(CH_3)_2NC_6H_4CHO$	,,	RCH=CHCOAr (89)	1924, 2019
	$4\text{-}CH_3OC_6H_4CH{=}CHCHO$	,,	RCH=CHCOAr (—)	2022
	Ferrocenecarboxaldehyde	,,	RCH=CHCOAr (30)	1975

4-NO_2	O_2N–(thiophene ring, S)–CHO	H_2SO_4	RCH=CHCOAr (41)	647
	O_2N–(furan ring, O)–CHO	NaOH	RCH=CHCOAr (35)	646
	(thiophene ring, S)–CHO	,,	RCH=CHCOAr (90)	2017, 1793
	(selenophene ring, Se)–CHO	,,	RCH=CHCOAr (77)	366
	(furan ring, O)–CHO	,,	RCH=CHCOAr (89)	2023, 646
	(pyridine ring, N)–CHO	,,	RCH=CHCOAr (89)	2023
	N(pyridine ring)–CHO	,,	RCH=CHCOAr (87)	2023
	2-FC_6H_4CHO	BF_3, CH_3CO_2H	RCH=CHCOAr (—)	1935
	2-$O_2NC_6H_4CHO$	Na_3PO_4	$RCHOHCH_2COAr$ (—)	361
	,,	$NaOC_2H_5$; also HCl	RCH=CHCOAr (18)	1948, 360, 1936
	3-$O_2NC_6H_4CHO$	$NaOC_2H_5$	RCH=CHCOAr (18)	1948, 360
	4-$O_2NC_6H_4CHO$	,,	RCH=CHCOAr (42)	1948, 360
	,,	HCl	RCH=CHCOAr (15)	357
	O_2N–(furan ring, O)–CH=CHCHO	H_2SO_4	RCH=CHCOAr (55)	646
	(thiophene ring, S)–CH=CHCHO	NaOH	RCH=CHCOAr (78)	647

Note: References 668–2359 are on pp. 403–438.

TABLE XVII. CONDENSATION OF ALDEHYDES WITH SUBSTITUTED ACETOPHENONES, CH_3COAr (*Continued*)

(R in the product is the group R in the aldehyde RCHO.)

Substituent(s) in Acetophenone	Aldehyde, RCHO	Catalyst	Product(s) (Yield, %)	Refs.
$4\text{-}NO_2$ (*contd.*)	C_6H_5CHO	$NaOCH_3$	RCH=CHCOAr (—)	1937, 360, 1926, 1939, 2021
	,,	HCl	RCH=CHCOAr (100)	349, 357
	$3\text{-}HOC_6H_4CHO$	,,	RCH=CHCOAr (60)	640
	$4\text{-}HOC_6H_4CHO$	,,	RCH=CHCOAr (40)	640, 2016
	2-Furyl-CH=CHCHO	H_2SO_4	RCH=CHCOAr (80)	646
	$3,4\text{-}(CH_2O_2)C_6H_3CHO$	NaOH	RCH=CHCOAr (70)	2023
	$4\text{-}CH_3OC_6H_4CHO$	HCl	RCH=CHCOAr (70)	357, 1937
	$4\text{-}(CH_3)_2NC_6H_4CHO$	Na_2CO_3	RCH=CHCOAr (81)	1810
	Quinoline-2-CHO	NaOH	$RCHOHCH_2COAr$ (—)	2024
	,,	,,	RCH=CHCOAr (85)	2023, 2024
	Quinoline-4-CHO	,,	RCH=CHCOAr (81)	2023
	$4\text{-}(CH_3)_2NC_6H_4CH{=}CHCHO$	Na_2CO_3	RCH=CHCOAr (72)	1810
	9-Formyl-3,4-benzacridine	NaOH	RCH=CHCOAr (75–95)	372
	10-Formyl-1,2-benzanthracene	,,	RCH=CHCOAr (75–95)	372
$2\text{-}HO\text{-}4\text{-}NO_2$	2-Furyl-CHO	,,	RCH=CHCOAr (91)	2025
	$2,4\text{-}Cl_2C_6H_3CHO$	HCl	RCH=CHCOAr (13)	357
	$2\text{-}BrC_6H_4CHO$	NaOH	RCH=CHCOAr (—)	357

	3-BrC_6H_4CHO	"	RCH=CHCOAr (—)	357
	4-BrC_6H_4CHO	"	RCH=CHCOAr (—)	357
	2-ClC_6H_4CHO	"	RCH=CHCOAr (—)	357
	3-ClC_6H_4CHO	"	RCH=CHCOAr (—)	357
	4-ClC_6H_4CHO	HCl	RCH=CHCOAr (11)	357
	2-FC_6H_4CHO	NaOH	RCH=CHCOAr (—)	357
	3-FC_6H_4CHO	"	RCH=CHCOAr (—)	357
	4-FC_6H_4CHO	"	RCH=CHCOAr (83)	357
	2-$O_2NC_6H_4CHO$	"	RCH=CHCOAr (56)	2026, 641
	3-$O_2NC_6H_4CHO$	"	RCH=CHCOAr (66)	2026, 641, 2021
	4-$O_2NC_6H_4CHO$	HCl	RCH=CHCOAr (3)	357, 641
	2-HO-3-$O_2NC_6H_3CHO$	NaOH	RCH=CHCOAr (—)	357
	C_6H_5CHO	"	RCH=CHCOAr (92)	2026, 357, 533, 641
	3-HOC_6H_4CHO	"	RCH=CHCOAr (—)	2028
	4-HOC_6H_4CHO	HCl	RCH=CHCOAr (27)	357, 2016, 2029
	3,4-$(HO)_2C_6H_3CHO$	"	RCH=CHCOAr (40)	2026
	4-NCC_6H_4CHO	NaOH	RCH=CHCOAr (74)	357
	4-$OHCC_6H_4CHO$	"	RCH=CHCOAr (—)	357
	3,4-$(CH_2O_2)C_6H_3CHO$	"	RCH=CHCOAr (—)	357
	2-$CH_3OC_6H_4CHO$	"	RCH=CHCOAr (58)	357
	4-$CH_3OC_6H_4CHO$	"	RCH=CHCOAr (—)	2016
	3-CH_3O-4-HOC_6H_3CHO	HCl	RCH=CHCOAr (40)	357
	$C_6H_5CH{=}CHCHO$	NaOH	RCH=CHCOAr (91)	2025
	2-$C_2H_5OC_6H_4CHO$	"	RCH=CHCOAr (—)	357
	3,4-$(CH_3O)_2C_6H_3CHO$	HCl	RCH=CHCOAr (32)	357
	4-*i*-$C_3H_7C_6H_4CHO$	NaOH	RCH=CHCOAr (—)	357
	1-$C_{10}H_7CHO$	"	RCH=CHCOAr (58)	357
	3,4-$(C_2H_5O)_2C_6H_3CHO$	"	RCH=CHCOAr (—)	357
2-HO-5-NO_2	2-thienyl-CHO (thiophene ring, S)	"	RCH=CHCOAr (60)	2006
	2-furyl-CHO (furan ring, O)	"	RCH=CHCOAr (100)	2025, 340, 2006

Note: References 668–2359 are on pp. 403–438.

TABLE XVII. Condensation of Aldehydes with Substituted Acetophenones, CH_3COAr (*Continued*)

(R in the product is the group R in the aldehyde RCHO.)

Substituent(s) in Acetophenone	Aldehyde, RCHO	Catalyst	Product(s) (Yield, %)	Refs.
2-HO-5-NO_2 (*contd.*)	Pyrrole-2-CHO (N-H)	$(CH_2)_5NH$	RCH=CHCOAr (33)	2006
	Pyridine-3-CHO	NaOH	RCH=CHCOAr (31)	2006
	2,4-$Cl_2C_6H_3CHO$	"	RCH=CHCOAr (85)	357
	2-BrC_6H_4CHO	"	RCH=CHCOAr (—)	357
	3-BrC_6H_4CHO	"	RCH=CHCOAr (—)	357
	4-BrC_6H_4CHO	"	RCH=CHCOAr (66)	357
	2-ClC_6H_4CHO	"	RCH=CHCOAr (61–75)	340, 357
	3-ClC_6H_4CHO	"	RCH=CHCOAr (58)	357
	4-ClC_6H_4CHO	"	RCH=CHCOAr (55)	357
	2-FC_6H_4CHO	"	RCH=CHCOAr (—)	357
	3-FC_6H_4CHO	"	RCH=CHCOAr (—)	357
	4-FC_6H_4CHO	"	RCH=CHCOAr (67)	357
	2-$O_2NC_6H_4CHO$	"	RCH=CHCOAr (75)	340, 641
	3-$O_2NC_6H_4CHO$	"	RCH=CHCOAr (—)	641
	4-$O_2NC_6H_4CHO$	"	RCH=CHCOAr (75)	357, 641
	2-HO-3-$O_2NC_6H_3CHO$	"	RCH=CHCOAr (13)	357
	2-HO-5-$O_2NC_6H_3CHO$	"	RCH=CHCOAr (—)	1983
	C_6H_5CHO	"	RCH=CHCOAr (97)	2026, 340, 357, 533, 641, 2006, 2021, 2027, 2030, 2031
	2-HOC_6H_4CHO	"	RCH=CHCOAr (80)	340, 2028
	3-HOC_6H_4CHO	"	RCH=CHCOAr (—)	2028
	4-HOC_6H_4CHO	HCl	RCH=CHCOAr (—)	2029, 357
	4-NCC_6H_4CHO	NaOH	RCH=CHCOAr (37)	357

	3,4-$(CH_2O_2)C_6H_3CHO$	,,	RCH=CHCOAr (39)	357, 2030
	2-$CH_3OC_6H_4CHO$	,,	RCH=CHCOAr (50)	357
	4-$CH_3OC_6H_4CHO$	,,	RCH=CHCOAr (80)	340, 2027, 2030
	3-CH_3O-4-HOC_6H_3CHO	,,	RCH=CHCOAr (1)	357
	,,	HCl	RCH=CHCOAr (23)	357
	C_6H_5CH=CHCHO	NaOH	RCH=CHCOAr (51)	2025
	2-$C_2H_5OC_6H_4CHO$	,,	RCH=CHCOAr (—)	357
	4-*i*-$C_3H_7C_6H_4CHO$	,,	RCH=CHCOAr (33)	357
	1-$C_{10}H_7CHO$	,,	RCH=CHCOAr (—)	357
	3,4-$(C_2H_5O)_2C_6H_3CHO$	,,	RCH=CHCOAr (24)	357
2-NO_2-3-HO	4-$O_2NC_6H_4CHO$	,,	RCH=CHCOAr (—)	2016
	3-HO-4-$O_2NC_6H_3CHO$	$NaOC_2H_5$	RCH=CHCOAr (63)	1948
	2-O_2N-5-HOC_6H_3CHO	,,	RCH=CHCOAr (72)	1948
	C_6H_5CHO	NaOH	RCH=CHCOAr (—)	2016
	4-$CH_3OC_6H_4CHO$	,,	RCH=CHCOAr (—)	2016
3-HO-4-NO_2	4-$O_2NC_6H_4CHO$	,,	RCH=CHCOAr (—)	2016
	3-HO-4-$O_2NC_6H_3CHO$	$NaOC_2H_5$	RCH=CHCOAr (66)	1948
	2-O_2N-5-HOC_6H_3CHO	,,	RCH=CHCOAr (60)	1948
	C_6H_5CHO	NaOH	RCH=CHCOAr (—)	2016
	4-HOC_6H_4CHO	HCl	RCH=CHCOAr (—)	2029
3-NO_2-4-HO	2,4-$Cl_2C_6H_3CHO$	NaOH	RCH=CHCOAr (—)	357
	2-BrC_6H_4CHO	,,	RCH=CHCOAr (—)	357
	3-BrC_6H_4CHO	,,	RCH=CHCOAr (—)	357
	4-BrC_6H_4CHO	,,	RCH=CHCOAr (—)	357
	2-HO-5-BrC_6H_3CHO	,,	RCH=CHCOAr (40)	1993
	2-ClC_6H_4CHO	,,	RCH=CHCOAr (—)	357
	3-ClC_6H_4CHO	,,	RCH=CHCOAr (47)	357
	2-FC_6H_4CHO	,,	RCH=CHCOAr (—)	357
	3-FC_6H_4CHO	,,	RCH=CHCOAr (—)	357
	4-FC_6H_4CHO	,,	RCH=CHCOAr (—)	357
	2-$O_2NC_6H_4CHO$	,,	RCH=CHCOAr (—)	641
	3-$O_2NC_6H_4CHO$	,,	RCH=CHCOAr (—)	641
	4-$O_2NC_6H_4CHO$	,,	RCH=CHCOAr (—)	641
	2-HO-3-$O_2NC_6H_3CHO$	,,	RCH=CHCOAr (—)	357
	3-HO-4-$O_2NC_6H_3CHO$	$NaOC_2H_5$	RCH=CHCOAr (53)	1948
	3-HO-6-$O_2NC_6H_3CHO$	,,	RCH=CHCOAr (84)	1948

Note: References 668–2359 are on pp. 403–438.

TABLE XVII. Condensation of Aldehydes with Substituted Acetophenones, CH_3COAr (*Continued*)

(R in the product is the group R in the aldehyde RCHO.)

Substituent(s) in Acetophenone	Aldehyde, RCHO	Catalyst	Product(s) (Yield, %)	Refs.
3-NO_2-4-HO (*contd.*)	3-O_2N-4-HOC_6H_3CHO	$NaOC_2H_5$	RCH=CHCOAr (27)	1948
	C_6H_5CHO	NaOH	RCH=CHCOAr (96)	2026, 533, 641, 2021
	2-HOC_6H_4CHO	,,	RCH=CHCOAr (—)	2028
	3-HOC_6H_4CHO	,,	RCH=CHCOAr (—)	2028
	4-HOC_6H_4CHO	HCl	RCH=CHCOAr (—)	2029
	4-NCC_6H_4CHO	NaOH	RCH=CHCOAr (—)	357
	3,4-$(CH_2O_2)C_6H_3CHO$	,,	RCH=CHCOAr (—)	357
	2-$CH_3OC_6H_4CHO$	,,	RCH=CHCOAr (39)	357
	4-$CH_3OC_6H_4CHO$	,,	RCH=CHCOAr (—)	2016
	3-CH_3O-4-HOC_6H_3CHO	HCl	RCH=CHCOAr (69)	357
	C_6H_5CH=CHCHO	NaOH	RCH=CHCOAr (—)	2025
	2-$C_2H_5OC_6H_4CHO$	,,	RCH=CHCOAr (—)	357
	3,4-$(CH_3O)_2C_6H_3CHO$	,,	RCH=CHCOAr (—)	357
	4-*i*-$C_3H_7C_6H_4CHO$	,,	RCH=CHCOAr (—)	357
	1-$C_{10}H_7CHO$	,,	RCH=CHCOAr (36)	357
	3,4-$(C_2H_5O)_2C_6H_3CHO$	,,	RCH=CHCOAr (15)	357
2-NO_2-5-HO	2,4-$Cl_2C_6H_3CHO$	,,	RCH=CHCOAr (—)	357
	2-BrC_6H_4CHO	,,	RCH=CHCOAr (—)	357
	3-BrC_6H_4CHO	,,	RCH=CHCOAr (—)	357
	4-BrC_6H_4CHO	,,	RCH=CHCOAr (—)	357
	2-ClC_6H_4CHO	,,	RCH=CHCOAr (—)	357
	3-ClC_6H_4CHO	,,	RCH=CHCOAr (69)	357
	4-ClC_6H_4CHO	,,	RCH=CHCOAr (56)	2028
	3-FC_6H_4CHO	,,	RCH=CHCOAr (—)	357
	4-FC_6H_4CHO	,,	RCH=CHCOAr (—)	357
	4-$O_2NC_6H_4CHO$	,,	RCH=CHCOAr (64)	2028, 2016
	2-HO-3-$O_2NC_6H_3CHO$	,,	RCH=CHCOAr (—)	357
	3-HO-4-$O_2NC_6H_3CHO$	$NaOC_2H_5$	RCH=CHCOAr (70)	1948
	2-O_2N-5-HOC_6H_3CHO	,,	RCH=CHCOAr (84)	1948
	C_6H_5CHO	,,	RCH=CHCOAr (80)	2028, 533, 2016
	3-HOC_6H_4CHO	,,	RCH=CHCOAr (40)	2028
	4-HOC_6H_4CHO	HCl	RCH=CHCOAr (—)	2016, 2029

	4-NCC_6H_4CHO	NaOH	RCH=CHCOAr (46)	357
	4-$OHCC_6H_4CHO$	,,	RCH=CHCOAr (—)	357
	3,4-$(CH_2O_2)C_6H_3CHO$	,,	RCH=CHCOAr (—)	357
	2-$CH_3OC_6H_4CHO$	,,	RCH=CHCOAr (69)	357
	4-$CH_3OC_6H_4CHO$	,,	RCH=CHCOAr (—)	2016
	$C_6H_5CH=CHCHO$	,,	RCH=CHCOAr (—)	2025
	2-$C_2H_5OC_6H_4CHO$	,,	RCH=CHCOAr (—)	357
	3,4-$(CH_3O)_2C_6H_3CHO$	,,	RCH=CHCOAr (—)	357
	1-$C_{10}H_7CHO$	,,	RCH=CHCOAr (—)	357
	3,4-$(C_2H_5O)_2C_6H_3CHO$	,,	RCH=CHCOAr (—)	357
2,4-$(HO)_2$-3-NO_2	3-$O_2NC_6H_4CHO$	KOH	RCH=CHCOAr (5)	2033
	C_6H_5CHO	,,	RCH=CHCOAr (20)	2033
	3,4-$(CH_2O_2)C_6H_3CHO$	,,	RCH=CHCOAr (25)	2033
	4-$CH_3C_6H_4CHO$	,,	RCH=CHCOAr (20)	2033
	2-$CH_3OC_6H_4CHO$	,,	RCH=CHCOAr (25)	2033
	3-$CH_3OC_6H_4CHO$	,,	RCH=CHCOAr (25)	2033
	4-$CH_3OC_6H_4CHO$	,,	RCH=CHCOAr (20)	2033
	1-$C_{10}H_7CHO$	,,	RCH=CHCOAr (5)	2033
2,4-$(HO)_2$-5-NO_2	2-HO-3,5-$Br_2C_6H_2CHO$	,,	RCH=CHCOAr (—)	1981
	C_6H_5CHO	,,	RCH=CHCOAr (15)	2033, 1981
	2-HOC_6H_4CHO	,,	RCH=CHCOAr (—)	1981
	3-Br-4-$CH_3OC_6H_3CHO$	,,	RCH=CHCOAr (—)	1981
	2-CH_3O-5-BrC_6H_3CHO	,,	RCH=CHCOAr (—)	1981
	3-CH_3O-4-HO-5-BrC_6H_2CHO	,,	RCH=CHCOAr (—)	1981
	2-$CH_3OC_6H_4CHO$	,,	RCH=CHCOAr (15)	2033, 1981
	4-$CH_3OC_6H_4CHO$	,,	RCH=CHCOAr (20)	2033, 1981
	3-CH_3O-4-HOC_6H_3CHO	,,	RCH=CHCOAr (—)	1981
2-HO	O_2N (S) CHO	HCl	RCH=CHCOAr (21)	2034, 2035
	,,	$NaOCH_3$	$RCHOHCH_2COAr$ (5)	2034
	O_2N (S) CHO	,,	$RCHOHCH_2COAr$ (15)	2034, 2035
	,,	H_2SO_4, CH_3CO_2H	RCH=CHCOAr (14)	2034

Note: References 668–2359 are on pp. 403–438.

TABLE XVII. Condensation of Aldehydes with Substituted Acetophenones, CH_3COAr (*Continued*)

(R in the product is the group R in the aldehyde RCHO.)

Substituent(s) in Acetophenone	Aldehyde, RCHO	Catalyst	Product(s) (Yield, %)	Refs.
2-HO (*contd.*)	4-Nitro-2-pyrrolecarboxaldehyde (O_2N, CHO, NH)	NaOH	RCH=CHCOAr (44)	2034, 2035
	5-Nitro-2-pyrrolecarboxaldehyde (O_2N, CHO, NH)	,,	RCH=CHCOAr (76)	2034, 2035
	2-Thiophenecarboxaldehyde	,,	RCH=CHCOAr (70)	2034–2037, 369
	,,	$NaOCH_3$	$RCHOHCH_2COAr$ (23)	2034
	2-Furaldehyde	NaOH	RCH=CHCOAr (72)	2034, 369, 1993, 2038
	2-Pyrrolecarboxaldehyde	$(CH_2)_5NH$	RCH=CHCOAr (25)	2034, 2035
	2-Pyridinecarboxaldehyde	KOH	RCH=CHCOAr (53)	2036, 2034, 2037
	,,	$NaOCH_3$	$RCHOHCH_2COAr$ (37)	2034, 369
	3-Pyridinecarboxaldehyde	NaOH	RCH=CHCOAr (60)	2034, 369, 371, 2036, 2037
	4-Pyridinecarboxaldehyde	,,	RCH=CHCOAr (17)	2034, 2037

Reactant	Catalyst	Product (Yield, %)	Refs.
,,	NaOH, $NaOCH_3$, or $(CH_2)_5NH$, CH_3CO_2H	$RCHOHCH_2COAr$ (27), RCH=CHCOAr (—)	2034, 369
2-HO-3,5-$Br_2C_6H_2CHO$	NaOH	RCH=CHCOAr (75)	1993
4-HO-3,5-$Br_2C_6H_2CHO$	,,	RCH=CHCOAr (45)	1993
2,3-$Cl_2C_6H_3CHO$	,,	RCH=CHCOAr (—)	2040
2,4-$Cl_2C_6H_3CHO$	,,	RCH=CHCOAr (—)	2040
2-HO-3,5-$Cl_2C_6H_2CHO$	,,	RCH=CHCOAr (60)	1993
2-BrC_6H_4CHO	KOH	RCH=CHCOAr (—)	2012
3-BrC_6H_4CHO	NaOH	RCH=CHCOAr (70)	1993
4-BrC_6H_4CHO	KOH	RCH=CHCOAr (—)	2012
2-HO-5-BrC_6H_3CHO	NaOH	RCH=CHCOAr (90)	1993
4-ClC_6H_4CHO	,,	RCH=CHCOAr (74)	2006
2-HO-5-ClC_6H_3CHO	,,	RCH=CHCOAr (85)	1993
2-FC_6H_4CHO	KOH	RCH=CHCOAr (—)	2012
3-FC_6H_4CHO	,,	RCH=CHCOAr (—)	2012
4-FC_6H_4CHO	,,	RCH=CHCOAr (—)	2012
2-IC_6H_4CHO	,,	RCH=CHCOAr (—)	2012
3-IC_6H_4CHO	,,	RCH=CHCOAr (—)	2012
4-IC_6H_4CHO	,,	RCH=CHCOAr (—)	2012
3-$O_2NC_6H_4CHO$	,,	RCH=CHCOAr (—)	2041, 2057
4-$O_2NC_6H_4CHO$	,,	RCH=CHCOAr (—), $RCHOHCH_2COAr$ (—)	2057
2-HO-5-$O_2NC_6H_3CHO$	KOH	RCH=CHCOAr (70)	1993
C_6H_5CHO	NaOH	RCH=CHCOAr (80)	2042–2044, 1953
2-HOC_6H_4CHO	KOH	RCH=CHCOAr (60)	1993, 348, 1953
3-HOC_6H_4CHO	NaOH	RCH=CHCOAr (50)	1993, 1999, 2039
4-HOC_6H_4CHO	KOH	RCH=CHCOAr (—)	348, 1999
2,4-$(HO)_2C_6H_3CHO$	,,	RCH=CHCOAr (—)	348
3,4-$(HO)_2C_6H_3CHO$	,,	RCH=CHCOAr (46)	2045
6-methylpyridine-2-carbaldehyde structure (CH_3, N, CHO)	,,	RCH=CHCOAr (37)	2036

Note: References 668–2359 are on pp. 403–438.

TABLE XVII. CONDENSATION OF ALDEHYDES WITH SUBSTITUTED ACETOPHENONES, CH_3COAr (*Continued*)

(R in the product is the group R in the aldehyde RCHO.)

Substituent(s) in Acetophenone	Aldehyde, RCHO	Catalyst	Product(s) (Yield, %)	Refs.
2-HO (*contd.*)	$4\text{-}NCC_6H_4CHO$	NaOH	RCH=CHCOAr (40–70)	1997, 2006
	$2,4\text{-}Cl_2\text{-}4\text{-}CH_3OC_6H_2CHO$	,,	RCH=CHCOAr (—)	2040
	$3,4\text{-}(CH_2O_2)C_6H_3CHO$	,,	RCH=CHCOAr (16)	527, 536, 2044
	$2\text{-}Br\text{-}4\text{-}CH_3C_6H_3CHO$	KOH	RCH=CHCOAr (70)	2046
	$2\text{-}Br\text{-}5\text{-}CH_3OC_6H_3CHO$	,,	RCH=CHCOAr (—)	1985
	$2\text{-}Cl\text{-}4\text{-}CH_3C_6H_3CHO$	NaOH	RCH=CHCOAr (—)	341
	$3\text{-}I\text{-}4\text{-}CH_3OC_6H_3CHO$	KOH	RCH=CHCOAr (85)	2047
	$2\text{-}CH_3O\text{-}5\text{-}IC_6H_3CHO$	,,	RCH=CHCOAr (85)	2047
	$2\text{-}CH_3C_6H_4CHO$	NaOH	RCH=CHCOAr (—)	2048, 1953
	$3\text{-}CH_3C_6H_4CHO$	,,	RCH=CHCOAr (—)	2048, 1953
	$4\text{-}CH_3C_6H_4CHO$	,,	RCH=CHCOAr (—)	2048, 1953
	$2\text{-}CH_3OC_6H_4CHO$	KOH	RCH=CHCOAr (—)	1953, 1985
	$3\text{-}CH_3OC_6H_4CHO$	,,	RCH=CHCOAr (—)	1953, 2049
	$4\text{-}CH_3OC_6H_4CHO$	NaOH	RCH=CHCOAr (54)	2006, 1953, 2039, 2044, 2050
	$2\text{-}HO\text{-}3\text{-}CH_3OC_6H_3CHO$	KOH	RCH=CHCOAr (—)	1953, 2039
	$3\text{-}HO\text{-}4\text{-}CH_3OC_6H_3CHO$	,,	RCH=CHCOAr (92)	2051
	$3\text{-}CH_3O\text{-}4\text{-}HOC_6H_3CHO$	,,	RCH=CHCOAr (—)	1953, 2052
	$4\text{-}C_2H_5OC_6H_4CHO$	NaOH	RCH=CHCOAr (—)	2050
	$2,3\text{-}(CH_3O)_2C_6H_3CHO$	KOH	RCH=CHCOAr (—)	1953, 2039
	$3,4\text{-}(CH_3O)_2C_6H_3CHO$	,,	RCH=CHCOAr (—)	1953, 527
	$4\text{-}(CH_3)_2NC_6H_4CHO$	$(CH_2)_5NH$	RCH=CHCOAr (20)	2006
	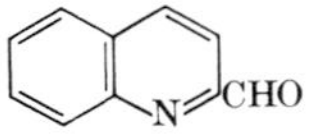 N CHO	KOH	RCH=CHCOAr (29)	2036, 2034
	,,	$NaOCH_3$	$RCHOHCH_2COAr$ (20)	2034

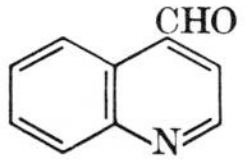 CHO, N	NaOH	RCH=CHCOAr (40)	2034
,,	$NaOCH_3$	$RCHOHCH_2COAr$ (10)	2034
2,6-$(CH_3)_2$-4-$CH_3OC_6H_2CHO$	NaOH	RCH=CHCOAr (33)	1609
1-$C_{10}H_7CHO$	KOH	RCH=CHCOAr (16)	2036
2-$C_{10}H_7CHO$	,,	RCH=CHCOAr (55)	2036
Ferrocenecarboxaldehyde	NaOH	RCH=CHCOAr (55)	1975
3,4-Dihydroxybenzaldehyde 4-*β*-*d*-glucoside	,,	RCH=CHCOAr (54)	2053
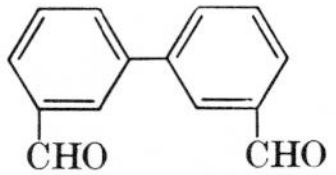 CHO, CHO	KOH	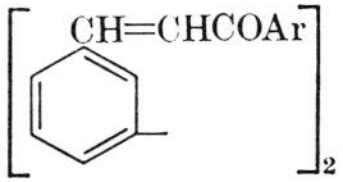 [CH=CHCOAr]$_2$ (66)	2047
4-$C_6H_5CH_2OC_6H_4CHO$	NaOH	RCH=CHCOAr (60)	2054
3-Hydroxy-4-methoxybenzaldehyde *β*-*d*-glucoside	Na_3PO_4, $Na_2B_4O_7$, NaOH	RCH=CHCOAr (9)	2055
3-Methoxy-4-hydroxybenzaldehyde *β*-*d*-glucoside	,,	RCH=CHCOAr (22)	2055
4-$C_6H_5CONHC_6H_4CHO$	NaOH	RCH=CHCOAr (—)	2057
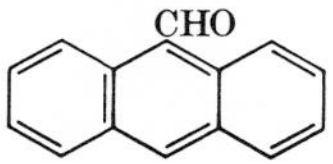 CHO	KOH	RCH=CHCOAr (58)	2036, 1978
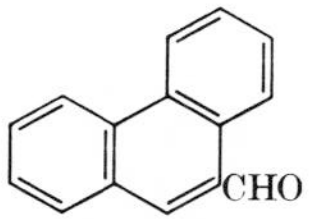 CHO	,,	RCH=CHCOAr (93)	2036
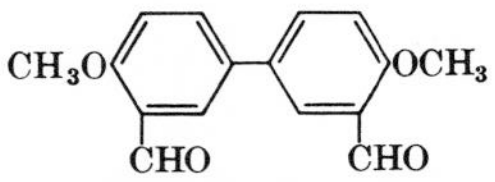 CH_3O, OCH_3, CHO, CHO	,,	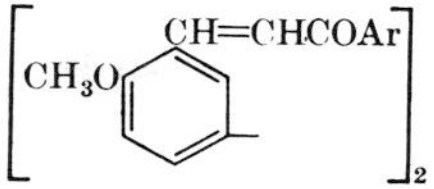 [CH_3O, CH=CHCOAr]$_2$ (59)	2047

Note: References 668–2359 are on pp. 403–438.

TABLE XVII. Condensation of Aldehydes with Substituted Acetophenones, CH_3COAr (*Continued*)

(R in the product is the group R in the aldehyde RCHO.)

Substituent(s) in Acetophenone	Aldehyde, RCHO	Catalyst	Product(s) (Yield, %)	Refs.
2-HO (*contd.*)	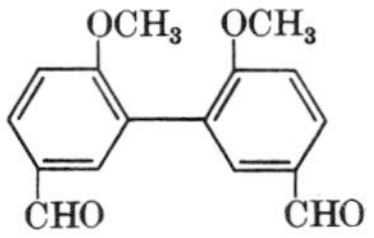 6,6'-Dimethoxybiphenyl-3,3'-dicarboxaldehyde (structure: two OCH_3, two CHO)	KOH	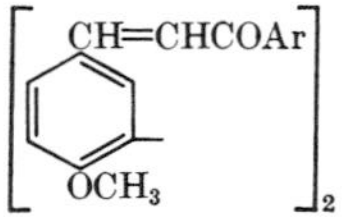 [biphenyl-diyl bis($CH=CHCOAr$), OCH_3]$_2$ (59)	2047
3-HO	C_6H_5CHO	NaOH	$RCH=CHCOAr$ (41)	353, 640, 1946
	3-HOC_6H_4CHO	HCl	$RCH=CHCOAr$ (30)	640
	4-HOC_6H_4CHO	,,	$RCH=CHCOAr$ (36)	640
	3,4,5-$(CH_3O)_3C_6H_2CHO$	$NaOCH_3$	$RCH=CHCOAr$ (—)	1971
	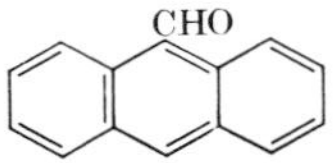 9-Anthraldehyde (structure: anthracene-CHO)	HCl	$RCH=CHCOAr$ (63)	1978
4-HO	$C(Cl_2)=CHCHO$	,,	$RCH=CHCOAr$ (80)	637
	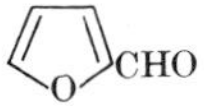 Furfural (structure: furan-CHO)	NaOH	$RCH=CHCOAr$ (—)	2056
	2-HO-3,5-$Br_2C_6H_2CHO$	,,	$RCH=CHCOAr$ (31)	549
	2-HO-5-BrC_6H_3CHO	,,	$RCH=CHCOAr$ (40)	1993
	3-ClC_6H_4CHO	HCl	$RCH=CHCOAr$ (38)	640
	4-ClC_6H_4CHO	,,	$RCH=CHCOAr$ (32)	640
	4-$O_2NC_6H_4CHO$	NaOH	$RCH=CHCOAr$ (—)	2016
	C_6H_5CHO	KOH	$RCH=CHCOAr$ (—)	1946
	,,	HCl	$RCH=CHCOAr$ (68)	349, 640
	3-HOC_6H_4CHO	,,	$RCH=CHCOAr$ (14)	640
	4-HOC_6H_4CHO	NaOH	$RCH=CHCOAr$ (52)	640, 348, 1572, 1948

	,,	HCl	RCH=CHCOAr (90)	357
	3,4-$(HO)_2C_6H_3CHO$	KOH	RCH=CHCOAr (23)	1948
	3-NCC_6H_4CHO	NaOH	RCH=CHCOAr (40–70)	1997
	4-NCC_6H_4CHO	,,	RCH=CHCOAr (40–70)	1997
	3-CH_3O-4-HO-5-BrC_6H_2CHO	CH_3CO_2H	RCH=CHCOAr (46)	1955
	3-CH_3O-4-HO-6-BrC_6H_2CHO	,,	RCH=CHCOAr (30)	1955
	4-$CH_3OC_6H_4CHO$	HCl	RCH=CHCOAr (68)	357, 1572, 2016
	3,4-$(CH_3O)_2C_6H_3CHO$	KOH	RCH=CHCOAr (—)	2058
	2-CH_3-4,5-$(CH_3O)_2C_6H_2CHO$	,,	RCH=CHCOAr (—)	2059
	[structure: anthracene with CHO]	HCl	RCH=CHCOAr (71)	1978
2,4-$(HO)_2$	2-ClC_6H_4CHO	KOH	RCH=CHCOAr (64)	2041
	3-ClC_6H_4CHO	,,	RCH=CHCOAr (33)	2041
	4-ClC_6H_4CHO	,,	RCH=CHCOAr (19)	2041
	2-$O_2NC_6H_4CHO$	,,	RCH=CHCOAr (0)	2041
	3-$O_2NC_6H_4CHO$	,,	RCH=CHCOAr (14)	2041
	4-$O_2NC_6H_4CHO$	,,	RCH=CHCOAr (0)	2041
	C_6H_5CHO	NaOH	RCH=CHCOAr (34–50)	340, 352, 1940,1953, 2060
	2-HOC_6H_4CHO	KOH	RCH=CHCOAr (—)	348, 1953, 2061
	3-HOC_6H_4CHO	,,	RCH=CHCOAr (40)	2062, 348, 1953, 1999, 2063
	4-HOC_6H_4CHO	,,	RCH=CHCOAr (27)	2064, 348, 2063
	3,4-$(HO)_2C_6H_3CHO$	,,	RCH=CHCOAr (18)	2060, 527, 1999, 2065
	3,4-$(CH_2O_2)C_6H_3CHO$	,,	RCH=CHCOAr (—)	2065
	2-$CH_3C_6H_4CHO$	,,	RCH=CHCOAr (20)	2066, 1953
	3-$CH_3C_6H_4CHO$	,,	RCH=CHCOAr (40)	2066, 1953
	4-$CH_3C_6H_4CHO$	,,	RCH=CHCOAr (50)	2066, 1953

Note: References 668–2359 are on pp. 403–438.

TABLE XVII. CONDENSATION OF ALDEHYDES WITH SUBSTITUTED ACETOPHENONES, CH_3COAr (*Continued*)

(R in the product is the group R in the aldehyde RCHO.)

Substituent(s) in Acetophenone	Aldehyde, RCHO	Catalyst	Product(s) (Yield, %)	Refs.
$2,4\text{-(HO)}_2$ (*contd.*)	$2\text{-}CH_3OC_6H_4CHO$	KOH	RCH=CHCOAr (—)	1953, 1985
	$3\text{-}CH_3OC_6H_4CHO$	"	RCH=CHCOAr (—)	1953, 1999
	$4\text{-}CH_3OC_6H_4CHO$	"	RCH=CHCOAr (53)	2064, 1953, 2067
	$2\text{-HO-3-}CH_3OC_6H_3CHO$	"	RCH=CHCOAr (64)	2068, 1953
	$3\text{-}CH_3O\text{-4-}HOC_6H_3CHO$	"	RCH=CHCOAr (31)	2062, 2066
	$4\text{-}C_2H_5OC_6H_4CHO$	"	RCH=CHCOAr (—)	348
	$2,3\text{-}(CH_3O)_2C_6H_3CHO$	"	RCH=CHCOAr (42)	2066, 1953
	$3,4\text{-}(CH_3O)_2C_6H_3CHO$	"	RCH=CHCOAr (—)	1953
	9-Anthraldehyde (CHO)	"	RCH=CHCOAr (40)	1978
	$3,4,5\text{-}(C_6H_5CH_2O)_3C_6H_2CHO$	"	RCH=CHCOAr (28)	2069
$2,5\text{-(HO)}_2$	$2\text{-}ClC_6H_4CHO$	"	RCH=CHCOAr (16)	2070
	$2\text{-HO-5-}BrC_6H_3CHO$	"	RCH=CHCOAr (—)	2070
	$3\text{-}O_2NC_6H_4CHO$	$POCl_3$	RCH=CHCOAr (22)	2070
	C_6H_5CHO	NaOH	RCH=CHCOAr (35)	2071, 2072
	$2\text{-}HOC_6H_4CHO$	KOH	RCH=CHCOAr (—)	348, 2071
	$4\text{-}HOC_6H_4CHO$	"	RCH=CHCOAr (—)	348, 2071
	$3,4\text{-}(CH_2O_2)C_6H_3CHO$	"	RCH=CHCOAr (25)	2070
	$4\text{-}CH_3OC_6H_4CHO$	NaOH	RCH=CHCOAr (—)	2071
	$3\text{-}CH_3O\text{-4-}HOC_6H_3CHO$	"	RCH=CHCOAr (—)	2071
	9-Anthraldehyde (CHO)	HCl	RCH=CHCOAr (—)	1978
$3,4\text{-(HO)}_2$	C_6H_5CHO	KOH	RCH=CHCOAr (65)	1948
	$4\text{-}HOC_6H_4CHO$	"	RCH=CHCOAr (45)	1948
	$3,4\text{-(HO)}_2C_6H_3CHO$	"	RCH=CHCOAr (55)	1948, 2073
	$3,4\text{-}(CH_2O_2)C_6H_3CHO$	"	RCH=CHCOAr (—)	2073
	$2\text{-HO-3,4-}(CH_3O)_2C_6H_2CHO$	"	RCH=CHCOAr (—)	348

2,3,4-$(HO)_3$	2-HO-5-BrC_6H_3CHO	NaOH	RCH=CHCOAr (40)	1993
	C_6H_5CHO	KOH	RCH=CHCOAr (61)	352, 348
	2-HOC_6H_4CHO	"	RCH=CHCOAr (45)	352, 348
	3-HOC_6H_4CHO	"	RCH=CHCOAr (25)	352
	4-HOC_6H_4CHO	"	RCH=CHCOAr (—)	348
	3,4-$(HO)_2C_6H_3CHO$	"	RCH=CHCOAr (39)	2045
2,3,6-$(HO)_3$	4-HOC_6H_4CHO	NaOH	RCH=CHCOAr (—)	2074
2,4,5-$(HO)_3$	3,4-$(HO)_2C_6H_3CHO$	KOH	RCH=CHCOAr (16)	2075
2,4,6-$(HO)_3$	"	NaOH	RCH=CHCOAr (39)	527
2-NH_2	C_6H_5CHO	"	RCH=CHCOAr (10)	1924
3-NH_2	"	"	RCH=CHCOAr (50)	1924
4-NH_2	CH_2O	None, heat	4-CH_2=$NC_6H_4COCH_2CH_2OH$ (23)	598, 359
	i-C_3H_7CHO	NaOH	RCH=CHCOAr (—)	1643
	[furan ring]CHO	KOH	4-RCH=NC_6H_4COCH=CHR (69)	2076, 359
	[pyridine ring]CHO	NaOH	RCH=CHCOAr (42)	545
	2-HO-3,5-$Br_2C_6H_2CHO$	"	RCH=CHCOAr (18)	549
	4-HO-3,5-$Br_2C_6H_2CHO$	"	RCH=CHCOAr (41)	549
	2-$O_2NC_6H_4CHO$	HCl	RCH=CHCOAr (—)	1936
	3-$O_2NC_6H_4CHO$	KOH	4-RCH=NC_6H_4COCH=CHR (—)	359
	4-$O_2NC_6H_4CHO$	HCl	RCH=CHCOAr (—)	2077
	"	KOH	4-RCH=NC_6H_4COCH=CHR (—)	359
	C_6H_5CHO	NaOH	RCH=CHCOAr (31)	1924, 359, 1934, 2078
	3-$H_2NC_6H_4CHO$	$NaOCH_3$	RCH=CHCOAr (68)	2004
	3-NCC_6H_4CHO	NaOH	RCH=CHCOAr (40–70)	1997
	4-NCC_6H_4CHO	"	RCH=CHCOAr (40–70)	1997
	3,4-$(CH_2O_2)C_6H_3CHO$	KOH	RCH=CHCOAr (—)	359
	4-$CH_3C_6H_4CHO$	"	4-RCH=NC_6H_4COCH=CHR (—)	359
	4-$CH_3OC_6H_4CHO$	"	4-RCH=NC_6H_4COCH=CHR (—)	359, 2078
	C_6H_5CH=CHCHO	"	4-RCH=NC_6H_4COCH=CHR (—)	359, 2078
	[quinoline ring]CHO	NaOH	RCH=CHCOAr (74)	2023, 2024
	4-*i*-$C_3H_7C_6H_4CHO$	KOH	4-RCH=NC_6H_4COCH=CHR (—)	359

Note: References 668–2359 are on pp. 403–438.

TABLE XVII. CONDENSATION OF ALDEHYDES WITH SUBSTITUTED ACETOPHENONES, CH_3COAr (*Continued*)

(R in the product is the group R in the aldehyde RCHO.)

Substituent(s) in Acetophenone	Aldehyde, RCHO	Catalyst	Product(s) (Yield, %)	Refs.
4-CN	$4\text{-}ClC_6H_4CHO$	NaOH	RCH=CHCOAr (40–70)	1997
	$2\text{-}O_2NC_6H_4CHO$	Na_3PO_4	$RCHOHCH_2COAr$ (—)	361
	,,	KOH	RCH=CHCOAr (—)	1936
	$4\text{-}NCC_6H_4CHO$	NaOH	RCH=CHCOAr (40–70)	1997
	$3,4\text{-}(CH_2O_2)C_6H_3CHO$	,,	RCH=CHCOAr (40–70)	1997
	$4\text{-}CH_3OC_6H_4CHO$	,,	RCH=CHCOAr (40–70)	1997, 2357
	$3,4\text{-}(CH_3O)_2C_6H_3CHO$	,,	RCH=CHCOAr (40–70)	1997, 2357
2-HO-4-CN	2-thiophenecarboxaldehyde (structure: thiophene ring, S, CHO)	,,	RCH=CHCOAr (53)	2006
	2-furaldehyde (structure: furan ring, O, CHO)	,,	RCH=CHCOAr (56)	2006
	pyrrole-2-carboxaldehyde (structure: pyrrole ring, N–H, CHO)	$(CH_2)_5NH$	RCH=CHCOAr (21)	2006
	pyridine-3-carboxaldehyde (structure: pyridine ring, N, CHO)	NaOH	RCH=CHCOAr (30)	2006
	C_6H_5CHO	,,	RCH=CHCOAr (38)	2006
$2\text{-}CH_3\text{-}3,5\text{-}(NO_2)_2$	,,	HCl, CH_3CO_2H	RCH=CHCOAr (67)	2079
$4\text{-}CO_2H$	$2\text{-}ClC_6H_4CHO$	NaOH	RCH=CHCOAr (60)	2080, 2350
	$2\text{-}O_2NC_6H_4CHO$	Na_3PO_4	$RCHOHCH_2COAr$ (—)	361
	,,	KOH	RCH=CHCOAr (—)	1936, 2350
	C_6H_5CHO	NaOH	RCH=CHCOAr (60)	2080, 2350
	$4\text{-}CH_3OC_6H_4CHO$	,,	RCH=CHCOAr (50)	2080, 2350
	$4\text{-}(CH_3)_2NC_6H_4CHO$	,,	RCH=CHCOAr (30)	2080, 2350
$3,4\text{-}(CH_2O_2)$	$3,4\text{-}(CH_2O_2)C_6H_3CHO$	,,	RCH=CHCOAr (90)	2081
$3\text{-}CO_2H\text{-}4\text{-}HO$	C_6H_5CHO	KOH	RCH=CHCOAr (55)	2082
	$2\text{-}HOC_6H_4CHO$	,,	RCH=CHCOAr (35)	2082

	3-HOC_6H_4CHO	,,	$RCH{=}CHCOAr$ (25)	2082
	4-HOC_6H_4CHO	,,	$RCH{=}CHCOAr$ (40)	2082
	2-$CH_3OC_6H_4CHO$	,,	$RCH{=}CHCOAr$ (45)	2082
	4-$CH_3OC_6H_4CHO$	,,	$RCH{=}CHCOAr$ (55)	2082
2-HO-5-CO_2H	C_6H_5CHO	,,	$RCH{=}CHCOAr$ (70)	2082
	3-HOC_6H_4CHO	,,	$RCH{=}CHCOAr$ (50)	2082
	4-HOC_6H_4CHO	,,	$RCH{=}CHCOAr$ (—)	2082
	3,4(CH_2O_2)C_6H_3CHO	NaOH	$RCH{=}CHCOAr$ (86)	2083
	2-$CH_3OC_6H_4CHO$	KOH	$RCH{=}CHCOAr$ (40)	2082
	4-$CH_3OC_6H_4CHO$	,,	$RCH{=}CHCOAr$ (55)	2082
4-CH_3O-5-Br	N CHO	,,	$RCH{=}CHCOAr$ (83)	370
2-HO-4-CH_3O-5-Br	2-Cl-4-$CH_3C_6H_3CHO$	NaOH	$RCH{=}CHCOAr$ (—)	341
	3-Cl-4-$CH_3C_6H_3CHO$	,,	$RCH{=}CHCOAr$ (—)	341
	2-$CH_3OC_6H_4CHO$	KOH	$RCH{=}CHCOAr$ (—)	1985
2-Cl-5-CH_3	2-$O_2NC_6H_4CHO$	,,	$RCH{=}CHCOAr$ (—)	1936
	C_6H_5CHO	NaOH	$RCH{=}CHCOAr$ (—)	2003
3-Cl-4-CH_3O	S CHO	,,	$RCH{=}CHCOAr$ (—)	2017
3-CH_3-4-F	4-$(CH_3)_2NC_6H_4CHO$	,,	$RCH{=}CHCOAr$ (—)	1995
3-NO_2-4-CH_3	2-$O_2NC_6H_4CHO$	Na_3PO_4	$RCHOHCH_2COAr$ (—)	361
	,,	HCl	$RCH{=}CHCOAr$ (—)	1936
3-NO_2-4-CH_3O	3,4-(CH_2O_2)C_6H_3CHO	KOH	$RCH{=}CHCOAr$ (—)	2084
2-HO-3-NO_2-5-CH_3	C_6H_5CHO	,,	$RCH{=}CHCOAr$ (—)	2085, 2043
	2-CH_3O-3-BrC_6H_3CHO	,,	$RCH{=}CHCOAr$ (—)	2085
	2-$CH_3OC_6H_4CHO$	,,	$RCH{=}CHCOAr$ (—)	2085
	3-CH_3O-4-HOC_6H_3CHO	—	$RCH{=}CHCOAr$ (—)	2043
	3,4-$(CH_3O)_2C_6H_3CHO$	—	$RCH{=}CHCOAr$ (—)	2043
2-HO-3-CH_3O-5-NO_2	C_6H_5CHO	NaOH	$RCH{=}CHCOAr$ (—)	2074
2-HO-3-NO_2-4-CH_3O	,,	KOH	$RCH{=}CHCOAr$ (60)	2033
	4-$CH_3OC_6H_4CHO$	,,	$RCH{=}CHCOAr$ (60)	2033, 2043

Note: References 668–2359 are on pp. 403–438.

TABLE XVII. CONDENSATION OF ALDEHYDES WITH SUBSTITUTED ACETOPHENONES, CH_3COAr (*Continued*)

(R in the product is the group R in the aldehyde RCHO.)

Substituent(s) in Acetophenone	Aldehyde, RCHO	Catalyst	Product(s) (Yield, %)	Refs.
2-HO-4-CH_3O-5-NO_2	2-HO-3,5-$Br_2C_6H_2CHO$	KOH	RCH=CHCOAr (—)	2086
	C_6H_5CHO	,,	RCH=CHCOAr (—)	2086
	2-HOC_6H_4CHO	,,	RCH=CHCOAr (—)	2086
	3,4-$(CH_2O_2)C_6H_3CHO$	,,	RCH=CHCOAr (—)	2087
	3-Br-4-$CH_3OC_6H_3CHO$	,,	RCH=CHCOAr (—)	2086
	2-CH_3O-5-BrC_6H_3CHO	,,	RCH=CHCOAr (—)	2086
	3-CH_3O-4-HO-5-BrC_6H_2CHO	,,	RCH=CHCOAr (—)	2086
	2-$CH_3OC_6H_4CHO$	,,	RCH=CHCOAr (—)	2086
	4-$CH_3OC_6H_4CHO$	,,	RCH=CHCOAr (—)	2086
	3-CH_3O-4-HOC_6H_3CHO	,,	RCH=CHCOAr (—)	2086
	3,4-$(CH_3O)_2C_6H_3CHO$	,,	RCH=CHCOAr (—)	2087
2-HO-3-NO_2-6-CH_3O	C_6H_5CHO	NaOH	RCH=CHCOAr (—)	2074
2-CH_3	,,	,,	RCH=CHCOAr (55)	1939
3-CH_3	,,	HCl	RCH=CHCOAr (65)	349, 1939
4-CH_3	$C(Cl_2)$=CHCHO	BaO	RCH=CHCOAr (45)	1908
	thiophene ring (S) with CHO	$NaOC_2H_5$	RCH=CHCOAr (86)	1914
	selenophene ring (Se) with CHO	$NaOCH_3$	RCH=CHCOAr (52)	1522, 366
	furan ring (O) with CHO	$NaOC_2H_5$	RCH=CHCOAr (—)	1860, 537
	2-HO-3,5-$Br_2C_6H_2CHO$	NaOH	RCH=CHCOAr (67)	549
	3,5-Br_2-4-HOC_6H_2CHO	,,	RCH=CHCOAr (50)	549
	2-HO-5-BrC_6H_3CHO	,,	RCH=CHCOAr (48)	549, 1965
	2-ClC_6H_4CHO	,,	RCH=CHCOAr (57)	353
	2-$O_2NC_6H_4CHO$	Na_3PO_4	$RCHOHCH_2COAr$ (—)	361

,,	NaOH	RCH=CHCOAr (—)	1938, 1936
$3\text{-}O_2NC_6H_4CHO$	,,	RCH=CHCOAr (—)	1938
$4\text{-}O_2NC_6H_4CHO$	,,	RCH=CHCOAr (—)	1938
C_6H_5CHO	HCl, $(CH_3CO)_2O$	RCH=CHCOAr (42)	353, 349, 538, 1938, 2088–2090
$2\text{-}HOC_6H_4CHO$	NaOH	RCH=CHCOAr (—)	2091
$3,4\text{-}(CH_2O_2)\text{-}6\text{-}BrC_6H_2CHO$	,,	RCH=CHCOAr (—)	2092
$3,4\text{-}(CH_2O_2)\text{-}6\text{-}ClC_6H_2CHO$	,,	RCH=CHCOAr (—)	2092
$4\text{-}OHCC_6H_4CHO$	KOH	RCH=CHCOAr (—)	544
,,	,,	$C_6H_4(CH{=}CHCOAr)_2\text{-}1,4$ (—)	544
$3,4\text{-}(CH_2O_2)C_6H_3CHO$	NaOH	RCH=CHCOAr (—)	1938, 2092
$3\text{-}Br\text{-}4\text{-}CH_3OC_6H_3CHO$	,,	RCH=CHCOAr (—)	2092
$2\text{-}CH_3O\text{-}5\text{-}BrC_6H_3CHO$	,,	RCH=CHCOAr (78)	549
$3\text{-}CH_3O\text{-}4\text{-}HO\text{-}5\text{-}BrC_6H_2CHO$	HCl, CH_3CO_2H	RCH=CHCOAr (19)	1955
$3\text{-}CH_3O\text{-}4\text{-}HO\text{-}6\text{-}BrC_6H_2CHO$	,,	RCH=CHCOAr (55)	1955
$3\text{-}Cl\text{-}4\text{-}CH_3OC_6H_3CHO$	NaOH	RCH=CHCOAr (—)	2092
$2\text{-}CH_3C_6H_4CHO$	,,	RCH=CHCOAr (30)	353
$3\text{-}CH_3C_6H_4CHO$	,,	RCH=CHCOAr (36)	353
$4\text{-}CH_3C_6H_4CHO$	$NaOCH_3$	RCH=CHCOAr (82)	2089, 353, 1959
$4\text{-}CH_3OC_6H_4CHO$	NaOH	RCH=CHCOAr (55)	353, 1959, 1964, 2092
(furan ring, O) $CH{=}C(CH_3)CHO$	Na_2CO_3	RCH=CHCOAr (41)	1766
$3\text{-}HO\text{-}4\text{-}CH_3OC_6H_3CHO$	NaOH	RCH=CHCOAr (—)	339
(ring structure: NH, CHO)	$(CH_2)_5NH$, 175°	RCH=CHCOAr (61)	600
(ring structure: O, CHO)	NaOH	RCH=CHCOAr (75–95)	1854

Note: References 668–2359 are on pp. 403–438.

TABLE XVII. Condensation of Aldehydes with Substituted Acetophenones, CH_3COAr (*Continued*)

(R in the product is the group R in the aldehyde RCHO.)

Substituent(s) in Acetophenone	Aldehyde, RCHO	Catalyst	Product(s) (Yield, %)	Refs.
$4\text{-}CH_3$ (*contd.*)	$2\text{-}C_2H_5O\text{-}3,5\text{-}Br_2C_6H_2CHO$	NaOH	RCH=CHCOAr (45)	549
	[furan ring, O]$(CH{=}CH)_2CHO$	Na_2CO_3	RCH=CHCOAr (92)	1766
	$2\text{-}C_2H_5O\text{-}5\text{-}BrC_6H_3CHO$	NaOH	RCH=CHCOAr (100)	549
	$4\text{-}(CH_3)_2NC_6H_4CHO$	,,	RCH=CHCOAr (—)	1800
	[indole ring: CHO, N, CH_3]	$(CH_2)_5NH$, 200°	RCH=CHCOAr (70)	600
	$3,4\text{-}(CH_2O_2)C_6H_3CH{=}CHCHO$	NaOH	RCH=CHCOAr (—)	1938
	$3,4\text{-}(C_2H_5O)_2C_6H_3CHO$	,,	RCH=CHCOAr (—)	1923
	$2\text{-}C_6H_5CH_2O\text{-}3,5\text{-}Br_2C_6H_2CHO$	,,	RCH=CHCOAr (—)	549
	$2\text{-}C_6H_5CH_2O\text{-}5\text{-}BrC_6H_3CHO$	,,	RCH=CHCOAr (71)	549
	[indole ring: CHO, C_6H_5, NH]	$(CH_2)_5NH$	RCH=CHCOAr (67)	1979
$2\text{-}CH_3Se$	C_6H_5CHO	$(CH_2)_5NH$, CH_3CO_2H	RCH=CHCOAr (67)	2093
$2\text{-}CH_3O$	,,	KOH	RCH=CHCOAr (—)	1953
	$4\text{-}NCC_6H_5CHO$	NaOH	RCH=CHCOAr (40–70)	1997
$4\text{-}CH_3O$	$C(Cl_2){=}CHCHO$	BaO	RCH=CHCOAr (56)	1908
	O_2N[thiophene ring, S]CHO	H_2SO_4, CH_3CO_2H	RCH=CHCOAr (40)	647

O_2N–[furan ring]–CHO	”	RCH=CHCOAr (90)	646
[thiophene ring]–CHO	NaOH	RCH=CHCOAr (98)	1913, 1793, 1914
[selenophene ring]–CHO	”	RCH=CHCOAr (75)	366
[furan ring]–CHO	”	RCH=CHCOAr (84)	1794, 1918, 2094
[pyridine ring]–CHO	”	RCH=CHCOAr (—)	1988
CH_3–[furan ring]–CHO	”	RCH=CHCOAr (—)	1988
2-HO-3,5-$Br_2C_6H_2CHO$	”	RCH=CHCOAr (58)	549
2,4-$Cl_2C_6H_3CHO$	”	RCH=CHCOAr (—)	1923
2-BrC_6H_4CHO	”	RCH=CHCOAr (34)	1929
2-HO-5-BrC_6H_3CHO	”	RCH=CHCOAr (60)	549
2-ClC_6H_4CHO	”	RCH=CHCOAr (64)	353, 1929
2-FC_6H_4CHO	BF_3, CH_3CO_2H	RCH=CHCOAr (—)	1935
4-FC_6H_4CHO	”	RCH=CHCOAr (—)	1935
2-$O_2NC_6H_4CHO$	Na_3PO_4	$RCHOHCH_2COAr$ (—)	361
	NaOH	RCH=CHCOAr (58)	1798
3-$O_2NC_6H_4CHO$	”	RCH=CHCOAr (96)	356
4-$O_2NC_6H_4CHO$	”	RCH=CHCOAr (90)	2095, 1798
O_2N–[furan ring]–CH=CHCHO	H_2SO_4, CH_3CO_2H	RCH=CHCOAr (25)	646
C_6H_5CHO	NaOH	RCH=CHCOAr (81)	353, 1994

Note: References 668–2359 are on pp. 403–438.

TABLE XVII. CONDENSATION OF ALDEHYDES WITH SUBSTITUTED ACETOPHENONES, CH_3COAr (*Continued*)

(R in the product is the group R in the aldehyde RCHO.)

Substituent(s) in Acetophenone	Aldehyde, RCHO	Catalyst	Product(s) (Yield, %)	Refs.
4-CH_3O (*contd.*)	2-HOC_6H_4CHO	NaOH	RCH=CHCOAr (—)	2097
	3-HOC_6H_4CHO	,,	RCH=CHCOAr (74)	353, 640
	4-HOC_6H_4CHO	,,	RCH=CHCOAr (—)	1572
	 O CH=CHCHO	,,	RCH=CHCOAr (88)	1794
	3-NCC_6H_4CHO	,,	RCH=CHCOAr (40–70)	1997
	4-NCC_6H_4CHO	,,	RCH=CHCOAr (40–70)	1997
	4-$OHCC_6H_4CHO$	KOH	RCH=CHCOAr (—)	544
	,,	,,	$C_6H_4(CH{=}CHCOAr)_2$-1,4 (—)	544
	3,4-$(CH_2O_2)C_6H_3CHO$	NaOH	RCH=CHCOAr (—)	1791, 2094
	3-CH_3O-4-HO-5-BrC_6H_2CHO	HCl, CH_3CO_2H	RCH=CHCOAr (67)	1955
	3-CH_3O-4-HO-6-BrC_6H_2CHO	,,	RCH=CHCOAr (22)	1955
	3-$CH_3OC_6H_4CHO$	$NaOCH_3$	RCH=CHCOAr (79)	2099
	4-$CH_3OC_6H_4CHO$	$NaOC_2H_5$	RCH=CHCOAr (92)	2100–2103, 353, 640, 1959, 1971, 2094
	2-HO-3-$CH_3OC_6H_3CHO$	NaOH	RCH=CHCOAr (61)	1966
	3-CH_3O-4-HOC_6H_3CHO	,,	RCH=CHCOAr (—)	1923
	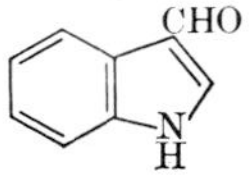 CHO N H	$(CH_2)_5NH$, 175°	RCH=CHCOAr (60)	600
	$C_6H_5CH{=}CHCHO$	NaOH	RCH=CHCOAr (—)	1791, 1798
	3-$CH_3CONHC_6H_4CHO$	,,	RCH=CHCOAr (76)	1798
	2-CH_3-4-$CH_3OC_6H_3CHO$	$NaOCH_3$	RCH=CHCOAr (80)	2104
	2,3-$(CH_3O)_2C_6H_3CHO$	,,	RCH=CHCOAr (—)	1971
	3,4-$(CH_3O)_2C_6H_3CHO$	NaOH	RCH=CHCOAr (—)	1929
	4-$(CH_3)_2NC_6H_4CHO$	,,	RCH=CHCOAr (40)	1798, 567
	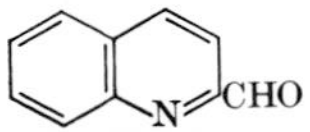 N CHO	$NaOCH_3$	RCH=CHCOAr (88)	2004, 2023, 2024

	[indole-3-CHO, N-CH$_3$]	$(CH_2)_5NH$, 200°	RCH=CHCOAr (55)	600
	4-i-$C_3H_7C_6H_4CHO$	NaOH	RCH=CHCOAr (—)	1923
	2-CH_3-4,5-$(CH_3O)_2C_6H_2CHO$	KOH	RCH=CHCOAr (—)	2059
	3,4,5-$(CH_3O)_3C_6H_2CHO$	$NaOCH_3$	RCH=CHCOAr (—)	1971
	4-$(CH_3)_2NC_6H_4CHO$	Na_2CO_3	RCH=CHCOAr (68)	1810
	3,4-$(C_2H_5O)_2C_6H_3CHO$	NaOH	RCH=CHCOAr (—)	1923
	2-CH_3-4-CH_3O-5-i-$C_3H_7C_6H_2CHO$	"	RCH=CHCOAr (—)	2105
	2-$C_6H_5CH_2O$-3,5-$Br_2C_6H_2CHO$	"	RCH=CHCOAr (88)	549
	2-$C_6H_5CH_2O$-5-BrC_6H_3CHO	"	RCH=CHCOAr (76)	549
	[indole-3-CHO, 2-C_6H_5, N-H]	$(CH_2)_5NH$	RCH=CHCOAr (78)	1979
2-HO-4-CH_3	[pyridine-2-CHO]	KOH	RCH=CHCOAr (—)	1990
	[pyridine-3-CHO]	"	RCH=CHCOAr (85)	371
	C_6H_5CHO	NaOH	RCH=CHCOAr (65)	2106
	4-$CH_3C_6H_4CHO$	"	RCH=CHCOAr (64)	2106
	4-$CH_3OC_6H_4CHO$	—	RCH=CHCOAr (—)	2043
	3,4-$(CH_3O)_2C_6H_3CHO$	NaOH	RCH=CHCOAr (—)	339
2-HO-5-CH_3	[pyridine-2-CHO]	KOH	RCH=CHCOAr (—)	371
	3-$O_2NC_6H_4CHO$	NaOH	RCH=CHCOAr (—)	339
	4-$O_2NC_6H_4CHO$	"	RCH=CHCOAr (—)	339
	C_6H_5CHO	KOH	RCH=CHCOAr (50)	529
	3-HOC_6H_4CHO	NaOH	RCH=CHCOAr (—)	339
	4-HOC_6H_4CHO	"	RCH=CHCOAr (60)	340, 339
	3,4-$(CH_2O_2)C_6H_3CHO$	KOH	RCH=CHCOAr (34)	529
	2-Br-4-$CH_3C_6H_3CHO$	"	RCH=CHCOAr (84)	2008
	2-CH_3-4-BrC_6H_3CHO	"	RCH=CHCOAr (—)	2008

Note: References 668–2359 are on pp. 403–438.

TABLE XVII. CONDENSATION OF ALDEHYDES WITH SUBSTITUTED ACETOPHENONES, CH_3COAr (*Continued*)
(R in the product is the group R in the aldehyde RCHO.)

Substituent(s) in Acetophenone	Aldehyde, RCHO	Catalyst	Product(s) (Yield, %)	Refs.
2-HO-5-CH_3 (*contd.*)	2-Cl-4-$CH_3C_6H_3CHO$	NaOH	RCH=CHCOAr (—)	341
	3-Cl-4-$CH_3C_6H_3CHO$	,,	RCH=CHCOAr (—)	341
	4-$CH_3OC_6H_4CHO$	,,	RCH=CHCOAr (70)	2107, 529, 2043
	2-HO-3-$CH_3OC_6H_3CHO$	KOH	RCH=CHCOAr (18)	529
	3-CH_3O-4-HOC_6H_3CHO	NaOH	RCH=CHCOAr (75)	340, 529
	C_6H_5CH=CHCHO	HCl, CH_3CO_2H	RCH=CHCOAr (—)	339
	2,4-$(CH_3)_2C_6H_3CHO$	NaOH	RCH=CHCOAr (—)	339
	3,4-$(CH_3O)_2C_6H_3CHO$	KOH	RCH=CHCOAr (60)	529
	2,4,6-$(CH_3)_3C_6H_2CHO$	NaOH	RCH=CHCOAr (—)	339
	2,4,6-$(CH_3O)_3C_6H_2CHO$	,,	RCH=CHCOAr (—)	339
	[9-anthryl]-CHO	KOH	RCH=CHCOAr (62)	2036
2-HO-3-CH_3O	2-HO-3-$CH_3OC_6H_3CHO$	NaOH	RCH=CHCOAr (22)	2108
	3-CH_3O-4-HOC_6H_3CHO	,,	RCH=CHCOAr (—)	2108
	3,4-$(CH_3O)_2C_6H_3CHO$	,,	RCH=CHCOAr (27)	2108, 2109
2-HO-4-CH_3O	[2-thienyl]-CHO	,,	RCH=CHCOAr (17)	2037
	[2-pyridyl]-CHO	,,	RCH=CHCOAr (16)	2037
	[3-pyridyl]-CHO	,,	RCH=CHCOAr (21)	2037

	4-Pyridinecarboxaldehyde (structure: pyridine ring, CHO)	,,	RCH=CHCOAr (8)	2037
	$2\text{-BrC}_6\text{H}_4\text{CHO}$	KOH	RCH=CHCOAr (—)	2046
	$\text{C}_6\text{H}_5\text{CHO}$	NaOH	RCH=CHCOAr (—)	2110, 2009
	$4\text{-HOC}_6\text{H}_4\text{CHO}$	KOH	RCH=CHCOAr (—)	348
	$3,4\text{-(CH}_2\text{O}_2\text{)C}_6\text{H}_3\text{CHO}$	NaOH	RCH=CHCOAr (—)	2110, 2094
	$2\text{-Br-4-CH}_3\text{C}_6\text{H}_3\text{CHO}$	KOH	RCH=CHCOAr (—)	2046
	$2\text{-Cl-4-CH}_3\text{C}_6\text{H}_3\text{CHO}$	NaOH	RCH=CHCOAr (—)	341
	$3\text{-Cl-4-CH}_3\text{C}_6\text{H}_3\text{CHO}$	,,	RCH=CHCOAr (—)	341
	$2\text{-CH}_3\text{OC}_6\text{H}_4\text{CHO}$	,,	RCH=CHCOAr (—)	2111, 1985
	$3\text{-CH}_3\text{OC}_6\text{H}_4\text{CHO}$	,,	RCH=CHCOAr (—)	2112
	$4\text{-CH}_3\text{OC}_6\text{H}_4\text{CHO}$	KOH	RCH=CHCOAr (—)	2113
	$3\text{-HO-4-CH}_3\text{OC}_6\text{H}_3\text{CHO}$	NaOH	RCH=CHCOAr (—)	339
	4-Hydroxybenzaldehyde *β-d*-glucoside	NaOH, $\text{Na}_2\text{B}_4\text{O}_7$	RCH=CHCOAr (3)	2055
$2\text{-HO-5-CH}_3\text{O}$	$2\text{-HO-5-BrC}_6\text{H}_3\text{CHO}$	KOH	RCH=CHCOAr (—)	2070
	$2\text{-ClC}_6\text{H}_4\text{CHO}$	,,	RCH=CHCOAr (—)	2070
	$3\text{-O}_2\text{NC}_6\text{H}_4\text{CHO}$	,,	RCH=CHCOAr (—)	2070
	$3,4\text{-(CH}_2\text{O}_2\text{)C}_6\text{H}_3\text{CHO}$	,,	RCH=CHCOAr (34)	2070
$2\text{-HO-6-CH}_3\text{O}$	$2\text{-HOC}_6\text{H}_4\text{CHO}$	NaOH	RCH=CHCOAr (86)	2114
	$2\text{-CH}_3\text{OC}_6\text{H}_4\text{CHO}$	,,	RCH=CHCOAr (—)	2115
$2\text{-CH}_3\text{O-4-HO}$	Furfural (structure: furan ring, O, CHO)	,,	RCH=CHCOAr (—)	2038
$3\text{-CH}_3\text{O-4-HO}$	$\text{C}_6\text{H}_5\text{CHO}$	KOH	RCH=CHCOAr (100)	2116
	$2\text{-HOC}_6\text{H}_4\text{CHO}$	,,	RCH=CHCOAr (—)	2052
	$3,4\text{-(CH}_3\text{O)}_2\text{C}_6\text{H}_3\text{CHO}$	,,	RCH=CHCOAr (48)	2116
$2,5\text{-(HO)}_2\text{-4-CH}_3\text{O}$	$3\text{-CH}_3\text{O-4-HOC}_6\text{H}_3\text{CHO}$	,,	RCH=CHCOAr (43)	346
$2\text{-CH}_3\text{O-4,6-(HO)}_2$	$\text{C}_6\text{H}_5\text{CHO}$	,,	RCH=CHCOAr (80)	2117
4-COCH_3	Furfural (structure: furan ring, O, CHO)	Na_2CO_3	$1,4\text{-(RCH=CHCO)}_2\text{C}_6\text{H}_4$ (—)	1801
	$2\text{-ClC}_6\text{H}_4\text{CHO}$	NaOH	$1,4\text{-(RCH=CHCO)}_2\text{C}_6\text{H}_4$ (32)	2080
	$\text{C}_6\text{H}_5\text{CHO}$	,,	$1,4\text{-(RCH=CHCO)}_2\text{C}_6\text{H}_4$ (46)	2080

Note: References 668–2359 are on pp. 403–438.

TABLE XVII. Condensation of Aldehydes with Substituted Acetophenones, CH_3COAr (*Continued*)
(R in the product is the group R in the aldehyde RCHO.)

Substituent(s) in Acetophenone	Aldehyde, RCHO	Catalyst	Product(s) (Yield, %)	Refs.
4-$COCH_3$ (*contd.*)	4-$CH_3OC_6H_4CHO$	NaOH	1,4-$(RCH{=}CHCO)_2C_6H_4$ (69)	2080
	$C_6H_5CH{=}CHCHO$	Na_2CO_3	1,4-$(RCH{=}CHCO)_2C_6H_4$ (—)	1801
3,4-$(CH_3O)_2$-ω-Br	C_6H_5CHO	HCl, CH_3CO_2H	$RCH{=}CBrCOAr$ (—)	1705
2,4-$(CH_3O)_2$-5-Br	2-CH_3O-5-BrC_6H_3CHO	KOH	$RCH{=}CHCOAr$ (—)	2119
	2-$CH_3OC_6H_4CHO$	"	$RCH{=}CHCOAr$ (—)	2119
3-CH_3CONH	2-$O_2NC_6H_4CHO$	"	$RCH{=}CHCOAr$ (—)	1936
	C_6H_5CHO	NaOH	$RCH{=}CHCOAr$ (63)	1924
4-CH_3CONH	2-ClC_6H_4CHO	$NaOC_2H_5$	$RCH{=}CHCOAr$ (69)	2120
	2-$O_2NC_6H_4CHO$	KOH	$RCH{=}CHCOAr$ (—)	1936
	3-$O_2NC_6H_4CHO$	$NaOCH_3$	$RCH{=}CHCOAr$ (80)	2004
	C_6H_5CHO	NaOH	$RCH{=}CHCOAr$ (84)	1924, 2078
	4-$CH_3OC_6H_4CHO$	"	$RCH{=}CHCOAr$ (—)	2078
	$C_6H_5CH{=}CHCHO$	"	$RCH{=}CHCOAr$ (—)	2078
	4-$(CH_3)_2NC_6H_4CHO$	"	$RCH{=}CHCOAr$ (39)	2121
2-HO-5-CH_3CONH	[thiophene ring, S]-CHO	"	$RCH{=}CHCOAr$ (57)	2006
	[furan ring, O]-CHO	"	$RCH{=}CHCOAr$ (64)	2006
	[pyrrole ring, NH]-CHO	"	$RCH{=}CHCOAr$ (22)	2006
	[pyridine ring, N]-CHO	"	$RCH{=}CHCOAr$ (30)	2006
	4-ClC_6H_4CHO	KOH	$RCH{=}CHCOAr$ (—)	2122
	2-$O_2NC_6H_4CHO$	NaOH	$RCH{=}CHCOAr$ (—)	2122
	4-$O_2NC_6H_4CHO$	"	$RCH{=}CHCOAr$ (Good)	2122

Substituent	Aldehyde	Catalyst	Product (Yield, %)	Refs.
	C_6H_5CHO	,,	RCH=CHCOAr (100)	2123, 2006, 2124
	2-HOC_6H_4CHO	KOH	RCH=CHCOAr (14)	2124, 2122
	3-HOC_6H_4CHO	,,	RCH=CHCOAr (37)	2124
	$OHCC_6H_4CHO$-1,4	NaOH	RCH=CHCOAr or $C_6H_4(CH{=}CHCOAr)_2$-1,4* (—)	2122
	3,4-$(CH_2O_2)C_6H_3CHO$	KOH	RCH=CHCOAr (36)	2124
	3-HO_2C-4-HOC_6H_3CHO	HCl	RCH=CHCOAr (65)	2124
	4-$CH_3C_6H_4CHO$	NaOH	RCH=CHCOAr (—)	2122
	4-$CH_3OC_6H_4CHO$	KOH	RCH=CHCOAr (33)	2124
3-NO_2-4-CH_3CONH	2-$O_2NC_6H_4CHO$	,,	RCH=CHCOAr (—)	1936
2,4-$(CH_3)_2$	,,	,,	RCH=CHCOAr (—)	1936
	2-HOC_6H_4CHO	NaOH	RCH=CHCOAr (—)	2048
	4-$CH_3OC_6H_4CHO$	,,	RCH=CHCOAr (—)	2048
2,5-$(CH_3)_2$	2-$O_2NC_6H_4CHO$	KOH	RCH=CHCOAr (—)	1936
	2-HOC_6H_4CHO	NaOH	RCH=CHCOAr (—)	2048
	4-$CH_3OC_6H_4CHO$	,,	RCH=CHCOAr (—)	2048
3,4-$(CH_3)_2$	2-$O_2NC_6H_4CHO$	KOH	RCH=CHCOAr (—)	1936
	,,	Na_3PO_4	$RCHOHCH_2COAr$ (—)	361
4-C_2H_5	C_6H_5CHO	$NaOCH_3$	RCH=CHCOAr (60)	2089, 349, 2090
2-CH_3Se-5-CH_3	,,	$(CH_2)_5NH$, CH_3CO_2H	RCH=CHCOAr (51)	2093
2-CH_3-4-CH_3O	2-CH_3-4-$CH_3OC_6H_3CHO$	$NaOCH_3$	RCH=CHCOAr (84)	2104
4-C_2H_5O	4-$CH_3OC_6H_4CHO$	$NaOC_2H_5$	RCH=CHCOAr (—)	2100
2-HO-4,6-$(CH_3)_2$	C_6H_5CHO	NaOH	RCH=CHCOAr (—)	339
	3-HOC_6H_4CHO	,,	RCH=CHCOAr (—)	339
	4-HOC_6H_4CHO	,,	RCH=CHCOAr (—)	339
	3-CH_3O-4-HOC_6H_3CHO	,,	RCH=CHCOAr (—)	339
	$C_6H_5CH{=}CHCHO$	,,	RCH=CHCOAr (—)	339
	2,4,6-$(CH_3)_3C_6H_2CHO$	,,	RCH=CHCOAr (—)	339
2,4-$(CH_3O)_2$	5-nitro-2-thiophenecarboxaldehyde (O_2N–thiophene–CHO)	CH_3CO_2H, $(CH_3CO)_2O$	RCH=CHCOAr (15)	647

Note: References 668–2359 are on pp. 403–438.

* The structure of the product was not established.

TABLE XVII. CONDENSATION OF ALDEHYDES WITH SUBSTITUTED ACETOPHENONES, CH_3COAr (*Continued*)

(R in the product is the group R in the aldehyde RCHO.)

Substituent(s) in Acetophenone	Aldehyde, RCHO	Catalyst	Product(s) (Yield, %)	Refs.
2,4-$(CH_3O)_2$ (*contd.*)	(2-thienyl)CHO	NaOH	RCH=CHCOAr (73)	1008
	(2-selenienyl)CHO	”	RCH=CHCOAr (61)	366
	(2-furyl)CHO	”	RCH=CHCOAr (—)	1794
	2-ClC_6H_4CHO	KOH	RCH=CHCOAr (—)	2041
	4-ClC_6H_4CHO	”	RCH=CHCOAr (—)	2041
	4-$O_2NC_6H_4CHO$	NaOH	RCH=CHCOAr (Good)	2095
	C_6H_5CHO	”	RCH=CHCOAr (—)	2125, 1940, 2096
	(2-thienyl)CH=CHCHO	”	RCH=CHCOAr (78)	1008
	2-HOC_6H_4CHO	KOH	RCH=CHCOAr (—)	571
	(2-furyl)CH=CHCHO	NaOH	RCH=CHCOAr (93)	1794
	3,4-$(CH_2O_2)C_6H_3CHO$	”	RCH=CHCOAr (—)	2125
	2-CH_3O-5-BrC_6H_3CHO	KOH	RCH=CHCOAr (—)	2119
	4-$CH_3C_6H_4CHO$	NaOH	RCH=CHCOAr (—)	2048
	2-$CH_3OC_6H_4CHO$	KOH	RCH=CHCOAr (—)	2119, 1940
	4-$CH_3OC_6H_4CHO$	NaOH	RCH=CHCOAr (—)	2125, 2101
	2-HO-3-$CH_3OC_6H_3CHO$	KOH	RCH=CHCOAr (80)	2068
	2-HO-5-$CH_3OC_6H_3CHO$	”	RCH=CHCOAr (82)	2068
	2-HO-6-$CH_3OC_6H_3CHO$	”	RCH=CHCOAr (82)	2068
	3-HO-4-$CH_3OC_6H_3CHO$	”	RCH=CHCOAr (74)	2068
	2,4-$(CH_3O)_2C_6H_3CHO$	”	RCH=CHCOAr (—)	2125

	2,5-$(CH_3O)_2C_6H_3CHO$	,,	RCH=CHCOAr (—)	2125
	3,4-$(CH_3O)_2C_6H_3CHO$	,,	RCH=CHCOAr (—)	2125, 2061
	4-$(CH_3)_2NC_6H_4CHO$	Na_2CO_3	RCH=CHCOAr (43)	1810
	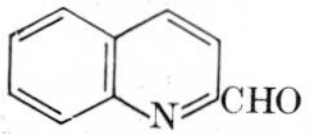 N CHO	NaOH	RCH=CHCOAr (—)	2024
	4-$(CH_3)_2NC_6H_4CH$=CHCHO	Na_2CO_3	RCH=CHCOAr (47)	1810
	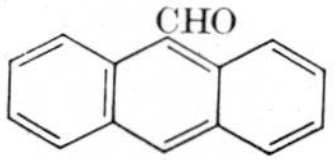 CHO	HCl	RCH=CHCOAr (36)	1978
2,5-$(CH_3O)_2$	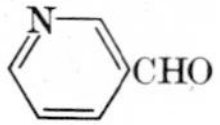 N CHO	NaOH	RCH=CHCOAr (—)	2126
	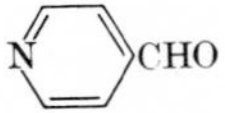 N CHO	$(CH_2)_5NH$, CH_3CO_2H	RCH=CHCOAr (49)	369
	2-HO-5-BrC_6H_3CHO	KOH	RCH=CHCOAr (—)	2070
	2-ClC_6H_4CHO	,,	RCH=CHCOAr (—)	2070
	3-$O_2NC_6H_4CHO$	,,	RCH=CHCOAr (—)	2070
	2-HOC_6H_4CHO	NaOH	RCH=CHCOAr (—)	2127
	3,4-$(CH_2O_2)C_6H_3CHO$	KOH	RCH=CHCOAr (—)	2070
	4-$CH_3OC_6H_4CHO$	,,	RCH=CHCOAr (—)	1922, 2101
	2,4-$(CH_3O)_2C_6H_3CHO$	NaOH	RCH=CHCOAr (—)	2125
	3,4-$(CH_3O)_2C_6H_3CHO$	,,	RCH=CHCOAr (—)	2125
2,6-$(CH_3O)_2$	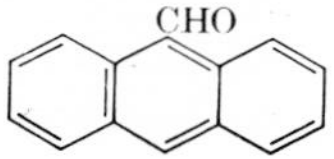 CHO	HCl	RCH=CHCOAr (70)	1978
3,4-$(CH_3O)_2$	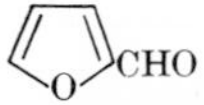 O CHO	NaOH	RCH=CHCOAr (—)	2125

Note: References 668–2359 are on pp. 403–438.

TABLE XVII. CONDENSATION OF ALDEHYDES WITH SUBSTITUTED ACETOPHENONES, CH_3COAr (*Continued*)

(R in the product is the group R in the aldehyde RCHO.)

Substituent(s) in Acetophenone	Aldehyde, RCHO	Catalyst	Product(s) (Yield, %)	Refs.
$3,4-(CH_3O)_2$ (*contd.*)	$2-O_2NC_6H_4CHO$	NaOH	RCH=CHCOAr (—)	2128
	$3-O_2NC_6H_4CHO$	,,	RCH=CHCOAr (—)	2128
	$4-O_2NC_6H_4CHO$	,,	RCH=CHCOAr (—)	2128
	C_6H_5CHO	,,	RCH=CHCOAr (—)	2125, 2058
	$2-HOC_6H_4CHO$	,,	RCH=CHCOAr (—)	2097
	$4-HOC_6H_4CHO$	KOH	RCH=CHCOAr (40)	2058
	$3,4-(CH_2O_2)C_6H_3CHO$	,,	RCH=CHCOAr (—)	2009
	$4-CH_3OC_6H_4CHO$	NaOH	RCH=CHCOAr (—)	2125, 2058, 2101
	$C_6H_5CH{=}CHCHO$	,,	RCH=CHCOAr (—)	2125
	$2,4-(CH_3O)_2C_6H_3CHO$	,,	RCH=CHCOAr (—)	2125
	$2,5-(CH_3O)_2C_6H_3CHO$	,,	RCH=CHCOAr (—)	2125
	$3,4-(CH_3O)_2C_6H_3CHO$	,,	RCH=CHCOAr (—)	2125, 2129, 2130
$3,5-(CH_3O)_2$	$3,4-(CH_2O_2)C_6H_3CHO$	$NaOCH_3$	RCH=CHCOAr (—)	1971
$2-HO-4-C_2H_5O$	C_6H_5CHO	NaOH	RCH=CHCOAr (—)	2131
	$4-CH_3OC_6H_4CHO$	,,	RCH=CHCOAr (—)	2173
	$3,4-(CH_2O_2)C_6H_3CHO$	,,	RCH=CHCOAr (—)	2131
	$3,4-(CH_3O)_2C_6H_3CHO$	KOH	RCH=CHCOAr (—)	2132
	$3-CH_3O-4-C_2H_5OC_6H_3CHO$	,,	RCH=CHCOAr (—)	2132
$2-HO-5-C_2H_5O$	C_6H_5CHO	NaOH	RCH=CHCOAr (—)	525, 528
	$2-HOC_6H_4CHO$	KOH	RCH=CHCOAr (—)	528
	$4-CH_3OC_6H_4CHO$	NaOH	RCH=CHCOAr (—)	526
	$2-C_2H_5OC_6H_4CHO$	KOH	RCH=CHCOAr (—)	528
$2-HO-3-CH_3O-5-CH_3$	C_6H_5CHO	NaOH	RCH=CHCOAr (11)	2134
$2-HO-4-CH_3O-6-CH_3$	,,	KOH	RCH=CHCOAr (—)	2135
	$4-HOC_6H_4CHO$	,,	RCH=CHCOAr (—)	2135
	$4-CH_3C_6H_4CHO$	,,	RCH=CHCOAr (Low)	2135
	$3,4-(CH_3O)_2C_6H_3CHO$	,,	RCH=CHCOAr (—)	339
$2-CH_3-4-CH_3O-5-HO$	$2-CH_3-4,5-(CH_3O)_2C_6H_2CHO$	,,	RCH=CHCOAr (—)	2059

2-HO-3,4-$(CH_3O)_2$	C_6H_5CHO	NaOH	RCH=CHCOAr	(—)	2009
	4-HOC_6H_4CHO	,,	RCH=CHCOAr	(45)	2136
	3,4-$(CH_2O_2)C_6H_3CHO$	,,	RCH=CHCOAr	(—)	2136, 2044
	2-$CH_3OC_6H_4CHO$	,,	RCH=CHCOAr	(—)	2115, 2137
	4-$CH_3OC_6H_4CHO$	,,	RCH=CHCOAr	(—)	2044, 2138
2-HO-3,5-$(CH_3O)_2$	C_6H_5CHO	,,	RCH=CHCOAr	(80)	2139, 2140
	4-$CH_3OC_6H_4CHO$	,,	RCH=CHCOAr	(85)	2139, 2140
	4-$C_6H_5CH_2OC_6H_4CHO$	,,	RCH=CHCOAr	(60)	2139
2-HO-3,6-$(CH_3O)_2$	[ring structure: N, CHO]	,,	RCH=CHCOAr	(—)	369, 2126
2-HO-4,5-$(CH_3O)_2$	[ring structure: O, CHO]	KOH	RCH=CHCOAr	(—)	2141
2-HO-4,6-$(CH_3O)_2$	3-$O_2NC_6H_4CHO$	NaOH	RCH=CHCOAr	(—)	2074
	C_6H_5CHO	,,	RCH=CHCOAr	(—)	2142, 340, 527
	2-HOC_6H_4CHO	,,	RCH=CHCOAr	(86)	2114, 340
	4-HOC_6H_4CHO	KOH	RCH=CHCOAr	(—)	2143
	3,4-$(CH_2O_2)C_6H_3CHO$	NaOH	RCH=CHCOAr	(—)	2142
	2-$CH_3OC_6H_4CHO$	KOH	RCH=CHCOAr	(—)	2114, 2115
	4-$CH_3OC_6H_4CHO$	NaOH	RCH=CHCOAr	(80)	340
	3-CH_3O-4-HOC_6H_3CHO	,,	RCH=CHCOAr	(60)	340
	2,4-$(CH_3O)_2C_6H_3CHO$	KOH	RCH=CHCOAr	(60)	2115, 2145
	3,4-$(CH_3O)_2C_6H_3CHO$	NaOH	RCH=CHCOAr	(41)	527, 2145
	4-$C_6H_5CH_2OC_6H_4CHO$	Na_2CO_3	RCH=CHCOAr	(—)	2146, 2347
2-HO-5,6-$(CH_3O)_2$	2-$CH_3OC_6H_4CHO$	KOH	RCH=CHCOAr	(—)	2147
	4-$CH_3OC_6H_4CHO$	,,	RCH=CHCOAr	(37)	346
	3,4-$(CH_3O)_2C_6H_3CHO$	,,	RCH=CHCOAr	(27)	346
4-HO-2,6-$(CH_3O)_2$	C_6H_5CHO	NaOH	RCH=CHCOAr	(—)	339
4-$(CH_3)_2N$	[ring structure: Se, CHO]	,,	RCH=CHCOAr	(43)	366

Note: References 668–2359 are on pp. 403–438.

TABLE XVII. Condensation of Aldehydes with Substituted Acetophenones, CH_3COAr (*Continued*)

(R in the product is the group R in the aldehyde RCHO.)

Substituent(s) in Acetophenone	Aldehyde, RCHO	Catalyst	Product(s) (Yield, %)	Refs.
4-$(CH_3)_2N$ (*contd.*)	[structure: O, CHO]	NaOH	RCH=CHCOAr (—)	2056
	4-$O_2NC_6H_4CHO$	Na_2CO_3	RCH=CHCOAr (91)	1810
	C_6H_5CHO	HCl	RCH=CHCOAr (76)	349
	4-$CH_3OC_6H_4CHO$	$NaOCH_3$	RCH=CHCOAr (—)	1934, 567
	C_6H_5CH=CHCHO	Na_2CO_3	RCH=CHCOAr (44)	1810
	2,4-$(CH_3O)_2C_6H_3CHO$	,,	RCH=CHCOAr (77)	1810
	4-$(CH_3)_2NC_6H_4CHO$	KOH	RCH=CHCOAr (92)	567, 1972
	[structure: N, CHO]	NaOH	RCH=CHCOAr (—)	2024
	4-$CH_3OC_6H_4CH$=CHCHO	Na_2CO_3	RCH=CHCOAr (30)	1810
	2,4,6-$(CH_3O)_3C_6H_2CHO$	,,	RCH=CHCOAr (47)	1810
	2,4-$(CH_3O)_2C_6H_3CH$=CHCHO	,,	RCH=CHCOAr (38)	1810
	4-$(CH_3)_2NC_6H_3CH$=CHCHO	,,	RCH=CHCOAr (17)	1810
	9-Formyl-3,4-benzacridine	NaOH	RCH=CHCOAr (75–95)	372
	10-Formyl-1,2-benzanthracene	,,	RCH=CHCOAr (75–95)	372
2-$CO_2C_2H_5$	C_6H_5CHO	,,	RCH=CHCOAr (82)	2148
2-(HO_2CCH_2O)-4-CH_3O	,,	,,	RCH=CHCOAr (—)	2149
	2-$CH_3OC_6H_4CHO$	,,	RCH=CHCOAr (—)	2149
	3-$CH_3OC_6H_4CHO$	,,	RCH=CHCOAr (—)	2149
	4-$CH_3OC_6H_4CHO$	,,	RCH=CHCOAr (—)	2149
	3,4-$(CH_3O)_2C_6H_3CHO$	,,	RCH=CHCOAr (—)	2149
1,3,5-$(CH_3)_3$	C_6H_5CHO	$NaOC_2H_5$	RCH=CHCOAr (—)	1607
	1,3,5-$(CH_3)_3C_6H_2CHO$	,,	RCH=CHCOAr (—)	1607
2,4,6-$(CH_3)_3$	CH_2O	K_2CO_3	$HOCH_2CH_2COA$ (71)	2150, 519
	4-BrC_6H_4CHO	NaOH	RCH=CHCOAr (91)	2151
	4-ClC_6H_4CHO	,,	RCH=CHCOAr (91)	2151
	3-$O_2NC_6H_4CHO$	,,	RCH=CHCOAr (100)	2152
	C_6H_5CHO	,,	RCH=CHCOAr (—)	2003
	4-$CH_3OC_6H_4CHO$	,,	RCH=CHCOAr (92)	2153

	2,4,6-$(CH_3)_3C_6H_2CHO$	$NaOC_2H_5$	RCH=CHCOAr	(90)	2154
4-n-C_3H_7	C_6H_5CHO	$NaOCH_3$	RCH=CHCOAr	(63)	2089, 2090
4-i-C_3H_7	,,	,,	RCH=CHCOAr	(65)	2155, 2090
	,,	HCl	RCH=CHCOAr	(68)	349
2-HO-3,4,6-$(CH_3)_3$	,,	NaOH	RCH=CHCOAr	(—)	2156
2-CH_3-4,5-$(CH_3O)_2$	3-CH_3-4-HOC_6H_3CHO	KOH	RCH=CHCOAr	(—)	2059
	3-CH_3O-4-$C_2H_5OC_6H_3CHO$	,,	RCH=CHCOAr	(—)	2059
	2-CH_3-4,5-$(CH_3O)_2C_6H_2CHO$	,,	RCH=CHCOAr	(—)	2059
2,3,4-$(CH_3O)_3$	3,4-$(CH_2O_2)C_6H_3CHO$	$NaOCH_3$	RCH=CHCOAr	(—)	1971, 2009, 2084
	4-$CH_3OC_6H_4CHO$	KOH	RCH=CHCOAr	(80)	2136, 2101
2,4,5-$(CH_3O)_3$	[furan ring, O]CHO	NaOH	RCH=CHCOAr	(—)	2094
	2-HOC_6H_4CHO	,,	RCH=CHCOAr	(—)	2157
	3,4-$(CH_2O_2)C_6H_3CHO$	,,	RCH=CHCOAr	(—)	2094
	4-$CH_3OC_6H_4CHO$	KOH	RCH=CHCOAr	(—)	2101
2,4,6-$(CH_3O)_3$	O_2N[thiophene ring, S]CHO	CH_3CO_2H, $(CH_3CO)_2O$	RCH=CHCOAr	(0)	647
	[thiophene ring, S]CHO	NaOH	RCH=CHCOAr	(84)	1008
	[selenophene ring, Se]CHO	,,	RCH=CHCOAr	(69)	366
	[thiophene ring, S]CH=CHCHO	,,	RCH=CHCOAr	(79)	1008
	2-HOC_6H_4CHO	KOH	RCH=CHCOAr	(50)	2158, 2115
	4-HOC_6H_4CHO	,,	RCH=CHCOAr	(—)	2143
	2-$CH_3OC_6H_4CHO$	,,	RCH=CHCOAr	(—)	2115
	4-$CH_3OC_6H_4CHO$	,,	RCH=CHCOAr	(—)	2101
	2,4-$(CH_3O)_2C_6H_3CHO$	NaOH	RCH=CHCOAr	(—)	2125, 2143
	2,5-$(CH_3O)_2C_6H_3CHO$	,,	RCH=CHCOAr	(—)	2125

Note: References 668–2359 are on pp. 403–438.

TABLE XVII. Condensation of Aldehydes with Substituted Acetophenones, CH_3COAr (*Continued*)

(R in the product is the group R in the aldehyde RCHO.)

Substituent(s) in Acetophenone	Aldehyde, RCHO	Catalyst	Product(s) (Yield, %)	Refs.
2,4,6-$(CH_3O)_3$ (*contd.*)	3,4-$(CH_3O)_2C_6H_3CHO$	NaOH	RCH=CHCOAr (—)	2125
	CHO	,,	RCH=CHCOAr (—)	2024
	4-$(CH_3)_2NC_6H_4CH$=CHCHO	Na_2CO_3	RCH=CHCOAr (83)	1810
3,4,5-$(CH_3O)_3$	3,4-$(CH_2O_2)C_6H_3CHO$	$NaOCH_3$	RCH=CHCOAr (—)	1971, 2009, 2084
2-HO-4-CH_3O-6-C_2H_5O	3,4-$(C_2H_5O)_2C_6H_3CHO$	NaOH	RCH=CHCOAr (—)	2159
2-HO-3-CH_3-4,6-$(CH_3O)_2$	3,4-$(CH_3O)_2C_6H_3CHO$	KOH	RCH=CHCOAr (50)	2160
2-HO-3,4,5-$(CH_3O)_3$	C_6H_5CHO	,,	RCH=CHCOAr (—)	2161
	4-$CH_3OC_6H_4CHO$	,,	RCH=CHCOAr (—)	2161
	3,4-$(CH_3O)_2C_6H_3CHO$	,,	RCH=CHCOAr (—)	2161
2-HO-3,4,6-$(CH_3O)_3$	2,4-$(CH_3O)_2C_6H_3CHO$	NaOH	RCH=CHCOAr (—)	2074
	2,6-$(CH_3O)_2C_6H_3CHO$	,,	RCH=CHCOAr (—)	2074
	3,4-$(CH_3O)_2C_6H_3CHO$	,,	RCH=CHCOAr (73)	2162
	2,4,6-$(CH_3O)_3C_6H_2CHO$	,,	RCH=CHCOAr (—)	2074
2-HO-4,5,6-$(CH_3O)_3$	C_6H_5CHO	NaOH	RCH=CHCOAr (—)	2156
	2-$C_6H_5CH_2O$-4-$CH_3OC_6H_3CHO$	KOH	RCH=CHCOAr (78)	2163
2,4-$(CH_3CO_2)_2$	CHO	HCl	RCH=CHCOAr (41)	1978
2,5-$(CH_3CO_2)_2$	,,	,,	RCH=CHCOAr (36)	1978
2-HO-3-CH_3O-5-$(CH_2$=$CHCH_2)$	C_6H_5CHO	KOH	RCH=CHCOAr (40)	2164
2-CH_3-5-*i*-C_3H_7	,,	NaOH	RCH=CHCOAr (—)	2003
4-*n*-C_4H_9	,,	$NaOCH_3$	RCH=CHCOAr (80)	2155
4-*i*-C_4H_9	,,	,,	RCH=CHCOAr (83)	2155
4-*t*-C_4H_9	,,	,,	RCH=CHCOAr (77)	2155, 349
2,4-$(C_2H_5O)_2$	2-HO-5-BrC_6H_3CHO	NaOH	RCH=CHCOAr (—)	2127
	C_6H_5CHO	,,	RCH=CHCOAr (—)	2165

2,5-$(C_2H_5O)_2$	2-HOC_6H_4CHO	,,	RCH=CHCOAr (—)	2127
2-HO-3-CH_3O-5-i-C_3H_7O	C_6H_5CHO	,,	RCH=CHCOAr (70)	2139
	4-$CH_3OC_6H_4CHO$	,,	RCH=CHCOAr (70)	2139
	4-i-$C_3H_7OC_6H_4CHO$	,,	RCH=CHCOAr (70)	2139
2-HO-4,6-$(C_2H_5O)_2$	C_6H_5CHO	,,	RCH=CHCOAr (—)	2142, 2166
	4-$CH_3OC_6H_4CHO$	KOH	RCH=CHCOAr (—)	2166
2,3,4,6-$(CH_3O)_4$	C_6H_5CHO	NaOH	RCH=CHCOAr (—)	2156
	3-CH_3-4,5-$(CH_3O)_2C_6H_2CHO$	KOH	RCH=CHCOAr (—)	2059
	2,3,4,6-$(CH_3O)_4C_6HCHO$	,,	RCH=CHCOAr (—)	2059
2-HO-3,4,5,6-$(CH_3O)_4$	4-$C_6H_5CH_2OC_6H_4CHO$	NaH	RCH=CHCOAr (—)	2167
4-$(CH_2)_4CH$	CH_3–(thiophene, S)–CHO	NaOH	RCH=CHCOAr (—	1988
	CH_3–(furan, O)–CHO	,,	RCH=CHCOAr (—)	1988
2,4,6-$(CH_3)_3$-3-CH_3CO	2-ClC_6H_4CHO	,,	1,3-$(RCH{=}CHCO)_2$-$C_6H(CH_3)_3$-2,4,6 (80)	2080
	4-ClC_6H_4CHO	,,	1,3-$(RCH{=}CHCO)_2$-$C_6H(CH_3)_3$-2,4,6 (85)	2080
	C_6H_5CHO	,,	1,3-$(RCH{=}CHCO)_2$-$C_6H(CH_3)_3$-2,4,6 (70)	2080
	4-$CH_3C_6H_4CHO$	,,	1,3-$(RCH{=}CHCO)_2$-$C_6H(CH_3)_3$-2,4,6 (30)	2080
	4-$CH_3OC_6H_4CHO$	,,	1,3-$(RCH{=}CHCO)_2$-$C_6H(CH_3)_3$-2,4,6 (35)	2080
	4-$(CH_3)_2NC_6H_4CHO$	,,	1,3-$(RCH{=}CHCO)_2$-$C_6H(CH_3)_3$-2,4,6 (60)	2080
4-n-C_5H_{11}	C_6H_5CHO	KOH	RCH=CHCOAr (51)	2155
4-i-$C_4H_9CH_2$	,,	$NaOCH_3$	RCH=CHCOAr (86)	2155
4-t-$C_4H_9CH_2$	,,	,,	RCH=CHCOAr (89)	2155
2-CH_3-4-CH_3O-5-i-C_3H_7	2,4-$Cl_2C_6H_3CHO$	NaOH	RCH=CHCOAr (80)	2168
	3,4-$Cl_2C_6H_3CHO$	,,	RCH=CHCOAr (80)	2168
	2-HO-3,5-$Cl_2C_6H_2CHO$	,,	RCH=CHCOAr (80)	2168

Note: References 668–2359 are on pp. 403–438.

TABLE XVII. CONDENSATION OF ALDEHYDES WITH SUBSTITUTED ACETOPHENONES, CH_3COAr (*Continued*)

(R in the product is the group R in the aldehyde RCHO.)

Substituent(s) in Acetophenone	Aldehyde, RCHO	Catalyst	Product(s) (Yield, %)	Refs.
2-CH_3-4-CH_3O-5-*i*-C_3H_7 (*contd.*)	2-ClC_6H_4CHO	NaOH	RCH=CHCOAr (80)	2168
	4-ClC_6H_4CHO	,,	RCH=CHCOAr (80)	2168
	C_6H_5CHO	KOH	RCH=CHCOAr (—)	2059
	2-HOC_6H_4CHO	,,	RCH=CHCOAr (—)	2059
	4-HOC_6H_4CHO	,,	RCH=CHCOAr (—)	2059
	3,4-$(CH_2O_2)C_6H_3CHO$	,,	RCH=CHCOAr (—)	2059
	4-$CH_3OC_6H_4CHO$	,,	RCH=CHCOAr (—)	2059
	3-CH_3O-4-$C_2H_5OC_6H_3CHO$	,,	RCH=CHCOAr (—)	2059
	2-CH_3-4,5-$(CH_3O)_2C_6H_2CHO$	,,	RCH=CHCOAr (—)	2059
	1-$C_{10}H_7CHO$	NaOH	RCH=CHCOAr (80)	2168
4-(4-$O_2NC_6H_4$)	C_6H_5CHO	$NaOC_2H_5$	RCH=CHCOAr (50)	1934
	4-$CH_3OC_6H_4CHO$	,,	RCH=CHCOAr (Good)	1934
4-(4-$O_2NC_6H_4O$)	C_6H_5CHO	$NaOCH_3$	RCH=CHCOAr (100)	2169
	4-$CH_3OC_6H_4CHO$	,,	RCH=CHCOAr (—)	2169
4-C_6H_5	2-thiophenecarboxaldehyde (thiophene ring, S, CHO)	NaOH	RCH=CHCOAr (51)	1845, 1829
	2-selenophenecarboxaldehyde (selenophene ring, Se, CHO)	,,	RCH=CHCOAr (100)	366
	2-furaldehyde (furan ring, O, CHO)	$NaOC_2H_5$	RCH=CHCOAr (50)	1823, 1860
	C_6H_5CHO	NaOH	RCH=CHCOAr (90)	1829, 540
	3,4-$(CH_2O_2)C_6H_3CHO$	,,	RCH=CHCOAr (100)	1829
	4-$CH_3OC_6H_4CHO$	KOH	RCH=CHCOAr (—)	540
	C_6H_5CH=CHCHO	,,	RCH=CHCOAr (—)	540
	n-C_5H_{11}-thiophene-CHO (thiophene ring, S)	NaOH	RCH=CHCOAr (—)	1829
	2-CH_3-4-HO-5-*i*-$C_3H_7C_6H_2CHO$	,,	RCH=CHCOAr (—)	2105
	2-CH_3-4-CH_3O-5-*i*-$C_3H_7C_6H_2CHO$	,,	RCH=CHCOAr (75)	2105

	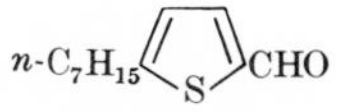 n-C_7H_{15} S CHO	,,	RCH=CHCOAr	(—)	1829
	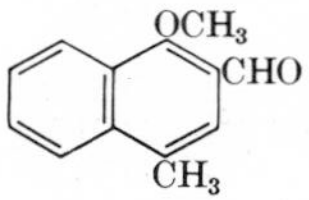 OCH_3 CHO CH_3	,,	RCH=CHCOAr	(—)	1829
	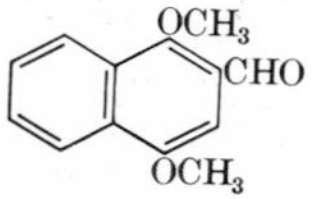 OCH_3 CHO OCH_3	,,	RCH=CHCOAr	(—)	1829
	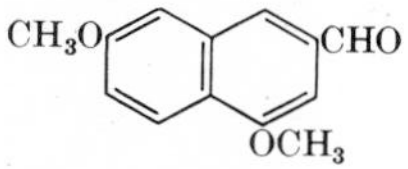 CH_3O CHO OCH_3	,,	RCH=CHCOAr	(—)	1829
	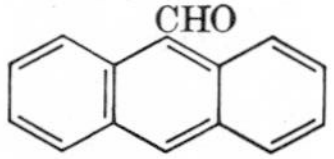 CHO	HCl	RCH=CHCOAr	(68)	1978
	3-CH_3O-4-n-$C_{12}H_{25}OC_6H_3CHO$	NaOH	RCH=CHCOAr	(—)	1829
4-C_6H_5S	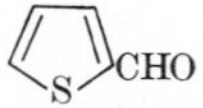 S CHO	,,	RCH=CHCOAr	(—)	2017
	C_6H_5CHO	$NaOCH_3$	RCH=CHCOAr	(—)	1934
	3,4-$(CH_2O_2)C_6H_3CHO$	,,	RCH=CHCOAr	(—)	1934
	4-$CH_3OC_6H_4CHO$	,,	RCH=CHCOAr	(—)	1934
4-C_6H_5O	C_6H_5CHO	$NaOC_2H_5$	RCH=CHCOAr	(—)	2169
4-(4-$H_2NC_6H_4$)	,,	$NaOCH_3$	RCH=CHCOAr	(—)	2169
4-$(CH_2)_5CH$	$(CH_3)_2C$=$CHCH_2CH_2$-$C(CH_3)$=CHCHO	NaOH	RCH=CHCOAr	(—)	1773

Note: References 668–2359 are on pp. 403–438.

TABLE XVII. CONDENSATION OF ALDEHYDES WITH SUBSTITUTED ACETOPHENONES, CH_3COAr (*Continued*)

(R in the product is the group R in the aldehyde RCHO.)

Substituent(s) in Acetophenone	Aldehyde, RCHO	Catalyst	Product(s) (Yield, %)	Refs.
2,4-$(HO)_2$-4-β-*d*-glucoside	3,4-$(CH_2O_2)C_6H_3CHO$	NaOH, $Na_2B_4O_7$	RCH=CHCOAr (10.5)	2055
	4-$CH_3OC_6H_4CHO$	,,	RCH=CHCOAr (20)	2055, 2170
	4-Hydroxy-3-methoxybenzaldehyde β-*d*-glucoside	,,	RCH=CHCOAr (18)	2055
	3-Hydroxy-4-methoxybenzaldehyde β-*d*-glucoside	NaOH	RCH=CHCOAr (54)	2053
4-*n*-C_6H_{13}	C_6H_5CHO	KOH	RCH=CHCOAr (57)	2155
2-HO-4-$[(C_2H_5)_2NCH_2CH_2O]$	4-HOC_6H_4CHO	,,	RCH=CHCOAr (25)	2171
2-$C_6H_5CO_2$	4-$C_6H_5CH_2OC_6H_4CHO$	HCl	RCH=CHCOAr (—)	355
	9-Anthraldehyde (structure: CHO)	,,	RCH=CHCOAr (68)	1978
	3-CH_3O-4-$C_6H_5CO_2C_6H_3CHO$	,,	RCH=CHCOAr (—)	355
2-HO-5-C_6H_5CO	3-$O_2NC_6H_4CHO$	KOH	RCH=CHCOAr (—)	2172
	C_6H_5CHO	,,	RCH=CHCOAr (—)	2172
	2-HOC_6H_4CHO	,,	RCH=CHCOAr (—)	2172
	3-HOC_6H_4CHO	,,	RCH=CHCOAr (—)	2172
	4-HOC_6H_4CHO	,,	RCH=CHCOAr (—)	2172
	3,4-$(CH_2O_2)C_6H_3CHO$	,,	RCH=CHCOAr (—)	2172
	3-Cl-4-$CH_3OC_6H_3CHO$	,,	RCH=CHCOAr (—)	2172
	4-$CH_3OC_6H_4CHO$	,,	RCH=CHCOAr (—)	2172
	3,4-$(CH_3O)_2C_6H_3CHO$	,,	RCH=CHCOAr (—)	2172
2-HO-4-$C_6H_5CH_2O$-5-Br	2-CH_3O-5-BrC_6H_3CHO	,,	RCH=CHCOAr (—)	2119
	2-$CH_3OC_6H_4CHO$	,,	RCH=CHCOAr (—)	2119
2-HO-4-$C_6H_5CH_2O$-5-NO_2	,,	,,	RCH=CHCOAr (—)	2085
4-(4-$CH_3OC_6H_4$)	2-Thiophenecarboxaldehyde (structure: S, CHO)	NaOH	RCH=CHCOAr (—)	2017

	C_6H_5CHO	,,	RCH=CHCOAr (—)	1995
	$4\text{-}CH_3OC_6H_4CHO$	,,	RCH=CHCOAr (—)	1995
$4\text{-}(4\text{-}CH_3C_6H_4O)$	C_6H_5CHO	$NaOCH_3$	RCH=CHCOAr (80)	2169
	$4\text{-}CH_3OC_6H_4CHO$	,,	RCH=CHCOAr (—)	2169
$2\text{-}HO\text{-}4\text{-}C_6H_5CH_2O$	C_6H_5CHO	NaOH	RCH=CHCOAr (—)	2173
	$2\text{-}CH_3O\text{-}5\text{-}BrC_6H_3CHO$	KOH	RCH=CHCOAr (—)	2119
	$2\text{-}CH_3OC_6H_4CHO$	,,	RCH=CHCOAr (—)	2119
	$3\text{-}CH_3OC_6H_4CHO$	NaOH	RCH=CHCOAr (70)	2054
	$4\text{-}CH_3OC_6H_4CHO$	KOH	RCH=CHCOAr (—)	2173
	$4\text{-}C_6H_5CH_2OC_6H_4CHO$	NaOH	RCH=CHCOAr (—)	2173
$4\text{-}n\text{-}C_7H_{15}$	C_6H_5CHO	$NaOCH_3$	RCH=CHCOAr (85)	2155
$4\text{-}(4\text{-}CH_3COC_6H_4S)$	,,	,,	$(4\text{-}RCH{=}CHCOC_6H_4)_2S$ (100)	1934
	$3,4\text{-}(CH_2O_2)C_6H_3CHO$	,,	$(4\text{-}RCH{=}CHCOC_6H_4)_2S$ (100)	1934
	$4\text{-}CH_3OC_6H_4CHO$	,,	$(4\text{-}RCH{=}CHCOC_6H_4)_2S$ (—)	1934
	$C_6H_5CH{=}CHCHO$	,,	$(4\text{-}RCH{=}CHCOC_6H_4)_2S$ (—)	1934
$4\text{-}(4\text{-}CH_3COC_6H_4Se)$	C_6H_5CHO	NaOH	$(4\text{-}RCH{=}CHCOC_6H_4)_2Se$ (75)	1934
	$3,4\text{-}(CH_2O_2)C_6H_3CHO$	,,	$(4\text{-}RCH{=}CHCOC_6H_4)_2Se$ (70)	1934
	$4\text{-}CH_3OC_6H_4CHO$	,,	$(4\text{-}RCH{=}CHCOC_6H_4)_2Se$ (70)	1934
	$C_6H_5CH{=}CHCHO$	,,	$(4\text{-}RCH{=}CHCOC_6H_4)_2Se$ (60)	1934
$4\text{-}(4\text{-}CH_3COC_6H_4O)$	C_6H_5CHO	$NaOCH_3$	$(4\text{-}RCH{=}CHCOC_6H_4)_2O$ (—)	2169
	$3,4\text{-}(CH_2O_2)C_6H_3CHO$	,,	$(4\text{-}RCH{=}CHCOC_6H_4)_2O$ (—)	1934
	$4\text{-}CH_3OC_6H_4CHO$	,,	$(4\text{-}RCH{=}CHCOC_6H_4)_2O$ (—)	2169
	$C_6H_5CH{=}CHCHO$	,,	$(4\text{-}RCH{=}CHCOC_6H_4)_2O$ (—)	1934
$2\text{-}HO\text{-}3,5,6\text{-}(CH_3O)_3\text{-}4\text{-}C_6H_5CH_2O$	$4\text{-}C_6H_5CH_2OC_6H_4CHO$	NaOH	RCH=CHCOAr (—)	2066, 2174
	$3\text{-}CH_3O\text{-}4\text{-}C_6H_5CH_2OC_6H_3CHO$	,,	RCH=CHCOAr (—)	2174
$4\text{-}n\text{-}C_8H_{17}$	C_6H_5CHO	$NaOCH_3$	RCH=CHCOAr (42)	2155
$4\text{-}n\text{-}C_9H_{19}$	,,	,,	RCH=CHCOAr (90)	2155
$2,4,6\text{-}(i\text{-}C_3H_7)_3$	,,	NaOH	RCH=CHCOAr (91)	2175
$2,4\text{-}(C_6H_5CO_2)_2$	$3,4\text{-}(HO)_2C_6H_3CHO$	HCl	RCH=CHCOAr (—)	2136
	$4\text{-}C_6H_5CO_2C_6H_4CHO$	,,	RCH=CHCOAr (—)	355
	$3\text{-}CH_3O\text{-}4\text{-}C_6H_5CO_2C_6H_3CHO$	,,	RCH=CHCOAr (98)	355
$2,5\text{-}(C_6H_5CO_2)_2$	$2\text{-}C_6H_5CO_2C_6H_4CHO$	,,	RCH=CHCOAr (82)	354
	$3\text{-}C_6H_5CO_2C_6H_4CHO$	,,	RCH=CHCOAr (67)	354
	$4\text{-}C_6H_5CO_2C_6H_4CHO$	,,	RCH=CHCOAr (78)	354

Note: References 668–2359 are on pp. 403–438.

TABLE XVII. CONDENSATION OF ALDEHYDES WITH SUBSTITUTED ACETOPHENONES, CH_3COAr (*Continued*)

(R in the product is the group R in the aldehyde RCHO.)

Substituent(s) in Acetophenone	Aldehyde, RCHO	Catalyst	Product(s) (Yield, %)	Refs.
2,5-$(C_6H_5CO_2)_2$ (*contd.*)	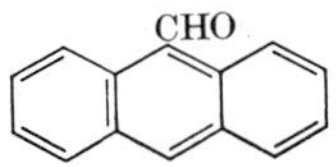 CHO	HCl	RCH=CHCOAr (55)	1978
	3-CH_3O-4-$C_6H_5CO_2C_6H_3CHO$	″	RCH=CHCOAr (88)	354
	2,4-$(C_6H_5CO_2)_2C_6H_3CHO$	″	RCH=CHCOAr (81)	354
	3,4-$(C_6H_5CO_2)_2C_6H_3CHO$	″	RCH=CHCOAr (78)	354
	2,4-$(C_6H_5CO_2)_2$-6-$CH_3C_6H_2CHO$	″	RCH=CHCOAr (42)	354
	1,3,5-$(C_6H_5CO_2)_3C_6H_2CHO$	″	RCH=CHCOAr (—)	354
2,6-$(C_6H_5CO_2)_2$	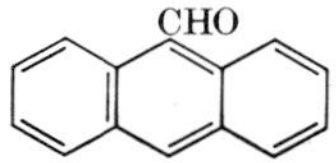 CHO	″	RCH=CHCOAr (—)	1978
2,4-$(C_6H_5CH_2O)_2$	C_6H_5CHO	NaOH	RCH=CHCOAr (—)	2176
	3,4-$(CH_3O)_2C_6H_3CHO$	″	RCH=CHCOAr (—)	2176
2,5-$(C_6H_5CH_2O)_2$	2,5-$(CH_3O)_2$-3-$O_2NC_6H_2CHO$	$NaOCH_3$	RCH=CHCOAr (—)	2177
2-HO-3-CH_3-4,6-$(C_6H_5CH_2O)_2$	C_6H_5CHO	KOH	RCH=CHCOAr (—)	2178
1,3,5-$(C_6H_5)_3$	″	$NaOC_2H_5$	RCH=CHCOAr (—)	1607
2,3,4-$(C_6H_5CO_2)_3$	4-$C_6H_5CO_2C_6H_4CHO$	HCl	RCH=CHCOAr (—)	355
	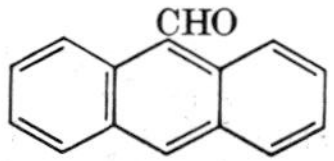 CHO	″	RCH=CHCOAr (—)	1978
	3-CH_3O-4-$C_6H_5CH_2OC_6H_3CHO$	″	RCH=CHCOAr (90)	355
	3,4-$(C_6H_5CH_2O)C_6H_3CHO$	″	RCH=CHCOAr (—)	2179
2,4,6-$(C_6H_5CO_2)_3$	4-$C_6H_5CO_2C_6H_4CHO$	″	RCH=CHCOAr (85)	355
	3-CH_3O-4-$C_6H_5CO_2C_6H_3CHO$	″	RCH=CHCOAr (90)	355
	3,4-$(C_6H_5CO_2)_2C_6H_3CHO$	″	RCH=CHCOAr (—)	2180

Note: References 668–2359 are on pp. 430–438.

TABLE XVIII. CONDENSATION OF ALDEHYDES WITH CARBOCYCLIC AROMATIC KETONES OTHER THAN ACETOPHENONES

(R in the product is the group R in the aldehyde RCHO.)

Ketone	Aldehyde, RCHO	Catalyst	Product(s) (Yield, %)	Refs.
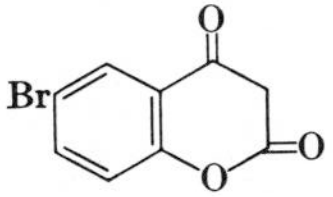 Br, O, O, =O	$(CH_3)_2NC_6H_4CHO$	$(CH_2)_5NH$, CH_3CO_2H	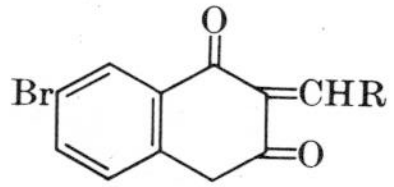 Br, O, =CHR, =O (55)	2181
	$(C_2H_5)_2NC_6H_4CHO$	,,	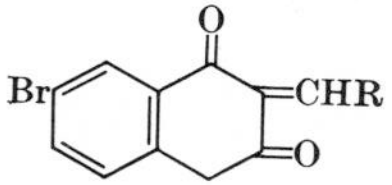 Br, O, =CHR, =O (42)	2181
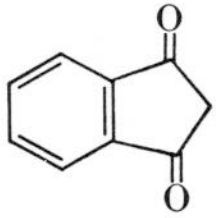 O, O	$C(Cl_2)$=CHCHO	None	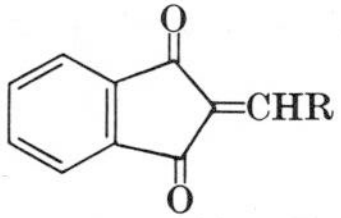 O, =CHR, O (21)	639
			A, R = $C(Cl_2)$=CH	
	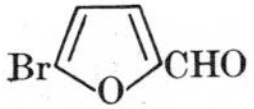 Br, O, CHO	H_2SO_4, CH_3CO_2H	A, R = 2-(C_4H_2BrO) (69)	2182
	 I, O, CHO	,,	A, R = 2-(C_4H_2IO) (89)	2182
	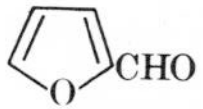 O, CHO	None, 110°	A, R = 2-(C_4H_3O) (—)	2183, 2184
	2,4-$Cl_2C_6H_3CHO$	$(CH_2)_5NH$	A, R = 2,4-$Cl_2C_6H_3$ (77)	2346
	4-BrC_6H_4CHO	$(CH_2)_5NH$, CH_3CO_2H	A, R = 4-BrC_6H_4 (74)	2184
	2-HOC_6H_4CHO	None, 110°	A, R = 2-HOC_6H_4 (—)	2183
	3-HOC_6H_4CHO	,,	A, R = 3-HOC_6H_4 (—)	2183
	4-HOC_6H_4CHO	,,	A, R = 4-HOC_6H_4 (—)	2183

Note: References 668–2359 are on pp. 403–438.

TABLE XVIII. CONDENSATION OF ALDEHYDES WITH CARBOCYCLIC AROMATIC KETONES OTHER THAN ACETOPHENONES (*Continued*)

(R in the product is the group R in the aldehyde RCHO.)

Ketone	Aldehyde, RCHO	Catalyst	Product(s) (Yield, %)	Refs.
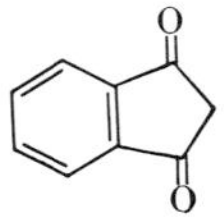 (*contd.*)	3,4-$(CH_2O_2)C_6H_3CHO$	NaOH	A, R = 3,4-$(CH_2O_2)C_6H_3$ (83)	299
	3-O_2N-4-$CH_3OC_6H_3CHO$	None, 130–140°	A, R = 3-O_2N-4-$CH_3OC_6H_3$ (60)	2185
	$C_6H_5CH{=}CHCHO$	None, 110°	A, R = $C_6H_5CH{=}CH$ (—)	2183
	2-$C_2H_5OC_6H_4CHO$	,,	A, R = 2-$C_2H_5OC_6H_4$ (—)	2183
	3-$C_2H_5OC_6H_4CHO$	,,	A, R = 3-$C_2H_5OC_6H_4$ (—)	2183
	4-$C_2H_5OC_6H_4CHO$	,,	A, R = 4-$C_2H_5OC_6H_4$ (—)	2183
	3,4-$(CH_3O)_2C_6H_3CHO$	NaOH	A, R = 3,4-$(CH_3O)_2C_6H_3$ (86)	299, 2184
	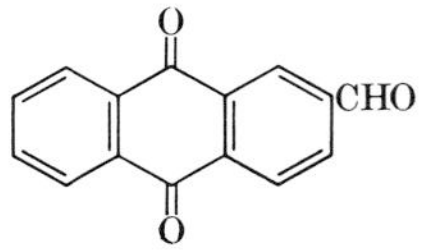 CHO	,,	A, R = 2-$(C_{14}H_7O_2)$ (—)	1977
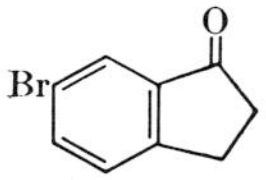 Br	C_6H_5CHO	,,	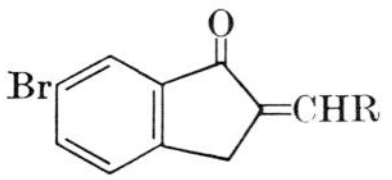 Br =CHR (—) B, R = C_6H_5	2186
	2-HOC_6H_4CHO	,,	B, R = 2-HOC_6H_4 (—)	2186
	3-HOC_6H_4CHO	,,	B, R = 3-HOC_6H_4 (—)	2186
	4-HOC_6H_4CHO	,,	B, R = 4-HOC_6H_4 (Poor)	2186
	,,	HCl	B, R = 4-HOC_6H_4 (—)	2186
	3,4-$(HO)_2C_6H_3CHO$	NaOH	B, R = 3,4-$(HO)_2C_6H_3$ (Very small)	2186
	,,	HCl	B, R = 3,4-$(HO)_2C_6H_3$ (—)	2186
	3,4-$(CH_2O_2)C_6H_3CHO$	NaOH	B, R = 3,4-$(CH_2O_2)C_6H_3$ (Good)	2186
	3-CH_3O-4-HOC_6H_3CHO	,,	B, R = 3-CH_3O-4-HOC_6H_3 (Poor)	2186

O	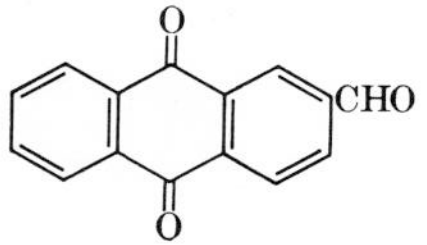 O, O, CHO	,,	B, R = 2-$(C_{14}H_7O_2)$ (—)	1977
	$C(Cl_2)$=CClCHO	H_2SO_4	O, =CHR (—)	289
			C, R = $C(Cl_2)$=CCl	
	$C(Cl_2)$=CHCHO	$(CH_2)_5NH$, CH_3CO_2H	C, R = $C(Cl_2)$=CH (60)	639
	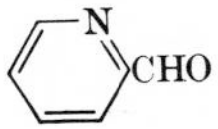 N, CHO	,,	C, R = 2-(C_5H_4N) (63–81)	363, 2187, 2333
	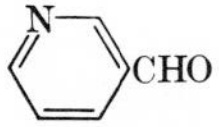 N, CHO	,,	C, R = 3-(C_5H_4N) (85)	363, 2187
	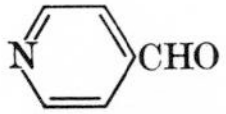 N, CHO	,,	C, R = 4-(C_5H_4N) (55)	363
	2-$O_2NC_6H_4CHO$	H_2SO_4; also KOH	C, R = 2-$O_2NC_6H_4$ (75)	643, 2188, 2189
	C_6H_5CHO	KOH	C, R = C_6H_5 (95)	643, 2190, 2345
	2-HOC_6H_4CHO	,,	C, R = 2-HOC_6H_4 (100)	2191, 2192
	3-HOC_6H_4CHO	NaOH	C, R = 3-HOC_6H_4 (100)	2192
	4-HOC_6H_4CHO	,,	C, R = 4-HOC_6H_4 (100)	2192
	2,4-$(HO)_2C_6H_3CHO$	HCl	C, R = 2,4-$(HO)_2C_6H_3$ (—)	2191, 571
	3,4-$(HO)_2C_6H_3CHO$	,,	C, R = 3,4-$(HO)_2C_6H_3$ (100)	2192
	3,4-(CH_2O_2)-6-$O_2NC_6H_2CHO$	$(CH_3CO)_2O$	C, R = 3,4-$(CH_2O_2)_2$-6-$O_2NC_6H_2$ (19)	2188
	3,4-$(CH_2O_2)C_6H_3CHO$	NaOH	C, R = 3,4-$(CH_2O_2)_2C_6H_3$ (100)	2192
	2-HO-4-$CH_3OC_6H_3CHO$	KOH	C, R = 2-HO-4-$CH_3OC_6H_3$ (—)	2192
	3-CH_3O-4-HOC_6H_3CHO	NaOH	C, R = 3-CH_3O-4-HOC_6H_3 (—)	2192

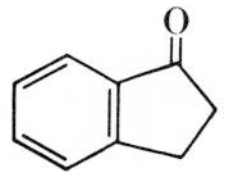

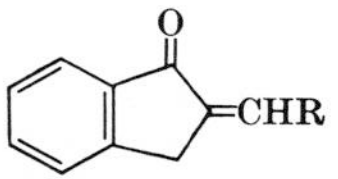

Note: References 668–2359 are on pp. 403–438.

TABLE XVIII. CONDENSATION OF ALDEHYDES WITH CARBOCYCLIC AROMATIC KETONES OTHER THAN ACETOPHENONES (*Continued*)

(R in the product is the group R in the aldehyde RCHO.)

Ketone	Aldehyde, RCHO	Catalyst	Product(s) (Yield, %)	Refs.
[1-indanone structure] (*contd.*)	3,4-$(CH_3O)_2$-6-$O_2NC_6H_2CHO$	$(CH_3CO)_2O$	C, R = 3,4-$(CH_3O)_2$-6-$O_2NC_6H_2$ (—)	2188
	2,4-$(CH_3O)_2C_6H_3CHO$	KOH	C, R = 2,4-$(CH_3O)_2C_6H_3$ (—)	2191
	4-$(CH_3)_2NC_6H_4CHO$	NaOH	C, R = 4-$(CH_3)_2NC_6H_4$ (100)	2192, 2345
[5-hydroxy-1-indanone structure, HO]	[2-pyridyl]CHO	$(CH_2)_5NH$, CH_3CO_2H	[HO-indanone]=CHR (41)	363
	[3-pyridyl]CHO	,,	[HO-indanone]=CHR (75)	363
	[4-pyridyl]CHO	,,	[HO-indanone]=CHR (62)	363
4-$ClC_6H_4COC_2H_5$	4-$CH_3OC_6H_4CHO$	Na_2CO_3	4-$ClC_6H_4COC(CH_3)$=CHR (—)	2194
2-HO-4-$O_2NC_6H_3$-COC_2H_5	[2-furyl]CHO	NaOH	2-HO-4-$O_2NC_6H_3COC(CH_3)$=CHR (—)	2193
	4-ClC_6H_4CHO	,,	2-HO-4-$O_2NC_6H_3COC(CH_3)$=CHR (—)	2193
	C_6H_5CHO	,,	2-HO-4-$O_2NC_6H_3COC(CH_3)$=CHR (—)	2193
2-HO-5-$O_2NC_6H_3$-COC_2H_5	[2-furyl]CHO	,,	2-HO-5-$O_2NC_6H_3COC(CH_3)$=CHR (—)	2193
	4-ClC_6H_4CHO	,,	2-HO-5-$O_2NC_6H_3COC(CH_3)$=CHR (—)	2193
	C_6H_5CHO	,,	2-HO-5-$O_2NC_6H_3COC(CH_3)$=CHR (—)	2193

$C_6H_5COC_2H_5$	CH_2O	Ion-exchange resin Amberlite IRA-400	$C_6H_5COCH(CH_3)CH_2OH$ (40)	337, 336, 2195, 2196
	$C(Cl_2)$=CClCHO	HCl	$C_6H_5COC(CH_3)$=CHR (77)	637
	i-C_3H_7CHO	$C_6H_5N(CH_3)$-MgBr	$C_6H_5COCH(CH_3)CHOHR$ (88)	177
	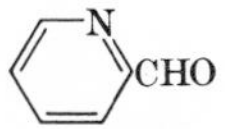 (pyridine-2-CHO)	NaOH	$C_6H_5COCH(CH_3)CHOHR$ (40)	545
	C_6H_5CHO	$C_6H_5N(CH_3)$-MgBr	$C_6H_5COCH(CH_3)CHOHR$ (—)	177
	"	HCl	$C_6H_5COC(CH_3)$=CHR (96)	342–345
	$C_6H_4(CHO)_2$-1,4	KOH	$C_6H_4[CH{=}C(CH_3)COC_6H_5]_2$-1,4 (13)	544
	4-$CH_3OC_6H_4CHO$	Na_2CO_3	$C_6H_5COC(CH_3)$=CHR (—)	2194
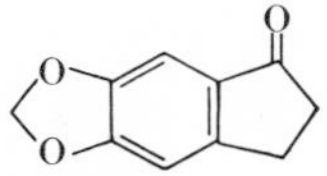 (structure)	2-$O_2NC_6H_4CHO$	$(CH_3CO)_2O$	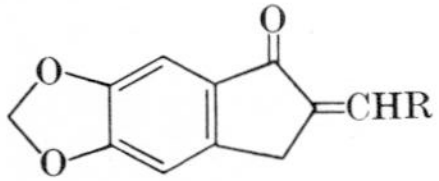 (structure)=CHR (36) D, R = 2-$O_2NC_6H_4$	2188
	2-HOC_6H_4CHO	KOH	D, R = 2-HOC_6H_4 (Good)	2191
	3,4-(CH_2O_2)-6-$O_2NC_6H_2CHO$	$(CH_3CO)_2O$	D, R = 3,4-(CH_2O_2)-6-$O_2NC_6H_2$ (34)	2188
	3,4-$(CH_3O)_2$-6-$O_2NC_6H_2CHO$	"	D, R = 3,4-$(CH_3O)_2$-6-$O_2NC_6H_2$ (—)	2188
	$(CH_3)_2NC_6H_4CHO$	KOH	D, R = $(CH_3)_2NC_6H_4$ (—)	2197
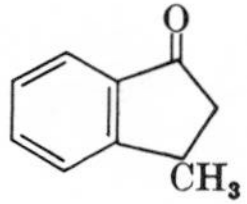 (structure, CH_3)	$C(Cl_2)$=CClCHO	BF_3	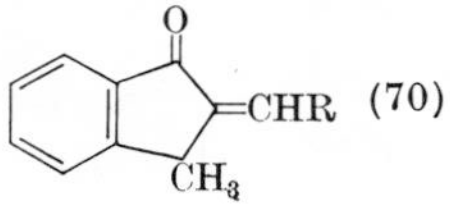 (structure, CH_3)=CHR (70)	289, 2345
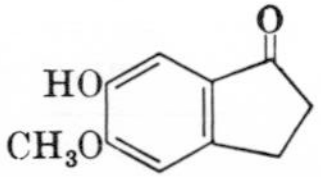 (structure, HO, CH_3O)	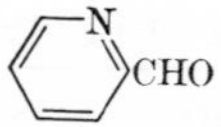 (pyridine-2-CHO)	$(CH_2)_5NH$, CH_3CO_2H	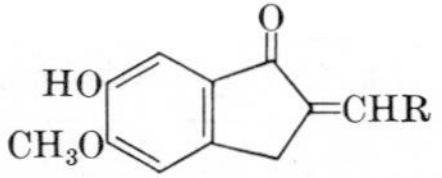 (structure, HO, CH_3O)=CHR (50)	363

Note: References 668–2359 are on pp. 403–438.

TABLE XVIII. Condensation of Aldehydes with Carbocyclic Aromatic Ketones Other than Acetophenones (*Continued*)

(R in the product is the group R in the aldehyde RCHO.)

Ketone	Aldehyde, RCHO	Catalyst	Product(s) (Yield, %)	Refs.
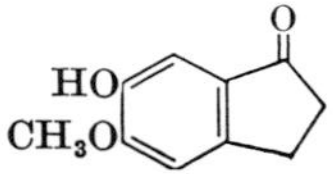 6-Hydroxy-5-methoxy-1-indanone (HO, CH_3O) (*contd.*)	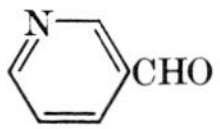 3-Pyridinecarboxaldehyde (CHO)	$(CH_2)_5NH$, CH_3CO_2H	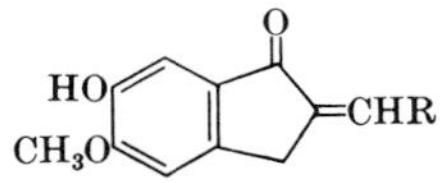 2-(=CHR)-6-hydroxy-5-methoxy-1-indanone (HO, CH_3O) (63)	363
	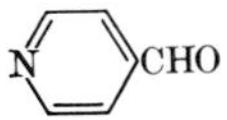 4-Pyridinecarboxaldehyde (CHO)	"	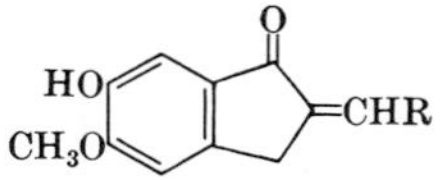 2-(=CHR)-6-hydroxy-5-methoxy-1-indanone (HO, CH_3O) (37)	363
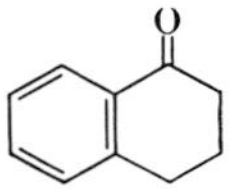 1-Tetralone	CH_2O	NaOH	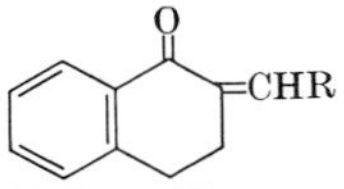 2-(=CHR)-1-tetralone (21) E, R = H	1822, 550, 2198
	$C(Cl_2){=}CClCHO$	H_2SO_4	E, R = $C(Cl_2){=}CCl$ (81)	639, 289
	$C(Cl_2){=}CHCHO$	HCl	E, R = $C(Cl_2){=}CH$ (70)	638
	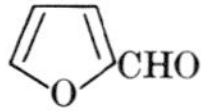 Furfural (CHO)	NaOH	E, R = 2-(C_4H_3O) (—)	2199
	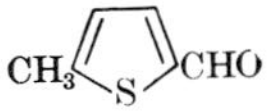 5-Methyl-2-thiophenecarboxaldehyde (CH_3, S, CHO)	"	E, R = 2-($CH_3C_4H_2S$) (—)	1988
	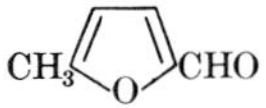 5-Methylfurfural (CH_3, O, CHO)	"	E, R = 2-($CH_3C_4H_2O$) (—)	1988
	2-$O_2NC_6H_4CHO$	$NaNH_2$	E, R = 2-$O_2NC_6H_4$ (12)	643
	"	H_2SO_4	E, R = 2-$O_2NC_6H_4$ (85)	643, 642
	C_6H_5CHO	KOH	E, R = C_6H_5 (88)	2200, 301, 1894, 2201, 2348
	2-$CH_3C_6H_4CHO$	"	E, R = 2-$CH_3C_6H_4$ (98)	301

	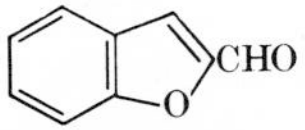 O, CHO	NaOH	E, R = 2-(C_8H_5O) (95)	1854
	2,4,6-$(CH_3)_3C_6H_2CHO$	KOH	E, R = 2,4,6-$(CH_3)_3C_6H_2$ (85)	301
	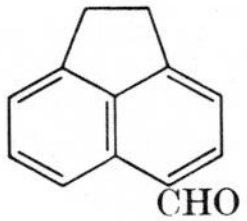 CHO	,,	E, R = 5-$C_{12}H_9$ (60)	2202
	Pyrene-1-carboxaldehyde	,,	E, R = 1-$C_{16}H_9$ (90)	2202
4-$ClC_6H_4COC_3H_7$-*i*	CH_2O	K_2CO_3	4-$ClC_6H_4COC(CH_2OH)(CH_3)_2$ (55)	2203
$C_6H_5COC_3H_7$-*n*	$C(Cl_2)$=CClCHO	HCl	$C_6H_5COC(C_2H_5)$=CHR (78)	637
	$CH_3(CH_2)_4CHO$	$C_6H_5N(CH_3)$-MgBr	$C_6H_5COCH(C_2H_5)CHOHR$ (40)	177
4-$CH_3OC_6H_4COC_2H_5$	4-$CH_3OC_6H_4CHO$	Na_2CO_3	4-$CH_3OC_6H_4COC(CH_3)$=CHR (—)	2194
O	O, CHO	KOH	O, CHR (98)	1848
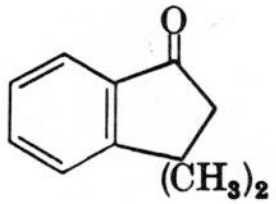 O, $(CH_3)_2$	C_6H_5CHO	NaOH	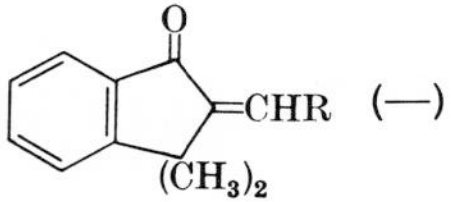 O, =CHR, $(CH_3)_2$ (—)	2204, 2205
CH_3O, O	2-$CH_3OC_6H_4CHO$	$NaOC_2H_5$	CH_3O, O, =CHR (—)	2206
			F, R = 2-$CH_3OC_6H_4$	
	3-$CH_3OC_6H_4CHO$	,,	F, R = 3-$CH_3OC_6H_4$ (—)	2206
	4-$CH_3OC_6H_4CHO$	,,	F, R = 4-$CH_3OC_6H_4$ (—)	2206

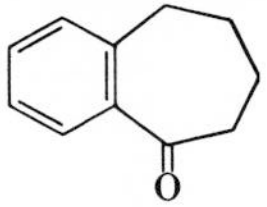

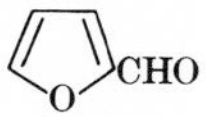

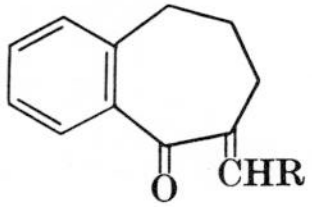

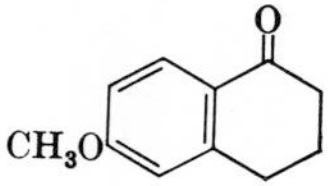

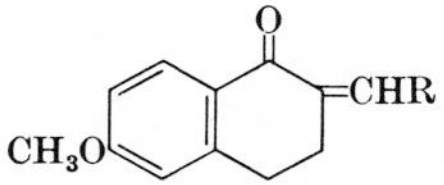

Note: References 668–2359 are on pp. 403–438.

TABLE XVIII. Condensation of Aldehydes with Carbocyclic Aromatic Ketones Other than Acetophenones (*Continued*)

(R in the product is the group R in the aldehyde RCHO.)

Ketone	Aldehyde, RCHO	Catalyst	Product(s) (Yield, %)	Refs.
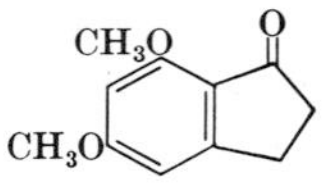 [structure: indanone with CH$_3$O, CH$_3$O, O]	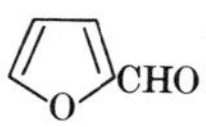 [structure: furan ring, CHO]	NaOH	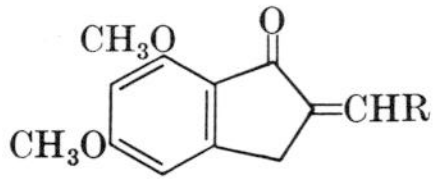 [structure: indanone with CH$_3$O, CH$_3$O, O, =CHR] (—)	2207
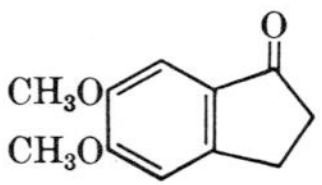 [structure: indanone with CH$_3$O, CH$_3$O, O]	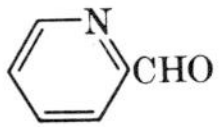 [structure: pyridine ring, N, CHO]	$(CH_2)_5NH$, CH_3CO_2H	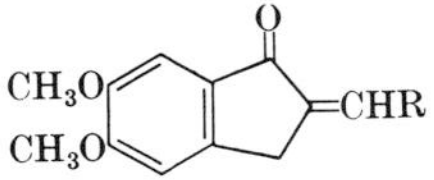 [structure: indanone with CH$_3$O, CH$_3$O, O, =CHR] (67) G, R = 2-(C_5H_4N)	363
	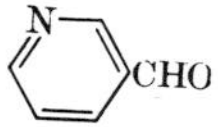 [structure: pyridine ring, N, CHO]	,,	G, R = 3-(C_5H_4N) (82)	363
	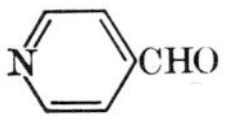 [structure: pyridine ring, N, CHO]	,,	G, R = 4-(C_5H_4N) (55)	363
	2-$O_2NC_6H_4CHO$	$(CH_3CO)_2O$	G, R = 2-$O_2NC_6H_4$ (37)	2188
	C_6H_5CHO	KOH	G, R = C_6H_5 (Good)	2191
	2-HOC_6H_4CHO	,,	G, R = 2-HOC_6H_4 (100)	2191, 571
	2,4-$(HO)_2C_6H_3CHO$	HCl	G, R = 2,4-$(HO)_2C_6H_3$ (—)	2191, 571
	3,4-(CH_2O_2)-6-$O_2NC_6H_2CHO$	$(CH_3CO)_2O$	G, R = 3,4-(CH_2O_2)-6-$O_2NC_6H_2$ (—)	2188
	3,4-$(CH_2O_2)C_6H_3CHO$	KOH	G, R = 3,4-$(CH_2O_2)C_6H_3$ (—)	2191
	2-$CH_3OC_6H_4CHO$	,,	G, R = 2-$CH_3OC_6H_4$ (Good)	2191
	3,4-$(CH_3O)_2$-6-$O_2NC_6H_2CHO$	$(CH_3CO)_2O$	G, R = 3,4-$(CH_3O)_2$-6-$O_2NC_6H_2$ (36)	2188
	3,4-$(CH_3O)_2C_6H_3CHO$	KOH	G, R = 3,4-$(CH_3O)_2C_6H_3$ (Good)	2191
$C_6H_5COC_4H_9$-*n*	CH_3CHO	$C_6H_5N(CH_3)$-MgBr	$C_6H_5COCH(C_3H_7$-*n*$)CHOHR$ (50)	177
2,4-$(CH_3O)_2C_6H_3$-COC_2H_5	4-$CH_3OC_6H_4CHO$	Na_2CO_3	2,4-$(CH_3O)_2C_6H_3COC(CH_3)$=CHR (—)	2194
4-$CH_3OC_6H_4COC_3H_7$-*n*	4-$CH_3OC_6H_4CHO$	HCl	4-$CH_3OC_6H_4COC(C_2H_5)$=CHR (—)	2102
4-$CH_3OC_6H_4COC_3H_7$-*i*	CH_2O	K_2CO_3	4-$CH_3OC_6H_4COC(CH_2OH)(CH_3)_2$ (17)	2203

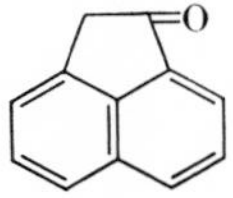

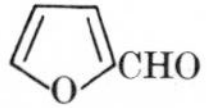

	2-Furyl-CHO	NaOH	H, R = 2-(C_4H_3O) (—)	1801, 2208
	3,4-$(CH_2O_2)C_6H_3CHO$	,,	H, R = 3,4-$(CH_2O_2)C_6H_3$ (—)	2208
	4-$CH_3OC_6H_4CHO$	,,	H, R = 4-$CH_3OC_6H_4$ (—)	2208
	3-CH_3O-4-HOC_6H_3CHO	,,	H, R = 3-CH_3O-4-HOC_6H_3 (—)	2208
	$C_6H_5CH{=}CHCHO$	,,	H, R = $C_6H_5CH{=}CH$ (—)	2208
	4-$(CH_3)_2NC_6H_4CHO$	,,	H, R = 4-$(CH_3)_2NC_6H_4$ (—)	2208
	2-HO-5-BrC_6H_3CHO	,,	I, R = 2-HO-5-BrC_6H_3 (—)	1993
	C_6H_5CHO	,,	I, R = C_6H_5 (—)	2209
	2-HOC_6H_4CHO	,,	I, R = 2-HOC_6H_4 (—)	1993
	4-$CH_3OC_6H_4CHO$	,,	I, R = 4-$CH_3OC_6H_4$ (—)	2210
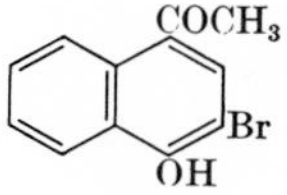	2-HO-5-BrC_6H_3CHO	,,	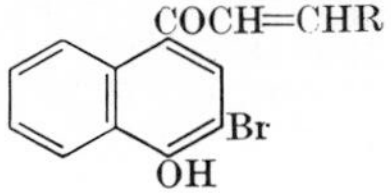 (15)	1993
	C_6H_5CHO	,,	J, R = C_6H_5 (18)	2211
	2-HOC_6H_4CHO	,,	J, R = 2-HOC_6H_4 (18)	2211
	3-HOC_6H_4CHO	,,	J, R = 3-HOC_6H_4 (18)	2211
	4-HOC_6H_4CHO	,,	J, R = 4-HOC_6H_4 (18)	2211
	4-$CH_3OC_6H_4CHO$	,,	J, R = 4-$CH_3OC_6H_4$ (18)	2211

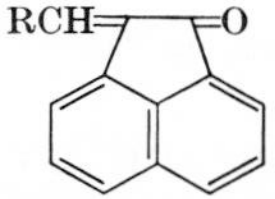

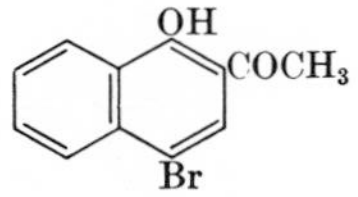

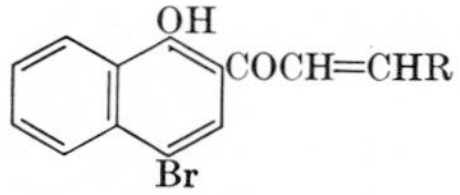

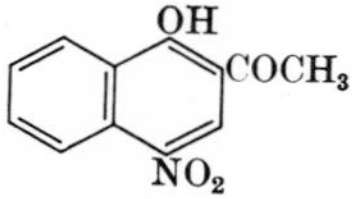

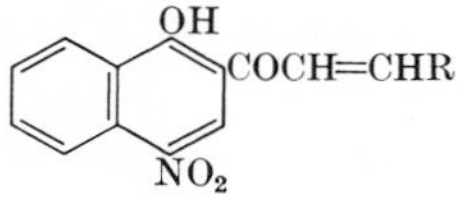

Note: References 668–2359 are on pp. 403–438.

TABLE XVIII. Condensation of Aldehydes with Carbocyclic Aromatic Ketones Other than Acetophenones (*Continued*)

(R in the product is the group R in the aldehyde RCHO.)

Ketone	Aldehyde, RCHO	Catalyst	Product(s) (Yield, %)	Refs.
$1\text{-}C_{10}H_7COCH_3$	$2\text{-}C_4H_3S\text{-}CHO$	$NaOC_2H_5$	$1\text{-}C_{10}H_7COCH{=}CHR$ (45)	1823, 1845
	$2\text{-}C_4H_3O\text{-}CHO$	,,	$1\text{-}C_{10}H_7COCH{=}CHR$ (68)	1823, 1860
	$4\text{-}O_2NC_6H_4CHO$	NaOH	$1\text{-}C_{10}H_7COCH{=}CHR$ (—)	1791
	C_6H_5CHO	,,	$1\text{-}C_{10}H_7COCH{=}CHR$ (—)	2212
	$3,4\text{-}(CH_2O_2)C_6H_3CHO$	,,	$1\text{-}C_{10}H_7COCH{=}CHR$ (—)	1744, 2212
	$4\text{-}CH_3C_6H_4CHO$	,,	$1\text{-}C_{10}H_7COCH{=}CHR$ (—)	2212
	$4\text{-}CH_3OC_6H_4CHO$	,,	$1\text{-}C_{10}H_7COCH{=}CHR$ (—)	2212
	$C_6H_5CH{=}CHCHO$	,,	$1\text{-}C_{10}H_7COCH{=}CHR$ (—)	1791
	$4\text{-}i\text{-}C_3H_7C_6H_4CHO$	,,	$1\text{-}C_{10}H_7COCH{=}CHR$ (—)	2212
	9-Formyl-3,4-benzacridine	,,	$1\text{-}C_{10}H_7COCH{=}CHR$ (75–95)	372
	10-Formyl-1,2-benzanthracene	,,	$1\text{-}C_{10}H_7COCH{=}CHR$ (75–95)	372
$2\text{-}C_{10}H_7COCH_3$	$C(Cl_2){=}CHCHO$	BaO	$2\text{-}C_{10}H_7COCH{=}CHR$ (25)	1908
	$5\text{-}Br\text{-}2\text{-}C_4H_2O\text{-}CHO$	NaOH	$2\text{-}C_{10}H_7COCH{=}CHR$ (—)	1644
	$2\text{-}C_4H_3S\text{-}CHO$	$NaOC_2H_5$	$2\text{-}C_{10}H_7COCH{=}CHR$ (57)	1823, 1845
	$2\text{-}C_4H_3O\text{-}CHO$	,,	$2\text{-}C_{10}H_7COCH{=}CHR$ (85)	1823, 1860
	$4\text{-}FC_6H_4CHO$	NaOH	$2\text{-}C_{10}H_7COCH{=}CHR$ (—)	1995
	C_6H_5CHO	,,	$2\text{-}C_{10}H_7COCH{=}CHR$ (—)	2212
	$3,5\text{-}(CH_3)_2\text{-}2\text{-}C_4HS\text{-}CHO$	,,	$2\text{-}C_{10}H_7COCH{=}CHR$ (—)	2017

	3-i-C_3H_7-4-CH_3O-6-$CH_3C_6H_2CHO$	,,	2-$C_{10}H_7COCH{=}CHR$ (75)	2105
	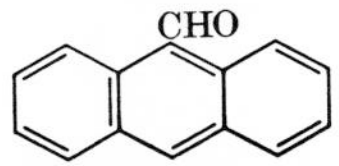 (structure: CHO)	HCl	2-$C_{10}H_7COCH{=}CHR$ (80)	1978
	Pyrene-3-carboxaldehyde	$NaOC_2H_5$	2-$C_{10}H_7COCH{=}CHR$ (100)	1615
	9-Formyl-3,4-benzacridine	NaOH	2-$C_{10}H_7COCH{=}CHR$ (75–95)	372
	10-Formyl-1,2-benzanthracene	,,	2-$C_{10}H_7COCH{=}CHR$ (75–95)	372
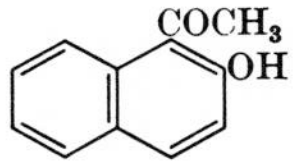 (structure: $COCH_3$, OH)	2-HO-5-BrC_6H_3CHO	,,	2-$HOC_{10}H_6COCH{=}CHR$-1 (80)	1993
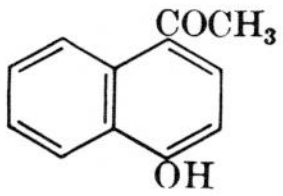 (structure: $COCH_3$, OH)	2-ClC_6H_4CHO	KOH	4-$HOC_{10}H_6COCH{=}CHR$-1 (78)	2213
	3-ClC_6H_4CHO	,,	4-$HOC_{10}H_6COCH{=}CHR$-1 (72)	2213
	4-ClC_6H_4CHO	,,	4-$HOC_{10}H_6COCH{=}CHR$-1 (76)	2213
	2-HOC_6H_4CHO	,,	4-$HOC_{10}H_6COCH{=}CHR$-1 (62)	2213
	4-$CH_3OC_6H_4CHO$	,,	4-$HOC_{10}H_6COCH{=}CHR$-1 (81)	2213
	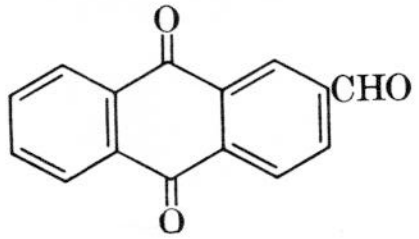 (structure: O, O, CHO)	NaOH	4-$HOC_{10}H_6COCH{=}CHR$-1 (—)	1977
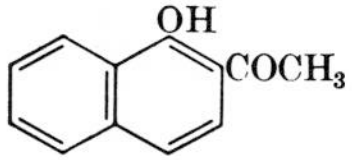 (structure: OH, $COCH_3$)	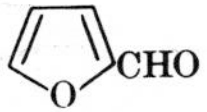 (structure: O, CHO)	,,	1-$HOC_{10}H_6COCH{=}CHR$-2 (—)	2214
	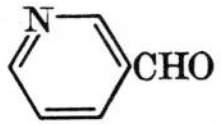 (structure: N, CHO)	KOH	1-$HOC_{10}H_6COCH{=}CHR$-3 (71)	371

Note: References 668–2359 are on pp. 403–438.

TABLE XVIII. CONDENSATION OF ALDEHYDES WITH CARBOCYCLIC AROMATIC KETONES OTHER THAN ACETOPHENONES (*Continued*)

(R in the product is the group R in the aldehyde RCHO.)

Ketone	Aldehyde, RCHO	Catalyst	Product(s) (Yield, %)	Refs.
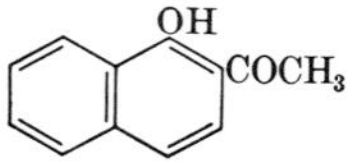 1-Hydroxy-2-naphthyl methyl ketone (structure: OH, $COCH_3$) (*contd.*)	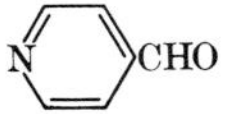 Pyridine-4-CHO (structure: N, CHO)	KOH	$1\text{-}HOC_{10}H_6COCH{=}CHR\text{-}2$ (73)	370
	$2\text{-}HO\text{-}5\text{-}BrC_6H_3CHO$	NaOH	$1\text{-}HOC_{10}H_6COCH{=}CHR\text{-}2$ (60)	1993
	C_6H_5CHO	,,	$1\text{-}HOC_{10}H_6COCH{=}CHR\text{-}2$ (—)	2215, 1929, 2044
	$2\text{-}HOC_6H_4CHO$	,,	$1\text{-}HOC_{10}H_6COCH{=}CHR\text{-}2$ (60)	1993
	$3,4\text{-}(CH_2O_2)C_6H_3CHO$	,,	$1\text{-}HOC_{10}H_6COCH{=}CHR\text{-}2$ (—)	2215, 1929, 2044
	$4\text{-}CH_3OC_6H_4CHO$	,,	$1\text{-}HOC_{10}H_6COCH{=}CHR\text{-}2$ (—)	1929, 2044, 2216
	$4\text{-}C_2H_5OC_6H_4CHO$	,,	$1\text{-}HOC_{10}H_6COCH{=}CHR\text{-}2$ (—)	2214
	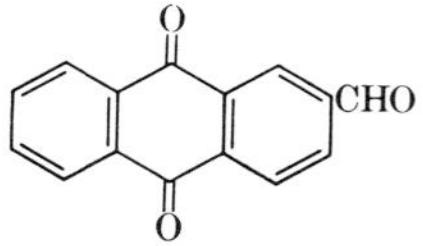 Anthraquinone-2-CHO (structure: O, O, CHO)	,,	$1\text{-}HOC_{10}H_6COCH{=}CHR\text{-}2$ (—)	1977
Acetylferrocene	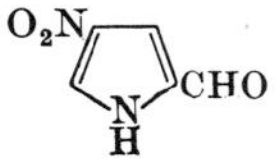 Nitropyrrole-2-CHO (structure: O_2N, N-H, CHO)	,,	$C_5H_5FeC_5H_4COCH{=}CHR$ (15)	1975
	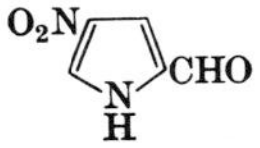 Nitropyrrole-2-CHO (structure: O_2N, N-H, CHO)	,,	$C_5H_5FeC_5H_4COCH{=}CHR$ (16)	1975
	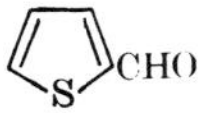 Thiophene-2-CHO (structure: S, CHO)	,,	$C_5H_5FeC_5H_4COCH{=}CHR$ (75)	1975

	2-Furyl-CHO (furan-2-carboxaldehyde)	,,	$C_5H_5FeC_5H_4COCH{=}CHR$ (65)	1975
	2-Pyrrolyl-CHO (pyrrole-2-carboxaldehyde, N–H)	,,	$C_5H_5FeC_5H_4COCH{=}CHR$ (19)	1975
	2-Pyridyl-CHO	,,	$C_5H_5FeC_5H_4COCH{=}CHR$ (50)	1975
	3-Pyridyl-CHO	,,	$C_5H_5FeC_5H_4COCH{=}CHR$ (44)	1975
	4-Pyridyl-CHO	,,	$C_5H_5FeC_5H_4COCH{=}CHR$ (32)	1975
	4-ClC_6H_4CHO	,,	$C_5H_5FeC_5H_4COCH{=}CHR$ (50)	1975
	4-$O_2NC_6H_4CHO$	,,	$C_5H_5FeC_5H_4COCH{=}CHR$ (50)	1975
	C_6H_5CHO	,,	$C_5H_5FeC_5H_4COCH{=}CHR$ (83)	1975, 2217, 2218
	4-NCC_6H_4CHO	,,	$C_5H_5FeC_5H_4COCH{=}CHR$ (40)	1975
	4-$CH_3OC_6H_4CHO$	,,	$C_5H_5FeC_5H_4COCH{=}CHR$ (30)	1975
	$C_6H_5CH{=}CHCHO$	,,	$C_5H_5FeC_5H_4COCH{=}CHR$ (40)	1975
	2-Quinolyl-CHO	,,	$C_5H_5FeC_5H_4COCH{=}CHR$ (22)	1975
	2-Naphthyl-CHO	,,	$C_5H_5FeC_5H_4COCH{=}CHR$ (35)	1975
	Ferrocenecarboxaldehyde	,,	$C_5H_5FeC_5H_4COCH{=}CHR$ (44)	1975
C_6H_5CO-(cyclopent-1-enyl)	4-ClC_6H_4CHO	$NaOCH_3$	C_6H_5CO-(cyclopentenyl)$=CHR$ (47)	2219

Note: References 668–2359 are on pp. 403–438.

TABLE XVIII. CONDENSATION OF ALDEHYDES WITH CARBOCYCLIC AROMATIC KETONES OTHER THAN ACETOPHENONES (*Continued*)

(R in the product is the group R in the aldehyde RCHO.)

Ketone	Aldehyde, RCHO	Catalyst	Product(s) (Yield, %)	Refs.
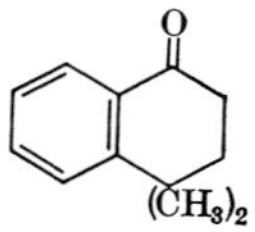 O, $(CH_3)_2$	4-Cl-2-$O_2NC_6H_3CHO$	CH_3CO_2H, H_2O, Fe, 100°	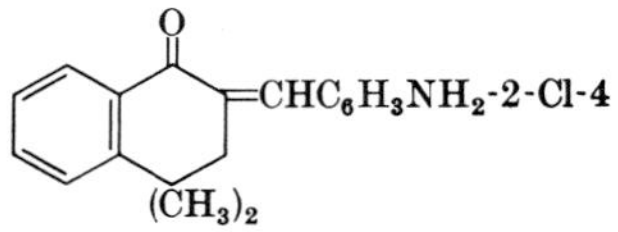 O, =$CHC_6H_3NH_2$-2-Cl-4, $(CH_3)_2$ (90)	645
	2-O_2N-5-ClC_6H_3CHO	"	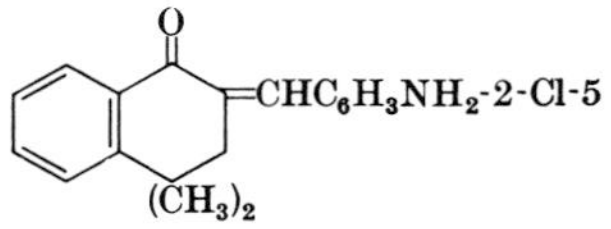 O, =$CHC_6H_3NH_2$-2-Cl-5, $(CH_3)_2$ (94)	645
	2-$O_2NC_6H_4CHO$	H_2SO_4, CH_3CO_2H	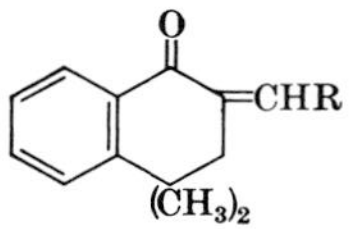 O, =CHR, $(CH_3)_2$ (94)	642
	C_6H_5CHO	KOH	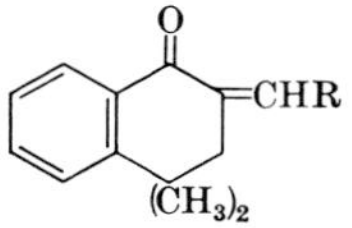 O, =CHR, $(CH_3)_2$ (76)	643, 2205
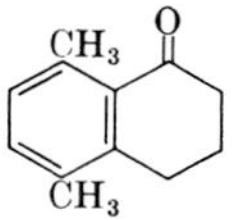 CH_3, O, CH_3	2-$O_2NC_6H_4CHO$	H_2SO_4, CH_3CO_2H	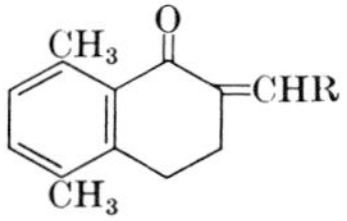 CH_3, O, =CHR, CH_3 (48)	644
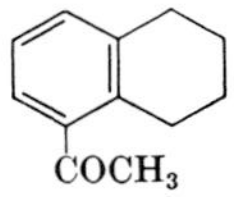 $COCH_3$	C_6H_5CHO	NaOH	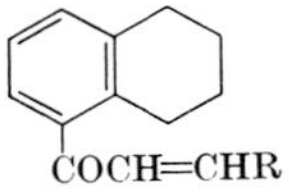 COCH=CHR (77)	2220

$COCH_3$	3-i-C_3H_7-4-CH_3O-6-CH_3-C_6H_2CHO	,,	$COCH{=}CHR$ (75)	2105
CH_3 O OCH_3	2-$OHCC_6H_4CHO$	KOH	CH_3 O $=CHR$ OCH_3 (—)	2221
CH_3O O CH_3O	CHO	NaOH	CH_3O O $=CHR$ CH_3O (54)	2207
CH_3O O CH_3O	2-$HO_2CC_6H_4CHO$	KOH	CH_3O O $=CHR$ CH_3O (65)	2221
2-HO-5-(C_2H_5CONH)-$C_6H_3COC_2H_5$	4-$O_2NC_6H_4CHO$	,,	2-HO-5-(C_2H_5CONH)C_6H_3-$COC(CH_3){=}CHR$ (—)	2222
2,4,6-$(CH_3)_3C_6H_2$-COC_2H_5	CH_2O	K_2CO_3	2,4,6-$(CH_3)_3COC_6H_2$-$COC(CH_3){=}CHR$ (70)	2150
	2,4,6-$(CH_3)_3C_6H_2CHO$	$NaOC_2H_5$	2,4,6-$(CH_3)_3COC_6H_2$-$COC(CH_3){=}CHR$ (51)	2154
O	C_6H_5CHO	KOH	O $=CHR$ (60)	2223
	CHO	,,	O $=CHR$ (60)	2223

Note: References 668–2359 are on pp. 403–438.

TABLE XVIII. Condensation of Aldehydes with Carbocyclic Aromatic Ketones Other than Acetophenones (*Continued*)

(R in the product is the group R in the aldehyde RCHO.)

Ketone	Aldehyde, RCHO	Catalyst	Product(s) (Yield, %)	Refs.
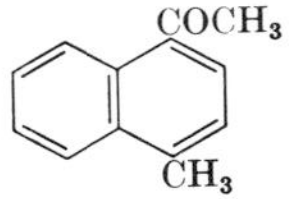 $COCH_3$; CH_3	2-$O_2NC_6H_4CHO$	HCl	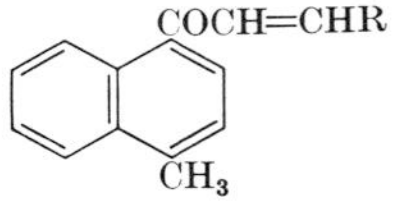 $COCH{=}CHR$; CH_3 (—)	1936
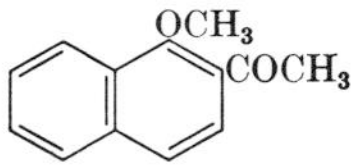 OCH_3; $COCH_3$	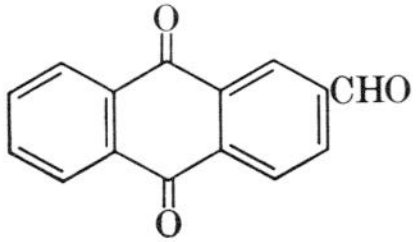 CHO; O; O	NaOH	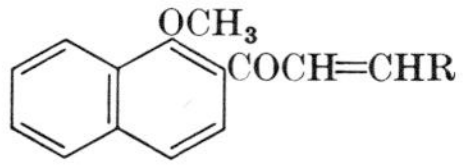 OCH_3; $COCH{=}CHR$ (—)	1977
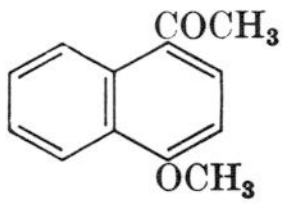 $COCH_3$; OCH_3	,,	,,	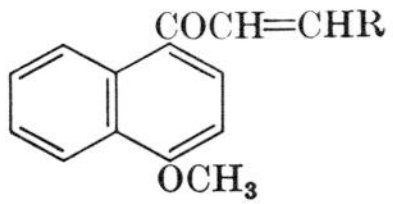 $COCH{=}CHR$; OCH_3 (—)	1977
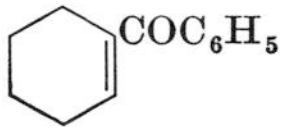 COC_6H_5	C_6H_5CHO	$NaOC_2H_5$	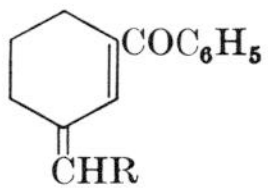 COC_6H_5; CHR (—)	316
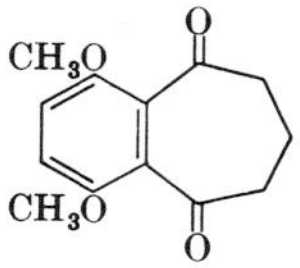 CH_3O; O; CH_3O; O	CH_3CHO	HCl	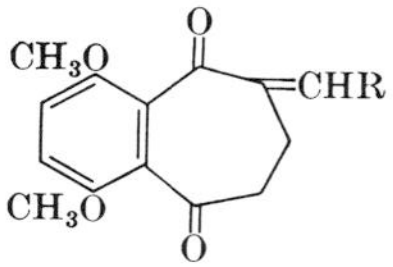 CH_3O; O; =CHR; CH_3O; O (63)	2224
	C_6H_5CHO	,,	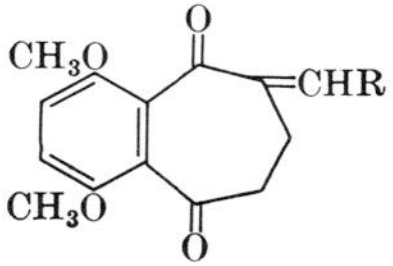 CH_3O; O; =CHR; CH_3O; O (58)	2224

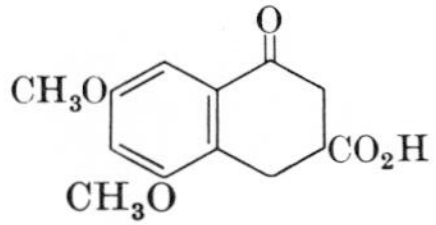 [structure: O; CH_3O; CH_3O; CO_2H]	2-Cl-5-$CH_3OC_6H_3CHO$	KOH	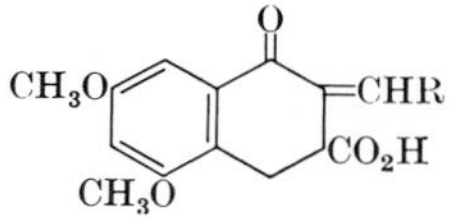 [structure: O; CH_3O; CH_3O; =CHR; CO_2H] (44)	2225
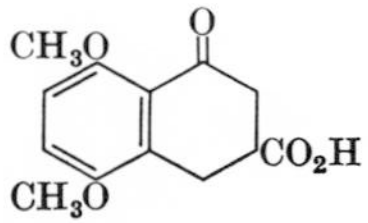 [structure: O; CH_3O; CH_3O; CO_2H]	2,4-$(CH_3O)_2C_6H_3CHO$	,,	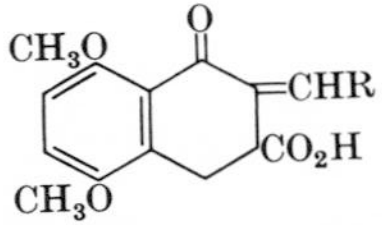 [structure: O; CH_3O; CH_3O; =CHR; CO_2H] (50)	2225
2,4,6-$(CH_3)_3C_6H_2$-$COCH{=}CHCH_3$	4-ClC_6H_4CHO	NaOH	2,4,6-$(CH_3)_3C_6H_2CO(CH{=}CH)_2R$ (45)	2226
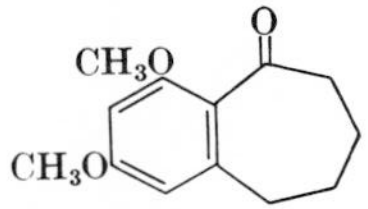 [structure: O; CH_3O; CH_3O]	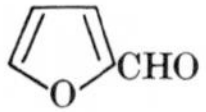 [structure: furan; CHO]	,,	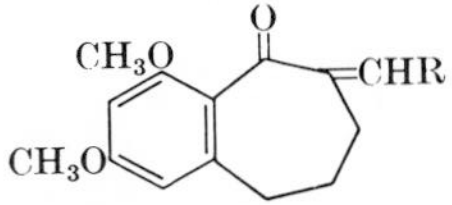 [structure: O; CH_3O; CH_3O; =CHR] (—)	2207
2,4,6-$(CH_3)_3C_6H_2$-COC_3H_7-*i*	CH_2O	K_2CO_3	2,4,6-$(CH_3)_3C_6H_2COC(CH_3)_2$-$CH_2OH$ (40)	2150
2,4,6-$(CH_3)_3C_6H_2$-$COCH_2CH_2OCH_3$	,,	,,	2,4,6-$(CH_3)_3C_6H_2$-$COC(CH_2OCH_3){=}CH_2$ (65)	519
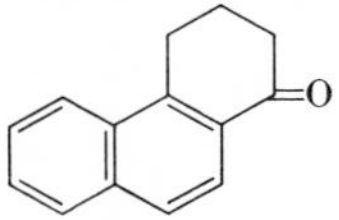 [structure: =O]	Pyrene-1-carboxaldehyde	KOH	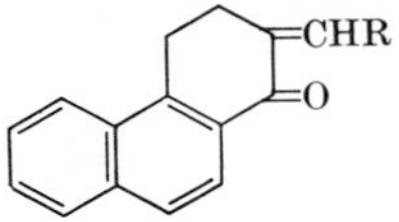 [structure: =CHR; =O] (60)	2202
	Phenanthrene-9-carboxaldehyde	,,	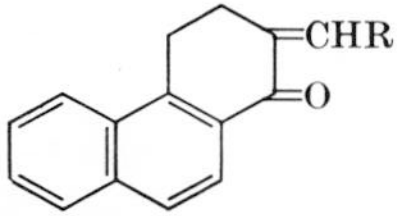 [structure: =CHR; =O] (—)	362

Note: References 668–2359 are on pp. 403–438.

TABLE XVIII. Condensation of Aldehydes with Carbocyclic Aromatic Ketones Other than Acetophenones (*Continued*)

(R in the product is the group R in the aldehyde RCHO.)

Ketone	Aldehyde, RCHO	Catalyst	Product(s) (Yield, %)	Refs.
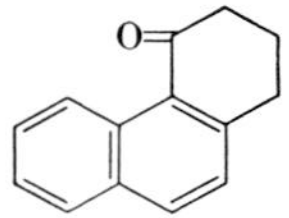 [structure: O=]	Pyrene-1-carboxaldehyde	KOH	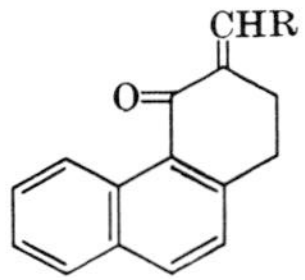 [structure: CHR, O=] (—)	2202
$C_6H_5COCH_2C_6H_5$	$C(Cl_2)$=CClCHO	H_2SO_4	$C_6H_5COCH(C_6H_5)CHOHR$ (31)	289
	2-ClC_6H_4CHO	HCl	$C_6H_5COC(C_6H_5)$=CHR (—)	2227
	4-$O_2NC_6H_4CHO$	NaOH	$C_6H_5COCH(C_6H_5)CHOHR$ (epimers) (75)	64, 2188
	C_6H_5CHO	$(CH_2)_5NH$	$C_6H_5COC(C_6H_5)$=CHR (77)	597, 548, 2228–2232
	4-$CH_3C_6H_4CHO$	HCl	$C_6H_5COC(C_6H_5)$=CHR (—)	2227
	4-$CH_3OC_6H_4CHO$	,,	$C_6H_5COC(C_6H_5)$=CHR (—)	2227
	3,4-$(CH_3O)_2C_6H_3CHO$	,,	$C_6H_5COC(C_6H_5)$=CHR (—)	2227
	4-*i*-$C_3H_7C_6H_4CHO$	,,	$C_6H_5COC(C_6H_5)$=CHR (—)	2227
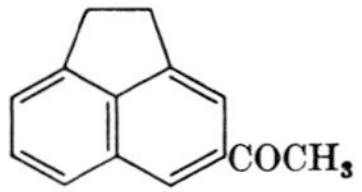 [structure: $COCH_3$]	9-Formyl-3,4-benzacridine	NaOH	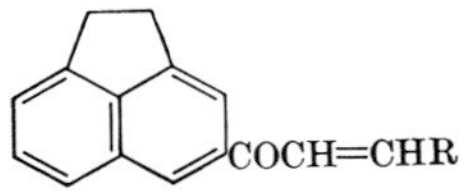 [structure: COCH=CHR] (75–95)	372
	10-Formyl-1,2-benzanthracene	,,	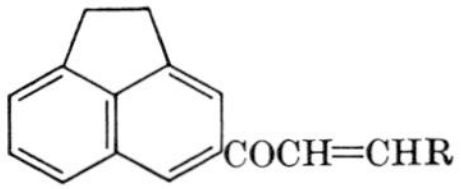 [structure: COCH=CHR] (75–95)	372
 [structure: $COCH_3$]	2-ClC_6H_4CHO	—	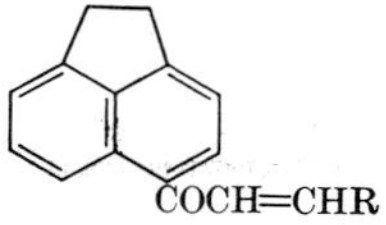 [structure: COCH=CHR] (—)	2233

	9-Formyl-3,4-benzacridine	NaOH	$COCH{=}CHR$	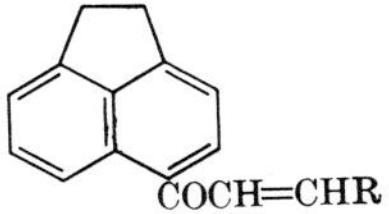 (75–95)	372
 $COCH_3$ (O)	C_6H_5CHO	KOH	$COCH{=}CHR$ (O)	(—)	2234
bis-Acetylferrocene	"	NaOH	$C_{10}H_8Fe(COCH{=}CHR)_2$	(75)	2235
CH_3O, $COCH_3$, CH_3O	S, CHO	"	CH_3O, $COCH{=}CHR$, CH_3O	(—)	1829
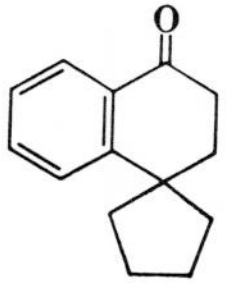 O	$2\text{-}O_2NC_6H_4CHO$	H_2SO_4, CH_3CO_2H	O, $=CHR$	(84)	644
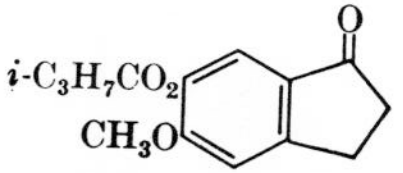 $i\text{-}C_3H_7CO_2$, CH_3O, O	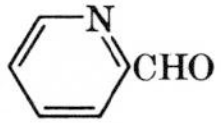 N, CHO	$(CH_2)_5NH$, CH_3CO_2H	$i\text{-}C_3H_7CO_2$, CH_3O, O, $=CHR$	(59)	363
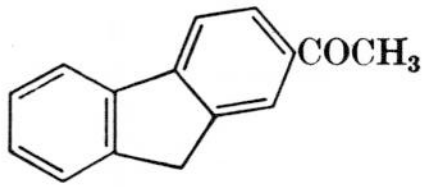 $COCH_3$	CH_2O	KOH	Polymer	(—)	2236

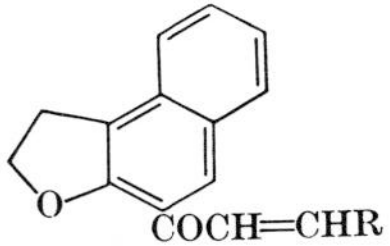

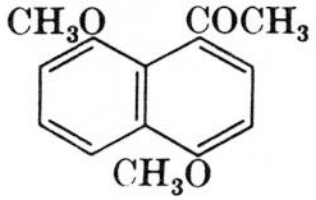

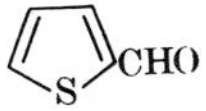

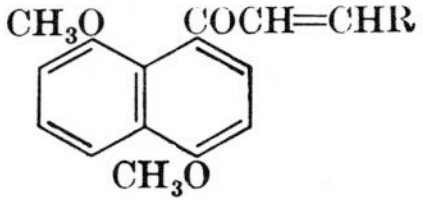

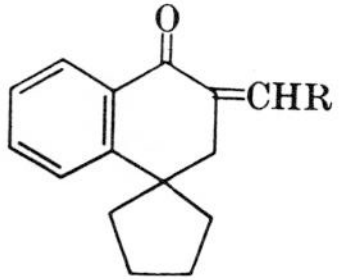

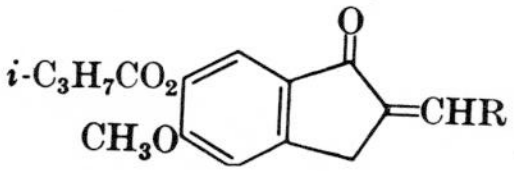

Note: References 668–2359 are on pp. 403–438.

TABLE XVIII. CONDENSATION OF ALDEHYDES WITH CARBOCYCLIC AROMATIC KETONES OTHER THAN ACETOPHENONES (*Continued*)

(R in the product is the group R in the aldehyde RCHO.)

Ketone	Aldehyde, RCHO	Catalyst	Product(s) (Yield, %)	Refs.
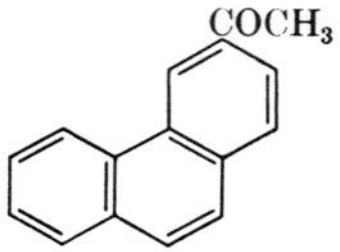	C_6H_5CHO	HCl	$3\text{-}C_{14}H_9COCH{=}CHR$ (60)	2238
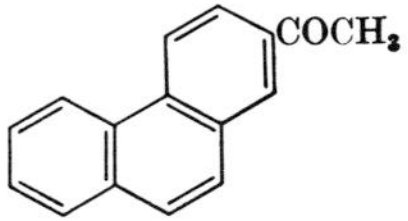	$4\text{-}O_2NC_6H_4CHO$	”	$2\text{-}C_{14}H_9COCH{=}CHR$ (60)	2238
	C_6H_5CHO	”	$2\text{-}C_{14}H_9COCH{=}CHR$ (60)	2238
$4\text{-}(CH_3)_2NC_6H_4CH_2\text{-}COC_6H_5$	CH_2O	$(CH_2)_5NH$	$4\text{-}(CH_3)_2NC_6H_4CH(CH_2OH)COC_6H_5$ (80)	598
$4\text{-}(CH_3)_2NC_6H_4COCH_2\text{-}C_6H_5$	”	”	$4\text{-}(CH_3)_2NC_6H_4COCH(CH_2OH)C_6H_5$ (60)	598
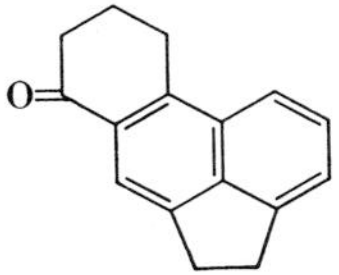	C_6H_5CHO	KOH	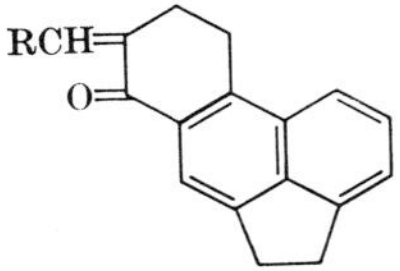 (70)	2202
	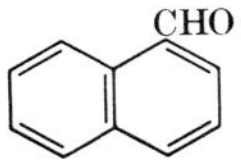	”	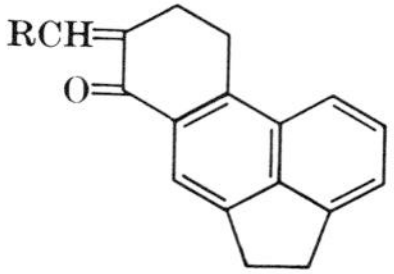 (65)	2202

$C_6H_5COCH_2COCH_2C_6H_5$	$2\text{-}HOC_6H_4CHO$	$(CH_2)_5NH$	$C_6H_5COCH_2COC(C_6H_5){=}CHR$ (63)	2237
$2,4,6\text{-}(CH_3)_3C_6H_2\text{-}CH_2COC_6H_5$	CH_2O	K_2CO_3	$2,4,6\text{-}(CH_3)_3C_6H_2C({=}CH_2)COC_6H_5$ (80)	2239
$2,4,6\text{-}(CH_3)_3C_6H_2\text{-}COCH_2C_6H_5$	$4\text{-}ClC_6H_4CHO$	NaOH	$2,4,6\text{-}(CH_3)_3C_6H_2COC(C_6H_5){=}CHR$ (83)	2240
	C_6H_5CHO	,,	$2,4,6\text{-}(CH_3)_3C_6H_2COC(C_6H_5){=}CHR$ (82)	2241
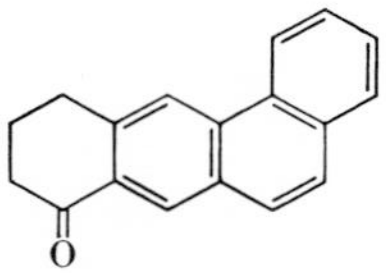 O	C_6H_5CHO	KOH	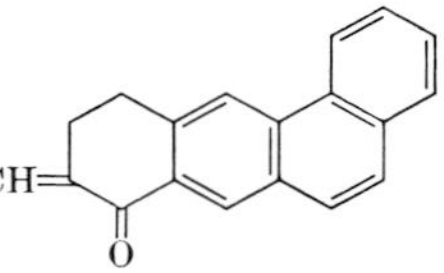 RCH= O (70)	362
	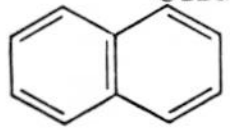 CHO	,,	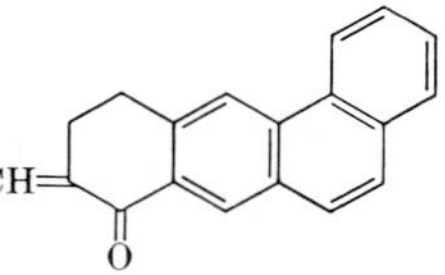 RCH= O (—)	362
	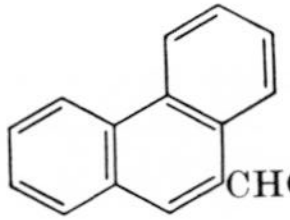 CHO	,,	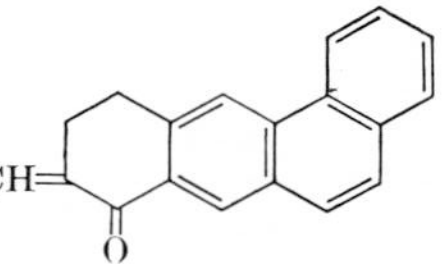 RCH= O (—)	362
$2,4,6\text{-}(CH_3)_3C_6H_2COCH_2\text{-}CH_2C_6H_5$	$4\text{-}ClC_6H_4CHO$	NaOH	$2,4,6\text{-}(CH_3)_3C_6H_2\text{-}COC(CH_2C_6H_5){=}CHR$ (—)	2242
$2,3,5,6\text{-}(CH_3)_4C_6HCOCH_2\text{-}C_6H_5$	,,	,,	$2,3,5,6\text{-}(CH_3)_4C_6HCOC(C_6H_5){=}CHR$ (81)	2240
	C_6H_5CHO	,,	$2,3,5,6\text{-}(CH_3)_4C_6HCOC(C_6H_5){=}CHR$ (70)	2240
$(C_6H_5COCH_2CH_2)_2$	,,	$NaOCH_3$	$[C_6H_5COC({=}CHR)CH_2]_2$ (—)	2243

Note: References 668–2359 are on pp. 403–438.

TABLE XVIII. CONDENSATION OF ALDEHYDES WITH CARBOCYCLIC AROMATIC KETONES OTHER THAN ACETOPHENONES (*Continued*)

(R in the product is the group R in the aldehyde RCHO.)

Ketone	Aldehyde, RCHO	Catalyst	Product(s) (Yield, %)	Refs.
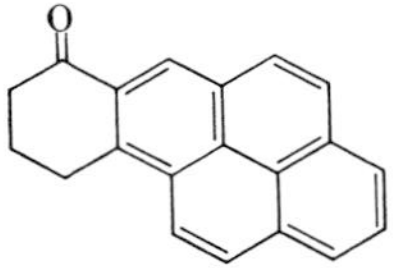	C_6H_5CHO	KOH	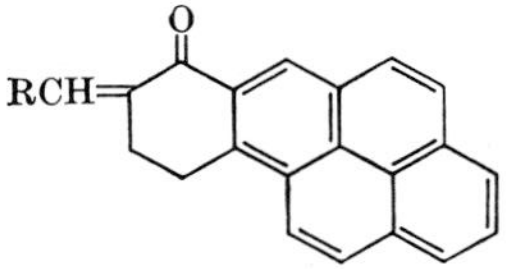 (80)	2202
	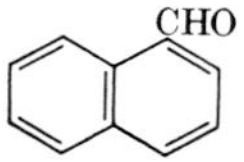 (1-naphthaldehyde, CHO)		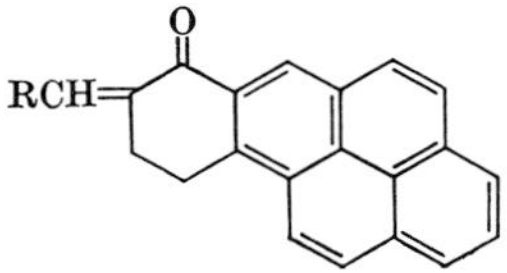 (—)	2202
2,4,6-$(CH_3)_3C_6H_2CH_2$-$COC_6H_2(CH_3)_3$-2,4,6	CH_2O	K_2CO_3	2,4,6-$(CH_3)_3C_6H_2C(CH_3)$=$C(OH)C_6H_2(CH_3)_3$-2,4,6 (50)	2244
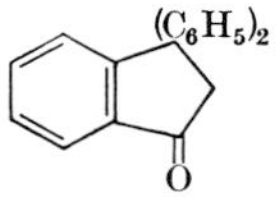 $(C_6H_5)_2$	2-ClC_6H_4CHO	NaOH	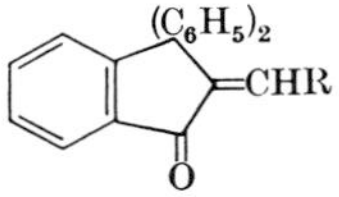 $(C_6H_5)_2$ =CHR (—) K, R = 2-ClC_6H_4 (*cis* and *trans*)	2247

	$4\text{-ClC}_6\text{H}_4\text{CHO}$	,,	K, R = $4\text{-ClC}_6\text{H}_4$ (—) (*cis* and *trans*)	2247
	$\text{C}_6\text{H}_5\text{CHO}$	,,	K, R = C_6H_5 (—) (*cis* and *trans*)	2247
	$2\text{-CH}_3\text{C}_6\text{H}_4\text{CHO}$	,,	K, R = $2\text{-CH}_3\text{C}_6\text{H}_4$ (—) (*cis* and *trans*)	2247
	$3\text{-CH}_3\text{C}_6\text{H}_4\text{CHO}$	,,	K, R = $3\text{-CH}_3\text{C}_6\text{H}_4$ (—) (*cis* and *trans*)	2247
	$2\text{-CH}_3\text{OC}_6\text{H}_4\text{CHO}$	,,	K, R = $2\text{-CH}_3\text{OC}_6\text{H}_4$ (—) (*cis* and *trans*)	2247
	$4\text{-CH}_3\text{OC}_6\text{H}_4\text{CHO}$	,,	K, R = $4\text{-CH}_3\text{OC}_6\text{H}_4$ (—) (*cis* and *trans*)	2247
	$2\text{-C}_2\text{H}_5\text{OC}_6\text{H}_4\text{CHO}$	,,	K, R = $2\text{-C}_2\text{H}_5\text{OC}_6\text{H}_4$ (—) (*cis* and *trans*)	2247
$2,4\text{-(C}_6\text{H}_5\text{CH}_2\text{O)}_2\text{C}_6\text{H}_3\text{-COCH}_2\text{OCH}_3$	$2\text{-HO-4,5-(CH}_2\text{O}_2\text{)C}_6\text{H}_2\text{CHO}$	HCl	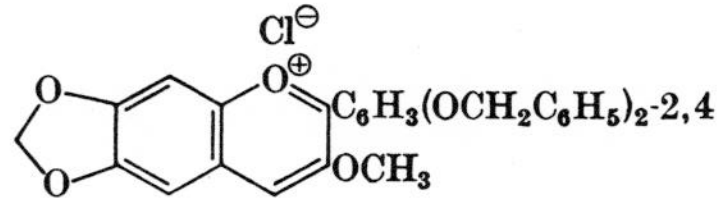 [structure: $\text{Cl}^{\ominus}$, $\text{O}^{\oplus}$, $\text{C}_6\text{H}_3(\text{OCH}_2\text{C}_6\text{H}_5)_2\text{-2,4}$, OCH_3] (—)	2245
$2,4,6\text{-(}i\text{-C}_3\text{H}_7\text{)}_3\text{C}_6\text{H}_2\text{-CH}_2\text{COC}_6\text{H}_5$	CH_2O	K_2CO_3	$2,4,6\text{-(}i\text{-C}_3\text{H}_7\text{)}_3\text{C}_6\text{H}_2\text{C(COC}_6\text{H}_5\text{)=CHR}$ (96)	2239
	$\text{C}_6\text{H}_5\text{CHO}$	NaOH	$2,4,6\text{-(}i\text{-C}_3\text{H}_7\text{)}_3\text{C}_6\text{H}_2\text{C(COC}_6\text{H}_5\text{)=CHR}$ (70)	2240
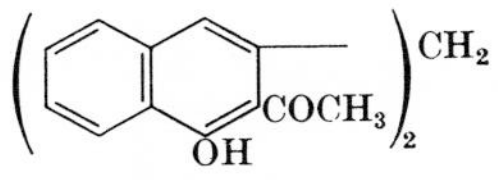 [structure: (naphthyl, OH, COCH_3)$_2$$\text{CH}_2$]	$3,4\text{-(CH}_2\text{O}_2\text{)C}_6\text{H}_3\text{CHO}$	,,	[structure: (naphthyl, OH, COCH=CHR)$_2$$\text{CH}_2$] (75)	2246

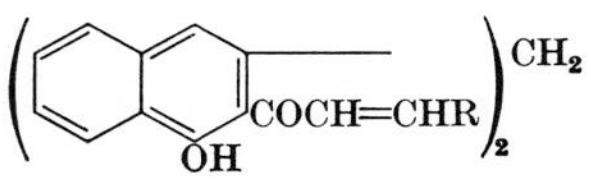

Note: References 668–2359 are on pp. 403–438.

TABLE XIX. CONDENSATION OF ALDEHYDES WITH HETEROCYCLIC KETONES

(R in the product is the group R in the aldehyde RCHO.)

Ketone and Aldehyde, RCHO	Catalyst	Product(s) (Yield, %)	Refs.
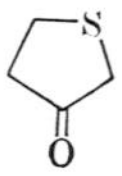 [structure: S, O] **and**			
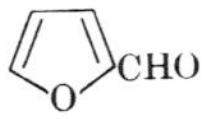 [structure: O] CHO	NaOH	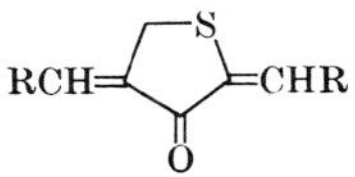 RCH= [structure: S, O] =CHR (99)	2248
C_6H_5CHO	"	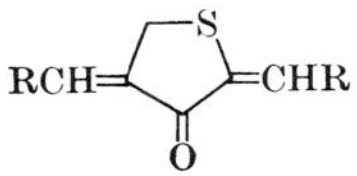 RCH= [structure: S, O] =CHR (46)	2248
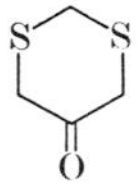 [structure: S, S, O] **and** C_6H_5CHO	$(CH_2)_5NH$	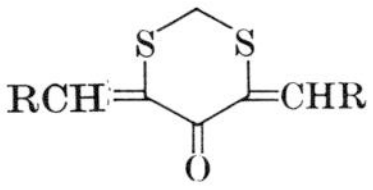 RCH= [structure: S, S, O] =CHR (78)	2249
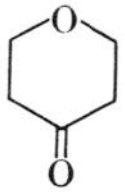 [structure: O, O] **and**			
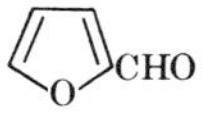 [structure: O] CHO	NaOH	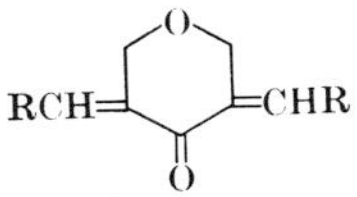 RCH= [structure: O, O] =CHR (92)	2250
		A, R = 2-(C_4H_3O)	
4-$O_2NC_6H_4CHO$	$(CH_2)_5NH$, CH_3CO_2H	A, R = 4-$O_2NC_6H_4$ (60)	599
C_6H_5CHO	NaOH	A, R = C_6H_5 (70)	2250, 599, 2251

 CHO	"	A, R = C_6H_{11} (—)	2250
3,4-$(CH_2O_2)C_6H_3CHO$	$(CH_2)_5NH$, CH_3CO_2H	A, R = 3,4-$(CH_2O_2)C_6H_3$ (55)	599
3-$CH_3C_6H_4CHO$	"	A, R = 3-$CH_3C_6H_4$ (33)	599
4-$CH_3C_6H_4CHO$	"	A, R = 4-$CH_3C_6H_4$ (50)	599, 2250
4-$CH_3OC_6H_4CHO$	"	A, R = 4-$CH_3OC_6H_4$ (42)	599, 2251
$C_6H_5CH{=}CHCHO$	NaOH	A, R = $C_6H_5CH{=}CH$ (65)	2250, 2251
4-$C_2H_5OC_6H_4CHO$	$(CH_2)_5NH$, CH_3CO_2H	A, R = 4-$C_2H_5OC_6H_4$ (56)	599
2,3-$(CH_3O)_2C_6H_3CHO$	"	A, R = 2,3-$(CH_3O)_2C_6H_3$ (50)	599
3,4-$(CH_3O)_2C_6H_3CHO$	"	A, R = 3,4-$(CH_3O)_2C_6H_3$ (58)	599
4-$(CH_3)_2NC_6H_4CHO$	"	A, R = 4-$(CH_3)_2NC_6H_4$ (42)	599
4-*i*-$C_3H_7C_6H_4CHO$	"	A, R = 4-*i*-$C_3H_7C_6H_4$ (69)	599
1-$C_{10}H_7CHO$	"	A, R = 1-$C_{10}H_7$ (58)	599
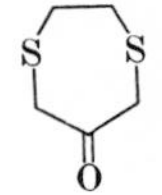 S, S, O and C_6H_5CHO	$(CH_2)_5NH$	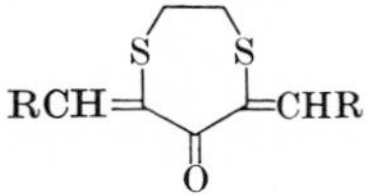 RCH=, S, S, =CHR, O (92)	2249
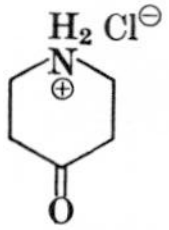 H_2 $Cl^{\ominus}$ N ⊕, O and C_6H_5CHO	HCl, CH_3CO_2H	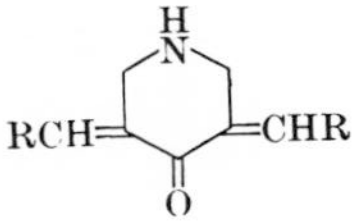 H N, RCH=, =CHR, O (58)	2252, 2253
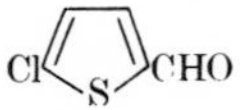 Br, S, $COCH_3$ and Cl, S, CHO	NaOH	 Br, S, COCH=CHR (—) B, R = 2-(ClC_4H_2S)	2017
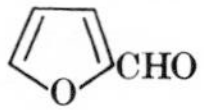 O, CHO	"	B, R = 2-(C_4H_3O) (—)	2017
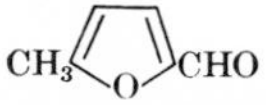 CH_3, O, CHO	"	B, R = 2-$(CH_3C_4H_2O)$ (—)	1988

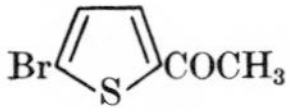

Note: References 668–2359 are on pp. 403–438.

TABLE XIX. Condensation of Aldehydes with Heterocyclic Ketones (*Continued*)
(R in the product is the group R in the aldehyde RCHO.)

Ketone and Aldehyde, RCHO	Catalyst	Product(s) (Yield, %)	Refs.
5-Br-2-thienyl-$COCH_3$ [Br–thiophene–$COCH_3$] (***contd.***) **and**			
$3,4\text{-}Cl_2C_6H_3CHO$	NaOH	B, R = $3,4\text{-}Cl_2C_6H_3$ (—)	2017
$4\text{-}ClC_6H_4CHO$	"	B, R = $4\text{-}ClC_6H_4$ (—)	2017
$4\text{-}FC_6H_4CHO$	"	B, R = $4\text{-}FC_6H_4$ (—)	2017
C_6H_5CHO	"	B, R = C_6H_5 (—)	2017
[2,5-dimethyl-3-thiophenecarboxaldehyde: CH_3, CH_3, CHO on thiophene]	"	B, R = 3-[$(CH_3)_2C_4HS$] (—)	2017
$4\text{-}CH_3OC_6H_4CHO$	"	B, R = $4\text{-}CH_3OC_6H_4$ (—)	2017
[2-benzofurancarboxaldehyde: benzofuran–CHO]	"	B, R = $2\text{-}(C_8H_5O)$ (75–95)	1854
$4\text{-}(CH_3)_2NC_6H_4CHO$	"	B, R = $4(CH_3)_2NC_6H_4$ (—)	2017
$1\text{-}C_{10}H_7CHO$	"	B, R = $1\text{-}C_{10}H_7$ (—)	2017
5-Cl-2-thienyl-$COCH_3$ [Cl–thiophene–$COCH_3$] **and**			
[5-nitro-2-furaldehyde: O_2N–furan–CHO]	H_2SO_4	[Cl–thiophene–]COCH=CHR **(—)** C, R = $2\text{-}(O_2NC_4H_2O)$	2144
[2-thiophenecarboxaldehyde: thiophene–CHO]	NaOH	C, R = $2\text{-}(C_4H_3S)$ (—)	2017
[2-furaldehyde: furan–CHO]	"	C, R = $2\text{-}(C_4H_3O)$ (—)	2017

[4-pyridyl]CHO (NC_5H_4CHO)	,,	C, R = 4-(C_5H_4N) (—)	1988
CH_3[thiophene, S]CHO	,,	C, R = 2-($CH_3C_4H_2S$) (—)	1988
CH_3[furan, O]CHO	,,	C, R = 2-($CH_3C_4H_2O$) (—)	1988
3,4-$Cl_2C_6H_3CHO$	,,	C, R = 3,4-$Cl_2C_6H_3$ (—)	2017
4-ClC_6H_4CHO	,,	C, R = 4-ClC_6H_4 (—)	2017
4-FC_6H_4CHO	,,	C, R = 4-FC_6H_4 (—)	2017
C_6H_5CHO	,,	C, R = C_6H_5 (—)	2017, 2254
[benzofuran, O]CHO	,,	C, R = 2-(C_8H_5O) (75–95)	1854
4-$(CH_3)_2NC_6H_4CHO$	,,	C, R = 4-$(CH_3)_2NC_6H_4$ (—)	2017
1-$C_{10}H_7CHO$	,,	C, R = 1-$C_{10}H_7$ (—)	2017
O_2N[thiophene, S]$COCH_3$ and			
O_2N[furan, O]CHO	H_2SO_4, CH_3CO_2H	O_2N[thiophene, S]COCH=CHR (32) D, R = 2-($O_2NC_4H_2O$)	1779
[furan, O]CHO	NaOH	D, R = 2-(C_4H_3O) (20)	1779
4-$O_2NC_6H_4CHO$	H_2SO_4, CH_3CO_2H	D, R = 2-$O_2NC_6H_4$ (90)	647
C_6H_5CHO	,,	D, R = C_6H_5 (60)	647
4-$CH_3OC_6H_4CHO$	,,	D, R = 4-$CH_3OC_6H_4$ (81)	647

Note: References 668–2359 are on pp. 403–438.

TABLE XIX. CONDENSATION OF ALDEHYDES WITH HETEROCYCLIC KETONES (*Continued*)

(R in the product is the group R in the aldehyde RCHO.)

Ketone and Aldehyde, RCHO	Catalyst	Product(s) (Yield, %)	Refs.
5-Nitro-2-thienyl-$COCH_3$ (*contd.*) and			
$C_6H_5CH{=}CHCHO$	H_2SO_4, CH_3CO_2H	D, R = $C_6H_5CH{=}CH$ (60)	647
2,4-$(CH_3O)_2C_6H_3CHO$	"	D, R = 2,4-$(CH_3O)_2C_6H_3$ (54)	647
2-Quinolyl-CHO	NaOH	D, R = 2-(C_9H_6N) (—)	2024
2,4,6-$(CH_3O)_3C_6H_2CHO$	H_2SO_4, CH_3CO_2H	D, R = 2,4,6-$(CH_3O)_3C_6H_2$ (50)	647
5-Nitro-2-furyl-$COCH_3$ and			
5-Nitro-2-thienyl-CHO	"	5-Nitro-2-furyl-COCH=CHR (51) E, R = 2-($O_2NC_4H_2S$)	1779
2-Thienyl-CHO	"	E, R = 2-(C_4H_3S) (70)	1779
4-$O_2NC_6H_4CHO$	"	E, R = 4-$O_2NC_6H_4$ (35)	646
C_6H_5CHO	"	E, R = C_6H_5 (50)	646
4-$CH_3OC_6H_4CHO$	"	E, R = 4-$CH_3OC_6H_4$ (33)	646
4-$O_2NC_6H_4CH{=}CHCHO$	"	E, R = 4-$O_2NC_6H_4CH{=}CH$ (45)	646
$C_6H_5CH{=}CHCHO$	"	E, R = $C_6H_5CH{=}CH$ (54)	646
2-Quinolyl-CHO	NaOH	E, R = 2-(C_9H_6N) (—)	2024
4-$CH_3OC_6H_4CH{=}CHCHO$	H_2SO_4, CH_3CO_2H	E, R = 4-$CH_3OC_6H_4CH{=}CH$ (25)	646
2,4,6-$(CH_3O)_3C_6H_2CHO$	"	E, R = 2,4,6-$(CH_3O)_3C_6H_2$ (50)	646

Reactants	Catalyst	Product (Yield, %)	Refs.
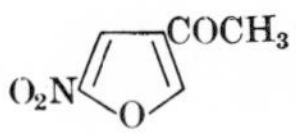 O_2N(furan, O)$COCH_3$ **and** 2,4-$(CH_3O)_2C_6H_3CHO$	,,	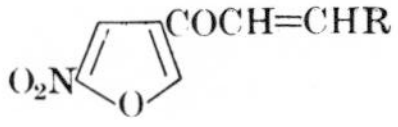 O_2N(furan, O)$COCH{=}CHR$ (30)	646
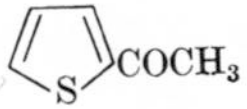 (thiophene, S)$COCH_3$ **and** $C(Cl_2){=}CClCHO$	H_2SO_4	 (thiophene, S)$COCH{=}CHR$ (80) F, R = $C(Cl_2){=}CCl$	289
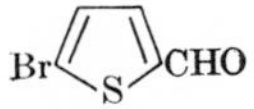 Br(thiophene, S)CHO	NaOH	F, R = 2-(BrC_4H_2S) (—)	2017
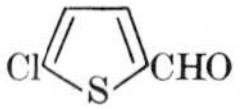 Cl(thiophene, S)CHO	,,	F, R = 2-(ClC_4H_2S) (—)	2017
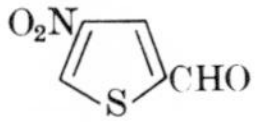 O_2N(thiophene, S)CHO	,,	F, R = 2-$(O_2NC_4H_2S)$ (55)	2034, 2035
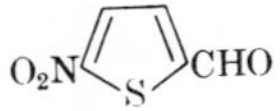 O_2N(thiophene, S)CHO	$NaOCH_3$	F, R = 2-$(O_2NC_4H_2S)$ (10)	2034
,,	H_2SO_4, CH_3CO_2H	F, R = 2-$(O_2NC_4H_2S)$ (20)	2034, 2035, 2144
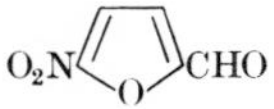 O_2N(furan, O)CHO	,,	F, R = 2-$(O_2NC_4H_2O)$ (12)	1779
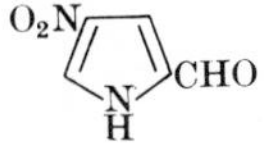 O_2N(pyrrole, NH)CHO	NaOH	F, R = 2-$(O_2NC_4H_2NH)$ (66)	2034, 2035
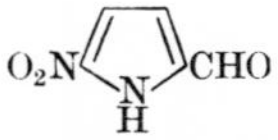 O_2N(pyrrole, NH)CHO	,,	F, R = 2-$(O_2NC_4H_2NH)$ (57)	2034, 2035

Note: References 668–2359 are on pp. 403–438.

TABLE XIX. CONDENSATION OF ALDEHYDES WITH HETEROCYCLIC KETONES (*Continued*)

(R in the product is the group R in the aldehyde RCHO.)

Ketone and Aldehyde, RCHO	Catalyst	Product(s) (Yield, %)	Refs.
2-Thienyl-$COCH_3$ (**_contd._**) and			
2-Thiophenecarboxaldehyde (thiophene-2-CHO)	NaOH	F, R = 2-(C_4H_3S) (90)	1845, 1030, 1520, 1823, 2035, 2255
Selenophene-2-CHO	$NaOCH_3$	F, R = 2-(C_4H_3Se) (56)	1915, 2351
Furan-2-CHO	KOH	F, R = 2-(C_4H_3O) (90)	1779, 1520, 1823, 2254, 2255
Pyrrole-2-CHO (N–H)	NaOH	F, R = 2-(C_4H_3NH) (83)	2034, 2035
OHC-(thiophene-2,5-diyl)-CHO	K_2CO_3	2-Thienyl-COCH=CH-(thiophene-2,5-diyl)-CH=CHCO-2-thienyl* (—)	365
OHC-(furan-2,5-diyl)-CHO	Na_2CO_3	2-Thienyl-COCH=CH-(furan-2,5-diyl)-CH=CHCO-2-thienyl* (76)	1921
5-CH_3-thiophene-2-CHO	NaOH	F, R = 2-($CH_3C_4H_2S$) (—)	2017
5-CH_3-furan-2-CHO	..	F, R = 2-($CH_3C_4H_2O$) (—)	1988
2,4-$Cl_2C_6H_3CHO$	..	F, R = 2,4-$Cl_2C_6H_3$ (82–95)	2254
3,4-$Cl_2C_6H_3CHO$	..	F, R = 3,4-$Cl_2C_6H_3$ (82–95)	2254, 2017
2-ClC_6H_4CHO	..	F, R = 2-ClC_6H_4 (82–95)	2254

4-ClC_6H_4CHO	"	F, R = 4-ClC_6H_4 (82–95)	2254
3-$O_2NC_6H_4CHO$	"	F, R = 3-$O_2NC_6H_4$ (93)	2023
4-$O_2NC_6H_4CHO$	"	F, R = 4-$O_2NC_6H_4$ (—)	2017, 1793
C_6H_5CHO	"	F, R = C_6H_5 (82–95)	2254, 1030, 1520, 1914, 1915, 2255, 2256
[thiophene ring, S]CH=CHCHO	"	F, R = 2-(C_4H_2S)CH=CH (—)	1793
[furan ring, O]CH=CHCHO	"	F, R = 2-(C_4H_2O)CH=CH (91)	1779
3-HOC_6H_4CHO	$NaOCH_3$	F, R = 3-HOC_6H_4 (—)	2257
4-HOC_6H_4CHO	NaOH	F, R = 4-HOC_6H_4 (32)	2258
3-$CF_3C_6H_4CHO$	"	F, R = 3-$CF_3C_6H_4$ (82–95)	2254
1,3-$C_6H_4(CHO)_2$	"	1,3-([thiophene ring, S]COCH=CH)$_2C_6H_4$ * (80)	1956
1,4-$C_6H_4(CHO)_2$	"	1,4-([thiophene ring, S]COCH=CH)$_2C_6H_4$ * (76)	1956, 1958
2-$CH_3OC_6H_4CHO$	"	F, R = 2-$CH_3OC_6H_4$ (83)	1914
4-$CH_3OC_6H_4CHO$	"	F, R = 4-$CH_3OC_6H_4$ (86)	1914, 1793
[benzofuran ring, O]CHO	"	F, R = 2-(C_8H_5O) (75–95)	1854
4-$O_2NC_6H_4CH{=}CHCHO$	"	F, R = 4-$O_2NC_6H_4CH{=}CH$ (40)	647
$C_6H_5CH{=}CHCHO$	"	F, R = $C_6H_5CH{=}CH$ (—)	1793

Note: References 668–2359 are on pp. 403–438.

* This is a complete structural formula.

TABLE XIX. CONDENSATION OF ALDEHYDES WITH HETEROCYCLIC KETONES (*Continued*)

(R in the product is the group R in the aldehyde RCHO.)

Ketone and Aldehyde, RCHO	Catalyst	Product(s) (Yield, %)	Refs.
2-Thienyl-$COCH_3$ (***contd.***) **and**			
2-Furyl-$(CH{=}CH)_2CHO$	NaOH	F, R = 2-$(C_4H_3O)CH{=}CHCH{=}CH$ (85)	1779
2,4-$(CH_3O)_2C_6H_3CHO$	"	F, R = 2,4-$(CH_3O)_2C_6H_3Cl$ (77)	1008
3,4-$(CH_3O)_2C_6H_3CHO$	"	F, R = 3,4-$(CH_3O)_2C_6H_3$ (—)	2017
Quinoline-2-CHO	"	F, R = 2-(C_9H_6N) (—)	2024
4-$CH_3OC_6H_4CH{=}CHCHO$	"	F, R = 4-$CH_3OC_6H_4CH{=}CH$ (—)	1793
2,4,6-$(CH_3O)_3C_6H_2CHO$	"	F, R = 2,4,6-$(CH_3O)_3C_6H_2$ (95)	1008
$(CH_3)_2C{=}CHCH_2CH_2C(CH_3){=}CHCHO$	"	F, R = $(CH_3)_2C{=}CHCH_2CH_2C(CH_3){=}CH$ (69)	1773
1-$C_{10}H_7CHO$	"	F, R = 1-$C_{10}H_7$ (—)	2017
Ferrocenecarboxaldehyde	"	F, R = $C_5H_5FeC_5H_4$ (69)	1975
2,4-$(CH_3O)_2C_6H_3CH{=}CHCHO$	"	F, R = 2,4-$(CH_3O)_2C_6H_3CH{=}CH$ (83)	1008
2-CH_3-4-HO-5-*i*-$C_3H_7C_6H_2CHO$	"	F, R = 2-CH_3-4-HO-5-*i*-$C_3H_7C_6H_2$ (75)	2105
2-CH_3-4-CH_3O-5-*i*-$C_3H_7C_6H_2CHO$	"	F, R = 2-CH_3-4-CH_3O-5-*i*-$C_3H_7C_6H_2$ (75)	2105
2-Selenienyl-$COCH_3$ **and**			
2-Thienyl-CHO	$NaOCH_3$	2-Selenienyl-$COCH{=}CHR$ (51)	1915, 2351
		G, R = 2-(C_4H_3S)	
2-Selenienyl-CHO	NaOH	G, R = 2-(C_4H_3Se) (—)	2351
OHC-2,5-furyl-CHO	Na_2CO_3	2-Selenienyl-$COCH{=}CH$-(2,5-furyl)-$CH{=}CHCO$-2-selenienyl* (78)	1921
C_6H_5CHO	NaOH	G, R = C_6H_5 (83)	366

$4\text{-}ClC_6H_4CHO$	,,	G, R = $4\text{-}ClC_6H_4$ (72)	366
$4\text{-}O_2NC_6H_4CHO$	,,	G, R = $4\text{-}O_2NC_6H_4$ (100)	366
$4\text{-}CH_3C_6H_4CHO$	,,	G, R = $4\text{-}CH_3C_6H_4$ (66)	366
$4\text{-}CH_3OC_6H_4CHO$	,,	G, R = $4\text{-}CH_3OC_6H_4$ (73)	366
$4\text{-}(CH_3)_2NC_6H_4CHO$	,,	G, R = $4\text{-}(CH_3)_2NC_6H_4$ (30)	366
$2,4\text{-}(CH_3O)_2C_6H_3CHO$	,,	G, R = $2,4\text{-}(CH_3O)_2C_6H_3$ (100)	366
$2,4,6\text{-}(CH_3O)_3C_6H_2CHO$	,,	G, R = $2,4,6\text{-}(CH_3O)_3C_6H_2$ (61)	366
$4\text{-}C_6H_5C_6H_4CHO$	,,	G, R = $4\text{-}C_6H_5C_6H_4$ (89)	366
2-(C_4H_3O)$COCH_3$ and			
5-O_2N-2-(C_4H_2S)CHO	,,	2-(C_4H_3O)COCH=CHR (24) H, R = 2-($O_2NC_4H_2S$)	1779
2-(C_4H_3S)CHO	,,	H, R = 2-(C_4H_3S) (70)	1779, 2255
2-(C_4H_3O)CHO	,,	H, R = 2-(C_4H_3O) (92)	1794, 2255
OHC-2,5-(C_4H_2O)-CHO	Na_2CO_3	2-(C_4H_3O)COCH=CH-2,5-(C_4H_2O)-CH=CHCO-2-(C_4H_3O)* (71)	1921
$3\text{-}O_2NC_6H_4CHO$	NaOH	H, R = $3\text{-}O_2NC_6H_4$ (92)	2023
$4\text{-}O_2NC_6H_4CHO$	,,	H, R = $4\text{-}O_2NC_6H_4$ (83)	646
C_6H_5CHO	,,	H, R = C_6H_5 (86)	1794, 1535, 1919, 2255
2-(C_4H_3S)CH=CHCHO	,,	H, R = 2-(C_4H_3S)CH=CH (93)	1779
$3\text{-}HOC_6H_4CHO$	$NaOCH_3$	H, R = $3\text{-}HOC_6H_4$ (—)	2257

Note: References 668–2359 are on pp. 403–438.

* This is a complete structural formula.

TABLE XIX. CONDENSATION OF ALDEHYDES WITH HETEROCYCLIC KETONES (*Continued*)

(R in the product is the group R in the aldehyde RCHO.)

Ketone and Aldehyde, RCHO	Catalyst	Product(s) (Yield, %)	Refs.
2-Furyl-$COCH_3$ (***contd.***) **and**			
2-Furyl-$CH{=}CHCHO$	NaOH	H, R = 2-$(C_4H_3O)CH{=}CH$ (80)	1794, 1535
1,3-$C_6H_4(CHO)_2$	"	1,3-(2-Furyl-$COCH{=}CH)_2C_6H_4$* (57)	1956
1,4-$C_6H_4(CHO)_2$	"	1,4-(2-Furyl-$COCH{=}CH)_2C_6H_4$* (51)	1956, 1958
2-$CH_3OC_6H_4CHO$	"	H, R = 2-$CH_3OC_6H_4$ (—)	1918
4-$CH_3OC_6H_4CHO$	"	H, R = 4-$CH_3OC_6H_4$ (92)	1794, 1918
3-$O_2NC_6H_4CH{=}CHCHO$	"	H, R = 3-$O_2NC_6H_4CH{=}CH$ (51)	646
$C_6H_5CH{=}CHCHO$	"	H, R = $C_6H_5CH{=}CH$ (96)	1794
2,4-$(CH_3O)_2C_6H_3CHO$	"	H, R = 2,4-$(CH_3O)_2C_6H_3$ (90)	1794
2-Quinolyl-CHO	"	H, R = 2-(C_9H_6N) (—)	2024
4-$CH_3OC_6H_4CH{=}CHCHO$	"	H, R = 4-$CH_3OC_6H_4CH{=}CH$ (80)	1794
$C_6H_5(CH{=}CH)_2CHO$	"	H, R = $C_6H_5(CH{=}CH)_2$ (35)	2260
2,4-$(CH_3O)_2C_6H_3CH{=}CHCHO$	"	H, R = 2,4-$(CH_3O)_2C_6H_3CH{=}CH$ (78)	1794
2-Pyrryl-$COCH_3$ **and**			

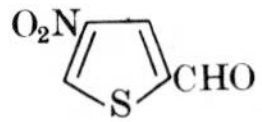 O_2N, S, CHO	..	 (pyrrole ring, NH)COCH=CHR (18) I, R = 2-($O_2NC_4H_2S$)	2034, 2035
	$NaOCH_3$	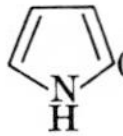 (pyrrole ring, NH)$COCH_2CHOHR$ (20)	2034
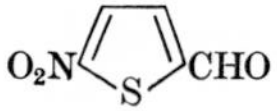 O_2N, S, CHO	NaOH	I, R = 2-($O_2NC_4H_2S$) (—)	2035
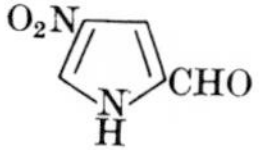 O_2N, NH, CHO	..	I, R = 2-($O_2NC_4H_2NH$) (20)	2034, 2035
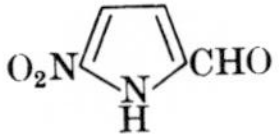 O_2N, NH, CHO	..	I, R = 2-($O_2NC_4H_2NH$) (52)	2034, 2035
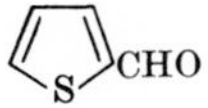 S, CHO	..	I, R = 2-(C_4H_3S) (55)	2034, 2035
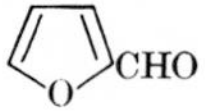 O, CHO	..	I, R = 2-(C_4H_3O) (100)	2259, 2034
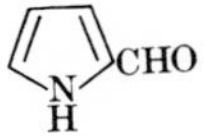 NH, CHO	..	I, R = 2-(C_4H_3NH) (40)	364, 2034, 2035, 2355
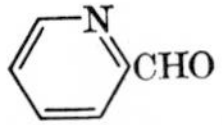 N, CHO	..	I, R = 2-(C_5H_4N) (50)	2034

Note: References 668–2359 are on pp. 403–438.

* This is a complete structural formula.

TABLE XIX. CONDENSATION OF ALDEHYDES WITH HETEROCYCLIC KETONES (*Continued*)

(R in the product is the group R in the aldehyde RCHO.)

Ketone and Aldehyde, RCHO	Catalyst	Product(s) (Yield, %)	Refs.
2-Pyrrolyl-$COCH_3$ (***contd.***) **and**			
3-Pyridyl-CHO	NaOH	I, R = 3-(C_5H_4N) (40)	2034
2-ClC_6H_4CHO	,,	I, R = 2-ClC_6H_4 (91)	1781
3-$O_2NC_6H_4CHO$	,,	I, R = 3-$O_2NC_6H_4$ (73)	1781
4-$O_2NC_6H_4CHO$	,,	I, R = 4-$O_2NC_6H_4$ (90)	1781
C_6H_5CHO	KOH	I, R = C_6H_5 (52)	364
2-Furyl-CH=CHCHO	Na_2CO_3	I, R = 2-(C_4H_3O)CH=CH (81)	1766
3,4-(CH_2O_2)C_6H_3CHO	KOH	I, R = 3,4-(CH_2O_2)C_6H_3 (75)	364
4-$CH_3C_6H_4CHO$	NaOH	I, R = 4-$CH_3C_6H_4$ (57)	1781
4-$CH_3OC_6H_4CHO$	,,	I, R = 4-$CH_3OC_6H_4$ (72)	1781
2-Furyl-CH=C(CH_3)CHO	Na_2CO_3	I, R = 2-(C_4H_3O)CH=C(CH_3) (50)	1766
C_6H_5CH=CHCHO	NaOH	I, R = C_6H_5CH=CH (—)	2259
2-Furyl-(CH=CH)$_2$CHO	Na_2CO_3	I, R = 2-(C_4H_3O)CH=CHCH=CH (72)	1766
4-(CH_3)$_2NC_6H_4CHO$	NaOH	I, R = 4-(CH_3)$_2NC_6H_4$ (60)	1781
Ferrocenecarboxaldehyde	,,	I, R = $C_5H_5FeC_5H_4$ (74)	1975
2,2-(CH_3)$_2$-tetrahydrofuran-3-one **and**			
i-C_3H_7CHO	,,	2,2-(CH_3)$_2$-5-(=CHR)-tetrahydrofuran-3-one (35)	2261, 2262

C_6H_5CHO	"	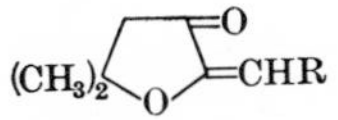 [structure: $(CH_3)_2$, O, =O, =CHR] (50)	2263
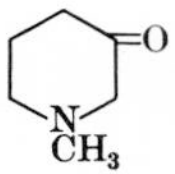 [structure: =O, N, CH_3] and C_6H_5CHO	KOH	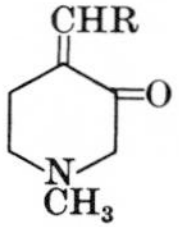 [structure: CHR, =O, N, CH_3] (25)	2264
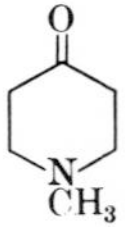 [structure: O, N, CH_3] and			
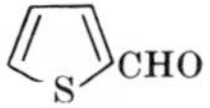 [structure: S] CHO	"	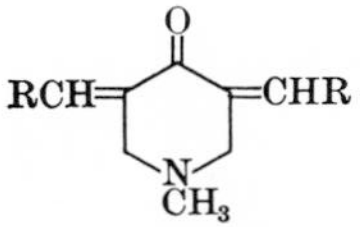 [structure: RCH=, O, =CHR, N, CH_3] (—) J, R = 2-(C_4H_3S)	2265
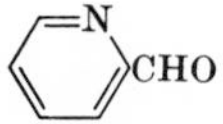 [structure: N] CHO	"	J, R = 2-(C_5H_4N) (—)	2265
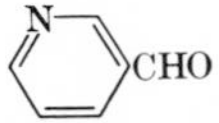 [structure: N] CHO	"	J, R = 3-(C_5H_4N) (—)	2265
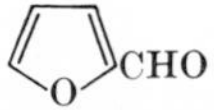 [structure: O] CHO	"	J, R = 2-(C_4H_3O) (—)	2265
2-ClC_6H_4CHO	"	J, R = 2-ClC_6H_4 (90)	2266, 2267
4-ClC_6H_4CHO	None, C_2H_5OH, rfx.	J, R = 4-ClC_6H_4 (89)	2267
4-FC_6H_4CHO	KOH	J, R = 4-FC_6H_4 (—)	2265
3-$O_2NC_6H_4CHO$	None, C_2H_5OH, rfx.	J, R = 3-$O_2NC_6H_4$ (82)	2267

Note: References 668–2359 are on pp. 403–438.

TABLE XIX. CONDENSATION OF ALDEHYDES WITH HETEROCYCLIC KETONES (*Continued*)

(R in the product is the group R in the aldehyde RCHO.)

Ketone and Aldehyde, RCHO	Catalyst	Product(s) (Yield, %)	Refs.
O, N, CH_3 (*contd.*) and			
$4\text{-}O_2NC_6H_4CHO$	None, C_2H_5OH, rfx.	J, R = $4\text{-}O_2NC_6H_4$ (100)	2267, 2268
C_6H_5CHO	KOH	J, R = C_6H_5 (44), O, CHOHR, N, CH_3 (36)	2266–2268
	,,	H, O, O, RCH=, NCH_3 †, N, CH_3, CHR (70)	2269
	,,	O, =CHR, N, CH_3 (61)	2266
$3,4\text{-}(CH_2O_2)C_6H_3CHO$	,,	J, R = $3,4\text{-}(CH_2O_2)C_6H_3$ (—)	2265
$2\text{-}CH_3C_6H_4CHO$	None, C_2H_5OH, rfx.	J, R = $2\text{-}CH_3C_6H_4$ (69)	2267
$4\text{-}CH_3C_6H_4CHO$	,,	J, R = $4\text{-}CH_3C_6H_4$ (63)	2267, 2265
$4\text{-}CH_3OC_6H_4CHO$	,,	J, R = $4\text{-}CH_3OC_6H_4$ (66)	2267, 2265
$C_6H_5CH{=}CHCHO$	KOH	J, R = $C_6H_5CH{=}CH$ (—)	2265
$4\text{-}CH_3CONHC_6H_4CHO$	,,	J, R = $4\text{-}CH_3CONHC_6H_4$ (—)	2265
$4\text{-}(CH_3)_2NC_6H_4CHO$	None, C_2H_5OH, rfx.	J, R = $4\text{-}(CH_3)_2NC_6H_4$ (58)	2267, 2265

Reactants	Catalyst	Product(s) (Yield, %)	Refs.
Quinoline-2-carboxaldehyde (C_9H_6N-2-CHO)	KOH	J, R = 2-(C_9H_6N) (39), 3-(RCH=)-1-methyl-4-piperidone (1-CH_3, 3-=CHR) (40)	2266
Quinoline-4-carboxaldehyde (C_9H_6N-4-CHO)	”	3-(RCH=)-1-methyl-4-piperidone (1-CH_3, 3-=CHR) (95)‡	2266
	”	J, R = 4-(C_9H_6N) (79)§	2266
4-i-$C_3H_7C_6H_4CHO$	None, C_2H_5OH, rfx.	J, R = 4-i-$C_3H_7C_6H_4$ (36)	2267
1-$C_{10}H_7CHO$	KOH	J, R = 1-$C_{10}H_7$ (88)	2266, 2265
4-$(CH_3)_2NC_6H_4CHO$	”	J, R = 4-$(CH_3)_2NC_6H_4$ (—)	2265
4-Methoxy-1-naphthaldehyde (OCH_3, CHO on naphthalene)	”	J, R = 1-($CH_3OC_{10}H_6$) (—)	2265
4-$C_6H_5C_6H_4CHO$	None, C_2H_5OH, rfx.	J, R = 4-$C_6H_5C_6H_4$ (—)	2267
Phenanthrene-9-carboxaldehyde	KOH	J, R = 9-$C_{14}H_9$ (—)	2265
2-Acetylpyridine (C_5H_4N-2-$COCH_3$) and			
Furfural (C_4H_3O-2-CHO)	$(C_2H_5)_2NH$	2-Pyridyl-COCH=CHR (—), K, R = 2-(C_4H_3O)	2270

Note: References 668–2359 are on pp. 403–438.

† This structure was suggested but not established.
‡ This was obtained with a 1:1 molar ratio of reactants.
§ This was obtained with a 2:1 molar ratio of aldehyde to ketone.

TABLE XIX. Condensation of Aldehydes with Heterocyclic Ketones (*Continued*)

(R in the product is the group R in the aldehyde RCHO.)

Ketone and Aldehyde, RCHO	Catalyst	Product(s) (Yield, %)	Refs.
2-Pyridyl-$COCH_3$ (*contd.*) and			
2-Pyridyl-CHO	NaOH	K, R = 2-(C_5H_4N) (40)	545
2,4-$Cl_2C_6H_3CHO$	,,	K, R = 2,4-$Cl_2C_6H_3$ (89)	534
2,5-$Cl_2C_6H_3CHO$	,,	K, R = 2,5-$Cl_2C_6H_3$ (89)	534
3,4-$Cl_2C_6H_3CHO$	,,	K, R = 3,4-$Cl_2C_6H_3$ (89)	534
2-ClC_6H_4CHO	,,	K, R = 2-ClC_6H_4 (99)	534
3-ClC_6H_4CHO	,,	K, R = 3-ClC_6H_4 (99)	534
2-Cl-4-HOC_6H_3CHO	,,	K, R = 2-Cl-4-HOC_6H_3 (95)	534
2-$O_2NC_6H_4CHO$	,,	2-Pyridyl-$COCH_2CHOHR$ (—)	2271
3-$O_2NC_6H_4CHO$	$NaOCH_3$	K, R = 3-$O_2NC_6H_4$ (96)	2004, 534
4-$O_2NC_6H_4CHO$	NaOH	K, R = 4-$O_2NC_6H_4$ (94)	534, 2270
C_6H_5CHO	,,	K, R = C_6H_5 (99)	534, 545, 2271
2-HOC_6H_4CHO	,,	K, R = 2-HOC_6H_4 (97)	534, 2270
4-HOC_6H_4CHO	,,	K, R = 4-HOC_6H_4 (90)	534
4-$HO_2CC_6H_4CHO$	,,	K, R = 4-$HO_2CC_6H_4$ (98)	534
1,3-$C_6H_4(CHO)_2$	,,	1,3-(2-Pyridyl-$COCH{=}CH)_2C_6H_4$* (58)	1956
1,4-$C_6H_4(CHO)_2$	,,	1,4-(2-Pyridyl-$COCH{=}CH)_2C_6H_4$* (68)	1956
3,4-$(CH_2O_2)C_6H_3CHO$	,,	K, R = **3,4-$(CH_2O_2)C_6H_3$** (91)	534, 2270

Reactants	Catalyst	Product (Yield, %)	Refs.
$2\text{-}CH_3OC_6H_4CHO$	,,	K, R = $2\text{-}CH_3OC_6H_4$ (98)	534
$4\text{-}CH_3OC_6H_4CHO$	,,	K, R = $4\text{-}CH_3OC_6H_4$ (91)	534, 2270
$2,3\text{-}(CH_3O)_2C_6H_3CHO$	,,	K, R = $2,3\text{-}(CH_3O)_2C_6H_3$ (98)	534
$2,4\text{-}(CH_3O)_2C_6H_3CHO$	,,	K, R = $2,4\text{-}(CH_3O)_2C_6H_3$ (98)	534
$2,5\text{-}(CH_3O)_2C_6H_3CHO$	,,	K, R = $2,5\text{-}(CH_3O)_2C_6H_3$ (98)	534
$3,4\text{-}(CH_3O)_2C_6H_3CHO$	,,	K, R = $3,4\text{-}(CH_3O)_2C_6H_3$ (92)	534, 2270
$4\text{-}(CH_3)_2NC_6H_4CHO$	,,	K, R = $4\text{-}(CH_3)_2NC_6H_4$ (60)	534
Ferrocenecarboxaldehyde	,,	K, R = $C_5H_5FeC_5H_4$ (45)	1975
3-Pyridyl-$COCH_3$ and			
2-Furyl-CHO	,,	3-Pyridyl-COCH=CHR (95) L, R = $2\text{-}(C_4H_3O)$	2272, 2270
$4\text{-}O_2NC_6H_4CHO$	$(C_2H_5)_2NH$	L, R = $4\text{-}O_2NC_6H_4$ (—)	2270
C_6H_5CHO	NaOH	L, R = C_6H_5 (22)	534, 2270
$2\text{-}HOC_6H_4CHO$	$(C_2H_5)_2NH$	L, R = $2\text{-}HOC_6H_4$ (—)	2270
$3\text{-}HOC_6H_4CHO$	NaOH	L, R = $3\text{-}HOC_6H_4$ (97)	534
$4\text{-}HOC_6H_4CHO$	,,	L, R = $4\text{-}HOC_6H_4$ (50)	534
$2\text{-}HO_2CC_6H_4CHO$	,,	L, R = $2\text{-}HO_2CC_6H_4$ (98)	534
$4\text{-}CH_3OC_6H_4CHO$	$(C_2H_5)_2NH$	L, R = $4\text{-}CH_3OC_6H_4$ (—)	2270
$3\text{-}CH_3O\text{-}4\text{-}HOC_6H_3CHO$	,,	L, R = $3\text{-}CH_3O\text{-}4\text{-}HOC_6H_3$ (—)	2270
$3,4\text{-}(CH_3O)_2C_6H_3CHO$	,,	L, R = $3,4\text{-}(CH_3O)_2C_6H_3$ (—)	2270
4-Pyridyl-$COCH_3$ and			
2-Furyl-CHO	NaOH	4-Pyridyl-COCH=CHR (77) M, R = $2\text{-}(C_4H_3O)$	2272, 2270

Note: References 668–2359 are on pp. 403–438.

* This is a complete structural formula.

TABLE XIX. CONDENSATION OF ALDEHYDES WITH HETEROCYCLIC KETONES (*Continued*)

(R in the product is the group R in the aldehyde RCHO.)

Ketone and Aldehyde, RCHO	Catalyst	Product(s) (Yield, %)	Refs.
3-Acetylpyridine (**contd.**) **and**			
3-ClC_6H_4CHO	NaOH	M, R = 3-ClC_6H_4 (88)	534
4-ClC_6H_4CHO	"	M, R = 4-ClC_6H_4 (99)	534
$2\text{-Cl-4-HOC}_6H_3CHO$	"	M, R = 2-Cl-4-HOC_6H_3 (98)	534
$2\text{-O}_2NC_6H_4CHO$	"	M, R = $2\text{-O}_2NC_6H_4$ (60)	534
$4\text{-O}_2NC_6H_4CHO$	"	M, R = $4\text{-O}_2NC_6H_4$ (60)	534
C_6H_5CHO	"	M, R = C_6H_5 (99)	534, 545, 2270
2-HOC_6H_4CHO	"	M, R = 2-HOC_6H_4 (90)	534, 2270
3-HOC_6H_4CHO	"	M, R = 3-HOC_6H_4 (97)	534
4-HOC_6H_4CHO	"	M, R = 4-HOC_6H_4 (86)	534
$1,4\text{-C}_6H_4(CHO)_2$	"	$1,4\text{-}(3\text{-pyridyl-COCH=CH})_2C_6H_4$* (33)	1956
$2\text{-HO}_2CC_6H_4CHO$	"	M, R = $2\text{-HO}_2CC_6H_4$ (98)	534
$3,4\text{-(CH}_2O_2)C_6H_3CHO$	"	M, R = $3,4\text{-(CH}_2O_2)C_6H_3$ (60)	534, 2270
$4\text{-CH}_3OC_6H_4CHO$	"	M, R = $4\text{-CH}_3OC_6H_4$ (99)	534, 2270
$3\text{-CH}_3O\text{-4-HOC}_6H_3CHO$	$(C_2H_5)_2NH$	M, R = $3\text{-CH}_3O\text{-4-HOC}_6H_3$ (—)	2270
$4\text{-(CH}_3)_2NC_6H_4CHO$	NaOH	M, R = $4\text{-(CH}_3)_2NC_6H_4$ (60)	534, 2270
Quinoline-2-CHO	"	M, R = $2\text{-(C}_9H_6N)$ (—)	2024
3-Quinuclidinone **and**			
CH_2O	K_2CO_3, H_2O	2,2-bis(hydroxymethyl)-3-quinuclidinone, $(CH_2OH)_2$* (21–54), 2,2-bis(hydroxymethyl)-3-quinuclidinol, $(CH_2OH)_2$* (15)	450

Reactants	Catalyst	Product (Yield, %)	Refs.
	K_2CO_3, H_2O, CH_3OH	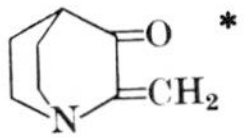 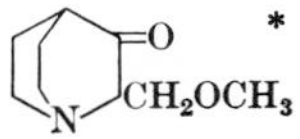 2-methylene-3-quinuclidinone (=O, =CH_2)* (54), 2-(methoxymethyl)-3-quinuclidinone (=O, CH_2OCH_3)* (16)	450
C_6H_5CHO	$(CH_2)_5NH$	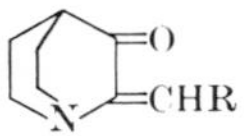 2-(=CHR)-3-quinuclidinone (=O, =CHR) (75)	2273, 2274
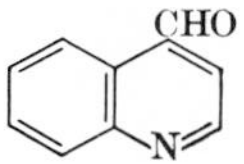 4-quinolinecarboxaldehyde (CHO, N)	$(CH_2)_5NH$, CH_3CO_2H	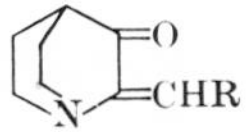 2-(=CHR)-3-quinuclidinone (=O, =CHR) (50)	2273
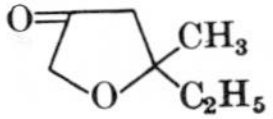 2-methyl-2-ethyltetrahydrofuran-3-one (O=, CH_3, C_2H_5) and			
CH_3CHO	NaOH	RCH= (4-oxo-5-CH_3-5-C_2H_5-tetrahydrofuran-2-ylidene) (38) N, R = CH_3	2261, 2262
C_2H_5CHO	"	N, R = C_2H_5 (17)	2261, 2262
n-C_3H_7CHO	"	N, R = n-C_3H_7 (22)	2261, 2262
i-C_3H_7CHO	"	N, R = i-C_3H_7 (35)	2261, 2262
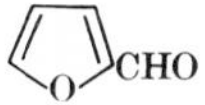 furfural (CHO)	"	N, R = 2-(C_4H_3O) (29)	2261, 2262
C_6H_5CHO	"	RCH= (ring with O=, =CHR, CH_3, C_2H_5) (33) O, R = C_6H_5	2261, 2262
n-$C_6H_{13}CHO$	"	N, R = n-C_6H_{13} (32)	2261, 2262
4-$CH_3OC_6H_4CHO$	"	N, R = 4-$CH_3OC_6H_4$ (46)‡	2261, 2262
	"	O, R = 4-$CH_3OC_6H_4$ (68)§	2261, 2262

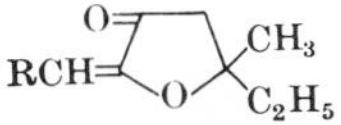

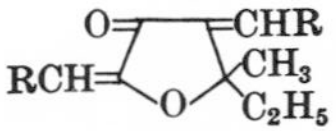

Note: References 668–2359 are on pp. 403–438.

* This is a complete structural formula.

‡ This was obtained with a 1:1 molar ratio of reactants.

§ This was obtained with a 2:1 molar ratio of aldehyde to ketone.

TABLE XIX. Condensation of Aldehydes with Heterocyclic Ketones (*Continued*)

(R in the product is the group R in the aldehyde RCHO.)

Ketone and Aldehyde, RCHO	Catalyst	Product(s) (Yield, %)	Refs.
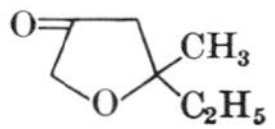 O= (ring) O, CH_3, C_2H_5 (***contd.***) **and**			
4-$(CH_3)_2NC_6H_4CHO$	NaOH	O, R = 4-$(CH_3)_2NC_6H_4$ (51)	2261, 2262
$(CH_3)_2C{=}CH(CH_2)_2CH(CH_3)CH_2CHO$	,,	N, R = $(CH_3)_2C{=}CH(CH_2)_2CH(CH_3)CH_2$ (49)	2261, 2262
4-$C_6H_5C_6H_4CHO$	,,	O, R = 4-$C_6H_5C_6H_4$ (55)	2261, 2262
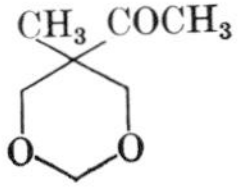 CH_3 $COCH_3$ (ring with O, O) **and** CH_2O	,,	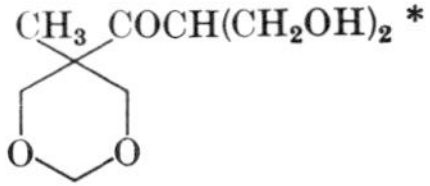 CH_3 $COCH(CH_2OH)_2$ * (ring with O, O) (—)	2275
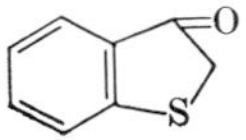 (ring with =O, S) **and** $C(Cl_2){=}CClCHO$	BF_3	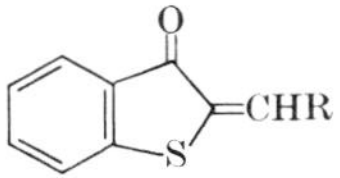 (ring with O, S) =CHR (80)	289
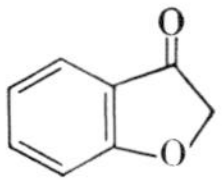 (ring with O, O) **and**			
3-CH_3O-4-HOC_6H_3CHO	HCl	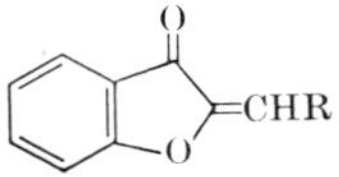 (ring with O, O) =CHR (55)	2052, 2342
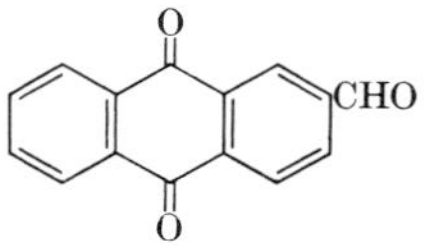 (ring with O, O) CHO	NaOH	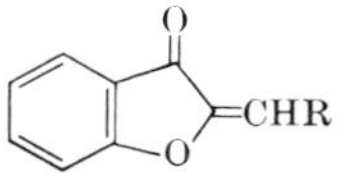 (ring with O, O) =CHR (—)	1977

Reactants	Catalyst	Product (Yield, %)	Refs.
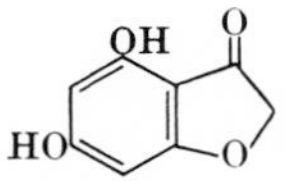 OH, O, HO, O (structure) and			
C_6H_5CHO	KOH	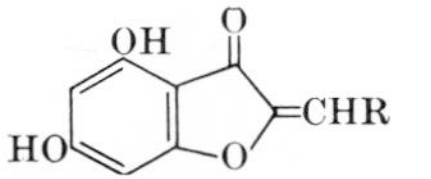 OH, O, HO, O (structure) $=CHR$ (—)	2165
		P, R = C_6H_5	
$2\text{-}CH_3OC_6H_4CHO$	,,	P, R = $2\text{-}CH_3OC_6H_4$ (—)	2276
$4\text{-}CH_3OC_6H_4CHO$	,,	P, R = $4\text{-}CH_3OC_6H_4$ (—)	2277
$3\text{-}CH_3O\text{-}4\text{-}HOC_6H_3CHO$	$(CH_3CO)_2O$, rfx.	P, R = $3\text{-}CH_3O\text{-}4\text{-}HOC_6H_3$‖ (60)	2278
$3\text{-}HO\text{-}4\text{-}CH_3OC_6H_3CHO$	,,	P, R = $3\text{-}HO\text{-}4\text{-}CH_3OC_6H_3$‖ (50)	2278
$3,4\text{-}(CH_3O)_2C_6H_3CHO$	KOH	P, R = $3,4\text{-}(CH_3O)_2C_6H_3$ (—)	2276
$3,4\text{-}(C_2H_5O)_2C_6H_3CHO$	NaOH	P, R = $3,4\text{-}(C_2H_5O)_2C_6H_3$ (—)	2159
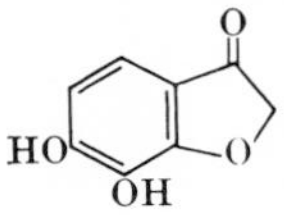 O, HO, O, OH (structure) and			
$3,4\text{-}(HO)_2C_6H_3CHO$	$(CH_3CO)_2O$	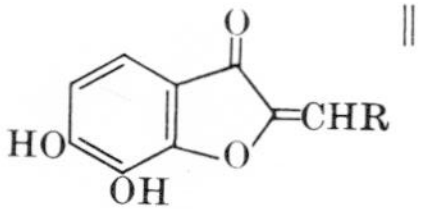 O, HO, O, OH (structure) $=CHR$ ‖ (—)	2279
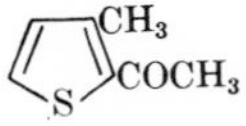 CH_3, S (structure) $COCH_3$ and C_6H_5CHO	NaOH	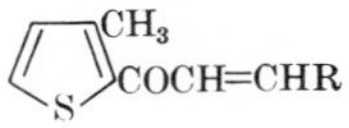 CH_3, S (structure) $COCH{=}CHR$ (82–95)	2254
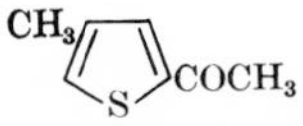 CH_3, S (structure) $COCH_3$ and C_6H_5CHO	,,	 CH_3, S (structure) $COCH{=}CHR$ (82–95)	2254

Note: References 668–2359 are on pp. 403–438.

* This is a complete structural formula.

‖ The product was initially isolated as a triacetate and was susbequently deacetylated.

TABLE XIX. CONDENSATION OF ALDEHYDES WITH HETEROCYCLIC KETONES (*Continued*)

(R in the product is the group R in the aldehyde RCHO.)

Ketone and Aldehyde, RCHO	Catalyst	Product(s) (Yield, %)	Refs.
5-CH_3-2-thienyl-$COCH_3$ and C_6H_5CHO	NaOH	5-CH_3-2-thienyl-$COCH{=}CHR$ (82–95)	2254
6-CH_3-3-pyridyl-$COCH_3$ and			
4-BrC_6H_4CHO	"	6-CH_3-3-pyridyl-$COCH{=}CHR$ (88) Q, R = 4-BrC_6H_4	367
2-ClC_6H_4CHO	"	Q, R = 2-ClC_6H_4 (88)	367
4-ClC_6H_4CHO	"	Q, R = 4-ClC_6H_4 (69)	367
2-$O_2NC_6H_4CHO$	"	Q, R = 2-$O_2NC_6H_4$ (27)	367
3-$O_2NC_6H_4CHO$	"	Q, R = 3-$O_2NC_6H_4$ (33)	367
4-$O_2NC_6H_4CHO$	"	Q, R = 4-$O_2NC_6H_4$ (78)	367
C_6H_5CHO	"	Q, R = C_6H_5 (49)	545
4-NCC_6H_4CHO	"	Q, R = 4-NCC_6H_4 (68)	367
4-$HO_2CC_6H_4CHO$	"	Q, R = 4-$HO_2CC_6H_4$ (44)	367
4-$(CH_3)_2NC_6H_4CHO$	"	Q, R = 4-$(CH_3)_2NC_6H_4$ (50)	367
4-$CH_3CONHC_6H_4CHO$	"	Q, R = 4-$CH_3CONHC_6H_4$ (68)	367
2,5-di(CH_3CO)pyrrole (CH_3CO–pyrrole(NH)–$COCH_3$) and 2-furyl-CHO	"	$RCH{=}CHCO$–pyrrole(NH)–$COCH{=}CHR$ (—)	2259
5-C_2H_5-2-thienyl-$COCH_3$ and 2-thienyl-CHO	"	5-C_2H_5-2-thienyl-$COCH{=}CHR$ (—)	2017
5-CH_3-3-CH_3-2-pyrrolyl(NH)-$COCH_3$ and			

[furan ring]CHO		[2,4-dimethylpyrrole ring: CH_3, CH_3, NH]$COCH{=}CHR$ (99)	2259
$C_6H_5CH{=}CHCHO$	"	[2,4-dimethylpyrrole ring: CH_3, CH_3, NH]$COCH{=}CHR$ (99)	2259
[pyrrole ring: CH_3, $COCH_3$, CH_3, NH] and [furan ring]CHO	"	[pyrrole ring: CH_3, $COCH{=}CHR$, CH_3, NH] (0)	2259
CH_3N[tropinone skeleton]$=O$ and			
C_6H_5CHO	KOH	CH_3N[tropinone skeleton: $=CHR$, $=O$, $=CHR$] (—)	2280
3,4-$(CH_2O_2)C_6H_3CHO$	"	CH_3N[tropinone skeleton: $=CHR$, $=O$, $=CHR$] (100)	2280
[tetrahydrofuran-3-one ring: $O{=}$, $(CH_3)_2$, O, $(CH_3)_2$] and			
i-C_3H_7CHO	"	[tetrahydrofuran-3-one ring: $O{=}$, $=CHR$, $(CH_3)_2$, O, $(CH_3)_2$] (27)	2281, 1355
		S, R = i-C_3H_7	
[thiophene ring]CHO	"	S, R = 2-(C_4H_3S) (100)	2282
[selenophene ring]CHO	"	S, R = 2-(C_4H_3Se) (100)	2282

Note: References 668–2359 are on pp. 403–438.

TABLE XIX. CONDENSATION OF ALDEHYDES WITH HETEROCYCLIC KETONES (*Continued*)

(R in the product is the group R in the aldehyde RCHO.)

Ketone and Aldehyde, RCHO	Catalyst	Product(s) (Yield, %)	Refs.
2,2,5,5-tetramethyltetrahydrofuran-3-one [$(CH_3)_2$, O, $(CH_3)_2$, C=O ring] (***contd.***) **and**			
furan-2-CHO	KOH	S, R = 2-(C_4H_3O) (85)	2281, 2282
5-CH_3-furan-2-CHO	"	S, R = 2-($CH_3C_4H_2O$) (100)	2282
C_6H_5CHO	"	S, R = C_6H_5 (97)	2283, 2284
2-$O_2NC_6H_4CHO$	"	S, R = 2-$O_2NC_6H_4$ (90)	2281
3-$O_2NC_6H_4CHO$	"	S, R = 3-$O_2NC_6H_4$ (43)	2281
2-HOC_6H_4CHO	"	S, R = 2-HOC_6H_4 (77)	2285
3,4-(CH_2O_2)C_6H_3CHO	"	S, R = 3,4-(CH_2O_2)C_6H_3 (95)	2281, 2284
2-$CH_3OC_6H_4CHO$	"	S, R = 2-$CH_3OC_6H_4$ (—)	2285
$C_6H_5CH{=}CHCHO$	"	S, R = $C_6H_5CH{=}CH$ (80)	2281
4-$(CH_3)_2NC_6H_4CHO$	"	S, R = 4-$(CH_3)_2NC_6H_4$ (74)	2281
CH_3-benzofuran-3(2H)-one **and**			
3-$O_2NC_6H_4CHO$	HCl, CH_3CO_2H	CH_3-2-(=CHR)-benzofuran-3(2H)-one (>70) T, R = 3-$O_2NC_6H_4$	339
4-HOC_6H_4CHO	"	T, R = 4-HOC_6H_4 (>70)	339
3,4-(CH_2O_2)C_6H_3CHO	KOH	T, R = 3,4-(CH_2O_2)C_6H_3 (—)	2276
2,4-$(CH_3)_2C_6H_3CHO$	HCl, CH_3CO_2H	T, R = 2,4-$(CH_3)_2C_6H_3$ (>70)	339
3,4-$(CH_3O)_2C_6H_3CHO$	KOH	T, R = 3,4-$(CH_3O)_2C_6H_3$ (—)	2276
2,4,6-$(CH_3)_3C_6H_2CHO$	HCl, CH_3CO_2H	T, R = 2,4,6-$(CH_3)_3C_6H_2$ (>70)	339
selenochroman-4-one (Se) **and** C_6H_5CHO	NaOH	3-(=CHR)-selenochroman-4-one (60)	2286

Reactants	Catalyst	Product (Yield, %)	Refs.
(structure) and			
C_6H_5CHO	HCl	(structure) =CHR (94) U, R = C_6H_5	2032, 2287–2289
3-HOC_6H_4CHO	"	U, R = 3-HOC_6H_4 (—)	2290
3,4-$(HO)_2C_6H_3CHO$	"	U, R = 3,4-$(HO)_2C_6H_3$ (45)	2291
3,4-$(CH_2O_2)C_6H_3CHO$	"	U, R = 3,4-$(CH_2O_2)C_6H_3$ (45)	2291
2-$CH_3OC_6H_4CHO$	"	U, R = 2-$CH_3OC_6H_4$ (82)	2032
3-$CH_3OC_6H_4CHO$	"	U, R = 3-$CH_3OC_6H_4$ (75)	2290
4-$CH_3OC_6H_4CHO$	"	U, R = 4-$CH_3OC_6H_4$ (88)	2032, 2291
3-CH_3O-4-HOC_6H_3CHO	"	U, R = 3-CH_3O-4-HOC_6H_3 (65)	2291
3,4-$(CH_3O)_2C_6H_3CHO$	"	U, R = 3,4-$(CH_3O)_2C_6H_3$ (99)	2032, 2290, 2291
CH_3O-(structure) and (structure)-CHO	NaOH	CH_3O-(structure) =CHR (—)	1977, 2343
(structure, OCH_3) and			
3-CH_3O-4-HOC_6H_3CHO	HCl	(structure, OCH_3) =CHR (77)	2109
3,4-$(CH_3O)_2C_6H_3CHO$	"	(structure, OCH_3) =CHR (—)	2109

Note: References 668–2359 are on pp. 403–438.

TABLE XIX. CONDENSATION OF ALDEHYDES WITH HETEROCYCLIC KETONES (*Continued*)

(R in the product is the group R in the aldehyde RCHO.)

Ketone and Aldehyde, RCHO	Catalyst	Product(s) (Yield, %)	Refs.
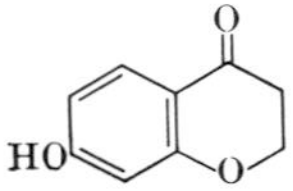 [structure] and			
$3,4\text{-}(HO)_2C_6H_3CHO$	HCl	[structure] (99) V, R = $3,4\text{-}(HO)_2C_6H_3$	2292, 2291
$3,4\text{-}(CH_2O_2)C_6H_3CHO$	"	V, R = $3,4\text{-}(CH_2O_2)C_6H_3$ (60)	2291
$4\text{-}CH_3OC_6H_4CHO$	$NaOC_2H_5$	V, R = $4\text{-}CH_3OC_6H_4$ (41)	2292
$3\text{-}CH_3O\text{-}4\text{-}HOC_6H_3CHO$	HCl	V, R = $3\text{-}CH_3O\text{-}4\text{-}HOC_6H_3$ (80)	2291
$3,4\text{-}(CH_3O)_2C_6H_3CHO$	$NaOC_2H_5$	V, R = $3,4\text{-}(CH_3O)_2C_6H_3$ (50)	2291
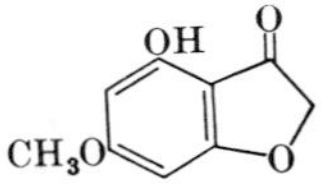 [structure] and $3,4\text{-}(HO)_2C_6H_3CHO$	$(CH_3CO)_2O$, rfx.	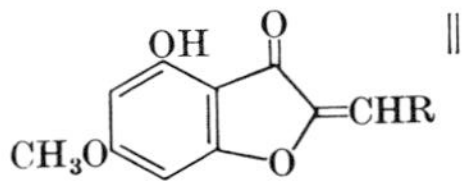 [structure] (66)	2278, 2293
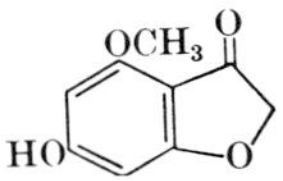 [structure] and $3,4\text{-}(HO)_2C_6H_3CHO$	"	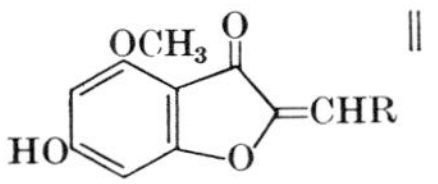 [structure] (43)	2278
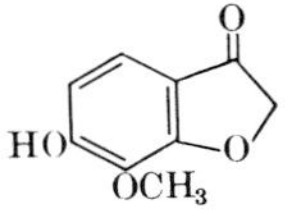 [structure] and $3,4\text{-}(HO)_2C_6H_3CHO$	$(CH_3CO)_2O$	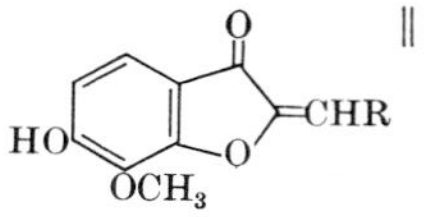 [structure] (—)	2294

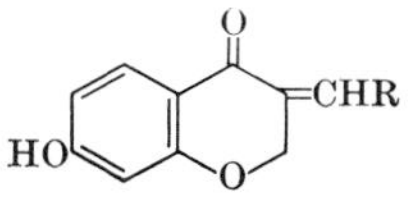

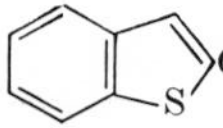

Reactants	Catalyst	Product(s) and Yield(s) (%)	Refs.
$COCH_3$ (S) and C_6H_5CHO	NaOH	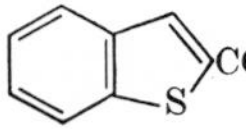 $COCH{=}CHR$ (S) (—)	1801
$COCH_3$ (S) and $3\text{-}O_2NC_6H_4CHO$	KOH	$COCH{=}CHR$ (S) (100) W, R = $3\text{-}O_2NC_6H_4$	2295
C_6H_5CHO	"	W, R = C_6H_5 (84)	2295
$3,4\text{-}(CH_2O_2)C_6H_3CHO$	"	W, R = $3,4\text{-}(CH_2O_2)C_6H_3$ (48)	2295
$2,3\text{-}(CH_3O)_2C_6H_3CHO$	"	W, R = $2,3\text{-}(CH_3O)_2C_6H_3$ (57)	2295
Ferrocenecarboxaldehyde	NaOH	W, R = $C_5H_5FeC_5H_4$ (25)	1975
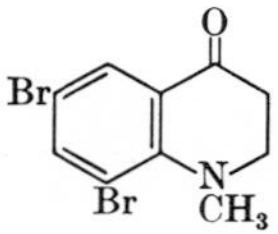 Br, O, Br, N, CH_3 and $4\text{-}(CH_3)_2NC_6H_4CHO$	$(CH_2)_5NH$	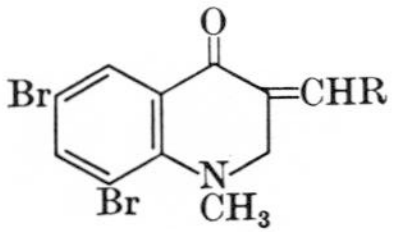 Br, O, $=CHR$, Br, N, CH_3 (—)	368
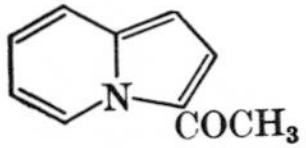 N, $COCH_3$ and C_6H_5CHO	NaOH	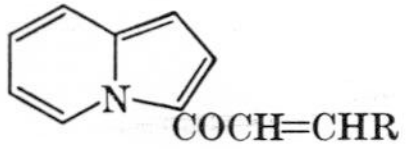 N, $COCH{=}CHR$ (—)	2296
 Br, O, N, CH_3 and $4\text{-}(CH_3)_2NC_6H_4CHO$	$(CH_2)_5NH$	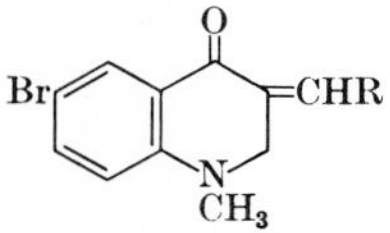 Br, O, $=CHR$, N, CH_3 (78)	368

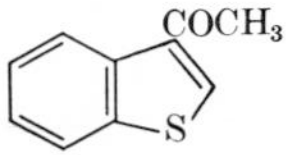

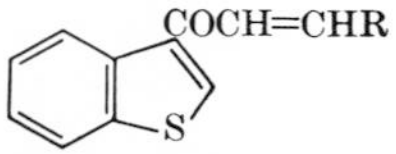

Note: References 668–2359 are on pp. 403–438.

‖ The product was initially isolated as a triacetate and was subsequently deacetylated.

TABLE XIX. Condensation of Aldehydes with Heterocyclic Ketones (*Continued*)

(R in the product is the group R in the aldehyde RCHO.)

Ketone and Aldehyde, RCHO	Catalyst	Product(s) (Yield, %)	Refs.
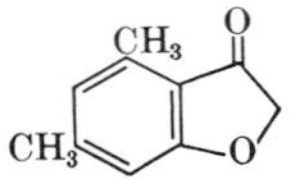 4,6-Dimethylcoumaranone [structure: CH_3, CH_3, O] and			
3-HOC_6H_4CHO	HCl, CH_3CO_2H	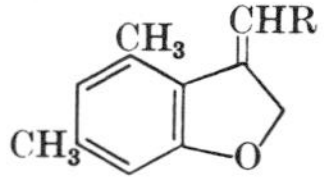 [structure: CH_3, CHR, CH_3, O] (—) X, R = 3-HOC_6H_4	339
4-HOC_6H_4CHO	"	X, R = 4-HOC_6H_4 (—)	339
4-$CH_3OC_6H_4CHO$	"	X, R = 4-$CH_3OC_6H_4$ (—)	339
$C_6H_5CH{=}CHCHO$	"	X, R = $C_6H_5CH{=}CH$ (—)	339
3,4-$(CH_3O)_2C_6H_3CHO$	"	X, R = 3,4-$(CH_3O)_2C_6H_3$ (—)	339
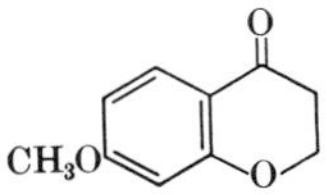 7-Methoxychromanone [structure: CH_3O, O] and			
2-BrC_6H_4CHO	NaOH	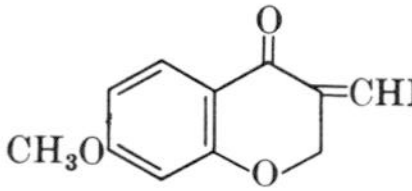 [structure: CH_3O, O, =CHR] (—) Y, R = 2-BrC_6H_4	1929
2-ClC_6H_4CHO	"	Y, R = 2-ClC_6H_4 (—)	1929
3,4-$(CH_3O)_2C_6H_3CHO$	HCl	Y, R = 3,4-$(CH_3O)_2C_6H_3$ (99)	2032, 2291
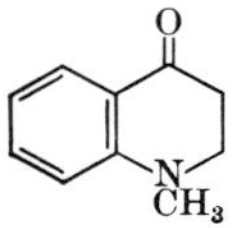 1-Methyl-4-quinolone [structure: O, N, CH_3] and			

C_6H_5CHO	NaOH	[structure: O, =CHR, N, CH_3]	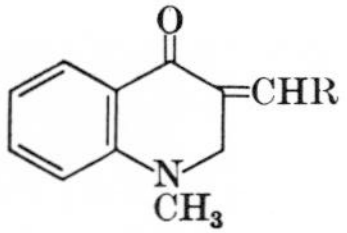 (75)	368
4-$(CH_3)_2NC_6H_4CHO$	$(CH_2)_5NH$	[structure: O, =CHR, N, CH_3]	(36)	368
[structure: CH_3CO, CH_3, CH_3, N, H, $COCH_3$] **and** [structure: O, CHO]	NaOH	[structure: CH_3CO, CH_3, CH_3, N, H, COCH=CHR]	(—)	2259
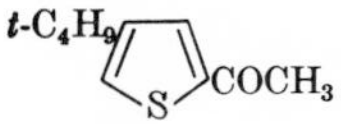 [structure: t-C_4H_9, S, $COCH_3$] **and** $(CH_3)_2C$=$CHCH_2CH_2C(CH_3)$=CHCHO	”	[structure: t-C_4H_9, S, COCH=CHR]	(67)	1773
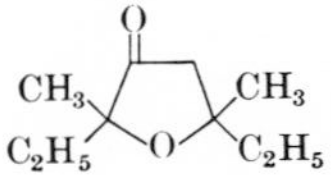 [structure: O, CH_3, C_2H_5, O, CH_3, C_2H_5] **and** 2-HOC_6H_4CHO	KOH	[structure: O, CHR, CH_3, C_2H_5, O, CH_3, C_2H_5]	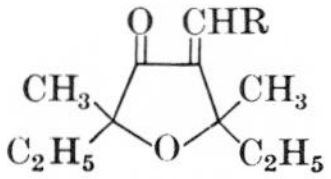 (15)	2285
4-$(CH_3)_2NC_6H_4CHO$	”	[structure: O, CHR, CH_3, C_2H_5, O, CH_3, C_2H_5]	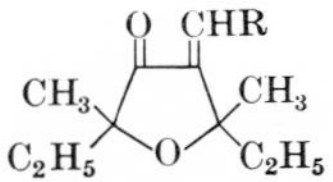 (100)	2282

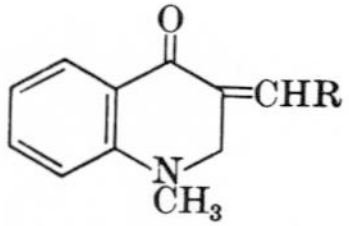

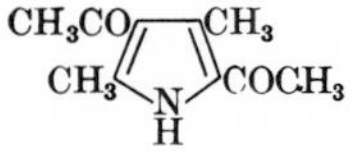

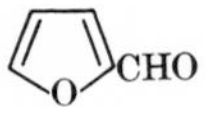

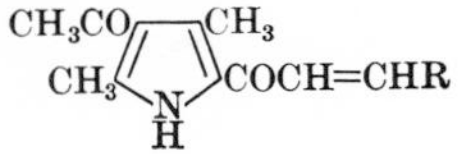

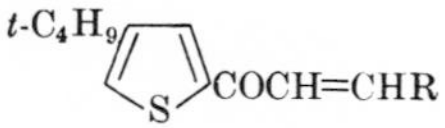

Note: References 668–2359 are on pp. 403–438.

TABLE XIX. CONDENSATION OF ALDEHYDES WITH HETEROCYCLIC KETONES (*Continued*)

(R in the product is the group R in the aldehyde RCHO.)

Ketone and Aldehyde, RCHO	Catalyst	Product(s) (Yield, %)	Refs.
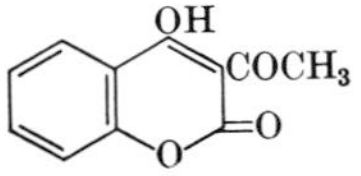 3-acetyl-4-hydroxycoumarin ($COCH_3$, OH, =O) and			
2-$O_2NC_6H_4CHO$	NaOH	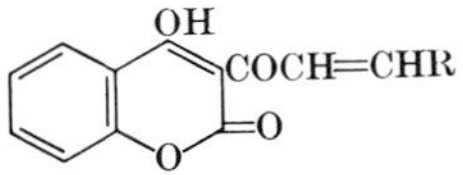 3-(COCH=CHR)-4-hydroxycoumarin (37) Z, R = 2-$O_2NC_6H_4$	2298
4-$O_2NC_6H_4CHO$	"	Z, R = 4-$O_2NC_6H_4$ (40)	2298
C_6H_5CHO	"	Z, R = C_6H_5 (—)	2298
3,4-$(CH_2O_2)C_6H_3CHO$	$NaOC_2H_5$	Z, R = 3,4-$(CH_2O_2)C_6H_3$ (60)	2299, 2298
4-$CH_3OC_6H_4CHO$	NaOH	Z, R = 4-$CH_3OC_6H_4$ (45)	2298
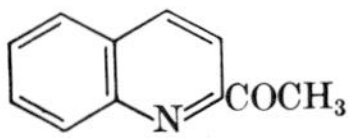 2-acetylquinoline ($COCH_3$) and			
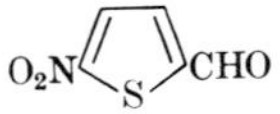 5-nitro-2-thiophenecarboxaldehyde (O_2N, S, CHO)	NaOH	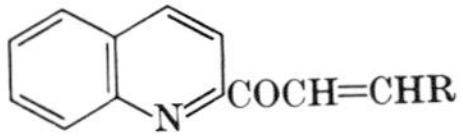 2-(COCH=CHR)quinoline (38) AA, R = 2-$(O_2NC_4H_2S)$	2301, 2356
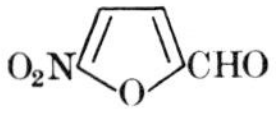 5-nitro-2-furaldehyde (O_2N, O, CHO)	"	AA, R = 2-$(O_2NC_4H_2O)$ (10)	2301, 2356
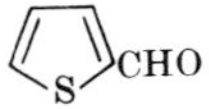 2-thiophenecarboxaldehyde (S, CHO)	"	AA, R = 2-(C_4H_3S) (94)	2301, 2356
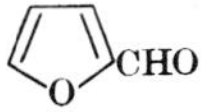 2-furaldehyde (O, CHO)	"	AA, R = 2-(C_4H_3O) (96)	2301, 2356

4-$O_2NC_6H_4CHO$	”	AA, R = 4-$O_2NC_6H_4$ (95)	2301, 2356
C_6H_5CHO	”	AA, R = C_6H_5 (96)	2301, 2356
4-$H_2NC_6H_4CHO$	”	AA, R = 4-$H_2NC_6H_4$ (50)	2301, 2356
4-$CH_3OC_6H_4CHO$	”	AA, R = 4-$CH_3OC_6H_4$ (95)	2301, 2356
$C_6H_5CH{=}CHCHO$	”	AA, R = $C_6H_5CH{=}CH$ (96)	2301, 2356
2,4-$(CH_3O)_2C_6H_3CHO$	”	AA, R = 2,4-$(CH_3O)_2C_6H_3$ (95)	2301, 2356
4-$(CH_3)_2NC_6H_4CHO$	”	AA, R = 4-$(CH_3)_2NC_6H_4$ (60)	2301, 2356
2,4,6-$(CH_3O)_3C_6H_2CHO$	”	AA, R = 2,4,6-$(CH_3O)_3C_6H_2$ (87)	2301, 2356
$COCH_3$ N **and** N CHO	$NaOCH_3$	COCH=CHR N (65)	2004
N $COCH_3$ **and** 2-ClC_6H_4CHO	NaOH	N COCH=CHR (93) BB, R = 2-ClC_6H_4	1770
C_6H_5CHO	”	BB, R = C_6H_5 (73)	1770
3,4-$(CH_2O_2)C_6H_3CHO$	”	BB, R = 3,4-$(CH_2O_2)C_6H_3$ (74)	1770
4-$CH_3OC_6H_4CHO$	”	BB, R = 4-$CH_3OC_6H_4$ (88)	1770
2,3-$(CH_3O)_2C_6H_3CHO$	”	BB, R = 2,3-$(CH_3O)_2C_6H_3$ (90)	1770
$COCH_3$ N **and** 2-ClC_6H_4CHO	NaOH	COCH=CHR N (71) CC, R = 2-ClC_6H_4	1770

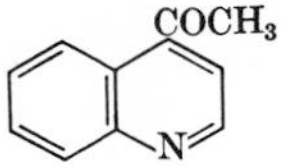

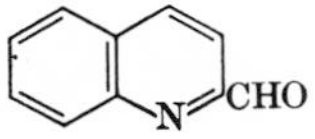

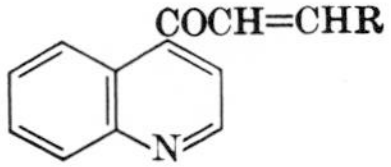

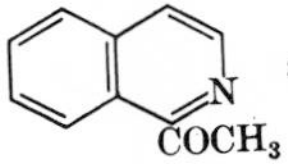

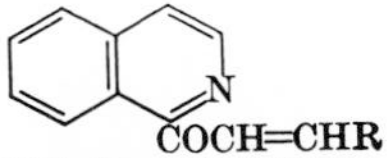

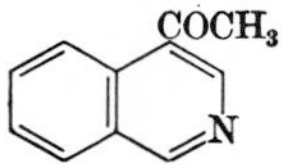

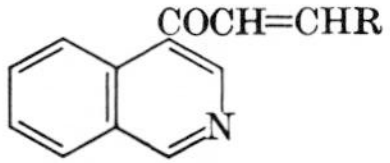

Note: References 668–2359 are on pp. 403–438.

TABLE XIX. CONDENSATION OF ALDEHYDES WITH HETEROCYCLIC KETONES (*Continued*)
(R in the product is the group R in the aldehyde RCHO.)

Ketone and Aldehyde, RCHO	Catalyst	Product(s) (Yield, %)	Refs.
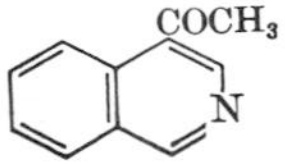 $COCH_3$, N (**contd.**) **and**			
C_6H_5CHO	$(CH_3)_2NH$	CC, R = C_6H_5 (52)	1770
$3,4\text{-}(CH_2O_2)C_6H_3CHO$	NaOH	CC, R = $3,4\text{-}(CH_2O_2)C_6H_3$ (95)	1770
$4\text{-}CH_3OC_6H_4CHO$	"	CC, R = $4\text{-}CH_3OC_6H_4$ (76)	1770
$2,3\text{-}(CH_3O)_2C_6H_3CHO$	"	CC, R = $2,3\text{-}(CH_3O)_2C_6H_3$ (66)	1770
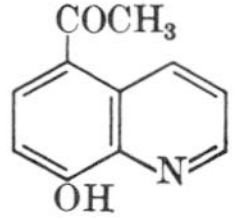 $COCH_3$, N, OH **and** $3,4\text{-}(CH_2O_2)C_6H_3CHO$	"	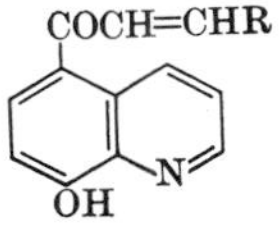 $COCH{=}CHR$, N, OH (94)	2297
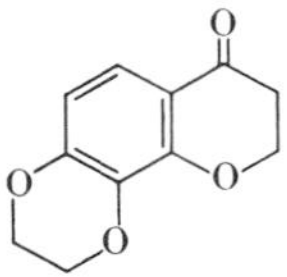 O, O, O, O **and** $3,4\text{-}(OCH_2CH_2O)C_6H_3CHO$	HCl, CH_3CO_2H	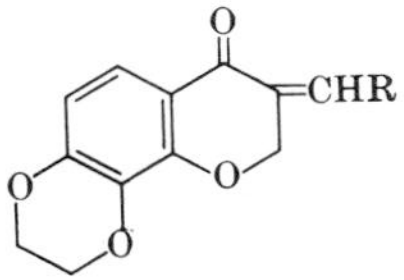 O, $=CHR$, O, O, O (75)	2302
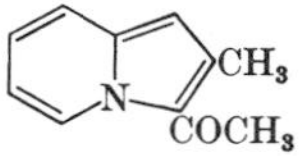 N, CH_3, $COCH_3$ **and** C_6H_5CHO	NaOH	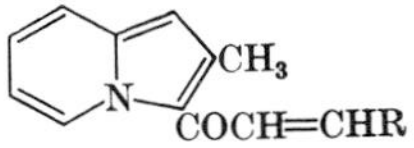 N, CH_3, $COCH{=}CHR$ (—)	2303
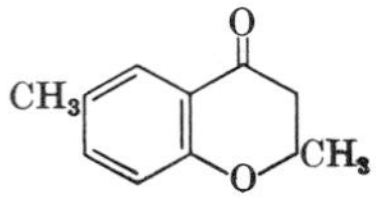 CH_3, O, O, CH_3 **and** C_6H_5CHO	HCl	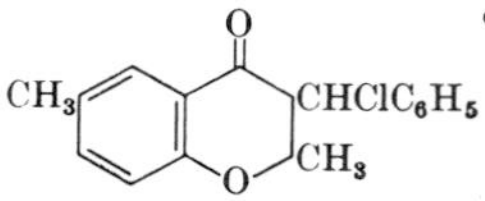 CH_3, O, $CHClC_6H_5$, O, CH_3 ¶ (84)	2304

Reactants	Catalyst	Product (Yield, %)	Refs.
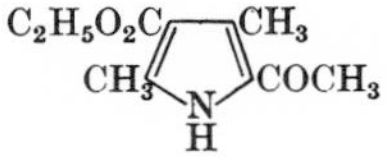 $C_2H_5O_2C$, CH_3, CH_3, N H, $COCH_3$ and (O) CHO 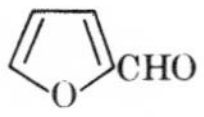	NaOH	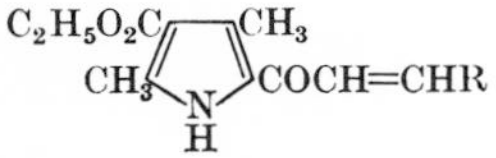 $C_2H_5O_2C$, CH_3, CH_3, N H, $COCH{=}CHR$ (—)	2259
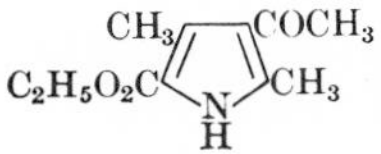 CH_3, $COCH_3$, $C_2H_5O_2C$, N H, CH_3 and			
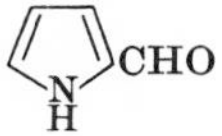 (N H) CHO	KOH	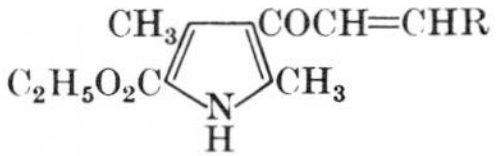 CH_3, $COCH{=}CHR$, $C_2H_5O_2C$, N H, CH_3 (65)	364
		DD, R = 2-(C_4H_3NH)	
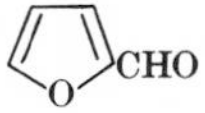 (O) CHO	"	DD, R = 2-(C_4H_3O) (69)	364, 2259
C_6H_5CHO	"	DD, R = C_6H_5 (66)	364
3,4-(CH_2O_2)C_6H_3CHO	"	DD, R = 3,4-(CH_2O_2)C_6H_3 (51)	364
4-$(CH_3)_2NC_6H_4CHO$	"	DD, R = $4(CH_3)_2NC_6H_4$ (65)	364
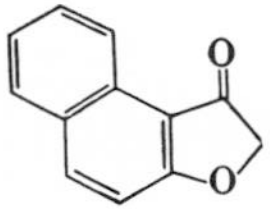 (O, O) and 4-$CH_3OC_6H_4CHO$	"	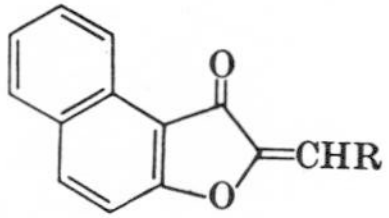 (O, O) =CHR (—)	2276
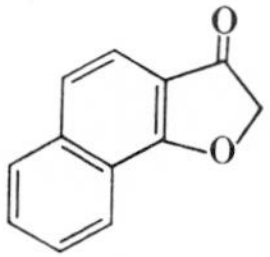 (O, O) and (O, O) CHO 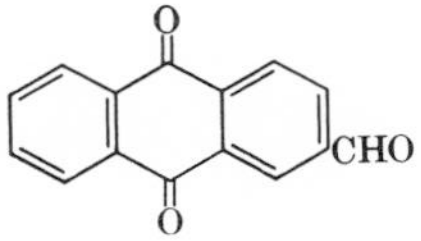	NaOH	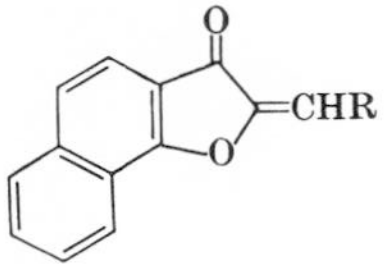 (O, O) =CHR (—)	1977

Note: References 668–2359 are on pp. 403–438.

¶ This product was converted quantitatively to the benzal derivative by heating at 180°.

TABLE XIX. CONDENSATION OF ALDEHYDES WITH HETEROCYCLIC KETONES (*Continued*)

(R in the product is the group R in the aldehyde RCHO.)

Ketone and Aldehyde, RCHO	Catalyst	Product(s) (Yield, %)	Refs.
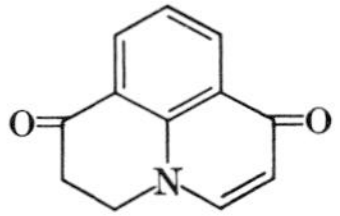 **and**			
C_6H_5CHO	$(CH_2)_5NH$	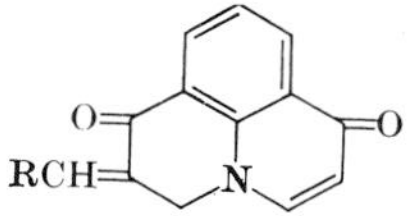 (—)	368
$4-(CH_3)_2NC_6H_4CHO$	"	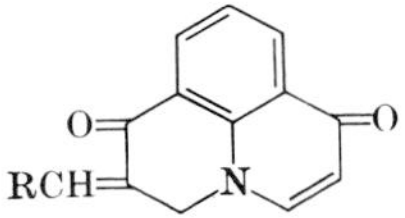 (—)	368
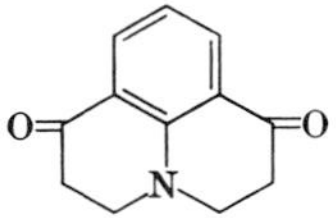 **and** 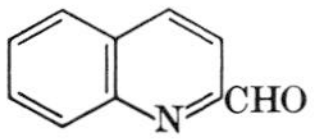	KOH	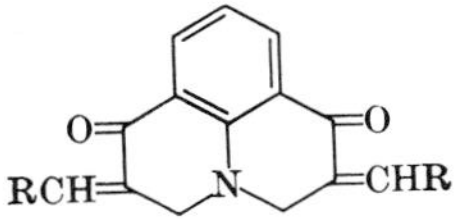 (25)	368, 2341
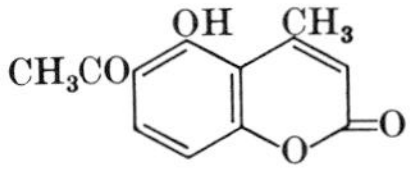 **and**			
C_6H_5CHO	NaOH	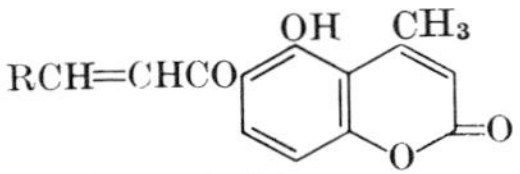 $RCH{=}CHCO$ (—)	2300
		EE, $R = C_6H_5$	
$2-HOC_6H_4CHO$	KOH	EE, $R = 2-HOC_6H_4$ (—)	2300
$3,4-(HO)_2C_6H_3CHO$	"	EE, $R = 3,4-(HO)_2C_6H_3$ (—)	2300
$4-CH_3OC_6H_4CHO$	"	EE, $R = 4-CH_3OC_6H_4$ (—)	2300
$3-CH_3O-4-HOC_6H_3CHO$	"	EE, $R = 3-CH_3O-4-HOC_6H_3$ (—)	2300

Reactants	Catalyst	Product (Yield, %)	Refs.
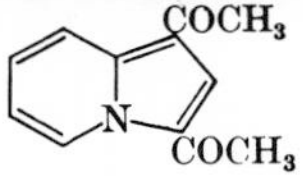 1,3-Diacetylindolizine ($COCH_3$, $COCH_3$) and			
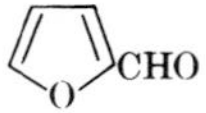 2-Furaldehyde (CHO)	NaOH	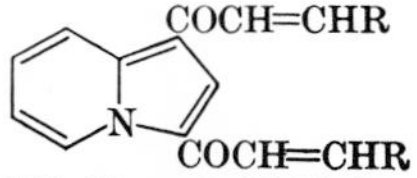 1,3-Bis(COCH=CHR)indolizine (—) FF, R = 2-(C_4H_3O)	2305
$2\text{-}O_2NC_6H_4CHO$	"	FF, R = $2\text{-}O_2NC_6H_4$ (—)	2306
$3\text{-}O_2NC_6H_4CHO$	"	FF, R = $3\text{-}O_2NC_6H_4$ (—)	2306
$4\text{-}O_2NC_6H_4CHO$	"	FF, R = $4\text{-}O_2NC_6H_4$ (—)	2306
C_6H_5CHO	"	FF, R = C_6H_5 (—)	2305
$3,4\text{-}(CH_2O_2)C_6H_3CHO$	"	FF, R = $3,4\text{-}(CH_2O_2)C_6H_3$ (—)	2305
$4\text{-}CH_3C_6H_4CHO$	"	FF, R = $4\text{-}CH_3C_6H_4$ (—)	2305
$C_6H_5CH{=}CHCHO$	"	FF, R = $C_6H_5CH{=}CH$ (—)	2305
$4\text{-}(CH_3)_2NC_6H_4CHO$	"	FF, R = $4\text{-}(CH_3)_2NC_6H_4$ (—)	2306
$4\text{-}i\text{-}C_3H_7C_6H_4CHO$	"	FF, R = $4\text{-}i\text{-}C_3H_7C_6H_4$ (—)	2305
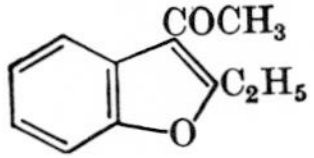 3-Acetyl-2-ethylbenzofuran ($COCH_3$, C_2H_5) and C_6H_5CHO	"	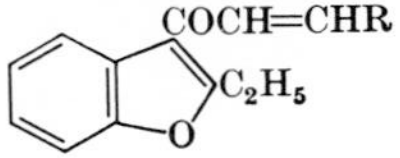 3-(COCH=CHR)-2-ethylbenzofuran (C_2H_5) (—)	1703
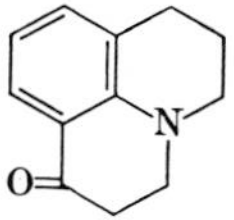 Julolidone (O, N) and			
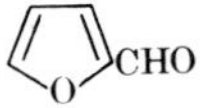 2-Furaldehyde (CHO)	"	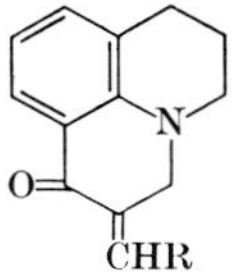 2-(=CHR) julolidone (O, N, CHR) (62)	2307
C_6H_5CHO	"	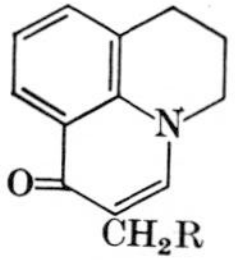 2-(CH_2R) julolid-2-en-1-one (O, N, CH_2R) (—)	2307

Note: References 668–2359 are on pp. 403–438.

TABLE XIX. Condensation of Aldehydes with Heterocyclic Ketones (*Continued*)

(R in the product is the group R in the aldehyde RCHO.)

Ketone and Aldehyde, RCHO	Catalyst	Product(s) (Yield, %)	Refs.
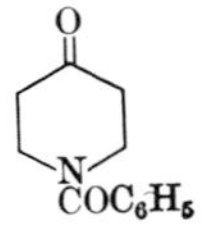 O, N, COC_6H_5 and			
$2\text{-}ClC_6H_4CHO$	HCl	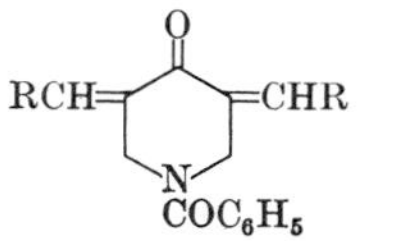 $RCH=$, O, $=CHR$, N, COC_6H_5 (—) GG, R = $2\text{-}ClC_6H_4$	358
$4\text{-}ClC_6H_4CHO$	,,	GG, R = $4\text{-}ClC_6H_4$ (73)	358
$3\text{-}O_2NC_6H_4CHO$	,,	GG, R = $3\text{-}O_2NC_6H_4$ (61)	358
$4\text{-}O_2NC_6H_4CHO$	,,	GG, R = $4\text{-}O_2NC_6H_4$ (52)	358
C_6H_5CHO	,,	GG, R = C_6H_5 (97)	358
$3,4\text{-}(CH_2O_2)C_6H_3CHO$	,,	GG, R = $3,4\text{-}(CH_2O_2)C_6H_3$ (99)	358
$2\text{-}CH_3OC_6H_4CHO$	,,	GG, R = $2\text{-}CH_3OC_6H_4$ (—)	358
$4\text{-}CH_3OC_6H_4CHO$	,,	GG, R = $4\text{-}CH_3OC_6H_4$ (91)	358
Br, O=, =O, N and C_6H_5CHO	$(CH_2)_5NH$	Br, O=, =O, $RCH=$, N, $=CHR$ (—)	368
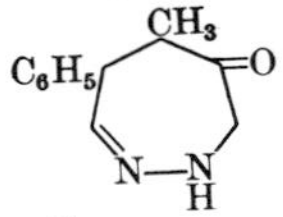 C_6H_5, CH_3, =O, N—N, H and C_6H_5CHO	$NaOC_2H_5$	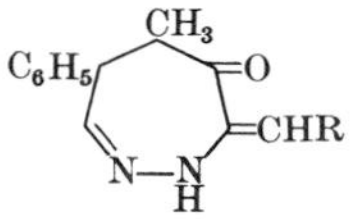 C_6H_5, CH_3, =O, $=CHR$, N—N, H (45)	2308
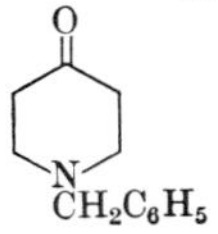 O, N, $CH_2C_6H_5$ and			

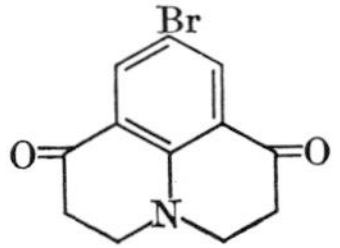

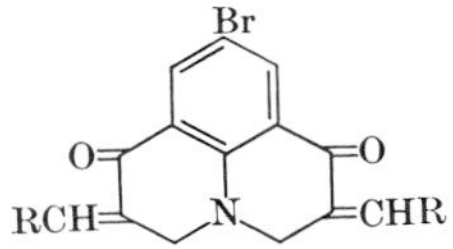

(thiophene ring, S) CHO	KOH	RCH=(ring, C=O, N–$CH_2C_6H_5$)=CHR (—) HH, R = 2-(C_4H_3S)	2265
(furan ring, O) CHO	”	HH, R = 2-(C_4H_3O) (—)	2265
(pyridine ring, N) CHO	”	HH, R = 2-(C_5H_4N) (—)	2265
CH_3 (furan ring, O) CHO	”	HH, R = 2-($CH_3C_4H_2O$) (—)	2265
4-FC_6H_4CHO	”	HH, R = 4-FC_6H_4 (—)	2265
C_6H_5CHO	”	HH, R = C_6H_5 (—)	2265
3,4-$(CH_2O_2)C_6H_3CHO$	”	HH, R = 3,4-$(CH_2O_2)C_6H_3$ (—)	2265
4-$CH_3OC_6H_4CHO$	”	HH, R = 4-$CH_3OC_6H_4$ (—)	2265
$C_6H_5CH{=}CHCHO$	”	HH, R = $C_6H_5CH{=}CH$ (—)	2265
4-$CH_3CONHC_6H_4CHO$	”	HH, R = 4-$CH_3CONHC_6H_4$ (—)	2265
4-$(CH_3)_2NC_6H_4CHO$	”	HH, R = 4-$(CH_3)_2NC_6H_4$ (—)	2265
1-$C_{10}H_7CHO$	”	HH, R = 1-$C_{10}H_7$ (—)	2265
(naphthalene ring) CHO, OCH_3	”	HH, R = 1-($CH_3OC_{10}H_6$) (—)	2265
(phenanthrene ring) CHO	”	HH, R = 9-($C_{14}H_9$) (—)	2265

Note: References 668–2359 are on pp. 403–438.

TABLE XIX. Condensation of Aldehydes with Heterocyclic Ketones (*Continued*)

(R in the product is the group R in the aldehyde RCHO.)

Ketone and Aldehyde, RCHO	Catalyst	Product(s) (Yield, %)	Refs.
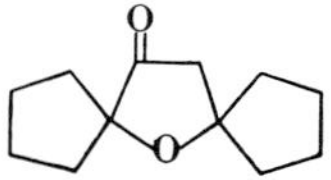 (dispiro ketone, O) and			
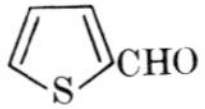 (S) CHO	KOH	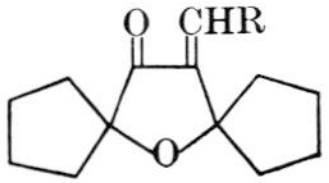 O CHR (O) (96) II, R = 2-(C_4H_3S)	2282
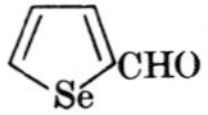 (Se) CHO	"	II, R = 2-(C_4H_3Se) (100)	2282
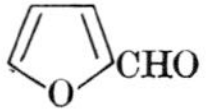 (O) CHO	"	II, R = 2-(C_4H_3O) (62)	2282
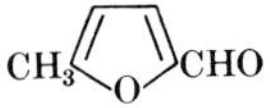 CH_3 (O) CHO	"	II, R = 2-($CH_3C_4H_2O$) (70)	2282
C_6H_5CHO	NaOH	II, R = C_6H_5 (43)	2309
4-$(CH_3)_2NC_6H_4CHO$	KOH	II, R = 4-$(CH_3)_2NC_6H_4$ (92)	2282
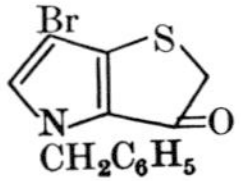 Br (S) (N, $CH_2C_6H_5$) =O and C_6H_5CHO	NaOH	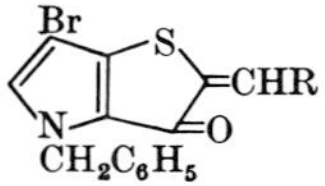 Br (S) =CHR (N, $CH_2C_6H_5$) =O (24)	2310
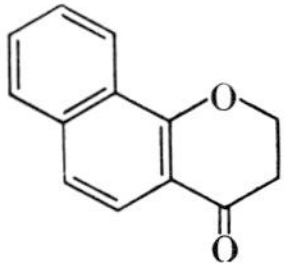 (O) (=O) and			

Reactants	Catalyst	Product (Yield, %)	Refs.
2-furaldehyde (furan-CHO)	,,	naphthopyranone =CHR (50) JJ, R = 2-(C_4H_3O)	1929
C_6H_5CHO	,,	JJ, R = C_6H_5 (50)	1929
3,4-$(CH_2O_2)C_6H_3CHO$	,,	JJ, R = 3,4-$(CH_2O_2)C_6H_3$ (25)	1929
N-$CH_2C_6H_5$ thienopyrrolone and C_6H_5CHO	,,	=CHR, N-$CH_2C_6H_5$ (73)	2310
$C_6H_5CH_2$-thienyl-$COCH_3$ and $(CH_3)_2C{=}CHCH_2CH_2C(CH_3){=}CHCHO$	,,	$C_6H_5CH_2$-thienyl-$COCH{=}CHR$ (65)	1773
2-acetyldibenzothiophene ($COCH_3$) and C_6H_5CHO	$NaOCH_3$	dibenzothiophene-$COCH{=}CHR$ (62)	2004
4-$(CH_3)_2NC_6H_4CHO$	,,	dibenzothiophene-$COCH{=}CHR$ (57)	2004

Note: References 668–2359 are on pp. 403–438

TABLE XIX. Condensation of Aldehydes with Heterocyclic Ketones (*Continued*)

(R in the product is the group R in the aldehyde RCHO.)

Ketone and Aldehyde, RCHO	Catalyst	Product(s) (Yield, %)	Refs.
Phenoxathiin-$COCH_3$ and			
2-Thienyl-CHO	$NaOC_2H_5$	Phenoxathiin-$COCH{=}CHR$ (81)	2312
		KK, R = 2-(C_4H_3S)	
2-Furyl-CHO	"	KK, R = 2-(C_4H_3O) (82)	2312
4-ClC_6H_4CHO	"	KK, R = 4-ClC_6H_4 (75)	2312
C_6H_5CHO	"	KK, R = C_6H_5 (83)	2312
4-$CH_3OC_6H_4CHO$	"	KK, R = 4-$CH_3OC_6H_4$ (90)	2312
4-$(CH_3)_2NC_6H_4CHO$	"	KK, R = 4-$(CH_3)_2NC_6H_4$ (70)	2312
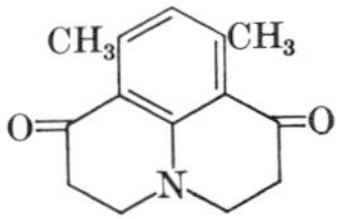 (CH_3, CH_3, O=, =O, N) and			
4-$(CH_3)_2NC_6H_4CHO$	KOH	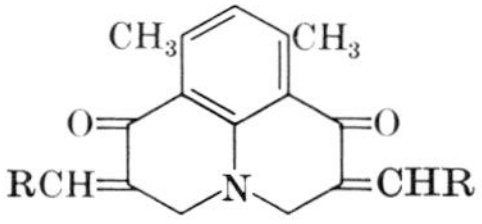 (CH_3, CH_3, O=, =O, RCH=, N, =CHR) (93)	368
2-Quinolyl-CHO	"	(CH_3, CH_3, O=, =O, RCH=, N, =CHR) (—)	368
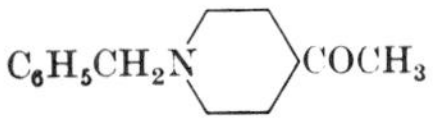 $C_6H_5CH_2N$ (piperidine) $COCH_3$ and CH_2O	NaOH	 $C_6H_5CH_2N$ (piperidine) (CH_2OH)($COCH_3$) (31)	450

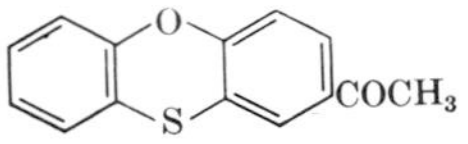

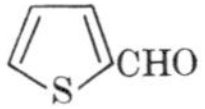

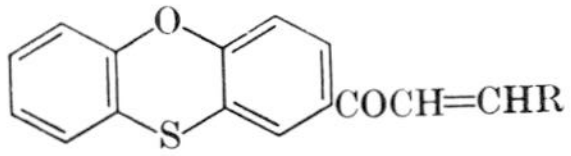

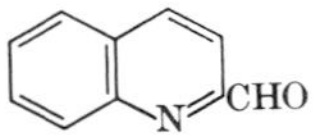

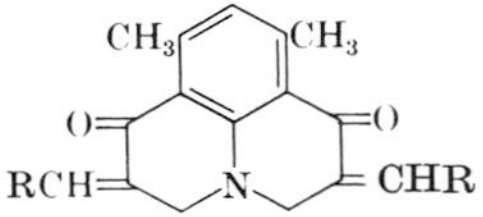

Reactants	Catalyst	Product (Yield, %)	Refs.
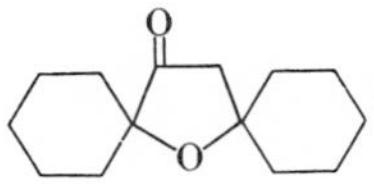 [structure] and			
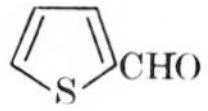 2-Thiophenecarboxaldehyde [structure]	KOH	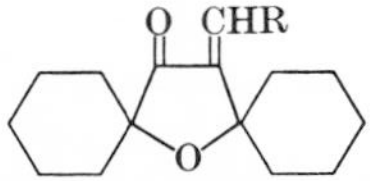 [structure] (94) LL, R = 2-(C_4H_3S)	2282
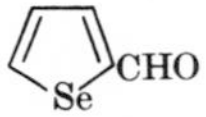 2-Selenophenecarboxaldehyde [structure]	"	LL, R = 2-(C_4H_3Se) (70)	2282
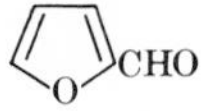 2-Furaldehyde [structure]	"	LL, R = 2-(C_4H_3O) (95)	2282
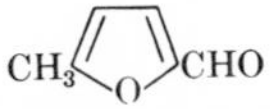 5-Methyl-2-furaldehyde [structure]	"	LL, R = 2-($CH_3C_4H_2O$) (92)	2282
C_6H_5CHO	NaOH	LL, R = C_6H_5 (55)	2309
4-$(CH_3)_2NC_6H_4CHO$	KOH	LL, R = 4-$(CH_3)_2NC_6H_4$ (90)	2282
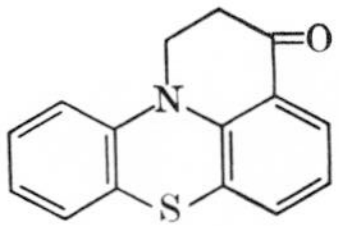 [structure] and			
4-ClC_6H_4CHO	None	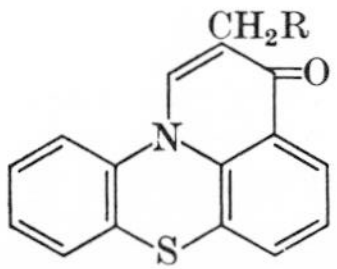 [structure] (—) MM, R = 4-ClC_6H_4	2313, 2328
3-$O_2NC_6H_4CHO$	"	MM, R = 3-$O_2NC_6H_4$ (—)	2313, 2328

Note: References 668–2359 are on pp. 403–438

TABLE XIX. CONDENSATION OF ALDEHYDES WITH HETEROCYCLIC KETONES (*Continued*)

(R in the product is the group R in the aldehyde RCHO.)

Ketone and Aldehyde, RCHO	Catalyst	Product(s) (Yield, %)	Refs.
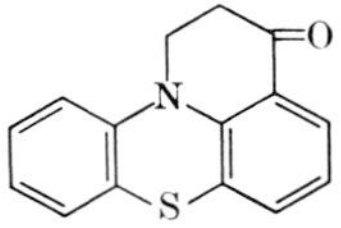 (*contd.*) and			
C_6H_5CHO	None	MM, R = C_6H_5 (—)	2313, 2328
$4\text{-}CH_3OC_6H_4CHO$	”	MM, R = $4\text{-}CH_3OC_6H_4$ (—)	2313, 2328
$4\text{-}(CH_3)_2NC_6H_4CHO$	”	MM, R = $4\text{-}(CH_3)_2NC_6H_4$ (—)	2313, 2328
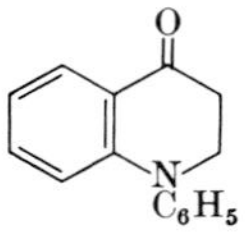 and 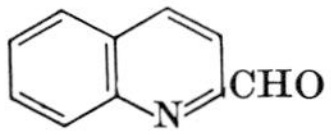	KOH	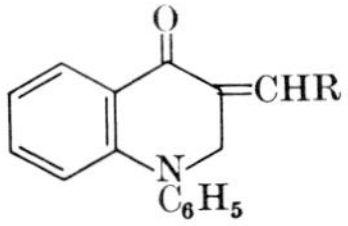 (80)	368
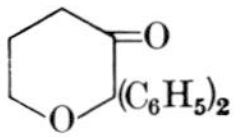 and 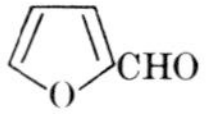	NaOH	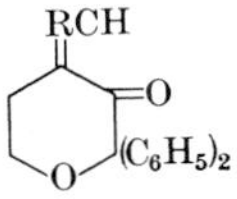 (—)	2314

Note: References 668–2359 are on pp. 403–438.

TABLE XX. Intramolecular Condensation of Ketoaldehydes

Ketoaldehyde	Catalyst	Product(s) (Yield, %)	Refs.
$CH_3CO(CH_2)_4CHO$*	NaOH	1-$COCH_3$-cyclopentene (73)	2315
$CH_3COCH(CH_3)(CH_2)_2CHO$†	$KOCH_3$	6-CH_3-cyclohex-2-enone (40)	447, 448
$CH_3COCH(CH_3)CH(CH_3)CHO$	CH_3CO_2Na	4,5-(CH_3)$_2$-cyclopent-2-enone (74)	2316
$CH_3COCH(CH_3)CH(CH_3)CH_2CHO$†	$KOCH_3$	5,6-(CH_3)$_2$-cyclohex-2-enone (35)	447, 448
$CH_3COC(CH_3)_2(CH_2)_2CHO$†	"	6,6-$(CH_3)_2$-cyclohex-2-enone (12)	447, 448
$CH_3COCH(C_2H_5)(CH_2)_2CHO$†	"	6-C_2H_5-cyclohex-2-enone (20)	447, 448

Note: References 668–2359 are on pp. 403–438.

* The ketoaldehyde was formed *in situ* from 2-benzyloxymethylene-1-carbethoxymethylcyclohexan-1-ol.

† The ketoaldehyde was formed *in situ* by Michael addition of the appropriate ketone to acrolein, α-methylacrolein, or crotonaldehyde.

TABLE XX. INTRAMOLECULAR CONDENSATION OF KETOALDEHYDES (*Continued*)

Ketoaldehyde	Catalyst	Product(s) (Yield, %)	Refs.
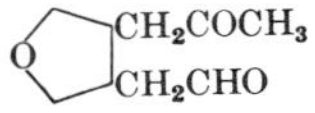 (tetrahydrofuran ring, O) CH_2COCH_3, CH_2CHO	$(CH_2)_5NH$, CH_3CO_2H; also KOH, KOC_4H_9-*t*, Al_2O_3	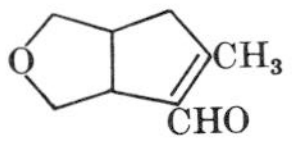 (bicyclic, O) CH_3, CHO (30)	438
$CH_3COC(CH_3)_2CH(CH_3)CH_2CHO$†	$KOCH_3$	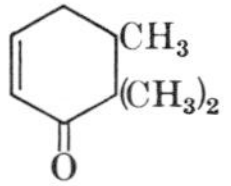 (cyclohexenone, O) CH_3, $(CH_3)_2$ (22)	447, 448
 CH_3CO(piperidine ring)NCH_2CHO‡	HCl	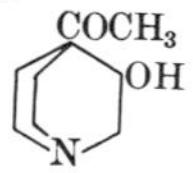 (bicyclic, N) $COCH_3$, OH (10)	450
$CH_3COCH(C_2H_5)CH(CH_3)CH_2CHO$†	$KOCH_3$	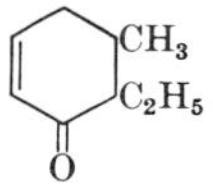 (cyclohexenone, O) CH_3, C_2H_5 (30)	447, 448
$CH_3CO(CH_2)_2CHCH_2CHO$ with $CH_3C{=}CH_2$ substituent	KOH	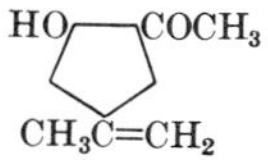 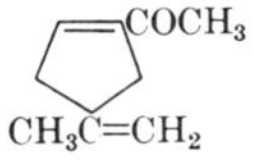 (cyclopentane) HO, $COCH_3$, $CH_3C{=}CH_2$ (25), (cyclopentene) $COCH_3$, $CH_3C{=}CH_2$ (25)	444
	$(CH_2)_5NH$, CH_3CO_2H	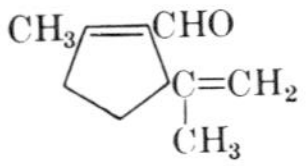 (cyclopentene) CH_3, CHO, $C{=}CH_2$, CH_3 (59)	445
$CH_3COCH(CO_2C_2H_5)(CH_2)_2CHO$	$NaOC_2H_5$	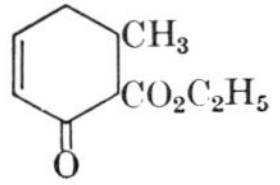 (cyclohexenone, O) CH_3, $CO_2C_2H_5$ (53)	2317

Reactant	Catalyst	Product(s) (Yield, %)	Refs.
$CH_3CO(CH_2)_2CH(C_3H_7\text{-}i)CH_2CHO$	KOH	$COCH_3$, $C_3H_7\text{-}i$ (—)	436, 437
	$(CH_2)_5NH$, CH_3CO_2H	CH_3, CHO, $C_3H_7\text{-}i$ (—)	436
O, CH_2CH_2CHO, $CO_2C_2H_5$	H_2SO_4	O, $CO_2C_2H_5$ I (—), O, $CO_2C_2H_5$ II (—) (I:II = 4.7)	451
O, CH_3, CH_2CH_2CHO, $CO_2C_2H_5$	HCl, CH_3CO_2H	HO, O, CH_3, $CO_2C_2H_5$ (80)	2318, 2358
	H_2SO_4	O, CH_3, $CO_2C_2H_5$ I (30), O, CH_3, $CO_2C_2H_5$ II (11)	2318, 451

Note: References 668–2359 are on pp. 403–438.

† The ketoaldehyde was formed *in situ* by Michael addition of the appropriate ketone to acrolein, α-methylacrolein, or crotonaldehyde.

‡ The ketoaldehyde was formed *in situ* from the diethyl acetal.

TABLE XX. INTRAMOLECULAR CONDENSATION OF KETOALDEHYDES (*Continued*)

Ketoaldehyde	Catalyst	Product(s) (Yield, %)	Refs.
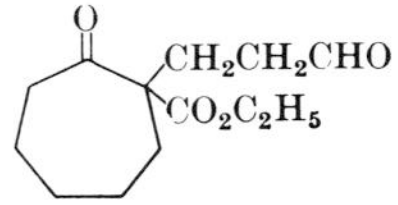 O, CH_2CH_2CHO, $CO_2C_2H_5$	H_2SO_4, then KOH	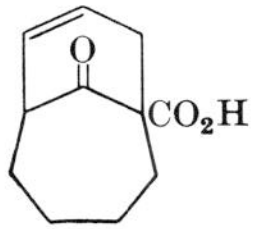 O, CO_2H (16)	2319
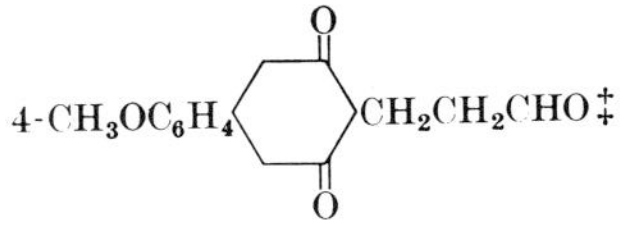 O, O, 4-$CH_3OC_6H_4$, CH_2CH_2CHO‡	HCl, CH_3CO_2H	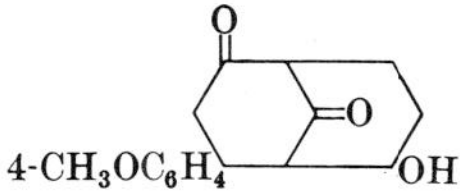 O, O, 4-$CH_3OC_6H_4$, OH (30)	2320
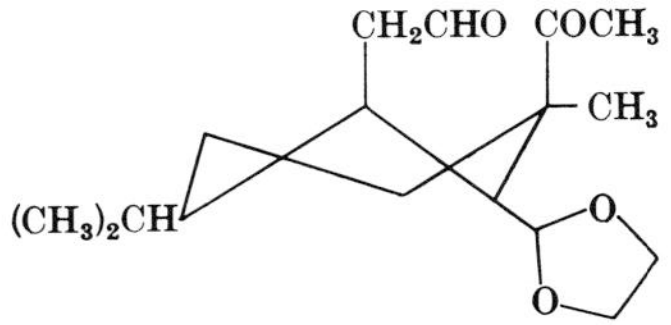 CH_2CHO, $COCH_3$, CH_3, $(CH_3)_2CH$, O, O	NaOH	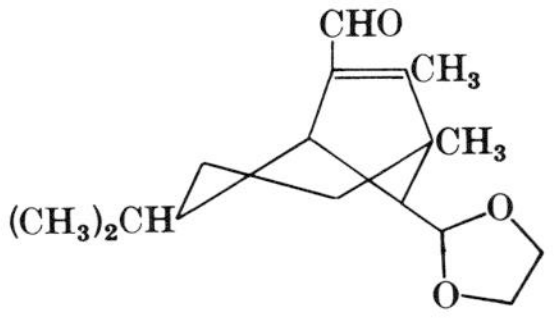 CHO, CH_3, CH_3, $(CH_3)_2CH$, O, O (50)	439
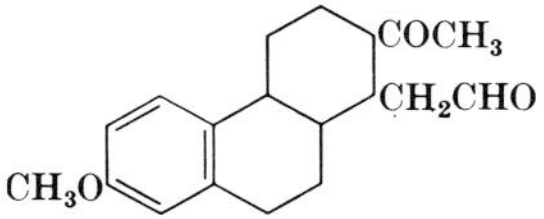 $COCH_3$, CH_2CHO, CH_3O	KOH	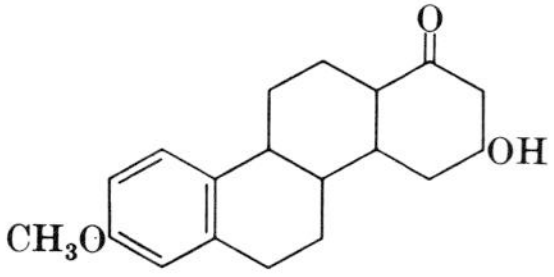 O, OH, CH_3O (90)	1316
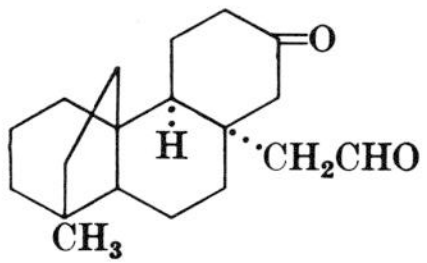 O, H, CH_2CHO, CH_3 §	H_2SO_4	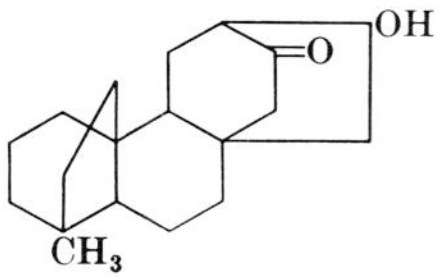 OH, O, CH_3 (45)	449

Reactant	Catalyst	Product(s) and Yield(s) (%)	Refs.
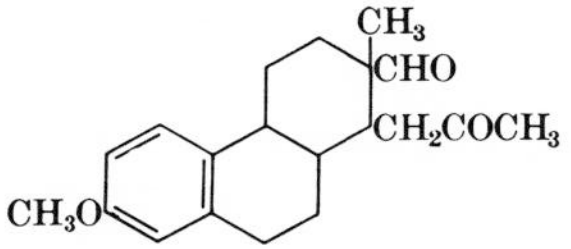 CH_3, CHO, CH_2COCH_3, CH_3O	CH_3CO_2Na, CH_3CO_2H	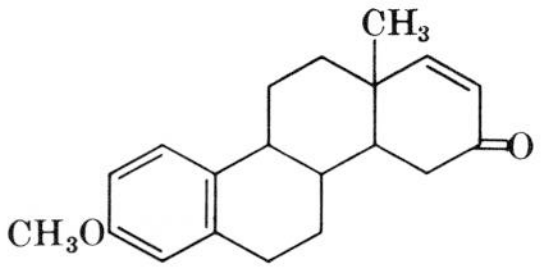 CH_3, O, CH_3O (—)	2321, 442
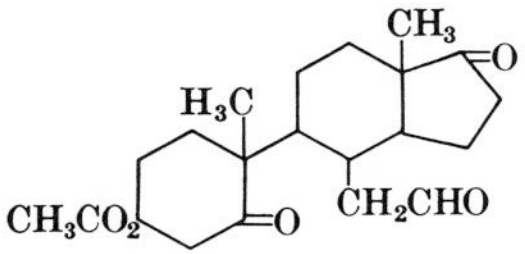 CH_3, O, H_3C, CH_3CO_2, O, CH_2CHO	Al_2O_3	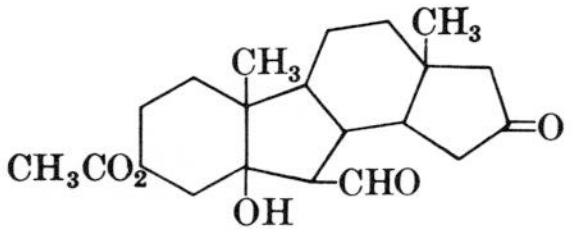 CH_3, CH_3, O, CH_3CO_2, OH, CHO (64)	441, 442
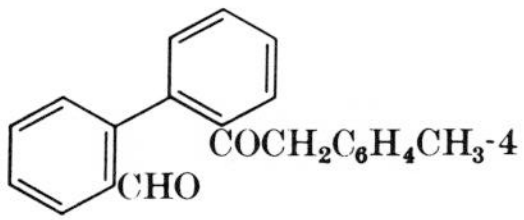 $COCH_2C_6H_4CH_3$-4, CHO	Florisil	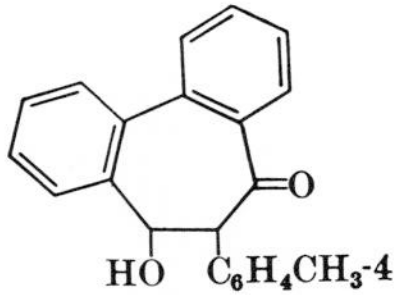 O, HO, $C_6H_4CH_3$-4 (epimers) (60)	70
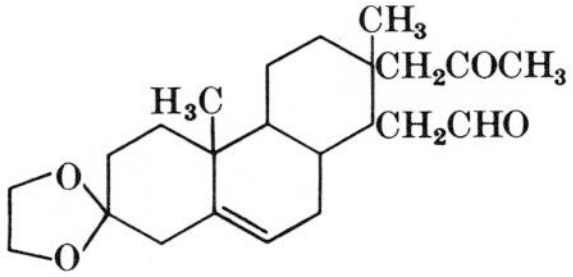 CH_3, CH_2COCH_3, H_3C, CH_2CHO, O, O	KOH	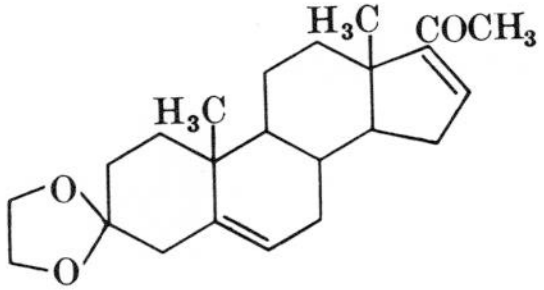 H_3C, $COCH_3$, H_3C, O, O (73)	434
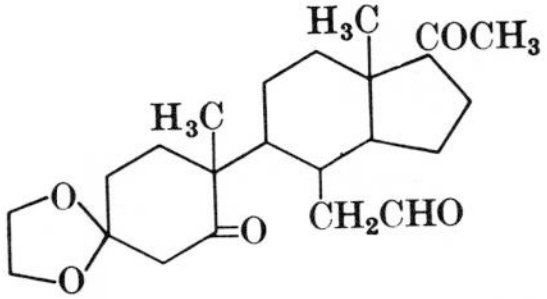 H_3C, $COCH_3$, H_3C, O, O, O, CH_2CHO	Al_2O_3	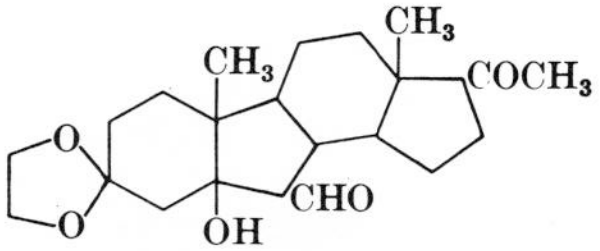 CH_3, CH_3, $COCH_3$, O, O, OH, CHO (—)	2322

Note: References 668–2359 are on pp. 403–438.

‡ The ketoaldehyde was formed *in situ* from the diethyl acetal.
§ The ketoaldehyde was formed *in situ* from the ketal.

TABLE XX. INTRAMOLECULAR CONDENSATION OF KETOALDEHYDES (*Continued*)

Ketoaldehyde	Catalyst	Product(s) (Yield, %)	Refs.
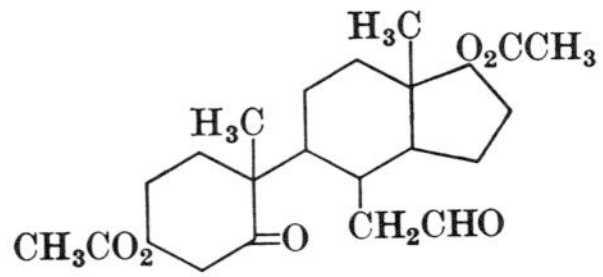 H_3C, O_2CCH_3, H_3C, CH_3CO_2, =O, CH_2CHO	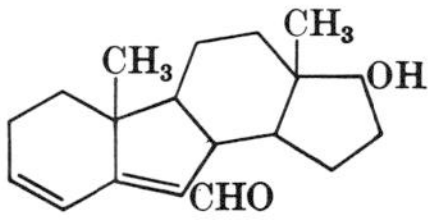 KOH	CH_3, CH_3, OH, CHO (25)	435
	Al_2O_3	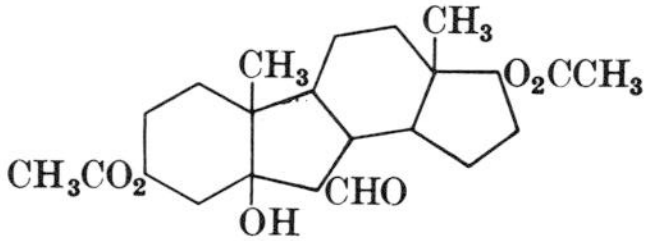 CH_3, CH_3, O_2CCH_3, CH_3CO_2, OH, CHO (68)	441, 442
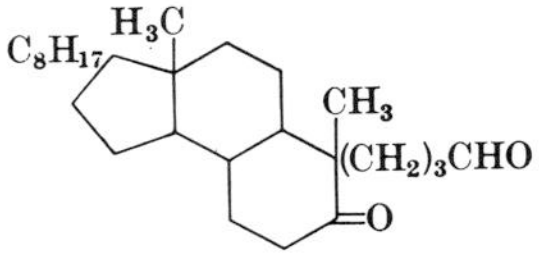 C_8H_{17}, H_3C, CH_3, $(CH_2)_3CHO$, =O	HCl, CH_3CO_2H 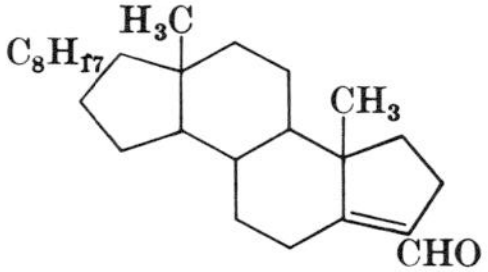	C_8H_{17}, H_3C, CH_3, CHO (90)	440
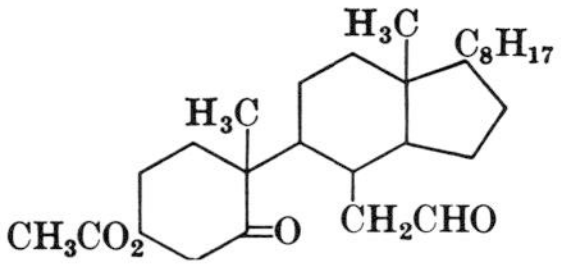 H_3C, C_8H_{17}, H_3C, CH_3CO_2, =O, CH_2CHO	Al_2O_3	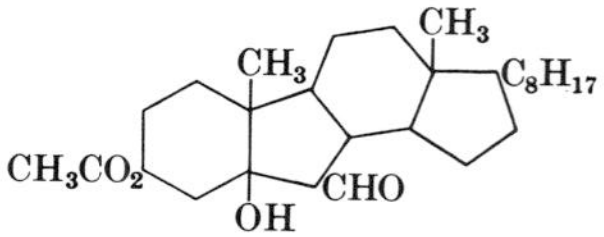 CH_3, CH_3, C_8H_{17}, CH_3CO_2, OH, CHO (74)	441, 442

Note: References 668–2359 are on pp. 403–438.

REFERENCES TO TABLES II–XX

[668] Magyar Ruggyantaárugyár R. T., Hung. pat. 135,586 [*C.A.*, **45,** 176 (1951)].

[669] C. J. Herrly, U.S. pat. 1,549,833 [*C.A.*, **19,** 3490 (1925)].

[670] M. Delépine, *Compt. Rend.*, **147,** 1316 (1908).

[671] A. Kekulé, *Ann.*, **162,** 77 (1872).

[672] W. Tischtschenko and M. Woronkow, *J. Russ. Phys. Chem. Soc.*, **38,** 547 (1906) [*Chem. Zentr.*, **77,** II, 1556 (1906)].

[673] Y. Miki, Jap. pat. 1578 (1953) [*C.A.*, **49,** 5522 (1955)].

[674] Usines de Melle, Brit. pat. 684,698 [*C.A.*, **48,** 2090 (1954)].

[675] G. S. Shaw, Can. pat. 281,951 [*C.A.*, **22,** 3498 (1928)].

[676] S. N. Ushakov and I. G. Shtan'ko, *Narodnuii Komissariat Tyazheloi Prom. SSSR, Leningrad, Plastmassui,* **1,** 309 (1935) [*C.A.*, **30,** 2173 (1936)].

[677] E. A. Shilov and G. I. Yakimov, *Sintet. Kauchuk,* (4), 7 (1934) [*C.A.*, **29,** 6798 (1935)].

[678] E. A. Shilov, *Zh. Prikl. Khim.*, **8,** 93 (1935) [*C.A.*, **29,** 6880 (1935)].

[679] E. J. Boake and L. W. E. Townsend, U.S. pat. 1,885, 221 [*C.A.*, **27,** 1005 (1933)].

[680] D. C. Hull, U.S. pat. 2,468,710 [*C.A.*, **43,** 5415 (1949)].

[681] Usines de Melle, Brit. pat. 630,904 [*C.A.*, **44,** 3518 (1950)].

[682] L. Alhéritière, U.S. pat. 2,489,608 [*C.A.*, **44,** 2017 (1950)].

[683] S. A. Miller and A. R. Hammond, U.S. pat. 2,517,013 [*C.A.*, **45,** 643 (1953)].

[684] H. M. Guinot and L. Alhéritière, Fr. pat. 942,981 [*C.A.*, **45,** 643 (1953)].

[685] H. W. Matheson, U.S. pat. 1,450,984 [*C.A.*, **17,** 1969 (1923)].

[686] R. P. Bell, *J. Chem. Soc.*, 1637 (1937).

[687] H. Hibbert, U.S. pat. 1,086,048 [*C.A.*, **8,** 1190 (1914)].

[688] E. I. du Pont de Nemours Co., Swiss pat. 64,932 [*C.A.*, **8,** 3490 (1914)].

[689] E. C. Craven and W. H. Gell, Brit. pat. 704,854 [*C.A.*, **49,** 4716 (1955)].

[690] H. Pauly and K. Feurstein, *Ber.*, **62,** 297 (1929).

[691] R. Ciola, *Anais. Assoc. Brasil. Quim.*, **20,** 63 (1961) [*C.A.*, **58,** 1327 (1963)].

[692] E Voyatzakis, *Ann. Fals. Fraudes*, **27,** 237 (1934) [*C.A.*, **28,** 5041 (1934)].

[693] R. B. Earle and L. P. Kyriakides, U.S. pat. 1,094,314 [*C.A.*, **8,** 2032 (1914)].

[694] H. Hammarsten, *Ann.*, **421,** 293 (1920).

[695] T. Ekecrantz, *Arkiv Kemi, Mineral. Geol.*, **4,** 1 (1912) [*Brit. Abstr.*, **I,** 788 (1912)].

[696] L. Alhéritière, U.S. pat. 2,442,280 [*C.A.*, **42,** 7788 (1950)].

[697] H. W. Matheson, U.S. pat. 1,151,113 [*C.A.*, **9,** 2795 (1915)].

[698] E. I. du Pont de Nemours Co., Fr. pat. 449,604 [*C.A.*, **7,** 2835 (1913)].

[699] A. K. Nowak, *Monatsh. Chem.*, **22,** 1140 (1901).

[700] Consortium für Elektrochem. Ind. Ges., Brit. pat. 19,463 [*C.A.*, **9,** 694 (1915)].

[701] Consortium für Elektrochem. Ind. Ges., Ger. pat. 269,996 [*C.A.*, **8,** 2222 (1914)].

[702] N. Grünstein, Brit. pat. 101,636 [*C.A.*, **11,** 195 (1917)].

[703] L. Alhéritière and G. Gobron, U.S. pat. 2,713,598 [*C.A.*, **49,** 12532 (1955)].

[704] J. E. Walker, U.S. pat. 2,400,274 [*C.A.*, **40,** 6094 (1946)].

[705] F. R. Charlton and D. A. Dowden, U.S. pat. 2,664,446 and Brit. pat. 638,207 [*C.A.*, **48,** 4583 (1954)].

[706] I. G. Farbenind. A.-G., Brit. pat. 439,195 and Fr. pat. 774,079 [*C.A.*, **30,** 2986 (1936); **29,** 2175 (1935)].

[707] J. N. Wickert, Fr. pat. 788,944 and Brit. pat. 438,848 [*C.A.*, **30,** 1811, 2986 (1936)].

[708] K. Billig, Ger. pat. 616,380 [*C.A.*, **29,** 7343 (1935)].

[709] E. J. Boake and L. W. E. Townsend, Brit. pat. 352,056 [*C.A.*, **26,** 5311 (1932)].

[710] L. Alhéritière, Ger. pat. 815,037 [*C.A.*, **51,** 12131 (1957)].

[711] M. Hori, *J. Agr. Chem. Soc. Japan*, **18,** 155 (1942) [*C.A.*, **45,** 4202 (1951)].

[712] K. Matsui, *J. Chem. Soc. Japan*, **64,** 1417 (1943) [*C.A.*, **41,** 3753 (1947)].

[713] B. Tollens, *Ber.*, **17,** 660 (1884).

[714] H. O. Mottern and V. F. Mistretta, U.S. pat. 2,530,987 [*C.A.*, **45,** 2969 (1953)].

[715] Y. A. Gorin and G. A. Sergicheva, *Zh. Obshch. Khim.*, **26,** 2444 (1956) [*C.A.*, **51,** 4992 (1957)].

[716] F. M. Scheidt, *J. Catalysis*, **3**, 372 (1964) [*C.A.*, **61**, 10580 (1964)].
[717] S. B. Newbury and W. S. Calkin, *Am. Chem. J.*, **12**, 523 (1890).
[718] J. H. Halpern, *Monatsh. Chem.*, **22**, 59 (1901).
[719] H. Hammarsten, *Ann.*, **420**, 262 (1919).
[720] W. R. Orndorff and S. B. Newbury, *Monatsh. Chem.*, **13**, 516 (1892).
[721] M. E. Puchot, *Ann. Chim. Phys.*, [6] **9**, 422 (1886).
[722] A. F. McLeod, *Am. Chem. J.*, **37**, 20 (1907).
[723] S. Zeisel and M. Neuwirth, *Ann.*, **433**, 121 (1923).
[724] P. Sabatier and G. Gaudion, *Compt. Rend.*, **166**, 632 (1918).
[725] V. Grignard and J. Rief, *Bull. Soc. Chim. France*, [4] **1**, 114 (1907).
[726] H. C. Chitwood and B. T. Freure, Brit. pat. 757, 812 [*C.A.*, **52**, 1205 (1958)].
[727] E. V. Budnitskaya, *Biokhimiya*, **6**, 146 (1941) [*C.A.*, **35**, 7938 (1941)].
[728] F. G. Fischer and A. Marschall, *Ber.*, **64**, 2825 (1931).
[729] W. Frankenburger, H. Hammerschmid, and G. Roessler, U.S. pat. 2,190,184 [*C.A.*, **34**, 4183 (1942)].
[730] A. Lieben and S. Zeisel, *Monatsh. Chem.*, **1**, 818 (1880).
[731] R. A. Heimsch and W. E. Weesner, U.S. pat. 3,148,218 [*C.A.*, **61**, 13196 (1964)].
[732] P. Mastagli and G. V. Durr, *Bull. Soc. Chim. France*, 268 (1955).
[733] E. Meier and H. Lauth, Ger. pat. 882,091 [*C.A.*, **52**, 11121 (1958)].
[734] F. J. Metzger, U.S. pat. 2,433,254 [*C.A.*, **43**, 3451 (1949)].
[735] Z. Czarny, *Chem. Stosowana, Ser. A*, **7**, 609 (1963) [*C.A.*, **62**, 6386 (1965)].
[736] A. Wurtz, *Compt. Rend.*, **76**, 1165 (1873).
[737] A. Wurtz, *Compt. Rend.*, **92**, 1438 (1881).
[738] A. Thalberg, *Monatsh. Chem.*, **19**, 154 (1898).
[739] Von O. Doebner and A. Weissenborn, *Ber.*, **35**, 1143 (1902).
[740] E. Hoppe, *Monatsh. Chem.*, **9**, 634 (1888).
[741] A. Lieben and S. Zeisel, *Monatsh. Chem.*, **4**, 10 (1883).
[742] W. Tischtschenko, *J. Russ. Phys. Chem. Soc.*, **38**, 482 (1906) [*Chem. Zentr.*, **77**, II, 1553 (1906)].
[743] C. D. Hurd and J. L. Abernethy, *J. Am. Chem. Soc.*, **63**, 1966 (1941).
[744] W. G. Berl and C. E. Feazel, *J. Am. Chem. Soc.*, **73**, 2054 (1951).
[745] C. Neuberg, *Ber.*, **35**, 2626 (1902).
[746] H. O. L. Fischer and E. Baer, *Helv. Chim. Acta*, **19**, 519 (1936).
[747] R. Nahum, *Compt. Rend.*, **243**, 849 (1956).
[748] I. Smedley, *J. Chem. Soc.*, **99**, 1627 (1911).
[749] H. Shingû and T. Okazaki, *Bull. Inst. Chem. Res. Kyoto Univ.*, **27**, 69 (1951) [*C.A.*, **47**, 2124 (1953)].
[750] H. Meerwein, B. v. Bock, B. Kirschnick, W. Lenz, and A. Migge, *J. Prakt. Chem.*, [2] **147**, 224 (1936).
[751] W. Langenbeck and O. Gödde, *Atti X° Congr. Intern. Chim.*, **3**, 230 (1939) [*C.A.*, **34**, 1970 (1940)].
[752] G. W. Seymour and V. S. Salvin, U.S. pat. 2,408,127 [*C.A.*, **41**, 771 (1949)].
[753] J. Schmitt and A. Obermeit, *Ann.*, **547**, 285 (1941).
[754] R. Kuhn and C. Grundmann, *Ber.*, **70**, 1318 (1937).
[755] K. Bernhauer and R. Drobnick, *Biochem. Z.*, **266**, 197 (1933) [*C.A.*, **28**, 740 (1934)].
[756] I. S. Hünig, *Ann.*, **569**, 198 (1950).
[757] Chemische Forschungs G.m.b.H., Ger. pat. 711, 732 [*C.A.*, **37**, 4161 (1943)].
[758] H. L. Du Mont and H. Fleischauer, *Ber.*, **71**, 1958 (1938).
[759] K. Bernhauer and G. Neubauer, *Biochem. Z.*, **251**, 173 (1932) [*C.A.*, **26**, 5907 (1932)].
[760] J. Jacques, *Ann. Chim. (Paris)*, [11] **20**, 363 (1945).
[761] P. Y. Blanc, *Helv. Chim. Acta*, **41**, 625 (1958).
[762] M. Delépine, *Ann. Chim. (Paris)*, [8] **20**, 389 (1910).
[763] C. Weizmann and S. F. Garrard, *J. Chem. Soc.*, **117**, 324 (1920).
[764] V. S. Batalin and S. E. Slavina, *Zh. Obshch. Khim.*, **7**, 202 (1937) [*C.A.*, **31**, 4267 (1937)].
[765] G. A. Raupenstrauch, *Monatsh. Chem.*, **8**, 108 (1887).
[766] R. Pummerer and J. Smidt, *Ann.*, **610**, 192 (1957).

[767] A. J. Hagemeyer and H. N. Wright, Ger. pat. 1,149,344 and Brit. pat. 922,826 [*C.A.*, **59,** 8596 (1963)].
[768] K. Billig, Ger. pat. 643,341 [*C.A.*, **31,** 4346 (1937)].
[769] R. N. Watts, W. J. Porter, J. A. Wingate, and J. A. Hanan, Fr. pat. 1,344,591 [*C.A.*, **60,** 10551 (1964)].
[770] F. Gregor, E. Pavlacka, and M. Marko, Czech. pat. 109,711 [*C.A.*, **60,** 15736 (1964)].
[771] T. Okada and T. Sumiya, Japan. pat. 3415 (1952) [*C.A.*, **48,** 3994 (1954)].
[772] V. Grignard and A. Vestermann, *Bull. Soc. Chim. France,* [4] **37,** 425 (1925).
[773] P. Herrmann, *Monatsh. Chem.*, **25,** 188 (1904).
[774] P. Richter, Czech. pat. 106,830 [*C.A.*, **60,** 2771 (1964)].
[775] V. F. Belyaev, *Sbortsiya iz Rastvorov Vysokopolimerami i Uglyami, Belorussk. Gos. Univ.*, **1961,** 143 [*C.A.*, **57,** 7080 (1962)].
[776] W. E. Heinz and A. P. MacLean, U.S. pat. 3,077,500 [*C.A.*, **59,** 1491 (1963)].
[777] A. Gorhan, *Monatsh. Chem.*, **26,** 73 (1905).
[778] R. Lombard and J. P. Stéphan, *Bull. Soc. Chim. France,* 1369 (1957).
[779] M. Backès, *Compt. Rend.*, **196,** 1674 (1933).
[780] J. Hagemeyer, U.S. pat. 2,829,169 [*C.A.*, **52,** 12897 (1958)].
[781] W. Fossek, *Monatsh. Chem.*, **4,** 663 (1883).
[782] M. Brauchbar, *Monatsh. Chem.*, **17,** 637 (1896).
[783] W. H. Perkin, Jr., *J. Chem. Soc.*, **43,** 90 (1883).
[784] A. Pfeiffer, *Ber.*, **5,** 699 (1872).
[785] A. Franke and L. Kohn, *Monatsh. Chem.*, **19,** 354 (1898).
[786] G. Urbain, *Bull. Soc. Chim. France,* [3] **13,** 1048 (1895).
[787] W. Fossek, *Monatsh. Chem.*, **2,** 614 (1881).
[788] E. H. Usherwood, *J. Chem. Soc.*, **123,** 1717 (1923).
[789] F. Urech, *Ber.*, **12,** 1744 (1879).
[790] F. J. Villani and F. F. Nord, *J. Am. Chem. Soc.*, **68,** 1674 (1946).
[791] L. Kohn, *Monatsh. Chem.*, **19,** 519 (1898).
[792] M. S. Oeconomides, *Bull. Soc. Chim. France,* [2] **36,** 209 (1881).
[793] W. V. McConnell and H. E. Davis, U.S. pat. 3,038,907 [*C.A.*, **57,** 12503 (1962)].
[794] M. Brauchbar and L. Kohn, *Monatsh. Chem.*, **19,** 16, 361 (1898).
[795] M. M. T. Plant, *J. Chem. Soc.*, 536 (1938).
[796] R. S. Raper, *J. Chem. Soc.*, **91,** 1831 (1907).
[797] L. Kohn, *Monatsh. Chem.*, **17,** 126 (1896).
[798] G. Bruylants, *Ber.*, **8,** 414 (1875).
[799] A. Kekulé, *Ber.*, **3,** 135 (1870).
[800] R. Fittig, *Ann.*, **117,** 68 (1861).
[801] E. H. Ingold, *J. Chem. Soc.*, **125,** 435 (1924).
[802] J. U. Nef, *Ann.*, **318,** 137 (1901).
[803] P. Schorigin, W. Issaguljanz, and A. Gussewa, *Ber.*, **66,** 1431 (1933).
[804] Y. Fujita, *J. Chem. Soc. Japan,* **58,** 305 (1937) [*C.A.*, **31,** 6192 (1937)].
[805] H. Rosinger, *Monatsh. Chem.*, **22,** 545 (1901).
[806] A. Lederer, *Monatsh. Chem.*, **22,** 536 (1901).
[807] V. Neustädter, *Monatsh. Chem.*, **27,** 879 (1906).
[808] W. Fossek, *Monatsh. Chem.*, **3,** 622 (1882).
[809] L. Kohn, *Monatsh. Chem.*, **18,** 189 (1897).
[810] M. Cihlar, *Monatsh. Chem.*, **25,** 149 (1904).
[811] J. Rainer, *Monatsh. Chem.*, **25,** 1035 (1904).
[812] R. Kitaoka, *Nippon Kagaku Zasshi,* **77,** 627 (1956) [*C.A.*, **52,** 322 (1958)].
[813] F. Büttner, *Ann.*, **583,** 184 (1953).
[814] R. R. Whetstone, U.S. pat. 2,481,377 and Brit. pat. 667,131 [*C.A.*, **46,** 7128 (1952); **44,** 2037 (1950)].
[815] H. R. Guest and B. W. Kiff, Brit. pat. 770,381 [*C.A.*, **51,** 13938 (1957)].
[816] H. R. Guest and B. W. Kiff, U.S. pat. 2,861,083 [*C.A.*, **53,** 12305 (1959)].
[817] R. B. Clayton, H. B. Henbest, and M. Smith, *J. Chem. Soc.*, 1982 (1957).

[818] J. English and G. W. Barber, *J. Am. Chem. Soc.*, **71**, 3310 (1949).

[819] A. Wohl and H. Schweitzer, *Ber.*, **39**, 890 (1906).

[820] A. Baeyer and H. von Liebeg, *Ber.*, **31**, 2106 (1898).

[821] A. Wohl and J. L. Losanitsch, *Ber.*, **40**, 4685 (1907).

[822] British Celanese Ltd., Brit. pat. 587,749 [*C.A.*, **42**, 216 (1948)].

[823] W. H. Perkin, Jr., *Ber.*, **16**, 1033 (1883).

[824] M. Métayer and G. de Bièvre-Gallin, *Compt. Rend.*, **225**, 635 (1947).

[825] G. Durr and P. Mastagli, *Compt. Rend.*, **235**, 1038 (1952).

[826] A. Bussy, *J. Prakt. Chem.*, [3] **37**, 92 (1854).

[827] W. Treibs and K. Krumbholz, *Chem. Ber.*, **85**, 1116 (1952).

[828] J. L. Ernst and N. L. Cull, U.S. pat. 2,810,762 [*C.A.*, **52**, 2894 (1958)].

[829] K. T. Potts and R. Robinson, *J. Chem. Soc.*, 2466 (1955).

[830] P. Baumgarten, *Ber.*, **68**, 1316 (1935).

[831] M. G. J. Beets and H. van Essen, *Rec. Trav. Chim.*, **74**, 98 (1955).

[832] J. v. Braun and F. Zobel, *Ber.*, **56**, 2139 (1923).

[833] E. V. Budnitskaya, *Tr. Komis. Analit. Khim. Akad. Nauk SSSR, Inst. Geokhim. i Analit. Khim.*, **6**, 197 (1955) [*C.A.*, **50**, 8967 (1956)].

[834] W. G. Young and S. Siegel, *J. Am. Chem. Soc.*, **66**, 354 (1944).

[835] H. E. French and D. M. Gallagher, *J. Am. Chem. Soc.*, **64**, 1497 (1942).

[836] J. Wolinsky, M. R. Slabaugh, and T. Gibson, *J. Org. Chem.*, **29**, 3740 (1964).

[837] H. Muxfeldt and G. Hardtmann, *Ann.*, **669**, 113 (1963).

[838] Farbwerke Hoechst A.-G., Brit. pat. 983,868 [*C.A.*, **63**, 4228 (1965)].

[839] S. Fujii, Japan. pat. 153,925 [*C.A.*, **43**, 3447 (1949)].

[840] H. Jackson and G. G. Jones, U.S. pat. 2,562,102 [*C.A.*, **46**, 3559 (1952)].

[841] J. A. Wyler, U.S. pat. 2,240,274 [*C.A.*, **35**, 5135 (1941).]

[842] R. F. Burghardt and R. H. Barth, U.S. pat. 2,401,749 [*C.A.*, **40**, 5254 (1946)].

[843] H. A. Bruson and J. J. Hewitt, U.S. pat. 2,998,458 [*C.A.*, **56**, 2330 (1962)].

[844] Celanese Corp. of America, Brit. pat. 757,564 [*C.A.*, **51**, 9678 (1957)].

[845] H. M. Spurlin, U.S. pat. 2,364,925 [*C.A.*, **39**, 4626 (1945)].

[846] A. Corbellini and A. Langini, *Giorn. Chim. Ind. ed Applicata*, **15**, 53 (1933) [*C.A.*, **27**, 4526 (1933)].

[847] M. O. Robeson, Ger. pat. 1,158,950 [*C.A.*, **60**, 15733 (1964)].

[848] F. O. Meissner and H. O. A. Meissner, Belg. pat. 626,845 [*C.A.*, **60**, 7916 (1964)].

[849] W. Fitzky, U.S. pat. 2,275,586 [*C.A.*, **36**, 4233 (1942)].

[850] H. J. Backer and H. B. J. Schurink, *Rec. Trav. Chim.*, **50**, 921 (1931).

[851] J. A. Wyler, U.S. pat. 2,206,379 [*C.A.*, **34**, 7301 (1940)].

[852] W. Pohl, Ger. pat. 845,194 and 857,804 [*C.A.*, **52**, 6397 (1958)].

[853] T. R. Paterson, U.S. pat. 2,011,589 [*C.A.*, **29**, 6610 (1935)].

[854] T. Sakai, Japan. pat. 94,210 [*C.A.*, **27**, 2697 (1933)].

[855] Deutsche Gold- und Silber-Scheideanstalt vorm. Roesseler, Fr. pat. 744,397 [*C.A.*, **27**, 3953 (1933)].

[856] J. A. Wyler, U.S. pat. 2,152,371 [*C.A.*, **33**, 5188 (1939)].

[857] W. Schemuth, Ger. pat. 1,048,567 [*C.A.*, **55**, 2489 (1961)].

[858] R. Carpentier and P. Lambert, Belg. pat. 565,088 and U.S. pat. 2,950,327 [*C.A.*, **53**, 7994 (1959)].

[859] A. Roche and G. Bourjol, Fr. pat. 1,173,735 [*C.A.*, **56**, 3356 (1962)].

[860] M. O. Robeson, Ger. pat. 1,158,950 [*C.A.*, **60**, 15733 (1964)].

[861] A. J. Poynton, U.S. pat. 2,978,514 [*C.A.*, **55**, 18598 (1961)].

[862] A. Roche, Fr. pat. 1,349,854 [*C.A.*, **62**, 9010 (1965)].

[863] E. Paltin and E. Teodoru, *Rev. Chim. (Bucharest)*, **15**, 546 (1964) [*C.A.*, **63**, 17875 (1965)].

[864] Manufactures de produits chimiques du Nord, Établissements Kuhlmann, Fr. pat. 962,381 [*C.A.*, **46**, 5615 (1952)].

[865] S. H. McAllister and E. F. Bullard, Brit. pat. 502,450 [*C.A.*, **33**, 6874 (1939)].

[866] A. V. Stepanov and M. Shchukina, *J. Russ. Phys. Chem. Soc.*, **58**, 840 (1926) [*C.A.*, **21**, 1094 (1927)].

[867] S. Ropuszynski and H. Matyschok, *Roczniki Chem.*, **39,** 1347 (1965) [*C.A.*, **64,** 3340 (1966)].

[868] S. Malinowski, H. Jedrzejewska, S. Basinski, Z. Lipski, and J. Moszczenska, *Roczniki Chem.*, **30,** 1129 (1956) [*C.A.*, **51,** 8650 (1957)].

[869] W. A. Smart, Brit. pat. 600,454 [*C.A.*, **42,** 7787 (1948)].

[870] T. B. Philip, H. M. Stanley, and W. A. Smart, Brit. pat. 573,573 [*C.A.*, **43,** 3026 (1949)].

[871] T. Ishikawa and T. Kamio, *Tokyo Kogyo Shikensho Hokoku*, **58,** 40 (1963) [*C.A.*, **61,** 4802 (1964)].

[872] A. F. MacLean, U.S. pat. 2,517,006 [*C.A.*, **45,** 639 (1951)].

[873] I. G. Farbenind. A.-G., Fr. pat. 847,370 [*C.A.*, **35,** 5127 (1941)].

[874] M. Gallagher and R. L. Harche, U.S. pat. 2,246,037 [*C.A.*, **35,** 5907 (1941)].

[875] M. Gallagher and R. L. Harche, U.S. pat. 2,245,582 [*C.A.*, **35,** 5907 (1941)].

[876] H. Schulz and H. Wagner, Ger. pat. 707,021 [*C.A.*, **36,** 1955 (1942)].

[877] Kodak-Pathé, Fr. pat. 830,750 [*C.A.*, **33,** 1343 (1939)].

[878] H. Schulz and H. Wagner, *Angew. Chem.*, **62,** 105 (1950).

[879] S. Malinowski, H. Jedrzejewska, S. Basinski, and Z. Lipski, *Roczniki Chem.*, **31,** 71 (1957) [*C.A.*, **51,** 14557 (1957)].

[880] S. Malinowski, S. Basinski, M. Olszewska, and H. Zieleniewska, *Roczniki Chem.*, **31,** 123 (1957) [*C.A.*, **51,** 14557 (1957)].

[881] H. Wagner, U.S. pat. 2,288,306 [*C.A.*, **37,** 201 (1943)].

[882] S. Malinowski and S. Basinski, *Przemysl Chem.*, **41,** 202 (1962) [*C.A.*, **57,** 5787 (1962)].

[883] T. Ichikawa and T. Kamio, *Yuki Gosei Kagaku Kyokai Shi*, **20,** 56 (1962) [*C.A.*, **56,** 11433 (1962)].

[884] S. Malinowski and S. Basinski, Pol. pat. 45,675 (1962) [*C.A.*, **59,** 9801 (1963)].

[885] G. L. Laemmle, J. G. Milligan, and W. J. Peppel, *Ind. Eng. Chem.*, **52,** 33 (1960).

[886] W. Winzer, K. H. Richter, and S. Neuber, Ger. (East) pat. 33,455 [*C.A.*, **63,** 13078 (1965)].

[887] M. Durand, J. M. Emeury, S. Marsaule, and M. A. Lanfrit, Fr. pat. 1,399,678 [*C.A.*, **63,** 16215 (1965)].

[888] M. O. Robeson, U.S. pat. 3,183,274 [*C.A.*, **63,** 4162 (1965)].

[889] M. O. Robeson, U.S. pat. 2,790,837 [*C.A.*, **51,** 15555 (1957)].

[890] M. M. Ketslakh, D. M. Rudkovskii, and F. A. Eppel, *Tr., Vses. Nauchn.-Issled. Inst. Neftekhim. Protsessov*, 178 (1960) [*C.A.*, **56,** 8542 (1962)].

[891] M. M. Brubaker and R. A. Jacobson, U.S. pat. 2,292,926 [*C.A.*, **37,** 890 (1943)].

[892] H. Hosaeus, *Ann.*, **276,** 75 (1893).

[893] H. Stetter and W. Böckmann, *Chem. Ber.*, **84,** 834 (1951).

[894] H. A. Poitras, J. E. Snow, and S. A. DeLorenzo, U.S. pat. 2,420,496 [*C.A.*, **41,** 5547 1947)].

[895] K. Koch and T. Zerner, *Monatsh. Chem.*, **22,** 443 (1901).

[896] F. Gresham, U.S. pat. 2,549,457 [*C.A.*, **45,** 8549 (1951)].

[897] T. Mitsui, M. Mitahara, and Y. Kiyatake, Japan. pat. 23,159 (1963) [*C.A.*, **60,** 2775 (1964)].

[898] T. Mitsui, M. Kitahara, and Y. Miyatake, *Rika Gaku Kenkyusho Hokoku*, **38,** 205, 217 (1962) [*C.A.*, **59,** 3762 (1963)].

[899] W. Friederich and W. Brün, *Ber.*, **63,** 2681 (1930).

[900] R. Pummerer, F. Aldebert, F. Büttner, F. Graser, E. Pirson, H. Rick and H. Sperber, *Ann.*, **583,** 161 (1953).

[901] P. Z. Bedoukian, *Am. Perfumer Essent. Oil Rev.*, **56,** 207 (1950) [*C.A.*, **45,** 2438 (1951)].

[902] W. E. Hanford and R. S. Schreiber, U.S. pat. 2,317,456 [*C.A.*, **37,** 5805 (1943)].

[903] R. Pummerer, E. Pirson, and H. Rick, Ger. pat. 821,204 [*C.A.*, **48,** 12170 (1954)].

[904] M. F. Shustakovskii and N. A. Keiko, *Dokl. Akad. Nauk SSSR*, **162,** 362 (1965) [*C.A.*, **63,** 5520 (1965)].

[905] H. Danziger and K. Haeseler, Ger. pat. 1,154,080 [*C.A.*, **60,** 14386 (1964)].

[906] Nobel-Bozel, Fr. pat. 1,299,029 [*C.A.*, **58,** 449 (1963)].

[907] Esso Research and Engineering Co., Brit. pat. 904,780 [*C.A.* **58,** 3317 (1963)].

[908] B. Audouze, P. Grancher, and M. Ortigues, Fr. pat. 1,255,982 [*C.A.*, **56,** 14082 (1962)].

[909] N. M. Bortnick, U.S. pat. 2,518,416 [*C.A.*, **45**, 1158 (1951)].
[910] M. Apel and B. Tollens, *Ber.*, **27**, 1087 (1894).
[911] M. Apel and B. Tollens, *Ann.*, **289**, 36 (1895).
[912] J. A. Wyler, U.S. pat. 2,786,083 [*C.A.*, **51**, 12132 (1957)].
[913] R. W. Shortridge, R. A. Craig, K. W. Greenlee, J. M. Derfer, and C. E. Boord, *J. Am. Chem. Soc.*, **70**, 946 (1948).
[914] J. F. Walker and N. Turnbull, U.S. pat. 2,135,063 [*C.A.*, **33**, 997 (1939)].
[915] P. Meyersberg, *Monatsh. Chem.*, **26**, 41 (1905).
[916] R. A. Jacobson, *J. Am. Chem. Soc.*, **67**, 1999 (1945).
[917] Y. Matsunaga, H. Ishida, and S. Takabe, Japan. pat. 609 (1964) [*C.A.*, **60**, 10548 (1964)].
[918] T. A. Favorskaya and Y. M. Portnyagin, *Zh. Obshch. Khim.*, **34**, 1065 (1964) [*C.A.*, **61**, 628 (1964)].
[919] L. Wessely, *Monatsh. Chem.*, **21**, 216 (1900).
[920] E. T. Stiller, S. A. Harris, J. Finkelstein, J. C. Keresztesy, and K. Folkers, *J. Am. Chem. Soc.*, **62**, 1785 (1940).
[921] J. H. Ford, *J. Am. Chem. Soc.*, **66**, 20 (1944).
[922] S. Sabetay and J. Bléger, *Bull. Soc. Chim. France*, [4] **47**, 885 (1930).
[923] J. W. Lynn, U.S. pat. 2,863,878 [*C.A.*, **53**, 7019 (1959)].
[924] H. N. Wright and H. J. Hagemeyer, Brit. pat. 881,842 [*C.A.*, **57**, 4545 (1962)].
[925] E. Späth and I. v. Szilágyi, *Ber.*, **76**, 949 (1943).
[926] V. P. Kravetz, *J. Russ. Phys. Chem. Soc.*, **45**, 1451 (1913) [*C.A.*, **8**, 325 (1914)].
[927] L. P. Kuhn and A. C. Duckworth, *J. Org. Chem.*, **24**, 1005 (1959).
[928] L. F. Theiling and R. J. Knopf, U.S. pat. 2,982,790 [*C.A.*, **56**, 326 (1962)].
[929] T. J. Prosser, *J. Org. Chem.*, **25**, 2039 (1960).
[930] V. Neustädter, *Ann.*, **351**, 294 (1906).
[931] Y. M. Slobodin, V. I. Grigor'eva, and Y. E. Shmulyakovskiĭ, *Zh. Obshch. Khim.*, **23**, 1665 (1953) [*C.A.*, **48**, 13637 (1954)].
[932] A. Franke, *Monatsh. Chem.*, **34**, 1893 (1913).
[933] Chemische Werke Hüls, A.-G., Brit. pat. 850,680 [*C.A.*, **55**, 7293 (1961)].
[934] M. M. Ketslakh, D. M. Rudkovskii, and F. A. Eppel, *Khim. Prom.*, (9), 666 (1962) [*C.A.*, **59**, 7361 (1963)].
[935] M. M. Ketslakh, D. M. Rudkovskii, and F. A. Eppel, *Okososinetz, Poluchemie Metodom Oksosinteza Al'degidov, Spiritov i Vtorichnykh Produktov naikh Osnove, Vses. Nauchn.-Issled. Inst. Neftekhim. Protsessov*, 156 (1963) [*C.A.*, **60**, 9133 (1964)].
[936] J. E. Snow, U.S. pat. 2,775,622 [*C.A.*, **51**, 12964 (1957)].
[937] A. Klemola and G. A. Nymann, *Acta Polytech. Scand., Chem. Met. Ser.*, (31), 16 (1964) [*C.A.*, **61**, 13176, 13272 (1964)].
[938] K. C. Brannock, U.S. pat. 2,912,447 [*C.A.*, **54**, 6114 (1960)].
[939] B. Weibull and M. Matell, *Acta Chem. Scand.*, **16**, 1062 (1962).
[940] E. G. Popova, I. A. Kuznetsova, and V. I. Zeifman, U.S.S.R. pat. 156,656 [*C.A.*, **60**, 6747 (1964)].
[941] T. Luessling, Ger. pat. 1,150,384 [*C.A.*, **59**, 13842 (1963)].
[942] R. Schaffer, *J. Am. Chem. Soc.*, **81**, 5452 (1959).
[943] C. Bernardy, M. Prillieux, and R. Muths, Fr. pat. 1,301,823 [*C.A.*, **59**, 9792 (1963)].
[944] M. Portelli and G. Luchi, *Ann. Chim. (Rome)*, **53**, 1441 (1963) [*C.A.*, **60**, 6736 (1964)].
[945] M. Portelli and G. Luchi, Brit. pat. 991,978 [*C.A.*, **63**, 8213 (1965)].
[946] H. A. Bruson and W. D. Niederhauser, U.S. pat. 2,418,290 [*C.A.*, **41**, 3813 (1947)].
[947] W. N. Cannon, *J. Org. Chem.*, **21**, 1519 (1956).
[948] H. A. Bruson, W. D. Niederhauser, and H. Iserson, U.S. pat. 2,417,100 [*C.A.*, **41**, 3819 (1947)].
[949] J. G. Burr, *J. Am. Chem. Soc.*, **73**, 5170 (1951).
[950] D. Bertin, H. Fritel, and L. Nedelec, Ger. pat. 1,159,439 [*C.A.*, **61**, 708 (1964)].
[951] G. Muller and J. Mathieu, Ger. pat. 1,145,611 [*C.A.*, **60**, 619 (1964)].
[952] D. Bertin, A. Locatelli, J. Mathieu, G. Muller, and H. Fritel, U.S. pat. 3,052,675 [*C.A.*, **58**, 2489 (1963)].
[953] A. Chwala and W. Bartek, *Monatsh. Chem.*, **82**, 652 (1951).

[954] H. Rosinger, *Monatsh. Chem.*, **28,** 947 (1907).

[955] F. X. Schmalzhofer, *Monatsh. Chem.*, **21,** 671 (1900).

[956] K. Matsuyo and S. Tsutsumi, Japan. pat. 2217 (1953) [*C.A.*, **49,** 1780 (1955)].

[957] R. Kuhn and M. Hoffer, *Ber.*, **64,** 1977 (1931).

[958] R. Kuhn, W. Badstübner, and C. Grundmann, *Ber.*, **69,** 98 (1936).

[959] R. Kuhn and M. Hoffer, *Ber.*, **63,** 2164 (1930).

[960] T. Reichstein and G. Trivelli, *Helv. Chim. Acta*, **15,** 1074 (1932).

[961] F. G. Fischer and O. Wiedemann, *Ann.*, **513,** 251 (1934).

[962] P. Baumgarten and G. Glatzel, *Ber.*, **59,** 2658 (1926).

[963] R. Decker and H. Holz, Ger. pat. 837,995 [*C.A.*, **51,** 15548 (1957)].

[964] M. Lilienfeld and S. Tauss, *Monatsh. Chem.*, **19,** 77 (1898).

[965] B. Eissler and A. Pollak, *Monatsh. Chem.*, **27,** 1129 (1906).

[966] A. Wogrinz, *Monatsh. Chem.*, **22,** 1 (1901).

[967] K. Adler, J. Haydn, K. Heimbach, and K. Neufang, *Ann.*, **586,** 110 (1954).

[968] Societé des Usines Chimiques Rhone-Poulenc, Fr. pat. 1,187,108 [*C.A.*, **56,** 354 (1962)].

[969] A. Einhorn, *Ber.*, **17,** 2026 (1884).

[970] R. Kuhn and A. Winterstein, *Helv. Chim. Acta*, **11,** 87 (1928).

[971] Y. Hirata, H. Nakata, K. Yamada, K. Okuhara, and T. Naito, *Tetrahedron*, **14,** 252 (1961).

[972] M. Lipp and F. Dallacker, *Chem. Ber.*, **90,** 1730 (1957).

[973] B. Rutovski and A. Korolev, *Trans. Sci. Chem-Pharm. Inst.* (*USSR*), 153 (1928) [*C.A.*, **23,** 4942 (1929)].

[974] L. Hough and J. K. N. Jones, *Nature*, **167,** 180 (1951).

[975] L. Hough and J. K. N. Jones, *J. Chem. Soc.*, 1122 (1951).

[976] A. Knorr and A. Weissenborn, U.S. pat. 1,716,822 [*C.A.*, **23,** 3714 (1929)].

[977] M. Mousseron-Canet and M. Mousseron, *Bull. Soc. Chim. France*, 391 (1956).

[978] Y. Obata and T. Yamanishi, *J. Agr. Chem. Soc. Japan*, **24,** 479 (1951) [*C.A.*, **46,** 11475 (1952)].

[979] R. Kuhn and K. Wallenfels, *Ber.*, **70,** 1331 (1937).

[980] J. W. Batty, A. Burawoy, I. M. Heilbron, W. E. Jones, and A. Lowe, *J. Chem. Soc.*, 755 (1937).

[981] J. Schreiber and C.-G. Wermuth, *Bull. Soc. Chim. France*, 2242 (1965).

[982] W. Morawetz, *Monatsh. Chem.*, **26,** 127 (1905).

[983] K. Michel and K. Spitzauer, *Monatsh. Chem.*, **22,** 1119 (1901).

[984] R. C. Fuson and R. E. Christ, *Science*, **84,** 294 (1936).

[985] B. G. Yasnitskii and E. B. Dol'berg, U.S.S.R. pat. 139,318 [*C.A.*, **56,** 7219 (1962)].

[986] J. A. Faust and M. Sahyun, U.S. pat. 3,094,561 [*C.A.*, **59,** 11330 (1963)].

[987] A. Baeyer and V. Drewsen, *Ber.*, **16,** 2205 (1883).

[988] C. F. Göhring, *Ber.*, **18,** 719 (1885).

[989] F. Kinkelin, *Ber.*, **18,** 483 (1885).

[990] C. F. Göhring, *Ber.*, **18,** 371 (1885).

[991] T. Nishimura, *Bull. Chem. Soc. Japan*, **25,** 54 (1952) [*C.A.*, **48,** 5143 (1954)].

[992] A. Scipioni, *Ann. Chim.* (*Rome*), **41,** 697 (1951) [*C.A.*, **46,** 8635 (1952)].

[993] S. Malinowski, W. Kiewlicz, and E. Soltys, *Roczniki Chem.*, **38,** 447 (1964).

[994] H. H. Richmond, U.S. pat. 2,529,186 [*C.A.*, **45,** 2979 (1951)].

[995] S. Kodama, *J. Chem. Soc. Japan*, **43,** 691 (1922) [*C.A.*, **17,** 2571 (1923)].

[996] A. Scipioni, Ital. pat. 481,245 [*C.A.*, **49,** 15966 (1955)].

[997] L. Chiozza, *Ann.*, **97,** 350 (1856).

[998] S. Malinowski, W. Kiewlicz, and E. Soltys, *Bull. Soc. Chim. France*, 439 (1963).

[999] D. Vorländer and E. Daehn, *Ber.*, **62,** 541 (1929).

[1000] D. Vorländer, E. Fischer, and K. Kunze, *Ber.*, **58,** 1284 (1925).

[1001] P. Friedlaender, *Ber.*, **15,** 2572 (1882).

[1002] A. Lüttringhaus and G. Schill, *Angew. Chem.*, **69,** 137 (1957).

[1003] A. Lüttringhaus and G. Schill, *Chem. Ber.*, **93,** 3048 (1960).

[1004] A. Landenburg and M. Scholtz, *Ber.*, **27,** 2958 (1894).

[1005] A. C. Cope, T. A. Liss, and D. S. Smith, *J. Am. Chem. Soc.*, **79,** 240 (1957).

[1006] M. Scholtz and A. Wiedemann, *Ber.*, **36,** 845 (1903).

[1007] K. Feurstein, *J. Prakt. Chem.*, [2] **143,** 174 (1935).

[1008] V. F. Lavrushin, S. V. Tsukerman, and V. M. Nikitchenko, *Zh. Obshch. Khim.*, **31,** 2845 (1961) [*C.A.*, **56,** 15456 (1962)].

[1009] H. Pauly and K. Wäscher, *Ber.*, **56,** 603 (1923).

[1010] W. König, W. Schramek, G. Rösch, and H. Arnold, *Ber.*, **61,** 2074 (1928).

[1011] J. Dorsky and W. M. Easter, U.S. pat. 2,976,321 [*C.A.*, **55,** 16485 (1961)].

[1012] H. Pauly and L. Strassberger, *Ber.*, **62,** 2277 (1929).

[1013] K. Bernhauer and I. Skudrzyk, *J. Prakt. Chem.*, [2] **155,** 310 (1940).

[1014] T. Hackhofer, *Monatsh. Chem.*, **22,** 95 (1901).

[1015] E. D. Laskina and T. A. Devitskaya, *Maslob.-Zhir. Prom.*, **29,** 23 (1963) [*C.A.*, **59,** 11313 (1963)].

[1016] A. Knorr and A. Weissenborn, U.S. pat. 1,844,013 [*C.A.*, **26,** 1940 (1932)].

[1017] R. Kuhn and A. Winterstein, *Helv. Chim. Acta*, **12,** 493 (1929).

[1018] W. M. Kraft, *J. Am. Chem. Soc.*, **70,** 3569 (1948).

[1019] M. J. Stritar, *Monatsh. Chem.*, **20,** 617 (1899).

[1020] A. Hildesheimer, *Monatsh. Chem.*, **22,** 497 (1901).

[1021] W. Subak, *Monatsh. Chem.*, **24,** 167 (1903).

[1022] B. N. Rutowski and A. I. Korolew, *J. Prakt. Chem.*, [2] **119,** 272 (1928).

[1023] P. E. Verkade and W. Meerburg, *Rec. Trav. Chim.*, **69,** 565 (1950).

[1024] A. Weizmann, *J. Am. Chem. Soc.*, **66,** 310 (1944).

[1025] A. P. Dunlop, U.S. pat. 2,993,912 and 2,993,913 [*C.A.*, **56,** 3454 (1962)].

[1026] W. C. Meuly, U.S. pat. 2,102,965 [*C.A.*, **32,** 1277 (1938)].

[1027] D. S. P. Eftax and A. P. Dunlop, *J. Org. Chem.*, **26,** 2106 (1961).

[1028] A. P. Dunlop and E. L. Washburn, U.S. pat. 2,527,714 [*C.A.*, **45,** 2715 (1951)].

[1029] H. Kappeler, U.S. pat. 1,873,599 [*C.A.*, **26,** 6081 (1932)].

[1030] R. E. Miller and F. F. Nord, *J. Org. Chem.*, **16,** 1720 (1951).

[1031] A. A. Ponomarev and M. D. Lipanova, *Zh. Obshch. Khim.*, **32,** 2535 (1962) [*C.A.*, **58,** 9066 (1963)].

[1032] G. Carrara, Brit. pat. 744,187 [*C.A.*, **51,** 1284 (1957)].

[1033] H. Saikachi, Y. Taniguchi, and H. Ogawa, *Yakugaku Zasshi*, **82,** 1262 (1962) [*C.A.*, **58,** 13887 (1963)].

[1034] H. Saikachi and S. Kimura, *J. Pharm. Soc. Japan*, **73,** 716 (1953) [*C.A.*, **48,** 7002 (1954)].

[1035] H. Keskin, R. E. Miller, and F. F. Nord, *J. Org. Chem.*, **16,** 199 (1951).

[1036] R. E. Miller and F. F. Nord, *J. Org. Chem.*, **16,** 1380 (1951).

[1037] G. Kimura and S. Chikaoka, *Yuki Gosei Kagaku Kyokai Shi*, **22,** 643 (1964) [*C.A.*, **61,** 9453 (1964)].

[1038] R. Takamoto and T. Hirohashi, *J. Pharm. Soc. Japan*, **48,** 446 (1928) [*C.A.*, **22,** 3409 (1928)].

[1039] H. Röhmer, *Ber.*, **31,** 281 (1898).

[1040] A. Hinz, G. Meyer, and G. Schücking, *Ber.*, **76,** 676 (1943).

[1041] W. König and K. Hey, Ger. pat. 330,358 [*C.A.*, **15,** 2102 (1921)].

[1042] W. König, *Ber.*, **58,** 2559 (1925).

[1043] J. Wiemann, *Bull. Soc. Chim. France*, [5] **2,** 1209 (1935).

[1044] H. Brunel, Fr. pat. 844,182 [*C.A.*, **34,** 7296 (1940)].

[1045] N. I. Shuĭkin and I. F. Bel'skiĭ, *Dokl. Akad. Nauk SSSR*, **137,** 622 (1961) [*C.A.*, **55,** 19889 (1961)].

[1046] A. K. Fukui and M. Takei, *Bull. Inst. Chem. Res. Kyoto Univ.*, **26,** 85 (1951) [*C.A.*, **49,** 5426 (1955)].

[1047] D. Ivanov, *Bull. Soc. Chim. France*, [4] **35,** 1658 (1924).

[1048] P. Mastagli, Z. Zafiriades, G. Durr, A. Floc'h, and G. Lagrange, *Bull. Soc. Chim. France*, [5] **20,** 693 (1953).

[1049] H. Saikachi and H. Ogawa, *J. Am. Chem. Soc.*, **80,** 3642 (1958).

[1050] W. J. Lantz and J. M. Walters, U.S. pat. 2,660,573 [*C.A.*, **48,** 3724 (1954)].

[1051] J. C. Schmidt, *Ber.*, **14,** 574 (1881).

[1052] K. Kulka, U.S. pat. 3,227,731 [*C.A.*, **64,** 8136 (1966)].

[1053] P. Mastagli, A. Floc'h, and G. Durr, *Compt. Rend.*, **235**, 1402 (1952).
[1054] I. K. Sarycheva, G. A. Serebrennikova, L. I. Mitrushkina, and N. A. Preobrazhenskii, *Zh. Obshch. Khim.*, **31**, 2190 (1961) [*C.A.*, **56**, 2366 (1962)].
[1055] J. B. Senderens, *Bull. Soc. Chim. France*, [4] **3**, 823 (1908); *Compt. Rend.*, **146**, 1211 (1908).
[1056] W. N. Ipatiev and A. D. Petrov, *Ber.*, **60**, 753 (1927).
[1057] W. N. Ipatiev and A. D. Petrov, *Ber.*, **59**, 2035 (1926).
[1058] A. D. Petrov, *Ber.*, **60**, 2548 (1927).
[1059] W. A. Bailey and W. H. Peterson, U.S. pat. 2,393,510 [*C.A.*, **40**, 2455 (1946)].
[1060] A. M. Kuliev, A. M. Levshina, and A. G. Zul'fugarova, *Azerb. Khim. Zh.*, (5), 29 (1959) [*C.A.*, **59**, 2638 (1963)].
[1061] I. G. Palmer and D. G. Cowan, Brit. pat. 921,510 [*C.A.*, **59**, 9801 (1963)].
[1062] J. Bredt and R. Rübel, *Ann.*, **299**, 160 (1897).
[1063] C. Völckel, *Ann.*, **82**, 63 (1852).
[1064] J. Hertkorn, Ger. pat. 258,057 [*C.A.*, **7**, 2836 (1913)].
[1065] R. Fittig, *Ann.*, **112**, 309 (1859).
[1066] R. Fittig, *Ann.*, **110**, 23 (1859).
[1067] C. Porlezza and V. Gatti, *Gazz. Chim. Ital.* **54**, 491 (1924) [*C.A.*, **19**, 35 (1925)].
[1068] R. Rieth and F. Beilstein, *Ann.*, **126**, 241 (1863).
[1069] D. Pavlov, *Ann.*, **188**, 138 (1877).
[1070] F. Bodroux and F. Taboury, *Bull. Soc. Chim. France*, [4] **3**, 829 (1908).
[1071] P. Sabatier and J. F. Durand, *Bull. Soc. Chim. France*, [4] **31**, 239 (1922).
[1072] P. P. Surmin, *Zh. Obshch. Khim.*, **5**, 1639 (1935) [*C.A.*, **30**, 3404 (1936)].
[1073] F. G. Klein and J. T. Banchero, *Ind. Eng. Chem.*, **48**, 1278 (1956).
[1074] S. Landa and V. Sesulka, *Chem. Listy*, **51**, 1159 (1957) [*C.A.*, **51**, 13814 (1957)].
[1075] W. R. Orndorff and S. W. Young, *Am. Chem. J.*, **15**, 249 (1893).
[1076] A. Pinner, *Ber.*, **15**, 586 (1882).
[1077] A W. Hofmann, *Ann.*, **71**, 121 (1849).
[1078] R. Fittig, *Ann.*, **141**, 129 (1867).
[1079] R. Fittig and W. H. Brueckner, *Ann.*, **147**, 42 (1868).
[1080] M. L. Tissier, *Ann. Chim. Phys.*, [6] **29**, 376 (1893).
[1081] G. Städeler, *Ann.*, **111**, 277 (1859).
[1082] P. C. Freer, *Ann.*, **278**, 116 (1894).
[1083] P. N. Raikow, *Chem. Ztg.*, **37**, 1455 (1913).
[1084] L. Bouveault and R. Locquin, *Ann. Chim. Phys.*, [8] **21**, 407 (1910).
[1085] A. E. Favorskiĭ and A. S. Onishchenko, *Zh. Obshch. Khim.*, **11**, 1111 (1941) [*C.A.*, **37**, 3735 (1943)].
[1086] K. Hirai and H. Hayashi, Ger. pat. 1, 161,872 [*C.A.*, **60**, 14390 (1964)].
[1087] R. Locquin, *Ann. Chim. (Paris)*, [9] **19**, 32 (1923).
[1088] N. A. Milas, U.S. pat. 2,997,486 [*C.A.*, **56**, 6061 (1962)].
[1089] J. Kordik, L. Risianova, and J. Husar, Czech. pat. 103,173 [*C.A.*, **59**, 9801 (1963)].
[1090] W. Heintz, *Ann.*, **169**, 114 (1873).
[1091] K. Schmitt, J. Disteldorf, and W. Baron, Ger. pat. 1,095,818 [*C.A.*, **56**, 5837 (1962)].
[1092] Hibernia-Chemie G.m.b.H., Belg. pat. 612,135 [*C.A.*, **58**, 3332 (1963)].
[1093] C. Loewig and S. Weidman, *J. Prakt. Chem.*, **21**, 54 (1840).
[1094] Societé industrielle des dérivés de l'acetylene, Brit. pat. 794, 344 [*C.A.*, **53**, 2119 (1959)].
[1095] V. Hancu, *Ber.*, **42**, 1052 (1909).
[1096] C. Mannich and V. H. Hancu, *Ber.*, **41**, 574 (1908).
[1097] G. Durr, *Compt. Rend.*, **236**, 1571 (1953).
[1098] A. Hoffman, Ger. pat. 229,678 [*C.A.*, **5**, 2535 (1911)].
[1099] A. Hoffman, *J. Am. Chem. Soc.*, **31**, 722 (1909).
[1100] W. Kerp and F. Müller, *Ann.*, **299**, 193 (1898).
[1101] W. Kerp, *Ann.*, **290**, 123 (1896).
[1102] J. Bredt and M. v. Rosenberg, *Ann.*, **289**, 1 (1896).
[1103] V. Grignard and M. Dubien, *Compt. Rend.*, **177**, 299 (1923).
[1104] V. Grignard, *Ann. Chim. Phys.*, [7] **24**, 433 (1901).

[1105] Y. Y. Tsmur, *Zh. Prikl. Khim.*, **34,** 1628 (1961) [*C.A.*, **55,** 27029 (1961)].

[1106] M. V. Gasselin, *Ann. Chim. Phys.*, [7] **3,** 5 (1894).

[1107] C. Weizmann, U.S. pat. 2,388, 101 [*C.A.*, **40,** 593 (1946)].

[1108] E. Louise, *Compt. Rend.*, **95,** 602 (1882).

[1109] C. Courtot and V. Ouperoff, *Compt. Rend.*, **191,** 416 (1930).

[1110] V. A. Zaitsev, K. P. Grinevitch, T. A. Balabina, I. M. Shevtsov, A. S. Sukhanova, R. L. Doroshkevitch, and I. V. Khvostov, U.S.S.R. pat. 104,868 [*C.A.*, **51,** 8777 (1957)].

[1111] O. Jacobsen, *Ann.*, **146,** 103 (1868).

[1112] M. Descudé, *Ann. Chim. Phys.*, [7] **29,** 486 (1903).

[1113] A. Baeyer, *Ann.*, **140,** 297 (1866).

[1114] L. Claisen, *Ann.*, **180,** 1 (1876).

[1115] P. C. Freer and A. Lachman, *Am. Chem. J.*, **19,** 887 (1897).

[1116] E. Knoevenagel and H. Beer, *Ber.*, **39,** 3457 (1906).

[1117] H. Midorikawa, *Rept. Sci. Res. Inst.* (*Tokyo*) **24,** 405 (1948) [*C.A.*, **45,** 4202 (1951)].

[1118] A. Mailhe and F. de Godon, *Bull. Soc. Chim. France,* [4] **21,** 61 (1917).

[1119] J. A. Mitchell and E. E. Reid, *J. Am. Chem. Soc.*, **53,** 330 (1931).

[1120] Esso Research and Engineering Co., Brit. pat. 909,941 [*C.A.*, **59,** 2652 (1963)].

[1121] M. Strell and E. Kopp, *Chem. Ber.*, **91,** 2854 (1958).

[1122] J. Colonge, *Bull. Soc. Chim. France,* [4] **45,** 200 (1929).

[1123] B. N. Rutovskiĭ, A. A. Berlin, and K. Zabyrina, *Zh. Obshch. Khim.*, **11,** 550 (1941) [*C.A.*, **35,** 6928 (1941)].

[1124] A. Franke and T. Köhler, *Ann.*, **433,** 314 (1923).

[1125] S. G. Powell and C. H. Secoy, *J. Am. Chem. Soc.*, **53,** 765 (1931).

[1126] J. E. Dubois, *Bull. Soc. Chim. France,* [5] **20,** C13 (1953).

[1127] H. Pariselle and Simon, *Compt. Rend.*, **173,** 86 (1921).

[1128] J. Colonge, *Bull. Soc. Chim. France,* [5] **1,** 1101 (1934).

[1129] V. I. Aksenova, *Uch. Zap. Saratovsk. Gos. Univ.*, (2), 92 (1939) [*C.A.*, **35,** 6238 (1941)].

[1130] O. Becker and J. F. Thorpe, *J. Chem. Soc.*, **121,** 1303 (1922).

[1131] F. Bodroux and F. Taboury, *Bull. Soc. Chim. France,* [4] **5,** 950 (1909).

[1132] F. Bodroux and F. Taboury, *Compt. Rend.*, **149,** 422 (1909).

[1133] R. V. Levina and N. P. Shusherina, *Zh. Obshch. Khim.*, **20,** 868 (1950) [*C.A.*, **44,** 9336 (1950)].

[1134] V. I. Esafov, V. M. Gulyakov, V. V. Kargopol'tseva, A. P. Kulakova, G. V. Razmyslov, and N. D. Toporov, *Zh. Obshch. Khim.*, **10,** 1973 (1940) [*C.A.*, **35,** 3958 (1941)].

[1135] B. Braun and H. Kittle, *Monatsh. Chem.*, **27,** 803 (1906).

[1136] J. Schramm, *Ber.*, **16,** 1581 (1883).

[1137] G. A. Razuavaev, S. V. Svetozarskii, E. N. Zil'berman, and K. L. Feller, *Tr. po Khim. i Khim. Technol.*, **4,** 611 (1961) [*C.A.*, **58,** 457 (1963)].

[1138] P. Barbier and G. Leser, *Bull. Soc. Chim. France,* [3] **31,** 278 (1904).

[1139] O. Jacobsen, *Ber.*, **7,** 1430 (1874).

[1140] A. P. Meshcheryakov and V. G. Glukhovtsev, *Izv. Akad. Nauk SSSR, Otd. Khim. Nauk,* 176 (1962) [*C.A.*, **57,** 11034 (1962)].

[1141] H. Beucker, Ger. pat. 740,425 [*C.A.*, **39,** 2430 (1943)].

[1142] H. Godchot and F. Taboury, *Bull. Soc. Chim. France,* [4] **13,** 16 (1913).

[1143] J. H. Burckhalter and P. Kurath, *J. Org. Chem.*, **24,** 990 (1959).

[1144] G. E. Goheen, *J. Am. Chem. Soc.*, **63,** 744 (1941).

[1145] O. Wallach, *Ann.*, **389,** 178 (1912).

[1146] O. Wallach, *Ber.*, **29,** 2965 (1896).

[1147] J. D. Cocker and T. G. Halsall, *J. Chem. Soc.*, 3441 (1957).

[1148] H. Christol, M. Mousseron, and R. Sallé, *Bull. Soc. Chim. France,* 556 (1958).

[1149] D. Varech, C. Ouannes, and J. Jacques, *Bull. Soc. Chim. France,* 1662 (1965).

[1150] J. Mleziva, *Chem. Listy,* **51,** 2364 (1957) [*C.A.*, **52,** 6206 (1958)].

[1151] F. Taboury and M. Godchot, *Compt. Rend.*, **169,** 62 (1919).

[1152] N. D. Zelinskii and N. I. Shuikin, *J. Russ. Phys. Chem. Soc.*, **62,** 1343 (1930) [*C.A.*, **25,** 2420 (1931)].

[1153] D. Ivanov and A. Spasov, *Bull. Soc. Chim. France,* [5] **2,** 1435 (1935).

[1154] H. Meerwein, *Ann.*, **405,** 155 (1914).
[1155] H. Meerwein, *Ann.*, **405,** 129 (1914).
[1156] O. Wallach, *Ber.*, **30,** 1094 (1897).
[1157] R. Mayer, *Ber.*, **89,** 1443 (1956).
[1158] R. Otte and H. v. Pechmann, *Ber.*, **22,** 2115 (1889).
[1159] G. A. R. Kon and E. Leton, *J. Chem. Soc.*, 2496 (1931).
[1160] J. B. Ekeley and M. S. Carpenter, *J. Am. Chem. Soc.*, **46,** 446 (1924).
[1161] J. B. Conant and A. H. Blatt, *J. Am. Chem. Soc.*, **51,** 1227 (1929).
[1162] J. Colonge and D. Joly, *Ann. Chim.* (*Paris*), [11] **18,** 286 (1943).
[1163] J. Colonge and D. Joly, *Ann. Chim.* (*Paris*), [11] **18,** 306 (1943).
[1164] F. C. Whitmore, J. S. Whitaker, W. A. Mosher, O. N. Breivik, W. R. Wheeler, C. S. Miner, L. H. Sutherland, R. B. Wagner, T. W. Clapper, C. E. Lewis, A. R. Lux, and A. H. Popkin, *J. Am. Chem. Soc.*, **63,** 643 (1941).
[1165] D. Ivanov and A. Spassov, *Bull. Soc. Chim. France*, [5] **2,** 816 (1935).
[1166] T. A. Favorskaya and Z. A. Shevchenko, *Zh. Obshch. Khim.*, **31**, 2526 (1961) [*C.A.*, **56,** 7258 (1962)].
[1167] I. N. Nazarov and S. G. Matsoyan, *Zh. Obshch. Khim.*, **27,** 2951 (1957) [*C.A.*, **52,** 8080 (1958)].
[1168] K. Conrow, *J. Org. Chem.*, **31,** 1050 (1966).
[1169] H. Stetter, E. Siehnhold, E. Klauke, and M. Coenen, *Chem. Ber.*, **86,** 1308 (1953).
[1170] J. Mleziva, *Chem. Listy*, **47,** 1354 (1953) [*C.A.*, **48,** 13642 (1954)].
[1171] T. L. Cairns, R. M. Joyce, and R. S. Schreiber, *J. Am. Chem. Soc.*, **70,** 1689 (1948).
[1172] E. L. Pelton, C. F. Starnes, and S. A. Shrader, *J. Am. Chem. Soc.*, **72,** 2039 (1950).
[1173] W. Treibs, *Ber.*, **61,** 683 (1928).
[1174] O. Wallach, M. Behnke, and F. Pauly, *Ann.*, **369,** 99 (1909).
[1175] J. Mleziva, *Chem. Listy*, **47,** 1031 (1953) [*C.A.*, **48,** 13642 (1954)].
[1176] P. Munk and J. Plesek, *Chem. Listy*, **51,** 771 (1957) [*C.A.*, **51,** 11261 (1957)].
[1177] S. V. Svetozarskiĭ, E. N. Zil'berman, and G. A. Razuvaev, *Zh. Obshch. Khim.*, **29,** 1454 (1959) [*C.A.*, **54,** 8668 (1960)].
[1178] K. Kunze, *Ber.*, **59,** 2085 (1926).
[1179] S. V. Svetozarskii, K. L. Feller, and E. N. Zil'berman, *Zh. Vses. Khim. Obshchestva im. D. I. Mendeleeva*, **8,** 113 (1963) [*C.A.*, **59,** 3788 (1963)].
[1180] G. R. Pettit and E. G. Thomas, *Chem. Ind.* (*London*), 1758 (1963).
[1181] O. Stichnoth, Ger. pat. 922,167 [*C.A.*, **51,** 17988 (1957)].
[1182] M. J. Astle and M. L. Pinns, *J. Org. Chem.*, **24,** 56 (1959).
[1183] A. D. Petrov, *Bull. Soc. Chim. France*, [4] **43,** 1272 (1928).
[1184] W. Hückel, O. Neunhoeffer, A. Gercke, and E. Frank, *Ann.*, **477,** 119 (1930).
[1185] N. D. Zelinskiĭ, N. I. Shuĭkin, and L. M. Fateev, *Zh. Obshch. Khim.*, **2,** 671 (1932) [*C.A.*, **27,** 2430 (1933)].
[1186] S. V. Svetozarskii, K. L. Feller, Y. Y. Samitov, E. N. Zil'berman, and G. A. Razuvaev, *Izv. Akad. Nauk SSSR, Otd. Khim. Nauk*, 121 (1964) [*C.A.*, **60,** 9225 (1964)].
[1187] H. Gault, L. Daltroff, and J. Eck-Tridon, *Bull. Soc. Chim. France*, [4] **12,** 952 (1945).
[1188] C. Mannich, *Ber.*, **40,** 153 (1907).
[1189] G. A. Razuvaev, E. N. Zil'berman, and S. V. Svetozarskiĭ, *Dokl. Akad. Nauk SSSR*, **131,** 850 (1960) [*C.A.*, **54,** 16439 (1960)].
[1190] J. Wiemann, B. Furth, and G. Dana, *Compt. Rend.*, **250,** 3674 (1960).
[1191] E. A. Braude and J. A. Coles, *J. Chem. Soc.*, 1425 (1952).
[1192] J. Wiemann, L. Martineau, and J. Tiquet, *Bull. Soc. Chim. France*, 1633 (1955).
[1193] J. Wiemann, L. T. Thuan, and J. M. Conia, *Bull. Soc. Chim. France*, 908 (1957).
[1194] J. Wiemann and L. T. Thuan, *Bull. Soc. Chim. France*, 696 (1957).
[1195] J. M. Conia, *Compt. Rend.*, **240,** 1545 (1955).
[1196] J. M. Conia and Y. R. Naves, *Compt. Rend.*, **250,** 356 (1960).
[1197] J. M. Conia, *Bull. Soc. Chim. France*, 690 (1954).
[1198] R. M. Acheson and R. Robinson, *J. Chem. Soc.*, 1127 (1952).
[1199] J. Colonge, R. Falcotet, and R. Gaumont, *Bull. Soc. Chim. France*, 211 (1958).
[1200] V. Grignard and J. Colonge, *Bull. Soc. Chim. Romania*, **15,** 5 (1933) [*C.A.*, **28,** 101 (1934)].

[1201] H. J. Shine and E. E. Turner, *Nature*, **158**, 170 (1946).

[1202] V. M. Tolstopyatov, *J. Russ. Phys. Chem. Soc.*, **62**, 1813 (1930) [*C.A.*, **25**, 3959 (1931)].

[1203] W. J. Hickenbottom and E. Schlücterer, *Nature*, **155**, 19 (1945).

[1204] V. M. Tolstopyatov, *J. Russ. Phys. Chem. Soc.*, **62**, 1813 (1930) [*C.A.*, **25**, 3959 (1931)].

[1205] T. A. Favorskaya and Z. A. Shevchenko, *Zh. Obshch. Khim.*, **32**, 46 (1962) [*C.A.*, **57** 16380 (1962)].

[1206] L. N. Stukanova, N. V. Zhdanova, V. I. Epishev, and A. A. Petrov, *Neftekhimiya*, **4**, 521 (1964) [*C.A.*, **61**, 13204 (1964)].

[1207] M. Godchot and F. Taboury, *Compt. Rend.* **169**, 1168 (1919).

[1208] T. A. Favorskaya and A. S. Lozhenitsyna, *Zh. Obshch. Khim.*, **33**, 2916 (1963) [*C.A.*, **60**, 1609 (1964)].

[1209] J. Reese, *Ber.*, **75**, 384 (1942).

[1210] M. Godchot and P. Brun, *Compt. Rend.*, **174**, 618 (1922).

[1211] T. A. Favorskaya, Z. A. Shevchenko, and T. A. Kuznetsova, *Zh. Obshch. Khim.*, **33**, 2909 (1963) [*C.A.*, **60**, 1675 (1964)].

[1212] K. Kulka, *Can. J. Chem.*, **42**, 2791 (1964).

[1213] C. Gastaldi and F. Cherchi, *Gazz. Chim. Ital.*, **45**, 251 (1915).

[1214] C. Gastaldi, *Gazz. Chim. Ital.*, **50**, 71 (1920).

[1215] M. Delacre, *Ann. Chim.* (*Paris*), [9] **2**, 63 (1914).

[1216] C. Porlezza and V. Gatti, *Gazz. Chim. Ital.*, **56**, 265 (1926).

[1217] N. O. Calloway and L. D. Green, *J. Am. Chem. Soc.*, **59**, 809 (1937).

[1218] W. Taylor, *J. Chem. Soc.*, 304 (1937).

[1219] G. Reddelien, *Ber.*, **46**, 2712 (1913).

[1220] G. F. Woods, *U.S. Dept. Com. Office Tech. Serv.*, AD 278,110 (1962) [*C.A.*, **60**, 5380 (1964)].

[1221] C. Engler and L. Dengler, *Ber.*, **26**, 1444 (1893).

[1222] W. Metzger and H. F. Schünemann, U.S. pat. 2,969,405 [*C.A.*, **55**, 12365 (1961)].

[1223] R. L. McLaughlin and J. W. Schick, U.S. pat. 3,023,245 [*C.A.*, **57**, 4599 (1962)].

[1224] W. N. Ipatiev and A. D. Petrov, *Ber.*, **60**, 1956 (1927).

[1225] W. C. Dovey and R. Robinson, *J. Chem. Soc.*, 1389 (1935).

[1226] F. Henrich and A. Wirth, *Monatsh. Chem.*, **25**, 423 (1904).

[1227] R. W. Roeske, D. B. Bright, R. L. Johnson, W. J. De Jarlais, R. W. Bush, and H. R. Snyder, *J. Am. Chem. Soc.*, **82**, 3128 (1960).

[1228] H. W. Moore and H. R. Snyder, *J. Org. Chem.*, **28**, 535 (1963).

[1229] M. Delacre, *Bull. Soc. Chim. France*, [4] **7**, 1041 (1910).

[1230] C. Engler and H. E. Berthold, *Ber.*, **7**, 1123 (1874).

[1231] G. Reddelien, *Ann.*, **388**, 165 (1912).

[1232] H. Hunsdiecker, *Ber.*, **75**, 455 (1942).

[1233] P. L. Southwick, E. P. Previc, J. Casanova, and E. H. Carlson, *J. Org. Chem.*, **21**, 1087 (1956).

[1234] L. Crombie and K. Mackenzie, *J. Chem. Soc.*, 4417 (1958).

[1235] F. S. Kipping, *J. Chem. Soc.*, **65**, 495 (1894).

[1236] E. D. Bergmann and R. Corett, *J. Org. Chem.*, **23**, 1507 (1958).

[1237] D. Ivanov and T. Ivanov, *Ber.*, **76**, 988 (1943).

[1238] L. A. Cort, R. G. Manders, and G. R. Pavlett, *J. Chem. Soc.*, 2844 (1964).

[1239] G. L. Buchanan, J. G. Hamilton, and R. A. Raphael, *J. Chem. Soc.*, 4606 (1963).

[1240] F. W. Semmler and K. Bartelt, *Ber.*, **41**, 866 (1908).

[1241] L. M. Mohunta and J. N. Ray, *J. Chem. Soc.*, 1328 (1934).

[1242] H. Paul and I. Wendel, *Ber.*, **90**, 1342 (1957).

[1243] M. V. Mavrov and V. F. Kucherov, *Izv. Akad. Nauk SSSR, Otd. Khim. Nauk*, 164 (1964) [*C.A.*, **60**, 9160 (1964)].

[1244] F. S. Kipping and W. H. Perkin, *J. Chem. Soc.*, **57**, 13 (1890).

[1245] H. Meerwein and J. Schäfer, *J. Prakt. Chem.*, [2] **104**, 289 (1922).

[1246] J. Lee, A. Ziering, S. D. Heineman, and L. Berger, *J. Org. Chem.*, **12**, 885 (1947).

[1247] D. B. Bright, *J. Am. Chem. Soc.*, **79**, 3200 (1957).

[1248] C. B. C. Boyce and J. S. Whitehurst, *J. Chem. Soc.*, 2022 (1959).

[1249] P. Wieland and K. Miescher, *Helv. Chim. Acta*, **33**, 2215 (1950).
[1250] P. De Mayo and H. Takeshita, *Can. J. Chem.*, **41**, 440 (1963).
[1251] D. J. Baisted and J. S. Whitehurst, *J. Chem. Soc.*, 4089 (1961).
[1252] A. M. Islam and R. A. Raphael, *J. Chem. Soc.*, 3151 (1955).
[1253] W. Hückel, A. Gercke, and A. Gross, *Ber.*, **66**, 563 (1933).
[1254] A. St. Pfau and P. Plattner, *Helv. Chim. Acta*, **19**, 858 (1936).
[1255] G. O. Schenck and K. H. Schulte-Elte, *Ann.*, **618**, 185 (1958).
[1256] M. W. Goldberg and P. Müller, *Helv. Chim. Acta*, **21**, 1699 (1938).
[1257] P. J. Ashworth, G. H. Whitham, and M. C. Whiting, *J. Chem. Soc.*, 4633 (1957).
[1258] M. Godchot, *Compt. Rend.*, **172**, 686 (1921).
[1259] W. Borsche and W. Menz, *Ber.*, **41**, 190 (1908).
[1260] N. S. Zefirov, P. V. Kostetskii, and Y. K. Yur'ev, *Zh. Obshch. Khim.*, **34**, 1069 (1964) [*C.A.*, **61**, 581 (1964)].
[1261] S. Julia, *Bull. Soc. Chim. France*, 780 (1954).
[1262] N. L. Wendler, H. L. Slates, and M. Tishler, *J. Am. Chem. Soc.*, **73**, 3816 (1951).
[1263] H. Hunsdiecker, *Ber.*, **75**, 460 (1942).
[1264] P. De Mayo, H. Takeshita, and A. B. M. A. Sattar, *Proc. Chem. Soc.*, 119 (1962).
[1265] J. A. Marshall and N. H. Anderson, *J. Org. Chem.*, **31**, 667 (1966).
[1266] H. Jäger and R. Keymer, *Arch. Pharm.*, **293**, 896 (1960) [*C.A.*, **55**, 3470 (1961)].
[1267] O. Wallach, *Ann.*, **408**, 202 (1915).
[1268] J. H. Amin, R. K. Razden, and S. C. Bhattacharaya, *Perfumery Essent. Oil Record*, **49**, 502 (1958) [*C.A.*, **53**, 8012 (1959)].
[1269] W. S. Johnson, S. Shulman, K. L. Williamson, and R. Pappo, *J. Org. Chem.*, **27**, 2015 (1962).
[1270] L. Crombie, S. H. Harper, and F. C. Newman, *J. Chem. Soc.*, 3963 (1956).
[1271] A. M. Islam and R. A. Raphael, *J. Chem. Soc.*, 4086 (1952).
[1272] A. M. Islam and M. T. Zemaity, *J. Am. Chem. Soc.*, **79**, 6023 (1957).
[1273] E. Knoevenagel, *Ann.*, **297**, 113 (1897).
[1274] A. L. Wilds and J. A. Johnson, *J. Am. Chem. Soc.*, **68**, 86 (1946).
[1275] P. Doyle, I. R. Maclean, R. D. H. Murray, W. Parker, and R. A. Raphael, *J. Chem. Soc.*, 1344 (1965).
[1276] A. S. Dreiding and A. J. Tomasewski, *J. Am. Chem. Soc.*, **77**, 411 (1955).
[1277] N. B. Haynes and C. J. Timmons, *J. Chem. Soc. Org.*, 224 (1966).
[1278] S. M. McElvain and P. H. Parker, *J. Am. Chem. Soc.*, **78**, 5312 (1956).
[1279] G. Ohloff, *Ann.*, **606**, 100 (1957).
[1280] H. Henecka, *Chem. Ber.*, **82**, 112 (1949).
[1281] R. L. Letsinger and J. D. Jamison, *J. Am. Chem. Soc.*, **83**, 193 (1961).
[1282] G. L. Buchanan, C. Maxwell, and W. Henderson, *Tetrahedron*, **21**, 3273 (1965).
[1283] V. Arkley, F. M. Dean, A. Robertson, and P. Sidisunthorn, *J. Chem. Soc.*, 2322 (1956).
[1284] W. Borsche and A. Fels, *Ber.*, **39**, 1922 (1906).
[1285] E. Ghera and F. Sondheimer, *Tetrahedron*, **21**, 977 (1965).
[1286] W. L. Meyer and B. S. Bielaski, *J. Org. Chem.*, **28**, 2896 (1963).
[1287] W. Reeve and E. Kiehlmann, *J. Org. Chem.* **31**, 2164 (1966).
[1288] P. S. Adamson, F. C. McQuillin, R. Robinson, and J. L. Simonsen, *J. Chem. Soc.*, 1576 (1937).
[1289] G. Büchi, O. Jeger, and L. Ruzicka, *Helv. Chim. Acta*, **31**, 241 (1948).
[1290] H. Dannenberg and S. Läufer, *Ber.*, **89**, 2242 (1956).
[1291] E. C. Kornfeld, E. J. Fornefeld, G. B. Kline, M. J. Mann, D. E. Morrison, R. G. Jones, and R. B. Woodward, *J. Am. Chem. Soc.*, **78**, 3087 (1956).
[1292] J. Bornstein and F. Nunes, *J. Org. Chem.*, **30**, 3324 (1965).
[1293] N. Barbulescu and M. Govela, *Analele Univ. "C. I. Parhon," Ser. Stiint. Nat.*, **10** (30), 151 (1961) [*C.A.*, **59**, 1506 (1963)].
[1294] J. Colonge and J. P. Kehlstadt, *Bull. Soc. Chim. France*, 504 (1955).
[1295] R. Anliker, A. S. Lindsey, D. E. Nettleton, and R. B. Turner, *J. Am. Chem. Soc.*, **79**, 220 (1957).

[1296] M. Palmade, P. Pesnelle, J. Streith, and G. Ourisson, *Bull. Soc. Chim. France,* 1950 (1963).

[1297] R. B. Bates, G. Büchi, T. Matsuura, and R. R. Shaffer, *J. Am. Chem. Soc.*, **82,** 2327 (1960).

[1298] J. Streith and G. Ourisson, *Bull. Soc. Chim. France,* 1960 (1963).

[1299] W. F. Newhall, S. A. Harris, F. W. Holly, E. L. Johnston, J. W. Richter, E. Walton, A. N. Wilson, and K. Folkers, *J. Am. Chem. Soc.* **77,** 5646 (1955).

[1300] P. Wieland, H. Ueberwasser, G. Anner, and K. Miescher, *Helv. Chim. Acta,* **36,** 376 (1953).

[1301] T. R. Marshall and W. H. Perkin, *J. Chem. Soc.*, **57,** 241 (1890).

[1302] M. Stoll and A. Rouvé, *Helv. Chim. Acta,* **30,** 2019 (1947).

[1303] B. R. T. Keene and K. Schofield, *J. Chem. Soc.*, 1080 (1958).

[1304] A. L. Wilds, *J. Am. Chem. Soc.*, **64,** 1421 (1942).

[1305] A. L. Wilds and C. H. Shunk, *J. Am. Chem. Soc.*, **72,** 2388 (1950).

[1306] T. Money, R. A. Raphael, A. I. Scott, and D. W. Young, *J. Chem. Soc.*, 3958 (1961).

[1307] E. Kloster-Jensen, E. Kováts, A. Eschenmoser, and E. Heilbronner, *Helv. Chim. Acta,* **39,** 1051 (1956).

[1308] L. J. Chinn and H. L. Dryden, *J. Org. Chem.*, **26,** 3904 (1961).

[1309] C. H. Shunk and A. L. Wilds, *J. Am. Chem. Soc.*, **71,** 3946 (1949).

[1310] A. L. Wilds and R. G. Werth, *J. Org. Chem.*, **17,** 1154 (1952).

[1311] B. K. Bhattacharyya, A. K. Bose, A. Chatterjee, and B. P. Sen, *J. Indian Chem. Soc.*, **41,** 479 (1964).

[1312] D. E. Clark, P. G. Holton, R. F. K. Meredith, A. C. Ritchie, T. Walker, and K. D. E. Whiting, *J. Chem. Soc.*, 2479 (1962).

[1313] A. L. Wilds and C. Djerassi, *J. Am. Chem. Soc.*, **68,** 1715 (1946).

[1314] O. Dann and H. Hofmann, *Ann.*, **667,** 116 (1963).

[1315] M. W. Goldberg and P. Müller, *Helv. Chim. Acta,* **28,** 831 (1940).

[1316] W. F. Johns, *J. Org. Chem.*, **26,** 4583 (1961).

[1317] H. Stobbe, *J. Prakt. Chem.*, [2] **86,** 209 (1912).

[1318] A. L. Wilds and W. J. Close, *J. Am. Chem. Soc.*, **68,** 83 (1946).

[1319] A. J. Birch and R. Robinson, *J. Chem. Soc.*, 503 (1944).

[1320] W. Nagata, Japan. pat. 11,619 (1963) [*C.A.*, **59,** 14074 (1963)].

[1321] L. Velluz, G. Muller, J. Mathieu, and A. Poittevin, *Compt. Rend.*, **252,** 4084 (1961).

[1322] Roussel-UCLAF, Belg. pat. 624,199 [*C.A.*, **60,** 12083 (1964)].

[1323] G. Nomine, J. Tessier, and A. Pierdet, Fr. pat. 1,360,155 [*C.A.*, **61,** 13383 (1964)].

[1324] W. Nagata, T. Teresawa, and T. Aoki, *Tetrahedron Letters,* (14), 865 (1963).

[1325] W. Nagata, T. Teresawa, and T. Aoki, U.S. pat. 3,190,879 [*C.A.*, **63,** 10032 (1965)].

[1326] J. L. Greene and H. D. Zook, *J. Am. Chem. Soc.*, **80,** 3629 (1958).

[1327] E. Buchta and K. Meyer, *Ber.*, **95,** 213 (1962).

[1328] C. Djerassi and T. T. Grossnickle, *J. Am. Chem. Soc.*, **76,** 1741 (1954).

[1329] E. Buchta and H. Krätzer, *Ber.*, **95,** 1820 (1962).

[1330] W. J. Wechter and G. Slomp, *J. Org. Chem.*, **27,** 2549 (1962).

[1331] H. H. Inhoffen and W. Bartmann, *Ann.*, **619,** 177 (1958).

[1332] J. M. H. Graves and H. J. Ringold, *Steroids, Suppl.* **1,** 23 (1965) [*C.A.*, **63,** 16406 (1965)].

[1333] E. Buchta and H. Krätzer, *Chem. Ber.*, **96,** 2093 (1963).

[1334] G. Nomine, R. Bocourt, and M. Vignau, U.S. pat. 3,085,098 [*C.A.*, **60,** 592 (1964)].

[1335] G. Nomine, A. Pierdet, R. Bucourt, and J. Tessier, U.S. pat. 3,138,617 [*C.A.*, **61,** 12061 (1964)].

[1336] D. Nasipuri, *J. Chem. Soc.*, 4192 (1958).

[1337] E. J. Corey, H. J. Hess, and S. Proskow, *J. Am. Chem. Soc.*, **85,** 3979 (1963).

[1338] R. C. Cookson and M. J. Nye, *J. Chem. Soc.*, 2009 (1965).

[1339] J. Fried, P. Grabowich, E. F. Sabo, and A. I. Cohen, *Tetrahedron,* **20,** 2297 (1964).

[1340] G. Snatzke and A. Nisar, *Ann.*, **683,** 159 (1965).

[1341] K. Dimroth, K. Wolf, and H. Kroke. *Ann.*, **678,** 183 (1964).

[1342] C. Armengaud, C. G. Wermuth, and J. Schreiber, *Compt. Rend.*, **254,** 2181 (1962).

[1343] E. Buchta and G. Satzinger, *Chem. Ber.*, **92,** 449 (1959).

[1344] M. Chaker and J. Schreiber, *Compt. Rend.*, **246,** 3646 (1958).

[1345] F. Ebel and O. Pesta, Ger. pat. 714,314 [*C.A.*, **38,** 1754 (1944)].

[1346] P. G. Stevens, *J. Am. Chem. Soc.*, **56,** 450 (1934).

[1347] C. V. Gheorghiu, *Bull. Sect. Sci. Acad. Roumaine*, **8,** 68 (1923) [*C.A.*, **17,** 2559 (1923)].

[1348] G. Vavon and A. Apchié, *Bull. Soc. Chim. France*, [4] **43,** 667 (1928).

[1349] G. O. Schenk, B. Brähler, and M. Cziesla, *Angew. Chem.*, **68,** 247 (1956).

[1350] J. R. Geigy A.-G., Neth. pat. appl. 6,408,223 [*C.A.*, **63,** 585 (1965)].

[1351] A. Behal, *Compt. Rend.*, **132,** 342 (1901).

[1352] J. Gore, *Bull. Soc. Chim. France*, 1710 (1964).

[1353] O. Wallach, *Ann.*, **300,** 267 (1898).

[1354] O. Wallach, *Ber.*, **29,** 2955 (1896).

[1355] E. Tamate, Japan. pat. 1346 (1951) [*C.A.*, **47,** 4917 (1953)].

[1356] E. Tamate, *Nippon Kagaku Zasshi*, **79,** 494 (1958) [*C.A.*, **54,** 4530 (1960)].

[1357] L. Geita and G. Vanags, *Latvijas PSR Zinatnu Akad. Vestis, Kim. Ser.*, 57 (1964) [*C.A.*, **61,** 3042 (1964)].

[1358] P. Cordier, *Compt. Rend.*, **205,** 918 (1937).

[1359] W. E. Hugh and G. A. R. Kon, *J. Chem. Soc.*, 2594 (1927).

[1360] J. Colonge, *Bull. Soc. Chim. France*, [5] **5,** 98 (1938).

[1361] A. Schönberg and K. Junghans, *Ber.*, **95,** 2137 (1962).

[1362] F. R. Japp and N. H. J. Miller, *J. Chem. Soc.*, **47,** 11, 35 (1885).

[1363] F. R. Japp and G. N. Lander, *J. Chem. Soc.*, **71,** 123 (1897).

[1364] F. R. Japp and N. H. J. Miller, *Ber.*, **18,** 179 (1885).

[1365] J. P. Collman, *J. Org. Chem.*, **26,** 3162 (1961).

[1366] J. Wiemann and Y. Dubois, *Compt. Rend.*, **253,** 1109 (1961).

[1367] J. Colonge and J. Dreux, *Compt. Rend.*, **231,** 1504 (1950).

[1368] F. D. Gunstone and R. M. Heggie, *J. Chem. Soc.*, 1437 (1952).

[1369] F. R. Japp and A. N. Meldrum, *J. Chem. Soc.*, **79,** 1024 (1901).

[1370] F. R. Japp and J. Knox, *J. Chem. Soc.*, **87,** 673 (1905).

[1371] S. Pietra and G. Tacconi, *Farmaco (Pavia) Ed. Sci.*, **13,** 893 (1958) [*C.A.*, **53,** 21875 (1959)].

[1372] P. Cordier and A. Haberzettl, *Compt. Rend.*, **254,** 699 (1962).

[1373] W. D. Garden and F. D. Gunstone, *J. Chem. Soc.*, 2650 (1952).

[1374] V. T. Esafov, L. I. Stashkov, L. B. Sirotkin, A. L. Suvorov, and E. G. Novikov, *Zh. Obshch. Khim.*, **29,** 845 (1959) [*C.A.*, **54,** 1394 (1960)].

[1375] H. G. Lindwall and J. S. Maclennan, *J. Am. Chem. Soc.*, **54,** 4739 (1932).

[1376] G. Kobayashi and S. Furukawa, *Chem. Pharm. Bull. (Tokyo)*, **12,** 1129 (1964) [*C.A.*, **62,** 539 (1965)].

[1377] M. Sy and G.-A. Thiault, *Bull. Soc. Chim. France*, 1308 (1965).

[1378] P. Cordier, *Compt. Rend.*, **202,** 1440 (1936).

[1379] G. H. Labib, *Compt. Rend.*, **259,** 1747 (1964).

[1380] C. F. H. Allen and H. B. Rosener, *J. Am. Chem. Soc.*, **49,** 2110 (1927).

[1381] F. R. Japp and F. Klingemann, *J. Chem. Soc.*, **57,** 662 (1890).

[1382] A. Chatterjee, R. C. Chatterjee, and D. K. Bhattacharyya, *J. Indian Chem. Soc.*, **34,** 855 (1957).

[1383] P. Cordier, *Compt. Rend.*, **225,** 388 (1947).

[1384] M. Kristensen-Reh, *Bull. Soc. Chim. France*, 882 (1956).

[1385] W. S. Johnson, J. Szmuszkovicz, and M. Miller, *J. Am. Chem. Soc.*, **72,** 3726 (1950).

[1386] S. K. Sen Gupta and B. K. Bhattacharyya, *J. Indian Chem. Soc.*, **36,** 273 (1959).

[1387] W. Dilthey and F. Quint, *J. Prakt. Chem.*, [2] **128,** 139 (1930).

[1388] H. Meerwein, H. Adam, and H. Buchloh, *Ber.*, **77,** 227 (1944).

[1389] F. Englehardt and J. Woellner, *Brennstoff-Chem.*, **44,** 178 (1963) [*C.A.*, **59,** 8582 (1963)].

[1390] I. G. Farbenind. A.-G., Brit. pat. 381,686 [*C.A.*, **27,** 3944 (1933)].

[1391] W. Flemming and H. D. von der Horst, Ger. pat. 544,887 [*C.A.*, **26,** 3521 (1932)].

[1392] W. Grimme and J. Wöllner, Ger. pat. 924,803 [*C.A.*, **52,** 3853 (1958)].

[1393] L. Claisen, *Ber.*, **25**, 3164 (1892).
[1394] J. Lincoln and J. G. N. Drewitt, U.S. pat. 2,395,414 and Brit. pat. 560,669 [*C.A.*, **40**, 3127 (1946)].
[1395] R. N. Haward, *J. Polymer Sci.*, **3**, 10 (1948).
[1396] Farbenfabriken vorm. F. Bayer and Co., Ger. pat. 223,207 [*C.A.*, **4**, 2980 (1910)].
[1397] Farbenfabriken vorm. F. Bayer and Co., Brit. pat. 19,087 [*C.A.*, **5**, 2153 (1911)].
[1398] J. Bolle, H. Jean, and T. Jullig, *Mem. Serv. Chim. État (Paris)*, **34**, 327 (1948) [*C.A.*, **44**, 6386 (1950)].
[1399] V. V. Chelintsev and G. I. Kuznetsova, *Zh. Obshch. Khim.*, **9**, 1901 (1939) [*C.A.*, **34**, 4387 (1940)].
[1400] J. Décombe, *Compt. Rend.*, **202**, 1685 (1936).
[1401] H. Dreyfus and J. G. N. Drewitt, U.S. pat. 2,378,988 [*C.A.*, **39**, 4626 (1945)].
[1402] H. Dreyfus and J. G. N. Drewitt, Brit. pat. 559,206 [*C.A.*, **40**, 1173 (1946)].
[1403] G. Merling and H. Köhler, U.S. pat. 981,668 [*C.A.*, **5**, 1192 (1911)].
[1404] W. M. Quattlebaum, U.S. pat. 2,064,564 [*C.A.*, **31**, 703 (1937)].
[1405] T. White, Brit. pat. 553,305 [*C.A.*, **38**, 4957 (1944)].
[1406] G. T. Morgan and E. L. Holmes, *J. Chem. Soc.*, 2667 (1932).
[1407] A. Müller, *Ber.*, **54**, 1142 (1921).
[1408] A. Werner, *Proc. Chem. Soc. (London)*, **20**, 196 (1904) [*Chem. Zentr.*, **76**, I, 221 (1905)].
[1409] T. White and R. N. Haward, *J. Chem. Soc.*, 25 (1943).
[1410] R. Gault and L. A. Germann, *Compt. Rend.*, **197**, 620 (1933).
[1411] I. G. Farbenind. A.-G., Brit. pat. 381,686 [*C.A.*, **27**, 3944 (1932)].
[1412] L. A. Germann, *Compt. Rend.*, **203**, 586 (1936).
[1413] G. Natta, U.S. pat. 2,378,573 [*C.A.*, **39**, 4090 (1945)].
[1414] C. Mannich and W. Brose, *Ber.*, **55**, 3155 (1922).
[1415] J. R. Roach, H. Wittcoff, and S. E. Miller, *J. Am. Chem. Soc.*, **69**, 2651 (1947).
[1416] R. V. Fedorova, M. I. Kogan, and O. D. Belova, *Tr. Vses. Nauchn.-Issled. Vitamin. Inst.*, **7**, 54 (1961) [*C.A.*, **59**, 1472 (1963)].
[1417] S. Malinowski, S. Basinski, and B. Polenska, *Roczniki Chem.*, **38**, 23 (1964).
[1418] P. Uschakoff, *J. Soc. Chim. Russe*, **29**, 113 (1897) [*Chem. Zentr.*, **68**, I, 1018 (1897)].
[1419] J. Wislicenus, T. Kircheisen, and E. Sattler, *Ber.*, **26**, 908 (1893).
[1420] A. M. Kuzin and N. A. Navraeva, *Biokhimiya*, **6**, 261 (1941) [*C.A.*, **35**, 7427 (1941)].
[1421] H. Pauly and H. v. Berg, *Ber.*, **34**, 2092 (1901).
[1422] H. Rupe and E. Hinterlach, *Ber.*, **40**, 4764 (1907).
[1423] A. Wohl and R. Maag, *Ber.*, **43**, 3280 (1910).
[1424] A. Dalgleish and R. N. Lacey, Brit. pat. 745,689 [*C.A.*, **51**, 1248 (1957)].
[1425] H. Albers, *Reichsamt Wirtschaftsausbau*, Prüf. Nr. **36**, (PB 52002), 15 (1940) [*C.A.*, **41**, 4461 (1947)].
[1426] C. Beyer, *J. Prakt. Chem.*, **32**, 125 (1885).
[1427] L. Knorr, *Ber.*, **20**, 1096 (1887).
[1428] E. R. Alexander and G. R. Coraor, *J. Am. Chem. Soc.*, **73**, 2721 (1951).
[1429] S. T. Young, J. B. Turner, and D. S. Tarbell, *J. Org. Chem.*, **28**, 928 (1963).
[1430] M. Julia and J. Bullot, *Bull. Soc. Chim. France*, 28 (1960).
[1431] F. Korte and H. Barkemeyer, *Chem. Ber.*, **90**, 2739 (1957).
[1432] H. B. Hill and J. Torrey, *Am. Chem. J.*, **22**, 89 (1899).
[1433] G. I. Fray, *Tetrahedron*, **14**, 161 (1961).
[1434] R. Robinson and G. I. Fray, Brit. pat. 979,925 [*C.A.*, **62**, 7656 (1965)].
[1435] J. Pastureau and M. Zamenhof, *Compt. Rend.*, **182**, 323 (1926).
[1436] P. Barbier and L. Bouveault, *Compt. Rend.*, **120**, 1269 (1895).
[1437] P. Barbier and L. Bouveault, *Compt. Rend.*, **120**, 1420 (1895).
[1438] V. V. Chelintsev and G. I. Kuznetsova, *Zh. Obshch. Khim.*, **9**, 1858 (1939) [*C.A.*, **34**, 4387 (1940)].
[1439] G. I. Kuznetsova, *Uch. Zap. Saratovsk. Gos. Univ.*, 12 (1938) [*C.A.*, **34**, 5823 (1940)].
[1440] G. I. Kuznetsova, *Uch. Zap. Saratovsk. Gos. Univ.*, (2), 80 (1939) [*C.A.*, **35**, 6236 (1951)].
[1441] H. Meerwein, *Ann.*, **358**, 71 (1908).
[1442] H. Nienburg, Ger. pat. 840,090 [*C.A.*, **45**, 2195 (1953)].

[1443] R. Fujii and K. Hoshiai, *J. Chem. Soc. Japan, Ind. Chem. Sect.*, **54,** 724 (1951) [*C.A.*, **48,** 4435 (1954)].

[1444] V. I. Esafov, *Zh. Obshch. Khim.*, **19,** 1115 (1949) [*C.A.*, **44,** 5795 (1950)].

[1445] S. G. Powell and D. A. Ballard, *J. Am. Chem. Soc.*, **60,** 1914 (1938).

[1446] R. Heilmann, G. de Gaudemaris, and P. Arnaud, *Compt. Rend.*, **242,** 2008 (1956).

[1447] A. Franke and L. Kohn, *Monatsh. Chem.*, **20,** 876 (1899).

[1448] N. Kizhner, *J. Russ. Phys. Chem. Soc.*, **45,** 987 (1913) [*C.A.*, **7,** 3965 (1913)].

[1449] R. Heilmann, *Bull. Soc. Chim. France*, [4] **49,** 75 (1931).

[1450] T. V. Nizovkina, I. M. Stroiman, N. M. Geller, G. M. Borovaya, and I. M. Saltykova, *Zh. Obshch. Khim.*, **34,** 3566 (1964) [*C.A.*, **62,** 5213 (1965)].

[1451] L. I. Zakharkin and L. P. Sorokina, *Izv. Akad. Nauk SSSR, Otd. Khim. Nauk*, 821 (1962) [*C.A.*, **57,** 12305 (1962)].

[1452] F. Dautwitz, *Monatsh. Chem.*, **27,** 773 (1906).

[1453] V. I. Anosov, N. N. Kozlov, M. A. Miropol'skaya, A. N. Nashatyrev, A. P. Savostin, V. G. Pines, G. I. Samokhvalov, and N. I. Fedotova, U.S.S.R., pat. 138,612 [*C.A.*, **56,** 8568 (1962)].

[1454] J. Colonge and A. Girantet, *Bull. Soc. Chim. France*, 1002 (1962).

[1455] I. V. Kamenskiĭ and N. V. Ungurean, *Zh. Prikl. Khim.*, **33,** 2121 (1960) [*C.A.*, **55,** 3549 (1961)].

[1456] I. G. Tishchenko, O. N. Bubel, and I. P. Zyat'kov, *Zh. Obshch. Khim.*, **33,** 2613 (1963) [*C.A.*, **60,** 483 (1964)].

[1457] G. Leser, *Bull. Soc. Chim. France*, [3] **17,** 108 (1897).

[1458] F. Tiemann and H. Tigges, *Ber.*, **33,** 559 (1900).

[1459] T. Nakayama, *J. Chem. Soc. Japan*, **59,** 224 (1938) [*C.A.*, **32,** 9061 (1938)].

[1460] F. Tiemann and P. Krüger, *Ber.*, **26,** 2115 (1893).

[1461] P. Barbier and L. Bouveault, *Compt. Rend.*, **118,** 198 (1894).

[1462] R. Heilmann, *Compt. Rend.*, **204,** 1345 (1937).

[1463] R. Heilmann, *Bull. Soc. Chim. France*, [5] **4,** 1072 (1937).

[1464] J. Colonge and P. Corbet, *Compt. Rend.*, **245,** 974 (1957).

[1465] H. G. Krey, Ger. (East) pat. 21,542 [*C.A.*, **56,** 5837 (1962)].

[1466] I. Heilbron, E. R. H. Jones, J. B. Toogood, and B. L. Weedon, *J. Chem. Soc.*, 1827 (1949).

[1467] J. Colonge and P. Jeltsch, *Bull. Soc. Chim. France*, 1288 (1963).

[1468] Yu. A. Arbuzov and E. I. Klimova, *Zh. Obshch. Khim.*, **32,** 3676 (1962) [*C.A.*, **58** 12416 (1963)].

[1469] I. G. Farbenind. A.-G., Brit. pat. 471,483 [*C.A.*, **32,** 1359 (1938)].

[1470] J. N. Wickert and C. A. Carter, U.S. pat. 2,088,017 [*C.A.*, **31,** 6770 (1937)].

[1471] A. Hinz, G. Meyer, and G. Schüking, *Ber.*, **76,** 676 (1943).

[1472] W. Surber, V. Theus, L. Colombi, and H. Schinz, *Helv. Chim. Acta*, **39,** 1299 (1956).

[1473] V. Theus and H. Schinz, *Helv. Chim. Acta*, **39,** 1290 (1956).

[1474] F. L. Breusch and F. Baykut, *Rev. Fac. Sci. Univ. Istanbul*, **16A,** 88 (1951) [*C.A.*, **46,** 3946 (1952)].

[1475] V. Grignard and F. Chambret, *Compt. Rend.*, **182,** 299 (1926).

[1476] K. Murakami, *Sci. Rept. Tohoku Imp. Univ.*, **18,** 639 (1929) [*C.A.*, **24,** 2426 (1930)].

[1477] H. Rupe and E. Willi, *Helv. Chim. Acta*, **15,** 842 (1932).

[1478] V. I. Esafov, I. F. Vladimiritsev, M. S. Kassikhinia, Z. S. Lisina, Z. S. Pronina, and I. I. Raikher, *Zh. Obshch. Khim.*, **13,** 814 (1943) [*C.A.*, **39,** 918 (1945)].

[1479] C. B. Clarke and A. R. Pinder, *J. Chem. Soc.*, 1967 (1958).

[1480] F. G. Fischer and H. Schulze, *Ber.*, **75,** 1467 (1942).

[1481] I. N. Nazarov, G. P. Kugatova, and G. A. Laumenskas, *Zh. Obshch. Khim.*, **27,** 2450 (1957) [*C.A.*, **52,** 7171 (1958)].

[1482] I. N. Nazarov, G. P. Kugatova, and V. V. Mozolis, *Zh. Obshch. Khim.*, **27,** 2635 (1957) [*C.A.*, **52,** 7167 (1958)].

[1483] D. Motskus and P. Kaikaris, *Lietuvos TSR Mokslu Akad. Darbai, Ser. B*, 95 (1964) [*C.A.*, **62,** 2717 (1965)].

[1484] O. Diels and K. Alder, Ger. pat. 545,398 [*C.A.*, **26,** 3075 (1932)].

[1485] I. G. Farbenind. A.-G., Brit. pat. 325,669 [*C.A.*, **24**, 4121 (1930)].

[1486] P. Karrer, C. Cochand, and N. Neuss, *Helv. Chim. Acta*, **29**, 1836 (1946).

[1487] T. Chaudron and R. Pallaud, *Compt. Rend.*, **249**, 2212 (1959).

[1488] J. L. Baas, A. Davies-Fidder, and H. O. Huisman, *Tetrahedron*, **22**, 259 (1966).

[1489] M. de Botton, *Compt. Rend.*, **256**, 2186 (1963).

[1490] E. C. Horning, M. G. Horning, and E. J. Platt, *J. Am. Chem. Soc.*, **71**, 1771 (1949).

[1491] International Flavors and Fragrances I.F.F., Neth. pat. 103,065 [*C.A.*, **62**, 13188 (1965)].

[1492] H. Hibbert and L. T. Cannon, *J. Am. Chem. Soc.*, **46**, 119 (1924).

[1493] V. Boulez, *Rev. Parfum.*, **7**, 380 (1927) [*C.A.*, **21**, 3599 (1927)].

[1494] Farbenfabriken, vorm. F. Bayer and Co., Ger. pat. 147,839 [*Frdl.*, **7**, 726 (1902–1904)].

[1495] Haarmann and Reimer, Ger. pat. 73,089 [*Frdl.*, **3**, 889 (1890–1894)].

[1496] Ping-Hsien Yeh, *Am. Perfumer*, **78**, 32 (1963) [*C.A.*, **59**, 3961 (1963)].

[1497] W. Stiehl, *J. Prakt. Chem.*, [2] **58**, 51 (1898).

[1498] F. Tiemann and P. Krüger, *Ber.*, **26**, 2675 (1893).

[1499] K. I. Bogacheva, Z. N. Bychkova, R. F. Shilina, E. F. Yakusheva, and E. P. Grigor'eva, *Tr. Vses. Nauchn.-Issled, Inst. Sintetich. i Natural'n. Dushistykh Veshchestv*, (5), 112 (1961) [*C.A.*, **58**, 4413 (1963)].

[1500] B. Tomek and J. Cvrtnik, Czech. pat. 100,648 [*C.A.*, **58**, 4429 (1963)].

[1501] A. M. Ille and P. Tipa, *Lucrarile Inst. Cercetari Aliment.*, **6**, 51 (1962–63) [*C.A.*, **61** 1895 (1964)].

[1502] Haarmann and Reimer, Ger. pat. 139,959 [*Frdl.*, **7**, 732 (1902–1904)].

[1503] L. Re and H. Schinz, *Helv. Chim. Acta*, **41**, 1710 (1958).

[1504] B. Willhalm, U. Steiner, and H. Schinz, *Helv. Chim. Acta*, **41**, 1359 (1958).

[1505] C. H. Eugster, E. Linner, A. H. Trivedi, and P. Karrer, *Helv, Chim. Acta*, **39**, 690 (1956).

[1506] Haarmann and Reimer, Ger. pat. 75,120 [*Frdl.*, **3**, 890 (1890–1894)].

[1507] H. Rupe and W. Lotz, *Ber.*, **36**, 2796 (1903).

[1508] E. H. Eschinasi and M. L. Cotter, *Tetrahedron Letters*, 3487 (1964).

[1509] M. Mousseron-Canet and C. Levallois, *Bull. Soc. Chim. France*, 993 (1963).

[1510] V. N. Belov, N. A. Daev, S. D. Kustova, K. V. Leets, S. S. Paddubnaya, N. I. Skvortsova, E. I. Shepelenkova, and A. K. Shumeiko, *Zh. Obshch. Khim.*, **27**, 1384 (1957) [*C.A.*, **52**, 3740 (1958)].

[1511] Soc. anon. M. Naef & Cie., Fr. pat. 744,345 [*C.A.*, **27**, 4031 (1933)].

[1512] V. E. Sibirtseva, G. V. Meleshkina, N. I. Skvortsova, and V. N. Belov, *Vopr. Khim. Terpenov i Terpenoidov, Akad. Nauk Lit. SSSR, Tr. Vses. Soveshch., Vilnyus*, 209 (1959) [*C.A.*, **55**, 16444 (1961)].

[1513] J. L. Baas, A. Davies-Fidder, F. R. Visser, and H. O. Huisman, *Tetrahedron*, **22**, 265 (1966).

[1514] H. O. Huisman, A. Smit, P. H. van Leeuwen, and J. H. van Rij, *Rec. Trav. Chim.*, **75**, 983 (1956).

[1515] G. Pappalardo, *Gazz. Chim. Ital.*, **89**, 540 (1959).

[1516] Z. N. Nazarova, *Zh. Obshch. Khim.*, **27**, 2931 (1957) [*C.A.*, **52**, 8115 (1958)].

[1517] Z. N. Nazarova and T. V. Ustimenko, *Zh. Obshch. Khim.*, **30**, 2017 (1960) [*C.A.*, **55**, 6463 (1961)].

[1518] H. Uota and T. Tanizaki, Japan. pat. 1678 (1964) [*C.A.*, **60**, 14474 (1964)].

[1519] G. Funatsukuri and M. Ueda, Japan. pat. 12,635 [*C.A.*, **61**, 16049 (1964)].

[1520] C. S. Marvel, J. M. Quinn, and J. S. Showell, *J. Org. Chem.*, **18**, 1730 (1953).

[1521] E. Grischkewitsch-Trochimowski and I. Mazurewitsch, *J. Russ. Phys. Chem. Soc.*, **44**, 570 (1912) [*Chem. Zentr.*, **83**, II, 1561 (1912)].

[1522] Yu. K. Yur'ev. N. N. Mezentsova, and V. E. Vas'kovskii, *Zh. Obshch. Khim.*, **27**, 3155 (1957) [*C.A.*, **52**, 9065 (1958)].

[1523] H. Haberland and E. Himmen, Ger. pat. 702,894 [*Chem. Zentr.*, **112**, I, 3194 (1941) [*C.A.*, **36**, 779 (1942)].

[1524] K. v. Auwers and H. Voss, *Ber.*, **42**, 4411 (1909).

[1525] L. Claisen and A. C. Ponder, *Ann.*, **223**, 137 (1884).

[1526] I. Kasiwagi, *Bull. Chem. Soc. Japan*, **1**, 90 (1926) [*C.A.*, **20**, 3005 (1926)].

[1527] A. Mangini and R. Andrisano, *Ann. Chim. Applicata*, **34**, 47 (1944) [*C.A.*, **40**, 7694 (1946)].
[1528] J. G. Schmidt, *Ber.*, **14**, 1459 (1881).
[1529] P. P. Surmin, *Zh. Obshch. Khim.*, **5**, 1642 (1935) [*C.A.*, **30**, 3430 (1936)].
[1530] Y. Terai and T. Tanaka, *Bull. Chem. Soc. Japan*, **29**, 822 (1956) [*C.A.*, **51**, 8004 (1957)].
[1531] V. I. Tikhonova, *Zh. Obshch. Khim.* **20**, 2213 (1950) [*C.A.*, **45**, 7100 (1951)].
[1532] A. N. Kost and D. V. Sibiryakova, *Zh. Obshch. Khim.*, **30**, 2920 (1960) [*C.A.*, **55**, 16517 (1961)].
[1533] A. D. Petrov, V. G. Glukhovstev, and S. V. Zakharova, *Dokl. Akad. Nauk SSSR*, **153**, 1346 (1963) [*C.A.*, **60**, 9225 (1964)].
[1534] N. I. Shuikin, I. F. Bel'skii, V. M. Shostakovskii, and R. A. Karakhanov, *Dokl. Akad. Nauk SSSR*, **151**, 1350 (1963) [*C.A.*, **59**, 13909 (1963)].
[1535] K. Alexander, L. S. Hafner, G. H. Smith, and L. E. Schniepp, *J. Am. Chem. Soc.*, **72**, 5506 (1950).
[1536] A. Isacescu, I. Gavat, C. Stoicescu, C. Vass, and I. Petrus, *Rev. Roumaine Chim.*, **10**, 219 (1965) [*C.A.*, **63**, 16284 (1965)].
[1537] Romania Ministry of Petroleum and Chemical Industry, Neth. pat. appl. 6,413,767 [*C.A.*, **63**, 18032 (1965)].
[1538] M. F. Stadnichuk, *Nauk. Zap. Chernivets'k. Derzh. Univ., Ser. Prirodn. Nauk*, **51**, 36 (1961) [*C.A.*, **62**, 9087 (1965)].
[1539] L. Claisen, *Ber.*, **14**, 2468 (1881).
[1540] E. Lubrzynska, *J. Chem. Soc.*, **109**, 1118 (1916).
[1541] W. Herz and J. Brasch, *J. Org. Chem.*, **23**, 1513 (1958).
[1542] C. S. Marvel and S. K. Stille, *J. Org. Chem.*, **22**, 1451 (1957).
[1543] J. Klosa, *Arch. Pharm.*, **289**, 177 (1956) [*C.A.*, **51**, 7373 (1957)].
[1544] D. Papa, E. Schwenk, F. Villani, and E. Klingsberg, *J. Am. Chem. Soc.*, **70**, 3356 (1948).
[1545] O. Neunhoeffer and D. Rosahl, *Chem. Ber.*, **86**, 226 (1953).
[1546] A. McGookin and I. M. Heilbron, *J. Chem. Soc.*, **125**, 2099 (1924).
[1547] I. M. Heilbron and A. B. Whitworth, *J. Chem. Soc.*, **123**, 238 (1923).
[1548] D. Vorländer, *Ann.*, **294**, 253 (1896).
[1549] S. v. Kostanecki and M. Schneider, *Ber.*, **29**, 1891 (1896).
[1550] P. I. Petrenko-Kritschenko and M. Bloch, *J. Prakt. Chem.*, [2] **60**, 154 (1899).
[1551] R. von Walther and W. Raetze, *J. Prakt. Chem.*, [5] **65**, 258 (1902).
[1552] I. Tanasescu and A. Georgescu, *Bull. Soc. Chim. France*, [4] **51**, 234 (1932).
[1553] J. van der Lee, *Rec. Trav. Chim.*, **47**, 920 (1928).
[1554] W. Kraszewski and B. Weicowna, *Roczniki Chem.*, **15**, 506 (1935) [*Chem. Zentr.*, **107**, I, 2540 (1936); *C.A.*, **30**, 2942 (1936)].
[1555] R. E. Corbett and C. L. Davey, *J. Chem. Soc.*, 296 (1955).
[1556] A. Baeyer and P. Becker, *Ber.*, **16**, 1968 (1883).
[1557] A. Baeyer, *Ann. Suppl.* **5**, 79 (1895).
[1558] L. Claisen and A. Claparède, *Ber.*, **14**, 2460 (1881).
[1559] L. Claisen, *Ann.*, **218**, 121 (1883).
[1560] V. I. Esafov and I. I. Raiker, *Zh. Obshch. Khim.*, **13**, 809 (1943) [*C.A.*, **39**, 918 (1945)].
[1561] P. Pfeiffer, B. Friedmann, Z. Goldberg, E. Pros, and V. Schwarzhopf, *Ann.*, **383**, 92 (1911).
[1562] F. Straus and F. Caspari, *Ber.*, **40**, 2689 (1907).
[1563] G. V. Austerweil and R. Pallaud, *J. Appl. Chem.* (*London*), **5**, 213 (1955).
[1564] C. D. Harries, *Ber.*, **24**, 3180 (1891).
[1565] F. Tiemann and A. Kees, *Ber.*, **18**, 1955 (1885).
[1566] H. Decker and H. Felser, *Ber.*, **41**, 2997 (1908).
[1567] R. Fabinyi, Ger. pat. 110,521 [*Chem. Zentr.*, **71**, II, 302 (1900)].
[1568] M. Miyano, S. Muraki, T. Kusunoki, T. Morita, and M. Matsui, *Nippon Nogeikagaku Kaishi*, **34**, 683 (1960) [*C.A.*, **59**, 13928 (1963)].
[1569] R. Kuhn, H. R. Hensel, and D. Weiser, *Ann.*, **611**, 83 (1958).
[1570] A. McGookin and D. J. Sinclair, *J. Chem. Soc.*, 1170 (1928).
[1571] T. Zincke and G. Mühlhausen, *Ber.*, **36**, 129 (1903).

[1572] D. Vorländer, *Ber.*, **58**, 118 (1925).

[1573] R. B. Woodward and E. C. Kornfeld, *J. Am. Chem. Soc.*, **70**, 2513 (1948).

[1574] J. Thiele and K. G. Falk, *Ann.*, **347**, 112 (1906).

[1575] P. Ruggli and E. Girod, *Helv. Chim. Acta*, **27**, 1464 (1944).

[1576] W. Löw, *Ann.*, **231**, 361 (1885).

[1577] A. Hamburger, *Monatsh. Chem.*, **19**, 427 (1898).

[1578] A. Kaufmann and R. Radovsevic, *Ber.*, **49**, 675 (1916).

[1579] A. Ya. Berlin and S. M. Sherlin, *Zh. Obshch. Khim.*, **18**, 1386 (1948) [*C.A.*, **43**, 2185 (1949)].

[1580] M. Faillebin, *Ann. Chim.* (*Paris*), [10] **4**, 410 (1925).

[1581] F. Haber, *Ber.*, **24**, 617 (1891).

[1582] E. Glaser and E. Tramer, *J. Prakt. Chem.*, [2] **116**, 331 (1927).

[1583] V. Hanzlik and A. Bianchi, *Ber.*, **32**, 2282 (1899).

[1584] A. Baeyer and V. Villiger, *Ber.*, **35**, 3013 (1902).

[1585] A. Baeyer and V. Villiger, *Ber.*, **35**, 1189 (1902).

[1586] A. Einhorn and J. P. Grabfield, *Ann.*, **243**, 362 (1888).

[1587] A. Ya. Berlin and T. P. Sycheva, *Zh. Obshch. Khim.*, **22**, 1998 (1952) [*C.A.*, **47**, 8681 (1953)].

[1588] T. Murai, *Sci. Rept. Tohoku Imp. Univ.*, **14**, 149 (1925) [*C.A.*, **19**, 2944 (1925)].

[1589] L. Francesconi and G. Cusmano, *Gazz. Chim. Ital.*, **38** (II), 70 (1908).

[1590] E. Profft and E. Wolf, *J. Prakt. Chem.*, [4] **19**, 192 (1963).

[1591] O. Gisvold, D. Buelow, and E. H. Carlson, *J. Am. Pharm. Assoc.*, **35**, 188 (1946).

[1592] S. Katsura, M. Indo, and H. Matsui, Japan. pat. 19,711 (1963) [*C.A.*, **60**, 3019 (1964)].

[1593] H. Nomura and S. Hotta, *Sci. Rept. Tohoku Imp. Univ.*, **14**, 119, 131 (1925) [*Chem. Zentr.*, **96**, II, 1744 (1925); *C.A.*, **19**, 2943 (1925)].

[1594] L. Diehl and A. Einhorn, *Ber.*, **18**, 2326 (1885).

[1595] L. Diehl and A. Einhorn, *Ber.*, **18**, 2320 (1885).

[1596] F. W. Hinrichsen and O. Lohse, *Ann.*, **336**, 337 (1904).

[1597] F. W. Hinrichsen and W. Triepel, *Ann.*, **336**, 196 (1904).

[1598] H. Stobbe and R. Haertel, *Ann.*, **370**, 99 (1909).

[1599] H. Ryan and G. Plunkett, *Proc. Roy. Irish Acad.*, **32**, 199 (1916) [*C.A.*, **10**, 1849 (1916)].

[1600] T. Imaki, K. Koike, S. Shimizu, and S. Takei, *J. Agr. Chem. Soc. Japan*, **20**, 289 (1944) [*C.A.*, **43**, 1743 (1949)].

[1601] F. Sachs and W. Lewin, *Ber.*, **35**, 3569 (1902).

[1602] B. Knott, Brit. pat. 584,381 [*C.A.*, **41**, 4729 (1947)].

[1603] P. Pfeiffer, *Ann.*, **412**, 308 (1917).

[1604] G. Lowe, F. G. Torto, and B. C. L. Weedon, *J. Chem. Soc.*, 1855 (1958).

[1605] V. Prelog, J. Führer, R. Hagenbach, and H. Frick, *Helv. Chim. Acta*, **30**, 113 (1947).

[1606] W. John and P. Günther, *Ber.*, **74**, 879 (1941).

[1607] E. P. Kohler and L. W. Blanchard, *J. Am. Chem. Soc.*, **57**, 367 (1935).

[1608] R. Beckstroem, *Arch. Pharm.*, **242**, 98 (1904) [*Chem. Zentr.*, **75**, I, 1008 (1904)].

[1609] C. T. Davis and T. A. Geissman, *J. Am. Chem. Soc.*, **76**, 3507 (1954).

[1610] C. S. Gibson, K. V. Hariharan, K. N. Menon, and J. L. Simonsen, *J. Chem. Soc.*, 2247 (1926).

[1611] R. Dickinson and I. M. Heilbron, *J. Chem. Soc.*, 14 (1927).

[1612] A. E. Chichibabin, S. Elgazin, and V. A. Lengold, *Bull. Soc. Chim. France*, [4] **43**, 238 (1928).

[1613] J. Gilbert, *J. Rech. Centre Natl. Rech. Sci. Lab. Bellevue* (*Paris*), (36), 271 (1956) [*C.A.*, **51**, 8702 (1957)].

[1614] F. Korte, E. Hackel, and H. Sieper, *Ann.*, **685**, 122 (1965).

[1615] M. Weizmann and E. Bograchov, *J. Am. Chem. Soc.*, **70**, 2829 (1948).

[1616] G. Morgan, N. J. L. Megson, and K. W. Pepper, *Chem. Ind.* (*London*), 885 (1938).

[1617] Rheinpreussen Akt.-Ges. für Bergbau und Chemie, Brit. pat. 774,934 [*C.A.*, **52**, 1213 (1958)].

[1618] J. Décombe, *Compt. Rend.*, **203**, 1077 (1936).

[1619] G. T. Morgan and C. F. Griffith, *J. Chem. Soc.*, 841 (1937).

[1620] E. F. Landau and E. P. Irany, *J. Org. Chem.*, **12,** 422 (1947).
[1621] J. Bolle, H. Jean, and T. Jullig, *Mem. Serv. Chim. etat (Paris)*, **34,** 321 (1948) [*C.A.*, **44,** 6386 (1950)].
[1622] J. Colonge and L. Cumet, *Bull. Soc. Chim. France*, [5] **14,** 838 (1947).
[1623] R. R. Dreisbach and G. B. Heusted, U.S. pat. 2,450,646 [*C.A.*, **43,** 1051 (1949)].
[1624] J. Dreux, *Bull. Soc. Chim. France*, 1443 (1954).
[1625] J. Woellner, U.S. pat. 3,024,249 [*C.A.*, **64,** 15894 (1966)].
[1626] B. N. Rutovskiĭ and A. Dmitrieva, *Zh. Prikl. Khim.*, **14,** 535 (1941) [*C.A.*, **36,** 3291 (1942)].
[1627] I. K. Sarycheva, G. A. Vorob'eva, L. G. Kucheryavenko, and N. A. Preobrazhenskiĭ, *Zh. Obshch. Khim.*, **21,** 2994 (1957) [*C.A.*, **52,** 8036 (1958)].
[1628] T. White, *J. Chem. Soc.*, 238 (1943).
[1629] T. White, Brit. pat. 551,219 [*C.A.*, **38,** 2350 (1944)].
[1630] G. S. Schaffel, U.S. pat. 3,009,005 [*C.A.*, **56,** 8557 (1962)].
[1631] J. Woellner and F. Engelhardt, Ger. pat. 1,102,714 [*C.A.*, **56,** 4619 (1962)].
[1632] E. M. McMahon, J. N. Roper, W. P. Utermohlen, R. H. Hasek, R. C. Harris, and J. H. Brant, *J. Am. Chem. Soc.*, **70,** 2971 (1948).
[1633] J. H. Brant and R. L. Hasche, U.S. pat. 2,245,567 [*C.A.*, **35,** 5907 (1941)].
[1634] E. P. Kohler, *Am. Chem. J.*, **38,** 511 (1907).
[1635] H. L. Fisher and F. D. Chittenden, *Ind. Eng. Chem.*, **22,** 869 (1930).
[1636] L. P. Kyrides, *J. Am. Chem. Soc.*, **55,** 3431 (1933).
[1637] R. B. Wagner, *J. Am. Chem. Soc.*, **71,** 3214 (1949).
[1638] G. F. Hennion, R. B. Davis, and D. E. Maloney, *J. Am. Chem. Soc.*, **71,** 2813 (1949).
[1639] L. E. Hinkel, E. E. Ayling, J. F. J. Dippy, and T. H. Angel, *J. Chem. Soc.*, 814 (1931).
[1640] H. B. Hill, C. A. Soch, and G. Oenslager, *Am. Chem. J.*, **24,** 1 (1901).
[1641] S. G. Powell, H. C. Murray, and M. N. Baldwin, *J. Am. Chem. Soc.*, **55,** 1153 (1933).
[1642] S. G. Powell, *J. Am. Chem. Soc.*, **46,** 2514 (1924).
[1643] H. Thoms and H. Kahre, *Arch. Pharm.*, **263,** 241 (1925) [*Chem. Zentr.*, **96,** II, 546 (1925); *C.A.*, **19,** 2474 (1925)].
[1644] I. F. Usmanov, I. V. Kamenskii, M. Tadzhieva, and A. Khidoyatov, *Uzbeksk. Khim. Zh.*, **8,** 48 (1964) [*C.A.*, **61,** 9373 (1964)].
[1645] S. Sukai, M. Kondo, A. Veno, and K. Momonki, Japan. pat. 12,637 (1964) [*C.A.*, **61,** 16049 (1964)].
[1646] H. Midorikawa, *Bull. Chem. Soc. Japan*, **26,** 460 (1953) [*C.A.*, **49,** 9607 (1955)].
[1647] I. Kasiwagi, *Bull. Chem. Soc. Japan*, **2,** 310 (1927) [*Chem. Zentr.*, **99,** I, 689 (1928)].
[1648] J. T. Thurston, U.S. pat. 2,385,314 [*C.A.*, **40,** 609 (1946)].
[1649] Z. V. Til, I. Markushina, K. Sopunar, and A. A. Ponomarev, *Zh. Obshch. Khim.*, **27,** 110 (1957) [*C.A.*, **51,** 12877 (1957)].
[1650] S. G. Powell and M. N. Baldwin, *J. Am. Chem. Soc.*, **58,** 1871 (1936).
[1651] J. E. Dubois and F. Weck, *Ann. Univ. Saraviensis*, I, 39 (1953).
[1652] J. N. Wickert and B. T. Freure, U.S. pat. 2,088,018 [*C.A.*, **31,** 6769 (1937)].
[1653] G. Heller, H. Lauth, and A. Buchwaldt, *Ber.*, **55,** 483 (1922).
[1654] E. Pavlovska, J. Lukac, and M. Borovicka, Czech. pat. 110,222 [*C.A.*, **61,** 9439 (1964)].
[1655] H. Midorikawa, *Bull. Chem. Soc. Japan*, **27,** 143 (1954) [*C.A.*, **50,** 293 (1956)].
[1656] H. Midorikawa, *Bull. Chem. Soc. Japan*, **27,** 131 (1954) [*C.A.*, **50,** 244 (1956)].
[1657] M. E. Egorova and M. A. Abramova, *Zh. Prikl. Khim.*, **24,** 1098 (1951) [*C.A.*, **46,** 7538 (1952)].
[1658] G. Massara, *Gazz. Chim. Ital.*, **63,** 199 (1933).
[1659] G. Massara, *Gazz. Chim. Ital.*, **67,** 440 (1937).
[1660] M. Métayer and N. Epinay, *Compt. Rend.*, **226,** 1095 (1948).
[1661] E. H. Woodruff and T. W. Conger, *J. Am. Chem. Soc.*, **60,** 465 (1938).
[1662] H. Haeussler and W. Schacht, *Chem. Ber.*, **83,** 129 (1950).
[1663] K. v. Auwers, *Ber.*, **45,** 2764 (1912).
[1664] H. Ryan and A. Devine, *Proc. Roy. Irish Acad.*, **32,** 208 (1916) [*C.A.*, **10,** 1850 (1916)].
[1665] H. Ryan and J. J. Lennon, *Proc. Roy. Dublin Soc.*, **19,** 121 (1928) [*C.A.*, **23,** 4471 (1929)].
[1666] C. V. Gheorghiu, *Bull. Soc. Chim. France*, [4] **53,** 1442 (1933).

[1667] A. McGookin and D. J. Sinclair, *J. Chem. Soc.*, **127,** 2539 (1925).
[1668] S. Ch. De, *J. Indian Chem. Soc.*, **4,** 137 (1927) [*Chem. Zentr.*, **98,** II, 1701 (1927)].
[1669] K. Iwamoto, *Bull. Chem. Soc. Japan*, **2,** 51 (1927) [*C.A.*, **21,** 1803 (1927)].
[1670] J. Eliasberg and P. Friedländer, *Ber.*, **25,** 1752 (1892).
[1671] R. Jacquier and S. Boyer, *Bull. Soc. Chim. France*, 8 (1955).
[1672] J. Thiele and E. Weitz, *Ann.*, **377,** 1 (1910).
[1673] H. Ryan and P. J. Cahill, *Proc. Roy. Irish Acad.*, **36B**, 334 (1924) [*C.A.*, **19,** 467 (1925)].
[1674] J. Harvey, I. M. Heilbron, and D. G. Wilkinson, *J. Chem. Soc.*, 423 (1930).
[1675] J. Ichikawa, *Sci. Rept. Tohoku Imp. Univ.*, **14,** 127 (1925) [*Chem. Zentr.*, **96,** II, 1744 (1925); *C.A.*, **19,** 2943 (1925)].
[1676] M. Scholtz, *Ber.*, **29,** 613 (1896).
[1677] Z. Zafiriadis, *Compt. Rend.*, **226,** 731 (1948).
[1678] I. G. Farbenind. A.-G., Brit. pat. 325,669 [*C.A.*, **24,** 4121 (1930)].
[1679] T. S. Warunis and P. Lekos, *Ber.*, **43,** 654 (1910).
[1680] E. Horiuchi and T. Takeshima, *Nat. Sci. Rept. Ochanomizu Univ. Tokyo*, **3,** 293 (1965) [*C.A.*, **63,** 13126 (1965)].
[1681] Y. R. Naves and P. Bachmann, *Helv. Chim. Acta*, **26,** 2151 (1943).
[1682] P. Coulin in Genf., Ger. pat. 150,771 [*Chem. Zentr.*, **75,** I, 1307 (1904)].
[1683] Haarmann and Reimer, Ger. pat. 127,424 [*Frdl.*, **6,** 1247 (1900–1902)].
[1684] Haarmann and Reimer, Ger. pat. 150,827 [*Chem. Zentr.*, **75,** I, 1379 (1904)].
[1685] H. Köster, *Ber.*, **77,** 553 (1944).
[1686] H. Köster, *J. Prakt. Chem.*, [2] **143,** 249 (1935).
[1687] H. Köster, *Chem. Ber.*, **80,** 248 (1947).
[1688] G. W. Pope and M. T. Bogert, *J. Org. Chem.*, **2,** 276 (1937).
[1689] Haarmann and Reimer, Ger. pat. 133,758 [*Chem. Zentr.*, **73,** II, 613 (1902)].
[1690] W. Ried and H. J. Schwencke, *Chem. Ber.*, **91,** 566 (1958).
[1691] C. D. Hurd, W. D. McPhee, and G. H. Morey, *J. Am. Chem. Soc.*, **70,** 329 (1948).
[1692] S. Olsen and G. Havre, *Acta Chem. Scand.*, **8,** 47 (1954).
[1693] L. I. Zakharkin and L. P. Sorokina, *Izv. Akad. Nauk SSSR, Otd. Khim. Nauk*, 2096 (1962) [*C.A.*, **58,** 9014 (1963)].
[1694] E. Friedmann, *Helv. Chim. Acta*, **14,** 783 (1931).
[1695] R. N. Sen and B. K. Sen, *J. Indian Chem. Soc.*, **11,** 411 (1934) [*C.A.*, **29,** 120 (1935)].
[1696] M. Strell and E. Kopp, *Chem. Ber.*, **91,** 1621 (1958).
[1697] M. Reimer, *J. Am. Chem. Soc.*, **48,** 2454 (1926).
[1698] M. Reimer, *J. Am. Chem. Soc.*, **53,** 3147 (1931).
[1699] A. A. Shamshurin and M. Karieva, *Tr. Uzbeksk. Gos. Univ.*, [N.S.] No. 25, *Khim.* No. 1, 23 (1941) [*C.A.*, **35,** 5875 (1941)].
[1700] E. Erlenmeyer, *Ber.*, **32,** 1450 (1899).
[1701] H. Kondo and K. Tanaka, *J. Pharm. Soc. Japan*, **55,** 209 [*C.A.*, **31,** 6647 (1937)].
[1702] I. Saikawa, S. Hirai, Y. Kodama, and A. Takai, Japan. pat. 19,450 (1964) [*C.A.*, **62,** 10411 (1965)].
[1703] M. Bisagni, N. P. Buu-Hoi, and R. Royer, *J. Chem. Soc.*, 3688 (1955).
[1704] V. Andreeva and M. M. Koton, *Zh. Obshch. Khim.*, **27,** 671 (1957) [*C.A.*, **51,** 16409 (1957)].
[1705] K. H. Bauer and F. Werner, *Ber.*, **55,** 2494 (1922).
[1706] C. Harries and W. S. Mills, *Ann.*, **330,** 247 (1904).
[1707] A. Dornow and W. Sassenberg, *Ann.*, **602,** 14 (1957).
[1708] N. J. Leonard, J. C. Little, and A. J. Kresge, *J. Am. Chem. Soc.*, **79,** 2642 (1957).
[1709] O. Diels, *Ann.*, **434,** 1 (1923).
[1710] O. Diels and E. Andersonn, *Ber.*, **44,** 883 (1911).
[1711] P. Karrer and N. Neuss, *Helv. Chim. Acta*, **28,** 1185 (1945).
[1712] P. Karrer and C. Cochand, *Helv. Chim. Acta*, **28,** 1181 (1945).
[1713] M. Kerfanto and J. P. Quentin, *Compt. Rend.*, **257,** 2660 (1963).
[1714] C. H. Tieman and M. H. Gold, *J. Org. Chem.*, **23,** 1856 (1958).
[1715] L. I. Smith and E. R. Rogier, *J. Am. Chem. Soc.*, **73,** 3840 (1951).
[1716] A. P. Meshcheryakov, V. G. Glukhovtstev, and N. N. Lemin, *Izv. Akad. Nauk SSSR, Otd. Khim. Nauk*, 1901 (1961) [*C.A.*, **56,** 8663 (1962)].

[1717] S. C. Bunce, H. J. Dorsman, and F. D. Popp, *J. Chem. Soc.*, 303 (1963).
[1718] R. P. Mariella and R. R. Raube, *J. Org. Chem.*, **18,** 282 (1953).
[1719] J. K. O'Loane, C. M. Combs, and R. L. Griffith, *J. Org. Chem.*, **29,** 1730 (1964).
[1720] P. Delest and R. Pallaud, *Compt. Rend.*, **246,** 1703 (1958).
[1721] H. Midorikawa, *Bull. Chem. Soc. Japan*, **27,** 213 (1954) [*C.A.*, **50,** 291 (1956)].
[1722] E. Knoevenagel and W. Faber, *Ber.*, **31,** 2773 (1898).
[1723] E. Knoevenagel and A. Herz, *Ber.*, **37,** 4483 (1904).
[1724] S. Olsen, *Acta Chem. Scand.*, **9,** 101 (1955).
[1725] S. Olsen, H. Balk, and K. Finholdt, *Ann.*, **635,** 52 (1960).
[1726] S. Olsen, A. Henriksen, H. Balk, and H. Russwurm, *Ann.*, **635,** 67 (1960).
[1727] S. Olsen and H. Russwurm, *Ann.*, **639,** 1 (1961).
[1728] F. Meingast, *Monatsh. Chem.*, **26,** 265 (1905).
[1729] A. Ludwig and E. A. Kehrer, *Ber.*, **24,** 2776 (1891).
[1730] E. A. Kehrer and W. Kleberg, *Ber.*, **26,** 345 (1893).
[1731] R. N. Sen and B. C. Roy, *J. Indian Chem. Soc.*, **7,** 401 (1930) [*C.A.*, **24,** 4763 (1930)].
[1732] S. H. Zaheer, I. K. Kacker, and N. S. Rao, *Chem. Ber.*, **89,** 351 (1956).
[1733] H. Erdmann, *Ann.*, **258,** 129 (1890).
[1734] W. S. Rapson and R. G. Shuttleworth, *J. Chem. Soc.*, 33 (1942).
[1735] H. Erdmann, *Ann.*, **254,** 182 (1889).
[1736] H. Erdmann, *Ber.*, **18,** 3441 (1885).
[1737] H. Kato and H. Ojima, *Aichi Gakugei Daigaku Kenkyu Hokoku, Shizen Kagaku* (7), 25 (1958) [*C.A.*, **53,** 14045 (1959)].
[1738] G. M. Picha, *J. Am. Chem. Soc.*, **75,** 3155 (1953).
[1739] W. Borsche, *Ber.*, **48,** 842 (1915).
[1740] J. Wöllner and F. Engelhardt, *Chem. Ber.*, **95,** 30 (1962).
[1741] H. Hunsdiecker, *Ber.*, **75,** 447 (1942).
[1742] C. D. Harries and P. Bromberger, *Ber.*, **35,** 3088 (1902).
[1743] I. M. Heilbron and F. Irving, *J. Chem. Soc.*, 931 (1929).
[1744] C. V. Gheorghiu and V. Matei, *Gazz. Chim. Ital.*, **73,** 65 (1943).
[1745] J. E. Dubois, R. Luft, and F. Weck, *Compt. Rend.*, **234,** 2289 (1952).
[1746] A. Lapworth and A. C. O. Hann, *J. Chem. Soc.*, **81,** 1485 (1902).
[1747] M. Winter, *Helv. Chim. Acta*, **44,** 2110 (1961).
[1748] H. Scheibler and A. Fischer, *Ber.*, **55,** 2903 (1922).
[1749] I. N. Nazarov, *Bull. Acad. Sci. URSS, Classe Sci. Chim.*, 107 (1948) [*C.A.*, **42,** 7737 (1948)].
[1750] Y. R. Naves and P. Ardizio, *Bull. Soc. Chim. France*, 1713 (1956).
[1751] H. Barbier, U.S. pat. 1,898,075 [*C.A.*, **27,** 2697 (1933)].
[1752] J. Wiemann, S.-L. T. Thuan, and D. Ramé, *Compt. Rend.*, **249,** 1529 (1959).
[1753] J. Wiemann, S.-L. T. Thuan, *Compt. Rend.*, **245,** 1552 (1957).
[1754] J. Wiemann and P. Lacroix, *Bull. Soc. Chim. France*, 2269 (1961).
[1755] J. Wiemann and J. Dupayrat, *Bull. Soc. Chim. France*, 757 (1961).
[1756] L. Durand, Huguenin and Co., Ger. pat. 118,288 [*Chem. Zentr.*, **72,** I, 711 (1901)].
[1757] R. Luft, *Compt. Rend. Congr. Soc. Savantes Paris Dept. Sect. Sci.*, **87,** 499 (1962) [*C.A.* **60,** 15725 (1964)].
[1758] J. E. Dubois and R. Luft, *Compt. Rend.*, **240,** 1540 (1955).
[1759] H. Midorikawa, *Bull. Chem. Soc. Japan*, **27,** 210 (1954) [*C.A.*, **50,** 290 (1956)].
[1760] H. Wachs and O. F. Hedenburg, *J. Am. Chem. Soc.*, **70,** 2695 (1948).
[1761] O. F. Hedenburg and H. Wachs, *J. Am. Chem. Soc.*, **70,** 2216 (1948).
[1762] J. C. Bardhan and D. N. Mukherji, *J. Chem. Soc.*, 4629 (1956).
[1763] H. Nomura and S. Hotta, *Sci. Rept. Tohoku Imp. Univ.*, **14,** 131 (1925) [*Chem. Zentr.*, **96,** II, 1744 (1925); *C.A.*, **19,** 2943 (1925)].
[1764] Carbide and Carbon Chemicals Corp., Fr. pat. 782,835 [*C.A.*, **29,** 6979 (1935)].
[1765] Z. Zafiriadis, *Compt. Rend.*, **228,** 250 (1949).
[1766] A. A. Ponomarev, Z. V. Til, and V. V. Zelenkova, *Zh. Obshch. Khim.*, **20,** 1085 (1950) [*C.A.*, **44,** 9403 (1950)].
[1767] J. Colonge and Y. Vaginag, *Compt. Rend.*, **260,** 203 (1965).

[1768] G. A. Hill, C. S. Spear, and J. S. Lachowicz, *J. Am. Chem. Soc.*, **45,** 1557 (1923).
[1769] A. A. Boon and F. J. Wilson, *J. Chem. Soc.*, **97,** 1751 (1910).
[1770] J. J. Padbury and H. G. Lindwall, *J. Am. Chem. Soc.*, **67,** 1268 (1945).
[1771] B. D. Vorländer and A. Knötzsch, *Ann.*, **294,** 317 (1897).
[1772] R. Ahmad and B. C. L. Weedon, *J. Chem. Soc.*, 3299 (1953).
[1773] M. Sy, *Compt. Rend.*, **243,** 1772 (1956).
[1774] J. Murai, *Sci. Rept. Tohoku Imp. Univ.*, **14,** 145 (1925) [*C.A.*, **19,** 2944 (1925)].
[1775] R. Luft, *Bull. Soc. Chim. France*, 181 (1957).
[1776] W. A. Mosher and J. C. Cox, *J. Am. Chem. Soc.*, **72,** 3701 (1950).
[1777] M. F. Ansell, M. A. Davis, J. W. Hancock, and W. J. Hickenbottom, *Chem. Ind. (London)*, **46,** 1483 (1955).
[1778] J. Redel and J. Boch, Fr. pat. 1,353,523 [*C.A.*, **61,** 694 (1964)].
[1779] S. V. Tsukerman, I. K. Gintse, and V. F. Lavrushin, *Zh. Obshch. Khim.*, **33,** 2383 (1963) [*C.A.*, **59,** 13919 (1963)].
[1780] N. Maxim and I. Copuzeaunu, *Bull. Soc. Chim. Roumania*, **16,** 117 (1934) [*C.A.*, **29,** 4355 (1935)].
[1781] N. Maxim and I. Copuzeaunu, *Bull. Soc. Chim. France*, [5] **3,** 2251 (1936).
[1782] N. S. Vul'fson, E. V. Savenkova, and L. D. Senyavina, *Zh. Obshch. Khim.*, **34,** 2743 (1964) [*C.A.*, **61,** 14627 (1964)].
[1783] H. Schinz and J.-P. Bourguin, *Helv. Chim. Acta.*, **25,** 1591 (1942).
[1784] Z. Zafiriadis, *Compt. Rend.*, **230,** 452 (1950).
[1785] H. Rupe and S. Wild, *Ann.*, **414,** 111 (1917).
[1786] B. Reichert and A. Lechner, *Arzneimittel-Forsch.*, **15,** 36 (1965) [*C.A.*, **62,** 10403 (1965)].
[1787] C. Harries, *Ann.*, **330,** 233, 257 (1904).
[1788] G. Goldschmiedt and G. Knöpfer, *Monatsh. Chem.*, **19,** 406 (1898).
[1789] G. Goldschmiedt and H. Krczmar, *Monatsh. Chem.*, **22,** 659 (1901).
[1790] T. N. Ghosh, A. Bosel, and A. Raychaudhuri, *J. Indian Chem. Soc.*, **37,** 93 (1960).
[1791] M. Scholtz and W. Meyer, *Ber.*, **43,** 1861 (1910).
[1792] J. Quarterman and T. S. Stevens, *J. Chem. Soc.*, 3292 (1955).
[1793] V. F. Lavrushin, S. V. Tsukerman, and V. M. Nikitchenko. *Ukr. Khim. Zh.*, **27,** 379 (1961) [*C.A.*, **56,** 3437 (1962)].
[1794] V. F. Lavrushin, S. V. Tsukerman, and S. I. Artemenko, *Zh. Obshch. Khim.*, **31,** 3037 (1961) [*C.A.*, **56,** 15454 (1962)].
[1795] A. Kauffmann and H. Burckhardt, *Ber.*, **46,** 3808 (1913).
[1796] S. v. Kostanecki and D. Maron, *Ber.*, **31,** 726 (1898).
[1797] F. Straus and O. Ecker, *Ber.*, **39,** 2977 (1906).
[1798] P. Pfeiffer, O. Angern, P. Backes, W. Fitz, E. Prahl, H. Rheinboldt, and W. Stoll, *Ann.*, **441,** 228 (1925).
[1799] W. Borsche, *Ann.*, **375,** 177 (1910).
[1800] P. Pfeiffer and H. Kleu, *Ber.*, **66,** 1704 (1933).
[1801] W. Neugebauer, M. Tomanek, and T. Scherer, U.S. pat. 2,768,077 [*C.A.*, **51,** 9388 (1957)].
[1802] C. R. Hauser and T. M. Harris, *J. Am. Chem. Soc.*, **80,** 6360 (1958).
[1803] C. M. Clark and J. D. A. Johnson, *J. Chem. Soc.*, 126 (1962).
[1804] J. A. Barltrop and N. A. J. Rogers, *J. Chem. Soc.*, 2566 (1958).
[1805] I. M. Heilbron and F. Irving, *J. Chem. Soc.*, 2323 (1928).
[1806] H. Carette, *Compt. Rend.*, **131,** 1225 (1900).
[1807] H. Thoms, *Ber. Deut. Pharm. Ges.*, **11,** 3 (1901) [*Chem. Zentr.*, **72,** I, 524 (1901)].
[1808] A. Roedig and E. Klappert, *Ann.*, **605,** 126 (1957).
[1809] H. Bauer and H. Dieterle, *Ber.*, **44,** 2697 (1911).
[1810] V. F. Lavrushin and V. P. Dzyuba, *Zh. Obshch. Khim.*, **33,** 2581 (1963) [*C.A.*, **60,** 443 (1964)].
[1811] I. Dory and Gy. Lanyi, *Acta Chim. Acad. Sci. Hung.*, **30,** 71 (1962) [*C.A.*, **58,** 564 (1963)].
[1812] E. Schmitz, *Ber.*, **46,** 2327 (1913).
[1813] F. Engelhardt and J. Woellner, Ger. pat. 1,110,624 [*C.A.*, **56,** 3355 (1962)].

[1814] D. Vorländer and K. Hobohm, *Ber.*, **29**, 1352 (1896).
[1815] R. F. Collins and M. Davis, *J. Chem. Soc.*, 1863 (1961).
[1816] E. Pace, *Atti Acad. Lincei*, [6] **9**, 778 (1929) [*C.A.*, **23**, 4941 (1929)].
[1817] W. Treibs and E. Lippmann, *Chem. Ber.*, **91**, 1999 (1958).
[1818] W. Davey and H. Gottfried, *J. Org. Chem.*, **26**, 3705 (1961).
[1819] G. Goldschmiedt and G. Knöpfer, *Monatsh. Chem.*, **20**, 734 (1899).
[1820] M. N. Tilichenko. *Zh. Obshch. Khim.*, **35**, 443 (1965) [*C.A.*, **63**, 527 (1965)].
[1821] H. W. Wanzlick and E. Peiler, *Chem. Ber.*, **89**, 1046 (1956).
[1822] J. Kenner, W. H. Ritchie, and F. S. Statham, *J. Chem. Soc.*, 1169 (1937).
[1823] R. Pallaud and F. Delaveau, *Compt. Rend.*, **237**, 1254 (1953).
[1824] R. Pallaud, *Chim. Anal.*, (Paris) **38**, 155 (1956) [*C.A.*, **50**, 12753 (1956)].
[1825] N. P. Buu-Hoi and N. D. Dat Xuong, *Compt. Rend.*, **251**, 2725 (1960).
[1826] A. Maccioni and E. Marongiu, *Gazz. Chim. Ital.*, **85**, 1570 (1955).
[1827] R. Pallaud and F. Delaveau, *Bull. Soc. Chim. France*, [5] **19**, 741 (1952).
[1828] D. Vorländer and K. Hobohm, *Ber.*, **29**, 1836 (1896).
[1829] N. P. Buu-Hoi and N. D. Dat Xuong, *Bull. Soc. Chim. France*, 758 (1958).
[1830] C. Mentzel, *Ber.*, **36**, 1499 (1903).
[1831] E. Braude and W. F. Forbes, *J. Chem. Soc.*, 1755 (1951).
[1832] W. Borsche and A. Geyer, *Ann.*, **393**, 29 (1912).
[1833] K. v. Auwers and F. Krollpfeiffer, *Ber.*, **48**, 1226 (1915).
[1834] N. P. Buu-Hoi, N. D. Dat Xuong, and N. V. Bac, *Compt. Rend.*, **258**, 1343 (1964).
[1835] R. Cornubert, M. André, M. de Deno, R. Joly, P. Louis, P. Robinet, and A. Strébel, *Bull. Soc. Chim. France*, [5] **5**, 513 (1938).
[1836] R. Cornubert and P. Louis, *Bull. Soc. Chim. France*, [5] **5**, 520 (1938).
[1837] R. Cornubert and C. Borrel, *Bull. Soc. Chim. France*, [4] **45**, 1148 (1929).
[1838] S. M. McElvain and E. J. Eisenbraun, *J. Am. Chem. Soc.*, **77**, 1599 (1955).
[1839] J. v. Braun, W. Keller, and K. Weissbach, *Ann.*, **490**, 179 (1931).
[1840] A. Maccioni and E. Marongiu, *Ann. Chim. (Rome)*, **46**, 252 (1956) [*C.A.*, **51**, 274 (1957)].
[1841] J. Colonge, *Bull. Soc. Chim. France*, 250 (1955).
[1842] F. Becke and K. Wick, Ger. pat. 1,085,520 [*C.A.*, **55**, 19827 (1961)].
[1843] F. A. Fries and F. Broich, Ger. pat. 956,948 [*C.A.*, **53**, 7056 (1959)].
[1844] I. A. Kaye and R. S. Matthews, *J. Org. Chem.*, **29**, 1341 (1964).
[1845] R. Pallaud and F. Delaveau, *Bull. Soc. Chim. France*, 1220 (1955).
[1846] V. G. Kharchenko, *Uch. Zap. Saratovsk. Gos. Univ.*, **75**, 71 (1962) [*C.A.*, **60**, 485 (1964)].
[1847] A. R. Poggi, A. Maccioni, and E. Marongiu, *Gazz. Chim. Ital.*, **84**, 528 (1954).
[1848] P. D. Gardner, C. E. Wulfman, and C. L. Osborn, *J. Am. Chem. Soc.*, **80**, 143 (1958).
[1849] A. C. Huitric and W. D. Kumler, *J. Am. Chem. Soc.*, **78**, 614 (1956).
[1850] A. C. Huitric and W. D. Kumler, *J. Am. Chem. Soc.*, **78**, 1147 (1956).
[1851] R. Poggi and P. Saltini, *Gazz. Chim. Ital.*, **62**, 678 (1932).
[1852] R. Poggi and V. Guastalla, *Gazz. Chim. Ital.*, **61**, 405 (1931).
[1853] K. Dimroth, *Ber.*, **71**, 1346 (1938).
[1854] V. T. Suu, N. P. Buu-Hoi, and N. Dat Xuong, *Bull. Soc. Chim. France*, 1875 (1962).
[1855] A. R. Poggi and M. Milletti, *Ann. Chim. Applicata*, **39**, 582 (1949) [*C.A.*, **46**, 2521 (1952)].
[1856] K. Dimroth and H. Jonsson, *Ber.*, **71**, 2658 (1938).
[1857] J. L. M. A. Schlatmann and E. Havinga, *Rec. Trav. Chim.*, **80**, 1101 (1961).
[1858] J. B. Aldersley, G. N. Burkhardt, A. E. Gillam, and N. C. Hindley. *J. Chem. Soc.*, 10 (1940).
[1859] N. Wolff, *Compt. Rend.*, **174**, 1469 (1924).
[1860] R. Pallaud and F. Delaveau, *Bull. Soc. Chim. France*, 35 (1955).
[1861] W. S. Johnson, *J. Am. Chem. Soc.*, **65**, 1320 (1943).
[1862] J. H. Brewster and J. E. Privett, *J. Am. Chem. Soc.*, **88**, 1419 (1966).
[1863] W. Borsche and K. Wunder, *Ann.*, **411**, 38 (1916).
[1864] A. Haller, *Compt. Rend.*, **136**, 1222 (1903).
[1865] M. L. Tétry, *Bull. Soc. Chim. France*, [3] **27**, 302 (1902).
[1866] O. Wallach, *Ann.*, **346**, 249 (1906).

[1867] A. R. Poggi and L. Sancio, *Gazz. Chim. Ital.*, **84,** 890 (1954).
[1868] M. S. R. Nair, H. H. Mathur, and S. C. Bhattacharyya, *J. Chem. Soc.*, 4154 (1964).
[1869] N. Milas and C. P. Priesing, *J. Am. Chem. Soc.*, **79,** 6295 (1957).
[1870] N. Milas and C. P. Priesing, *J. Am. Chem. Soc.*, **81,** 397 (1959).
[1871] I. T. Harrison and B. Lythgoe, *J. Chem. Soc.*, 837 (1958).
[1872] F. Johnson, W. D. Gurowitz, and N. A. Starkovsky, *Tetrahedron Letters*, 1167 (1962).
[1873] R. H. Burnell, *J. Chem. Soc.*, 1307 (1958).
[1874] F. Bohlmann and J. Riemann, *Chem. Ber.*, **97,** 1515 (1964).
[1875] V. Boekelheide, *J. Am. Chem. Soc.*, **69,** 790 (1947).
[1876] N. P. Buu-Hoi, T. Ba Loc, and N. Dat Xuong, *Compt. Rend.*, **244,** 477 (1957).
[1877] W. Parker, R. Ramage, and R. A. Raphael, *J. Chem. Soc.*, 1558 (1962).
[1878] Dainippon Pharmaceutical Co., Ltd., Japan. pat. 17,516 [*C.A.*, **62,** 5232 (1965)].
[1879] W. Treibs and P. Grossman, *Chem. Ber.*, **92,** 273 (1959).
[1880] N. J. Leonard and G. C. Robinson, *J. Am. Chem. Soc.*, **75,** 2143 (1953).
[1881] N. J. Leonard, L. A. Miller, and J. W. Berry, *J. Am. Chem. Soc.*, **79,** 1482 (1957).
[1882] R. Cornubert, R. Joly, and A. Strébel, *Bull. Soc. Chim. France*, [5] **5,** 1501 (1938).
[1883] R. Cornubert, M. André, and R. Joly, *Bull. Soc. Chim. France*, [5] **6,** 265 (1939).
[1884] F. Mattu and M. R. Manco, *Ann. Chim.* (*Rome*), **46,** 1173 (1956) [*C.A.*, **51,** 8036 (1957)].
[1885] O. Wallach, *Ann.*, **313,** 345, 365 (1900).
[1886] H. W. Wanzlick, M. Lehmann-Horchler, and S. Mohrmann, *Chem. Ber.*, **90,** 2521 (1957).
[1887] N. Maxim, I. Zugrăvescu, and N. Teodorescu, *Soc. Romania Sect. Soc. Române Stiinte, Bul. Chim. Pura Apl.*, [2] **3A,** 24 (1941–1942) [*C.A.*, **38,** 5496 (1944)].
[1888] J. Read and H. G. Smith, *J. Chem. Soc.*, **119,** 779 (1921).
[1889] J. Read and H. G. Smith, *J. Chem. Soc.*, **120,** 574 (1922).
[1890] J. C. Bardhan and R. N. Adhya, *J. Chem. Soc.*, 260 (1956).
[1891] J. C. Earl and J. Read, *J. Chem. Soc.*, 2072 (1926).
[1892] N. Wolff, *Compt. Rend.*, **172,** 1357 (1923).
[1893] A. Haller, *Compt. Rend.*, **128,** 1270 (1899).
[1894] O. Wallach, *Ann.*, **437,** 148 (1924).
[1895] G. V. Boyd, *J. Chem. Soc.*, 1978 (1958).
[1896] R. Huisgen, E. Rauenbusch, and G. Seidl, *Chem. Ber.*, **90,** 1958 (1957).
[1897] R. D. Haworth and A. F. Turner, *J. Chem. Soc.*, 1240 (1958).
[1898] W. S. Johnson, D. K. Bannerjee, W. P. Schneider, C. D. Gutsche, W. E. Shelburg, and L. J. Chinn, *J. Am. Chem. Soc.*, **74,** 2832 (1952).
[1899] P. Kurath and W. Cole, *J. Org. Chem.*, **26,** 4592 (1961).
[1900] R. G. Christiansen, R. O. Clinton, and J. W. Dean, U.S. pat. 3,163,641 [*C.A.*, **62,** 7834 (1965)].
[1901] W. L. Meyer, D. D. Cameron, and W. S. Johnson, *J. Org. Chem.*, **27,** 1130 (1962).
[1902] R. Hanna, C. Sandris, and C. Ourisson, *Bull. Soc. Chim. France*, 1454 (1959).
[1903] B. A. Shoulders, A. W. Kwie, W. Klyne, and P. D. Gardner, *Tetrahedron*, **21,** 2973 (1965).
[1904] L. G. Heeringa and M. G. J. Beets, *Rec. Trav. Chim.*, **76,** 213 (1957).
[1905] C. M. van Marle and B. Tollens, *Ber.*, **36,** 1351 (1903).
[1906] B. Tollens, *Ber.*, **37,** 1435 (1904).
[1907] A. Terada, *Nippon Kagaku Zasshi*, **81,** 606 (1960) [*C.A.*, **56,** 1446 (1962)].
[1908] M. Julia and J. Bullot, *Bull. Soc. Chim. France*, 1689 (1959).
[1909] W. Koenigs and E. Wagstaffe, *Ber.*, **26,** 554 (1893).
[1910] K. Kulka, R. J. Eiserle, J. A. Rogers, and F. W. Richter, *J. Org. Chem.*, **25,** 270 (1960).
[1911] R. Anet, *J. Org. Chem.*, **26,** 246 (1961).
[1912] W. J. Raich and C. S. Hamilton, *J. Am. Chem. Soc.*, **79,** 3800 (1957).
[1913] N. S. Emerson and T. M. Patrick, *J. Org. Chem.*, **14,** 790 (1949).
[1914] G. A. Hanson, *Bull. Soc. Chim. Belges*, **67,** 712 (1958) [*C.A.*, **53,** 17995 (1959)].
[1915] Yu. K. Yur'ev, N. N. Magdesieva, and V. V. Titov, *Zh. Obshch. Khim.*, **34,** 1078 (1964) [*C.A.*, **61,** 648 (1964)].
[1916] N. L. Drake and H. W. Gilbert, *J. Am. Chem. Soc.*, **52,** 4965 (1930).

[1917] F. W. Semmler and E. Ascher, *Ber.*, **42**, 2355 (1909).

[1918] G. A. Hanson, *Bull. Soc. Chim. Belges*, **67**, 91 (1958) [*C.A.*, **52**, 20111 (1958)].

[1919] A. P. Terent'ev, R. A. Gracheva, and L. M. Volkova, *Dokl. Akad. Nauk SSSR*, **140**, 610 (1961) [*C.A.*, **56**, 4709 (1962)].

[1920] I. V. Kamenskii, V. I. Itinskii, and N. E. Teplov, U.S.S.R. pat. 138,739 [*C.A.*, **56**, 7280 (1962)].

[1921] K. Y. Novitskii, V. P. Volkov, and Y. K. Yur'ev, *Zh. Obshch. Khim.*, **32**, 399 (1962) [C.A., **58**, 494 (1963)].

[1922] P. R. Shah and N. M. Shah, *Current Sci. (India)*, **31**, 12 (1962) [*C.A.*, **58**, 5585 (1963)].

[1923] C. K. Bradsher, F. C. Brown, and W. B. Blue, *J. Am. Chem. Soc.*, **71**, 3570 (1949).

[1924] W. Davey and J. R. Gwilt, *J. Chem. Soc.*, 1008 (1957).

[1925] J. C. S. Stevens, *J. Chem. Soc.*, 2107 (1930).

[1926] C. Weygand, E. Bauer, H. Günther, and W. Heynemann, *Ann.*, **459**, 99 (1927).

[1927] S. v. Kostanecki and E. Oppelt, *Ber.*, **29**, 244 (1896).

[1928] A. P. A. Rossi, L. H. Werner, W. L. Bencze, and G. de Stevens, Belg. pat. 641,254 [*C.A.*, **63**, 9956 (1965)].

[1929] P. Pfeiffer, E. Kalckbrenner, W. Kunze, and K. Levin, *J. Prakt. Chem.*, [2] **119**, 109 (1928).

[1930] J. F. G. Dippy and R. H. Lewis, *Rec. Trav. Chim.*, **56**, 1000 (1937).

[1931] C. L. Bickel, *J. Am. Chem. Soc.*, **68**, 865 (1946).

[1932] E. Weitz and A. Scheffer, *Ber.*, **54**, 2327 (1921).

[1933] J. T. Eaton, D. B. Black, and R. C. Fuson, *J. Am. Chem. Soc.*, **56**, 687 (1934).

[1934] W. Dilthey, L. Neuhaus, E. Reis, and W. Schommer, *J. Prakt. Chem.*, [2] **124**, 81 (1930).

[1935] Z. Csuros and G. Deák, *Acta Chim. Acad. Sci. Hung.*, **17**, 439 (1958) [*C.A.*, **53**, 17053 (1959)].

[1936] I. Tanasescu and A. Baciu, *Bull. Soc. Chim. France*, [5] **4**, 1742 (1937).

[1937] W. Dilthey, L. Neuhaus, and W. Schommer, *J. Prakt. Chem.*, [2] **123**, 235 (1929).

[1938] R. Sorge, *Ber.*, **35**, 1065 (1902).

[1939] C. Weygand and F. Schächer, *Ber.*, **68**, 227 (1935).

[1940] G. V. Jadhav and V. G. Kulkarni, *Current Sci. (India)*, **20**, 42 (1951) [*C.A.*, **45**, 10215 (1951)].

[1941] R. J. W. LeFevre, P. J. Markham, and J. Pearson, *J. Chem. Soc.*, 344 (1933).

[1942] H. Wieland, *Ber.*, **37**, 1148 (1904).

[1943] E. A. Braude and E. S. Waight, *J. Chem. Soc.*, 419 (1953).

[1944] D. S. Breslow and C. R. Hauser, *J. Am. Chem. Soc.*, **62**, 2385 (1940).

[1945] S. v. Kostanecki and G. Rossbach, *Ber.*, **29**, 1488 (1896).

[1946] S. v. Kostanecki and J. Tambor, *Ber.*, **32**, 1921 (1899).

[1947] F. Straus and H. Grindel, *Ann.*, **439**, 276 (1924).

[1948] P. Klinke and H. Gibian, *Chem. Ber.*, **94**, 26 (1961).

[1949] F. G. Pond, O. P. Maxwell, and G. M. Norman, *J. Am. Chem. Soc.*, **21**, 955 (1899).

[1950] C. Harries and G. Busse, *Ber.*, **29**, 375 (1896).

[1951] H. Bablich and S. v. Kostanecki, *Ber.*, **29**, 233 (1896).

[1952] E. Bargellini and L. Bini, *Gazz. Chim. Ital.*, **41**, 435 (1911).

[1953] D. N. Dhar, *J. Indian Chem. Soc.*, **37**, 799 (1960).

[1954] L. C. Raiford and G. V. Gundy, *J. Am. Chem. Soc.*, **54**, 1191 (1932).

[1955] L. C. Raiford and G. V. Gundy, *J. Org. Chem.*, **3**, 265 (1938).

[1956] W. Davey and D. H. Maass, *J. Chem. Soc.*, 4386 (1963).

[1957] N. B. Sunshine and G. F. Woods, *J. Org. Chem.*, **28**, 2517 (1963).

[1958] Z. S. Ariyan and B. Mooney, *J. Chem. Soc.*, 1519 (1962).

[1959] H. Stobbe and K. Bremer, *J. Prakt. Chem.*, [2] **123**, 1 (1929).

[1960] M. Giua, *Gazz. Chim. Ital.*, **46** (I), 289 (1916).

[1961] H. Stobbe and F. J. Wilson, *J. Chem. Soc.*, **97**, 1722 (1910).

[1962] G. A. Holmberg and J. Axberg, *Acta Chem. Scand.*, **17**, 967 (1963).

[1963] F. J. Pond and A. S. Schoffstall, *J. Am. Chem. Soc.*, **22**, 666 (1900).

[1964] A. D. Petrov, *Ber.*, **63**, 898 (1930).

[1965] W. Feuerstein and S. v. Kostanecki, *Ber.*, **31**, 710 (1898).

[1966] P. Pfeiffer, G. Armbruster, P. Backes, and H. Oberlin, *J. Prakt. Chem.*, [2] **108**, 341 (1924).

[1967] M. Scholtz, *Ber.*, **28**, 1726 (1895).

[1968] Z. Zafiriadis, *Compt. Rend.*, **228**, 852 (1949).

[1969] J. N. Chatterjea and K. Prasad, *J. Sci. Ind. Res. (India)*, **14B**, 383 (1955) [*C.A.*, **50**, 13908 (1956)].

[1970] D. S. Noyce and M. J. Jorgenson, *J. Am. Chem. Soc.*, **84**, 4312 (1962).

[1971] R. G. Christiansen, R. R. Brown, A. S. Hay, A. Nickon, and R. B. Sandin, *J. Am. Chem. Soc.*, **77**, 948 (1955).

[1972] A. Treibs and H. Bader, *Chem. Ber.*, **90**, 790 (1957).

[1973] W. C. Baird and R. L. Shriner, *J. Am. Chem. Soc.*, **86**, 3142 (1964).

[1974] S. Bodforss, *Ber.*, **52**, 142 (1919).

[1975] J. Boichard, J-P. Monin, and J. Tirouflet, *Bull. Soc. Chim. France*, 851 (1963).

[1976] J. O. Hawthorne, E. L. Mihelic, M. S. Morgan, and M. H. Wilt, *J. Org. Chem.*, **28**, 2831 (1963).

[1977] W. Jacob, *Helv. Chim. Acta*, **4**, 782 (1921).

[1978] A. Russel and W. B. Happoldt, *J. Am. Chem. Soc.*, **64**, 1101 (1942).

[1979] R. C. Blume and H. G. Lindwall, *J. Org. Chem.*, **11**, 185 (1946).

[1980] K. C. Joshi and A. K. Jauhar, *J. Indian Chem. Soc.*, **39**, 463 (1962).

[1981] V. G. Kulkarni and G. V. Jadhav, *J. Indian Chem. Soc.*, **31**, 746 (1954).

[1982] H. R. Trivedi and M. N. Vakharia, *J. Univ. Bombay*, **30**, 83 (1961–62) [*C.A.*, **60**, 5382 (1964)].

[1983] R. Kuhn and H. R. Hensel, *Chem. Ber.*, **86**, 1333 (1953).

[1984] W. Dahse, *Ber.*, **41**, 1619 (1908).

[1985] H. P. Vandrewalla and G. V. Jadhav, *Proc. Indian Acad. Sci.*, **28A**, 125 (1948) [*C.A.*, **44**, 3981 (1950)].

[1986] K. C. Joshi and A. J. Jauhar, *Indian J. Chem.*, **1**, 477 (1963).

[1987] D. B. Ghadawala and G. C. Amin, *Sci. Cult. (Bombay)*, **21**, 268 (1955) [*C.A.*, **50**, 11295 (1956)].

[1988] N. P. Buu-Hoi, N. Dat Xuong, and T. L. Trieu, *Bull. Soc. Chim. France*, 584 (1961).

[1989] S. S. Tiwari and A. Singh, *J. Indian Chem. Soc.*, **38**, 931 (1961).

[1990] A. C. Annigeri and S. Siddappa, *J. Karnatak Univ.*, **9–10**, 21 (1964–1965), [*C.A.*, **64**, 9676 (1966)].

[1991] B. C. Jha and G. C. Amin, *Tetrahedron*, **2**, 241 (1958).

[1992] R. P. Barnes and N. F. Payton, *J. Am. Chem. Soc.*, **58**, 1300 (1936).

[1993] E. Schraufstätter and S. Deutsch, *Chem. Ber.*, **81**, 489 (1948).

[1994] C. F. H. Allen and G. H. Frame, *Can. J. Res.*, **6**, 605 (1932).

[1995] N. P. Buu-Hoi and M. Sy, *Bull. Soc. Chim. France*, 219 (1958).

[1996] M. Brandstätter, *Monatsh. Chem.*, **80**, 1 (1949).

[1997] J. R. Merchant and A. S. U. Choughuley, *Chem. Ber.*, **95**, 1792 (1962).

[1998] F. C. Chen and K. T. Chang, *J. Taiwan Pharm. Assoc.*, **4**, 38 (1952) [*C.A.*, **49**, 3175 (1955)].

[1999] M. Vandewalle and A. Vandendooren, *Bull. Soc. Chim. Belges*, **71**, 131 (1962) [*C.A.*, **57**, 15061 (1962)].

[2000] B. J. Ghiya and M. G. Marathey, *J. Indian Chem. Soc.*, **37**, 739 (1960).

[2001] C. S. Marvel, W. R. Peterson, H. K. Inskip, J. E. McCorkle, W. K. Taft, and B. G. Labbe, *Ind. Eng. Chem.*, **45**, 1532 (1953).

[2002] F. Straus and A. Ackermann, *Ber.*, **42**, 1804 (1909).

[2003] C. F. H. Allen, J. B. Normington, and C. V. Wilson, *Can. J. Res.*, **11**, 382 (1934).

[2004] H. Gilman and L. F. Cason, *J. Am. Chem. Soc.*, **72**, 3469 (1950).

[2005] M. M. Shah and S. R. Parikh, *J. Indian Chem. Soc.*, **36**, 726 (1959).

[2006] P. L. Cheng, P. Fournari, and J. Tirouflet, *Bull. Soc. Chim. France*, 2248 (1963).

[2007] F. Kunckell and A. Fürstenberg, *Ber.*, **44**, 3654 (1911).

[2008] S. S. Vernekar and S. Rajogopal, *Rec. Trav. Chim.*, **81**, 710 (1962).

[2009] G. Bargellini and L. Monti, *Gazz. Chim. Ital.*, **44**, 25 (1914) [*C.A.*, **9**, 447 (1915)].

[2010] F. Bergmann, A. Kalmus, and S. Vromen, *J. Am. Chem. Soc.*, **77**, 2494 (1955).

[2011] E. A. Brown and J. Chinn, U.S. pat. 3,101,346 [*C.A.*, **59,** 11355 (1963)].

[2012] F. C. Chen and C. H. Yang, *J. Taiwan Pharm. Assoc.*, **3,** 39 (1951) [*C.A.*, **49,** 2432 (1955)].

[2013] N. B. Mulchandani and W. M. Shah, *Chem. Ber.*, **93,** 1913 (1960).

[2014] R. P. Barnes, J. H. Graham, and M. A. Salim Qureshi, *J. Org. Chem.*, **28,** 2890 (1963).

[2015] C. Engler and K. Dorant, *Ber.*, **28,** 2497 (1895).

[2016] G. Sipos, I. Dobo, and B. Czukor, *Acta Univ. Szeged. Acta Phys. Chem.*, **8,** 160 (1962) [*C.A.*, **59,** 5059 (1963)].

[2017] N. P. Buu-Hoi, N. Dat Xuong, and M. Sy, *Bull. Soc. Chim. France,* 1646 (1956).

[2018] I. Tanasescu and E. Tanasescu, *Bull. Soc. Chim. France*, [5] **3,** 865 (1936).

[2019] H. Rupe and D. Wasserzug, *Ber.*, **34,** 3527 (1901).

[2020] R. P. Barnes and L. B. Dodson, *J. Am. Chem. Soc.*, **65,** 1585 (1943).

[2021] T. Széll and A. Bajusz, *Magy. Kem. Folyoirat,* **60,** 5 (1954) [*C.A.*, **52,** 5352 (1958)].

[2022] M. Giua, *Gazz. Chim. Ital.*, **55,** 567 (1925) [*C.A.*, **20,** 749 (1926)].

[2023] Z. S. Ariyan and H. Suschitzky, *J. Chem. Soc.*, 2242 (1961).

[2024] S. V. Tsukerman, K-S. Chang, and V. F. Lavrushin, *Zh. Obshch. Khim.*, **34,** 832 (1964) [*C.A.*, **60,** 15826 (1964)].

[2025] T. Széll, *Chem. Ber.*, **93,** 1928 (1960).

[2026] T. Széll, *Chem. Ber.*, **91,** 2609 (1958).

[2027] G. S. Chhaya, P. L. Trivedi, and G. V. Jadhav, *J. Univ. Bombay,* **25,** 8 (1957) [*C.A.*, **52,** 14598 (1958)].

[2028] G. Sipos, T. Széll, and I. Várnai, *Acta Univ. Szeged. Acta Phys. Chem.*, **6,** 109 (1960) [*C.A.*, **55,** 14377 (1961)].

[2029] G. Sipos and F. Sirokman, *Nature,* **202,** 489 (1964).

[2030] M. Christian and G. C. Amin, *Chem. Ber.*, **90,** 1287 (1957).

[2031] T. Széll and G. Sipos, *Ann.*, **641,** 113 (1961).

[2032] O. Dann and H. Hoffmann, *Chem. Ber.*, **95,** 1446 (1962).

[2033] S. Seshardi and P. L. Trivedi, *J. Org. Chem.*, **22,** 1633 (1957).

[2034] A. Corvaisier, *Bull. Soc. Chim. France,* 528 (1962).

[2035] J. Tirouflet and A. Corvaisier, *Compt. Rend.*, **250,** 1276 (1960).

[2036] W. B. Raut and S. H. Wender, *J. Org. Chem.*, **25,** 50 (1960).

[2037] P. F. Devitt, A. Timoney, and M. A. Vickars, *J. Org. Chem.*, **26,** 4941 (1961).

[2038] Y. Otsuka, *J. Chem. Soc. Japan,* **65,** 539 (1944) [*C.A.*, **41,** 3797 (1947)].

[2039] D. N. Dhar, *J. Indian Chem. Soc.*, **38,** 823 (1961).

[2040] N. Gudi, S. Hiremath, V. Badiger, and S. Rajagopal, *Arch. Pharm.*, **295,** 16 (1962) [*C.A.*, **57,** 7154 (1962)].

[2041] D. N. Dhar, *J. Org. Chem.*, **25,** 1247 (1960).

[2042] K. Freudenberg and L. Orthner, *Ber.*, **55,** 1748 (1922).

[2043] P. P. Trakroo and A. J. Mukhedkar, *J. Indian Chem. Soc.*, **41,** 595 (1964).

[2044] H. S. Mahal and K. Venkataraman, *J. Chem. Soc.*, 616 (1933).

[2045] E. F. Kurth, *J. Am. Chem. Soc.*, **61,** 861 (1939).

[2046] S. Y. Ambekar, S. H. Dandeganoker, S. D. Jolad, and S. Rajagopal, *J. Indian Chem. Soc.*, **40,** 1041 (1963).

[2047] K. P. Mathai and S. Sethna, *J. Indian Chem. Soc.*, **41,** 347 (1964).

[2048] D. N. Dhar, *J. Indian Chem. Soc.*, **37,** 363 (1960).

[2049] A. Gutzeit and S. v. Kostanecki, *Ber.*, **38,** 933 (1905).

[2050] F. Herstein and S. v. Kostanecki, *Ber.*, **32,** 318 (1899).

[2051] L. Hoerhammer, H. Wagner, H. Roesler, M. Keckei Sen, and L. Farkas, *Tetrahedron,* **21,** 969 (1965).

[2052] B. Sen, *J. Am. Chem. Soc.*, **74,** 3445 (1952).

[2053] L. Reichel and J. Marchand, *Ber.*, **76,** 1132 (1943).

[2054] T. H. Simpson and J. L. Beton, *J. Chem. Soc.*, 4065 (1954).

[2055] L. Reichel and R. Schickle, *Ber.*, **76,** 1134 (1943).

[2056] S-T. Chen, P-C. Kan, and Yu-C. Hu, *Yao Hsueh Hsueh Pao,* **10** (2), 97 (1963) [*C.A.*, **59,** 13914 (1963)].

[2057] S. Fujise, R. Fujii, T. Maeda, K. Takagi, and S. Nakamura, *J. Chem. Soc. Japan*, **74**, 827 (1953) [*C.A.*, **49**, 3952 (1955)].

[2058] A. v. Wacek and E. David, *Ber.*, **70**, 190 (1937).

[2059] C. Kuroda and T. Nakamura, *Sci. Papers Inst. Phys. Chem. Res.* (*Tokyo*), **18**, 61 (1932) [*Chem. Zentr.*, **103**, I, 2170 (1932); *C.A.*, **26**, 2442 (1932)].

[2060] I. Z. Saiyad, D. R. Nadkarni, and T. S. Wheeler, *J. Chem. Soc.*, 1737 (1937).

[2061] A. Göschke and J. Tambor, *Ber.*, **45**, 186 (1912).

[2062] D. N. Dhar, *J. Indian Chem. Soc.*, **37**, 496 (1960).

[2063] J. Tambor, *Ber.*, **49**, 1704 (1916).

[2064] D. R. Nadkarni and T. S. Wheeler, *J. Chem. Soc.*, 1320 (1938).

[2065] A. Göschke and J. Tambor, *Ber.*, **44**, 3502 (1911).

[2066] D. N. Dhar and J. B. Lal, *J. Org. Chem.*, **23**, 1159 (1958).

[2067] J. Shinoda, *J. Pharm. Soc. Japan*, **48**, 214 (1928) [*C.A.*, **22**, 2947 (1928)].

[2068] J. B. Lal, *J. Indian Chem. Soc.*, **16**, 296 (1939) [*C.A.*, **34**, 1005 (1940)].

[2069] R. O. Clinton and T. A. Geissman, *J. Am. Chem. Soc.*, **65**, 85 (1943).

[2070] N. B. Mulchandani and W. M. Shah, *Chem. Ber.*, **93**, 1918 (1960).

[2071] G. N. Vyas and N. M. Shah, *J. Indian Chem. Soc.*, **26**, 273 (1949) [*C.A.*, **44**, 3981 (1950)].

[2072] P. R. Shah and N. M. Shah, *Chem. Ber.*, **97**, 1453 (1964).

[2073] G. J. Martin, J. M. Beiler, and S. Avakien, U.S. pat. 2,769,817 [*C.A.*, **51**, 14815 (1957)].

[2074] B. Cummins, D. M. X. Donnelly, J. F. Eades, H. Fletcher, F. O. Cinneide, E. M. Philbin, J. Swirski, T. S. Wheeler, and R. K. Wilson, *Tetrahedron*, **19**, 499 (1963).

[2075] M. K. Seikel, A. L. Haines, and H. D. Thompson, *J. Am. Chem. Soc.*, **77**, 1196 (1955).

[2076] D. H. Marrian, P. B. Russell, and A. R. Todd, *J. Chem. Soc.*, 1419 (1947).

[2077] F. Bergel, A. L. Morrison, and N. Rinderknecht, *J. Chem. Soc.*, 659 (1950).

[2078] M. Giua and E. Bagiella, *Gazz. Chim. Ital.*, **51** (II), 116 (1921) [*C.A.*, **16**, 557 (1922)].

[2079] R. C. Fuson and G. Munn, *J. Am. Chem. Soc.*, **71**, 1116 (1949).

[2080] P. Pfeiffer, K. Kollbach, and E. Haack, *Ann.*, **460**, 138 (1928).

[2081] K. W. Gopinath, T. R. Govindachari, and N. Viswanathan, *Tetrahedron*, **14**, 322 (1961).

[2082] G. N. Vyas and N. M. Shah, *J. Indian Chem. Soc.*, **28**, 41, 75 (1951) [*C.A.*, **46**, 1506 (1952)].

[2083] T. Nagano and K. Matsumura, *J. Am. Chem. Soc.*, **75**, 6237 (1953).

[2084] V. J. Harding, *J. Chem. Soc.*, **105**, 2790 (1914).

[2085] F. A. Atchabba, P. L. Trivedi, and G. V. Jadhav, *J. Univ. Bombay*, **25**, 1 (1957) [*C.A.*, **52**, 14599 (1958)].

[2086] V. G. Kulkarni and G. V. Jadhav, *J. Indian Chem. Soc.*, **32**, 97 (1955).

[2087] V. G. Kulkarni and G. V. Jadhav, *J. Univ. Bombay*, **23A**, 14 (1955) [*C.A.*, **51**, 11301 (1957)].

[2088] C. Weygand, *Ber.*, **57**, 413 (1924).

[2089] C. Weygand and A. Matthes, *Ann.*, **449**, 31 (1926).

[2090] J. Michel, *Bull. Soc. Chim. Belges*, **48**, 105 (1939) [*C.A.*, **33**, 7650 (1939)].

[2091] S. v. Kostanecki and J. Tambor, *Ber.*, **29**, 237 (1896).

[2092] S. M. Nadkarni, A. M. Warriar, and T. S. Wheeler, *J. Chem. Soc.*, 1798 (1937).

[2093] J. Gosselck and E. Wolters, *Chem. Ber.*, **95**, 1237 (1962).

[2094] V. Tognazzi, *Gazz. Chim. Ital.*, **54**, 697 (1924) [*C.A.*, **19**, 823 (1925)].

[2095] H. Kauffmann, *Ber.*, **54**, 795 (1921).

[2096] S. Matsueda, K. Sannohe, and Y. Saito, *Bull. Chem. Soc. Japan*, **36**, 1528 (1963) [*C.A.*, **60**, 5379 (1964)].

[2097] F. Zwayer and S. v. Kostanecki, *Ber.*, **41**, 1335 (1908).

[2098] F. R. Brown and W. Cummings, *J. Chem. Soc.*, 4302 (1958).

[2099] B. J. F. Hudson, *J. Chem. Soc.*, 754 (1946).

[2100] E. Rohrmann, R. G. Jones, and H. A. Shonle, *J. Am. Chem. Soc.*, **66**, 1856 (1944).

[2101] C. Kuroda and T. Matsukuma, *Sci. Papers Inst. Phys. Chem. Res.* (*Tokyo*), **18**, 51 (1932) [*Chem. Zentr.*, **103**, I, 2170 (1932); *C.A.*, **26**, 2442 (1932)].

[2102] R. C. Tallman and A. H. Stuart, U.S. pat. 2,400,033 [*C.A.*, **40**, 4484 (1946)].

[2103] F. Strauss, *Ann.*, **374**, 139 (1910).

[2104] B. J. F. Hudson and E. Walton, *J. Chem. Soc.*, 85 (1946).

[2105] R. Royer, P. Demerseman, A. Cheutin, and M. Hubert-Habart, *Bull. Soc. Chim. France*, 304 (1957).
[2106] F. D. Cramer and G. H. Elschnig, *Chem. Ber.*, **89**, 1 (1956).
[2107] K. v. Auwers and L. Anschütz, *Ber.*, **54**, 1543 (1921).
[2108] H. E. Smith and M. C. Paulson, *J. Am. Chem. Soc.*, **76**, 4486 (1954).
[2109] H. Richtzenhain and B. Alfredsson, *Chem. Ber.*, **89**, 378 (1956).
[2110] T. Emilewicz and S. v. Kostanecki, *Ber.*, **32**, 309 (1899).
[2111] S. v. Kostanecki and A. V. Szlagier, *Ber.*, **37**, 4155 (1904).
[2112] S. v. Kostanecki and A. Widmer, *Ber.*, **37**, 4159 (1904).
[2113] A. Sonn, *Ber.*, **54**, 358 (1921).
[2114] T. H. Simpson and W. B. Whalley, *J. Chem. Soc.*, 166 (1955).
[2115] N. Narasimachari, R. Rajagopalan, and T. R. Seshadri, *Proc. Indian Acad. Sci.*, **36A**, 231 (1952) [*C.A.*, **48**, 2698 (1954)].
[2116] K. Kratzl and H. Däubner, *Ber.*, **77**, 519 (1944).
[2117] G. Zemplén, R. Bognar, and J. Mechner, *Ber.*, **77**, 99 (1944).
[2118] M. Stefanovic, D. Miljkovic, M. Miljkovic, A. Jokic, and B. Stipanovic, *Tetrahedron Letters*, 3891 (1966).
[2119] H. P. Vandrewala and G. V. Jadhav, *J. Univ. Bombay*, **16**, 43 (1948) [*C.A.*, **43**, 1742 (1949)].
[2120] H. Gilman, L. Fullhart, and L. F. Cason, *J. Org. Chem.*, **21**, 826 (1956).
[2121] J. Dewar, *J. Chem. Soc.*, 619 (1944).
[2122] F. Kunckell and M. Hammerschmidt, *Ber.*, **46**, 2676 (1913).
[2123] F. Kunckell, *Ber.*, **37**, 2826 (1904).
[2124] A. A. Raval and N. M. Shah, *J. Org. Chem.*, **21**, 1408 (1956).
[2125] H. Kauffmann and F. Kieser, *Ber.*, **46**, 3788 (1913).
[2126] J. Koo, *J. Org. Chem.*, **26**, 4185 (1961).
[2127] J. Tambor, *Ber.*, **44**, 3215 (1911).
[2128] H. S. Mehra and K. B. L. Mathur, *J. Indian Chem. Soc.*, **33**, 618 (1956).
[2129] S. Tamura, K. Okuna, H. Akabori, and K. Kanezaki, *J. Agr. Chem. Soc. Japan*, **27**, 318 (1953) [*C.A.*, **50**, 6402 (1956)].
[2130] T. H. Marquardt, *Helv. Chim. Acta*, **48**, 1476 (1965).
[2131] T. Emilewicz and S. v. Kostanecki, *Ber.*, **31**, 696 (1898).
[2132] S. v. Kostanecki and A. Rozycki, *Ber.*, **32**, 2257 (1899).
[2133] S. v. Kostanecki and F. W. Osius, *Ber.*, **32**, 321 (1899).
[2134] M. F. Browne and R. L. Shriner, *J. Org. Chem.*, **22**, 1320 (1957).
[2135] P. B. Mahajani, S. S. Lele, and S. Sethna, *J. Maharaja Sayajirao Univ. Baroda*, **3**, 41 (1954) [*C.A.*, **50**, 324 (1956)].
[2136] A. Russell, *J. Chem. Soc.*, 218 (1934).
[2137] S. S. Cohen and S. v. Kostanecki, *Ber.*, **37**, 2627 (1904).
[2138] S. S. Cohen and O. Schleifenbaum, *Ber.*, **37**, 2631 (1904).
[2139] T. H. Simpson, *J. Org. Chem.*, **28**, 2107 (1963).
[2140] T. H. Simpson, *Chem. Ind.* (*London*), **1955**, 1672.
[2141] G. B. Marini-Bettolo, *Gazz. Chim. Ital.*, **71**, 635 (1941).
[2142] S. v. Kostanecki and J. Tambor, *Ber.*, **32**, 2260 (1899).
[2143] W. Mosimann and J. Tambor, *Ber.*, **49**, 1700 (1916).
[2144] CIBA Ltd., Brit. pat. 929,450 [*C.A.*, **60**, 1704 (1964)].
[2145] S. v. Kostanecki and J. Tambor, *Ber.*, **37**, 792 (1904).
[2146] H. S. Mahal and K. Venkataraman, *J. Chem. Soc.*, 569 (1936).
[2147] L. Farkas and M. Nogradi, *Chem. Ber.*, **98**, 164 (1965).
[2148] G. A. Hanson and P. Tarte, *Bull. Soc. Chim. Belges*, **66**, 619 (1947) [*C.A.*, **52**, 7244 (1958)].
[2149] J. Abelin and S. v. Kostanecki, *Ber.*, **43**, 2157 (1910).
[2150] R. C. Fuson, W. E. Ross, and C. H. McKeever, *J. Am. Chem. Soc.*, **61**, 414 (1939).
[2151] R. C. Fuson, E. W. Maynert, and W. J. Shenk, *J. Am. Chem. Soc.*, **67**, 1939 (1945).
[2152] R. P. Barnes and A. S. Spriggs, *J. Am. Chem. Soc.*, **67**, 134 (1945).
[2153] R. P. Barnes and H. Delaney, *J. Am. Chem. Soc.*, **65**, 2155 (1943).

[2154] R. C. Fuson and J. S. Meek, *J. Org. Chem.*, **10,** 551 (1945).

[2155] C. Weygand and L. Mensdorf, *Ber.*, **68,** 1825 (1935).

[2156] J. Chopin, D. Molho, H. Pachéco, and C. Mentzer, *Bull. Soc. Chim. France*, 192 (1957).

[2157] J. Reigrodski and J. Tambor, *Ber.*, **43,** 1964 (1910).

[2158] N. M. Cullinane and D. Philpott, *J. Chem. Soc.*, 1761 (1929).

[2159] F. É. King, T. J. King, and D. W. Rustige, *J. Chem. Soc.*, 1192 (1962).

[2160] N. R. Bannerjee and T. R. Seshadri, *J. Sci. Ind. Res. (India)*, **13B,** 598 (1954) [*C.A.*, **50,** 316 (1956)].

[2161] G. Bargellini and A. Oliverio, *Ber.*, **75,** 2083 (1942).

[2162] K. V. Rao and T. R. Seshadri, *Proc. Indian Acad. Sci.*, **24A,** 375 (1946) [*C.A.*, **41,** 2735 (1947)].

[2163] L. Hörhammer, H. Wagner, H. Rösler, E. Graf, and L. Farkas, *Chem. Ber.*, **97,** 2857 (1964).

[2164] J. C. Pew, *J. Am. Chem. Soc.*, **77,** 2831 (1955).

[2165] L. Kesselkaul and S. v. Kostanecki, *Ber.*, **29,** 1886 (1896).

[2166] C-T. Chang, *Formosan Sci.*, **16,** 117 (1962) [*C.A.*, **59,** 2760 (1963)].

[2167] G. H. Stout and V. F. Stout, *Tetrahedron*, **14,** 296 (1961).

[2168] R. Royer, *Bull. Soc. Chim. France*, [5] **20,** 412 (1953).

[2169] W. Dilthey, E. Bach, H. Grütering, and E. Hausdörfer, *J. Prakt. Chem.*, [2] **117,** 337 (1927).

[2170] H. Pachéco and A. Grouiller, *Bull. Soc. Chim. France*, 2937 (1965).

[2171] H. Grasshof, Ger. pat. 1,174,311 [*C.A.*, **61,** 11933 (1964)].

[2172] V. G. Joshi and G. C. Amin, *J. Indian Chem. Soc.*, **38,** 159 (1961).

[2173] H. S. Mahal, H. S. Rai, and K. Venkataraman, *J. Chem. Soc.*, 866 (1935).

[2174] H. H. Lee and C. H. Tan, *J. Chem. Soc., Suppl.* 2, 6255 (1964).

[2175] R. C. Fuson and A. I. Rachlin, *J. Am. Chem. Soc.*, **67,** 2055 (1945).

[2176] J. Chopin and P. Durual, *Bull. Soc. Chim. France*, 3350 (1965).

[2177] E. R. Blout and M. S. Simon, U.S. pat. 3,131,219 [*C.A.*, **61,** 3028 (1964)].

[2178] S. C. Bharara, R. N. Goel, A. C. Jain, and T. R. Seshadri, *Indian J. Chem.*, **2,** 399 (1964).

[2179] A. Russel and J. Todd, *J. Chem. Soc.*, 1506 (1934).

[2180] A. Russel and J. Todd, *J. Chem. Soc.*, 1066 (1934).

[2181] P. K. Jesthi and M. K. Rout, *Indian J. Chem.*, **3,** 461 (1965).

[2182] L. Geita and G. Vanags, *Latvigjas PSR Zinatnu Akad. Vestis, Kim. Ser.*, 439 (1962) [*C.A.*, **59,** 7458 (1963)].

[2183] S. v. Kostanecki and L. Laczkowski, *Ber.*, **30,** 2138 (1897).

[2184] J. Stradins, E. Ermane, T. Dumpis, J. Linabergs, and G. Vanags, *Zh. Organ. Khim.* **1,** 388 (1965) [*C.A.*, **62,** 16028 (1965)].

[2185] V. Grinsteins and A. Sausins, *Latvijas PSR Zinatnu Akad. Vestis, Kim. Ser.*, 605 (1963) [*C.A.*, **61,** 3037 (1964)].

[2186] W. Klobski and S. v. Kostanecki, *Ber.*, **31,** 720 (1898).

[2187] J. Sam, J. D. England, and D. W. Alwani, *J. Med. Chem.*, **7,** 732 (1964).

[2188] G. Singh and J. N. Ray, *J. Indian Chem. Soc.*, **7,** 637 (1930) [*C.A.*, **25,** 1253 (1931)].

[2189] A. Hassner and D. R. Fitchum, *Tetrahedron Letters*, 1991 (1966).

[2190] F. S. Kipping, *J. Chem. Soc.*, **65,** 480 (1894).

[2191] W. H. Perkin, Jr. and R. Robinson, *J. Chem. Soc.*, **91,** 1073 (1907).

[2192] W. Feuerstein, *Ber.*, **34,** 412 (1901).

[2193] T. Széll, *J. Prakt. Chem.*, [4] **17,** 346 (1962).

[2194] F. M. K/Serho, Neth. pat. appl. 6,400,267 [*C.A.*, **62,** 564 (1965)].

[2195] M. N. Tilichenko and N. K. Astakova, *Dokl. Akad. Nauk SSSR* **74,** 951 (1950) [*C.A.*, **45,** 4679 (1951)].

[2196] J. Manta, *J. Prakt. Chem.*, [2] **142,** 11 (1935).

[2197] E. Hannig, *Pharmazie*, **20,** 762 (1965) [*C.A.*, **64,** 6609 (1966)].

[2198] J. Décombe, *Compt. Rend.*, **226,** 1991 (1948).

[2199] D. A. Peak, R. Robinson, and J. Walker, *J. Chem. Soc.*, 752 (1936).

[2200] A. Hassner, N. H. Cromwell, and S. J. Davis, *J. Am. Chem. Soc.*, **79,** 230 (1957).

[2201] J. van Alphen and G. Drost, *Rec. trav. chim.*, **69,** 1080 (1950).

[2202] G. Saint-Ruf, N. P. Buu-Hoi, and P. Jacquignon, *J. Chem. Soc.*, 48 (1958); 3237 (1959).

[2203] T. I. Temnikova and N. A. Oshueva, *Zh. Obshch. Khim.*, **33**, 1403 (1963) [*C.A.*, **59**, 11314 (1963)].

[2204] B. D. Pearson, R. P. Ayer, and N. H. Cromwell, *J. Org. Chem.*, **27**, 3038 (1962).

[2205] D. N. Kevill, E. D. Weiler, and N. H. Cromwell, *J. Org. Chem.*, **29**, 1276 (1964).

[2206] K. W. Bentley and W. C. Firth, *J. Chem. Soc.*, 2403 (1955).

[2207] R. Huisgen, G. Seidl, and I. Wimmer, *Ann. Chem.*, **677**, 21 (1964).

[2208] R. de Fazi, *Gazz. Chim. Ital.*, **54**, 658 (1924) [*C.A.*, **19**, 821 (1925)].

[2209] H. A. Torrey and C. M. Brewster, *J. Am. Chem. Soc.*, **31**, 1322 (1909).

[2210] A. P. Khanolkar and T. S. Wheeler, *J. Chem. Soc.*, 2118 (1938).

[2211] B. J. Ghiya and M. G. Marathey, *J. Indian Chem. Soc.*, **38**, 331 (1961).

[2212] A. Maccioni and E. Marongiu, *Ann. Chim. (Rome)*, **50**, 1806 (1960) [*C.A.*, **55**, 17588 (1961)].

[2213] G. K. Trivedi and K. B. Raut, *LABDEV*, **2**, 66 (1964) [*C.A.*, **60**, 15794 (1964)].

[2214] D. Alperin and S. v. Kostanecki, *Ber.*, **32**, 1037 (1899).

[2215] S. v. Kostanecki, *Ber.*, **31**, 705 (1898).

[2216] E. Keller and S. v. Kostanecki, *Ber.*, **32**, 1034 (1899).

[2217] C. R. Hauser and J. K. Lindsay, *J. Org. Chem.*, **22**, 482 (1957).

[2218] M. D. Rausch and L. E. Coleman, Jr., *J. Org. Chem.*, **23**, 107 (1958).

[2219] R. C. Fuson, R. Johnson, and W. Cole, *J. Am. Chem. Soc.*, **60**, 1594 (1938).

[2220] R. Granger and H. Orzalési, *Bull. Soc. Chim. France*, 986 (1958).

[2221] A. L. Pandit and A. B. Kulkarni, *J. Sci. Ind. Res. (India)*, **19B**, 138 (1960) [*C.A.*, **55**, 2584 (1961)].

[2222] A. A. Raval and V. M. Thakor, *J. Indian Chem. Soc.*, **38**, 421 (1961).

[2223] N. P. Buu-Hoi and P. Jacquignon, *J. Org. Chem.*, **24**, 126 (1959).

[2224] A. J. S. Sorrie and R. H. Thomson, *J. Chem. Soc.*, 2233 (1955).

[2225] B. D. Hosangadi, A. L. Pandit, and A. B. Kulkarni, *Indian J. Chem.*, **2**, 235 (1964).

[2226] R. C. Fuson and F. N. Baumgartner, *J. Am. Chem. Soc.*, **70**, 3255 (1948).

[2227] A. Klages and F. Tetzner, *Ber.*, **35**, 3965 (1902).

[2228] H. Stobbe and K. Niedenzu, *Ber.*, **34**, 3897 (1901).

[2229] E. Knoevenagel and R. Weissgerber, *Ber.*, **26**, 441 (1893).

[2230] A. Klages and E. Knoevenagel, *Ber.*, **26**, 447 (1893).

[2231] E. Knoevenagel and R. Weissgerber, *Ber.*, **26**, 436 (1893).

[2232] E. P. Kohler and E. M. Nygaard, *J. Am. Chem. Soc.*, **52**, 4128 (1930).

[2233] K. Fleischer and P. Wolff, *Ber.*, **53**, 925 (1920).

[2234] P. Cagniant and D. Cagniant, *Bull. Soc. Chim. France*, 931 (1955).

[2235] T. A. Mashburn, C. E. Cain, and C. R. Hauser, *J. Org. Chem.*, **25**, 1982 (1960).

[2236] C. Astorino and F. Riccardi, *Ric. Sci. Rend., Ser. A.*, **8**, 193 (1965) [*C.A.*, **63**, 14781 (1965)].

[2237] A. B. E. Lovett and E. Roberts, *J. Chem. Soc.*, 1975 (1928).

[2238] J. Asselineau and A. Willemart, *Bull. Soc. Chim. France*, [5] **14**, 116 (1947).

[2239] R. C. Fuson, L. J. Armstrong, D. H. Chadwick, J. W. Kneisley, S. P. Rowland, W. J. Shenk, and Q. F. Soper, *J. Am. Chem. Soc.*, **67**, 386 (1945).

[2240] R. C. Fuson and R. E. Foster, *J. Am. Chem. Soc.*, **65**, 913 (1943).

[2241] R. C. Fuson, D. J. Byers, C. A. Sperati, R. E. Foster, and P. F. Warfield, *J. Org. Chem.*, **10**, 69 (1945).

[2242] R. C. Fuson, E. H. Hess, and H. S. Killam, *J. Org. Chem.*, **23**, 645 (1958).

[2243] R. C. Fuson and R. F. Heitmiller, *J. Am. Chem. Soc.*, **77**, 174 (1955).

[2244] R. C. Fuson, J. Corse, and C. H. McKeever, *J. Am. Chem. Soc.*, **62**, 3250 (1940).

[2245] L. Jurd, *J. Pharm. Sci.*, **54**, 1221 (1965) [*C.A.*, **63**, 13227 (1965)].

[2246] A. E. I. Sammour, *Tetrahedron*, **20**, 1067 (1964).

[2247] J. A. Bevan, P. E. Gagon, and I. D. Rae, *Can. J. Chem.*, **43**, 2612 (1965).

[2248] R. B. Woodward and R. H. Eastman, *J. Am. Chem. Soc.*, **68**, 2229 (1946).

[2249] E. G. Howard and R. V. Lindsey, *J. Am. Chem. Soc.*, **82**, 158 (1960).

[2250] R. Cornubert, R. Delmas, S. Monteil, and J. Viriot, *Bull. Soc. Chim. France*, [5] **17**, 36 (1950).

[2251] W. Borsche, *Ber.*, **48**, 682 (1915).
[2252] G. M. Kuettel and S. M. McElvain, *J. Am. Chem. Soc.*, **53**, 2692 (1931).
[2253] L. Ruzicka and V. Fornasir, *Helv. Chim. Acta*, **3**, 806 (1920).
[2254] H. H. Szmant and A. J. Basso, *J. Am. Chem. Soc.*, **73**, 4521 (1951).
[2255] C. Weygand and F. Strobelt, *Ber.*, **68**, 1839 (1935).
[2256] H. Brunswig, *Ber.*, **19**, 2890 (1886).
[2257] D. H. Deutsch and E. N. Garcia, U.S. pat. 2,754,299 [*C.A.*, **51**, 4437 (1957)].
[2258] A. M. Packman and N. Rubin, *Am. J. Pharm.*, **134**, 35 (1962) [*C.A.*, **59**, 1517 (1963)].
[2259] C. Finzi and E. Vecchi, *Gazz. Chim. Ital.*, **47** (II), 10 (1917) [*C.A.*, **12**, 1175 (1918)].
[2260] V. F. Lavrushin, S. V. Tsukerman, and S. I. Artemenko, *Zh. Obshch. Khim.*, **32**, 2551 (1962) [*C.A.*, **58**, 4046 (1963)].
[2261] J. Huet, *Bull. Soc. Chim. France*, 1670 (1965).
[2262] J. Huet and J. Dreux, *Compt. Rend.*, **255**, 2453 (1962).
[2263] J. Colonge and R. Gaumont, *Bull. Soc. Chim. France*, 939 (1959).
[2264] S. M. McElvain and J. F. Vozza, *J. Am. Chem. Soc.*, **71**, 896 (1949).
[2265] N. P. Buu-Hoi, O. Roussel, and P. Jacquignon, *Bull. Soc. Chim. France*, 3096 (1964).
[2266] S. M. McElvain and K. Rorig, *J. Am. Chem. Soc.*, **70**, 1820 (1948).
[2267] N. J. Leonard and D. M. Locke, *J. Am. Chem. Soc.*, **77**, 1852 (1955).
[2268] D. R. Howton, *J. Org. Chem.*, **10**, 277 (1945).
[2269] S. M. McElvain and P. H. Parker, *J. Am. Chem. Soc.*, **77**, 492 (1955).
[2270] L. Krasnec, J. Durinda, and L. Szucs, *Chem. Zvesti*, **15**, 558 (1961) [*C.A.*, **56**, 12847 (1962)].
[2271] C. Engler and A. Engler, *Ber.*, **35**, 4061 (1902).
[2272] A. P. Terentev, R. A. Gracheva, N. N. Presbrazhenskaya, and L. M. Volkova, *Zh. Obshch. Khim.*, **33**, 4006 (1963) [*C.A.*, **60**, 9242 (1964)].
[2273] G. R. Clemo and E. Hoggarth, *J. Chem. Soc.*, 1241 (1939).
[2274] V. Braschler, C. A. Grob, and A. Kaiser, *Helv. Chim. Acta*, **46**, 2646 (1963).
[2275] V. Grimme and J. Wöllner, Ger. pat. 1,052,385 [*C.A.*, **55**, 5349 (1961)].
[2276] D. M. Fitzgerald, J. F. O'Sullivan, E. M. Philbin, and T. S. Wheeler, *J. Chem. Soc.* 860 (1955).
[2277] A. Sonn, *Ber.*, **50**, 1262 (1917).
[2278] L. Farkas, M. Nogradi, and L. Pallos, *Chem. Ber.*, **97**, 1044 (1964).
[2279] L. Farkas, L. Pallos, and M. Nogradi, *Magy. Kem. Folyoirat*, **71**, 270 (1965) [*C.A.*, **63**, 9901 (1965)].
[2280] R. Robinson, *J. Chem. Soc.*, **111**, 762 (1917).
[2281] E. Tamate, *Nippon Kagaku Zasshi*, **78**, 1293 (1957) [*C.A.*, **54**, 476 (1960)].
[2282] I. K. Korobitsyna, G. V. Marinova, and Y. K. Yur'ev, *Zh. Obshch. Khim.*, **31**, 2131 (1961) [*C.A.*, **56**, 436 (1962)].
[2283] E. Tamate, *Nippon Kagaku Zasshi*, **80**, 942, 1047 (1959) [*C.A.*, **55**, 4470 (1961)].
[2284] D. C. Rowlands, U.S. pat. 2,776,239 [*C.A.*, **51**, 4639 (1957)].
[2285] I. K. Korobitsyna, C-L. Yin, and Y. K. Yur'ev, *Zh. Obshch. Khim.*, **31**, 2548 (1961) [*C.A.*, **56**, 8673 (1962)].
[2286] M. Renson, *Bull. Soc. Chim. Belges*, **73**, 483 (1964) [*C.A.*, **61**, 8278 (1964)].
[2287] P. Pfeiffer, K. Grimm, and H. Schmidt, *Ann.*, **564**, 208 (1949).
[2288] F. Arndt and G. Kallner, *Ber.*, **57**, 202 (1924).
[2289] S. G. Powell, *J. Am. Chem. Soc.*, **45**, 2708 (1923).
[2290] P. Pfeiffer and E. Döring, *Ber.*, **71**, 279 (1938).
[2291] P. Pfeiffer, E. Breith, and H. Hoyer, *J. Prakt. Chem.*, [2] **129**, 31 (1931).
[2292] P. Pfeiffer, H. Oberlin, and E. Konermann, *Ber.*, **58**, 1947 (1925).
[2293] L. Farkas, M. Nogradi, and L. Pallos, *Tetrahedron Letters*, 1999 (1963).
[2294] L. Farkas, L. Pallos, and M. Nogradi, *Magy. Kem. Folyoirat*, **71**, 272 (1965) [*C.A.*, **63**, 9901 (1965)].
[2295] C. Hansch and H. G. Lindwall, *J. Org. Chem.*, **10**, 381 (1945).
[2296] M. Scholtz, *Ber.*, **45**, 1718 (1912).
[2297] K. Matsumura and C. Sone, *J. Am. Chem. Soc.*, **53**, 1490 (1931).

[2298] A. Mustafa, O. H. Hsihmat, S. M. A. D. Zayed, and A. A. Nawar, *Tetrahedron*, **19**, 1831 (1963).

[2299] G. G. Badcock, F. M. Dean, A. Robertson, and W. B. Whalley, *J. Chem. Soc.*, 903 (1950).

[2300] N. M. Shah, *J. Univ. Bombay*, **11**, 109 (1942) [*C.A.*, **37**, 2000 (1943)].

[2301] S. V. Tsukerman, K. S. Chang, and V. F. Lavrushin, *Zh. Obshch. Khim.*, **34**, 2881 (1964) [*C.A.*, **62**, 520 (1965)].

[2302] W. H. Perkin, Jr., A. Pollard, and R. Robinson, *J. Chem. Soc.*, 49 (1937).

[2303] E. T. Borrows, D. O. Holland, and J. Kenyon, *J. Chem. Soc.*, 1069 (1946).

[2304] A. Schönberg and E. Singer, *Chem. Ber.*, **94**, 241 (1961).

[2305] M. Scholtz, *Ber.*, **45**, 734 (1912).

[2306] M. Scholtz and W. Fraude, *Ber.*, **46**, 1069 (1913).

[2307] H. Rapoport and J. R. Tretter, *J. Org. Chem.*, **23**, 248 (1958).

[2308] R. K. Bly, E. C. Zoll, and J. A. Moore, *J. Org. Chem.*, **29**, 2128 (1964).

[2309] I. K. Korobitsyna, Y. K. Yur'ev, A. Cheburkov, and E. M. Lukina, *Zh. Obshch, Khim.*, **25**, 734 (1955) [*C.A.*, **50**, 2536 (1956)].

[2310] E. T. Holmes and H. R. Snyder, *J. Org. Chem.*, **29**, 2725 (1964).

[2311] S. M. McElvain and P. H. Parker, Jr., *J. Am. Chem. Soc.*, **78**, 5312 (1956).

[2312] N. Barbulescu, O. Maior, and A. Gioaba, *Rev. Chim.* (*Bucharest*), **15**, 330 (1964) [*C.A.* **61**, 10677 (1964)].

[2313] G. Sunagawa and T. Ichii, *Yakugaku Zasshi*, **82**, 987 (1962) [*C.A.*, **58**, 5665 (1963)].

[2314] K. R. Huffman and D. S. Tarbell, *J. Am. Chem. Soc.*, **80**, 6341 (1958).

[2315] A. S. Dreiding and A. J. Tomasewski, *J. Am. Chem. Soc.*, **76**, 6388 (1954).

[2316] W. Keller-Schierlein, M. L. Mihailovic, and V. Prelog, *Helv. Chim. Acta*, **41**, 220 (1958).

[2317] F. Bohlmann and K. Prezewowsky, *Chem. Ber.*, **97**, 1176 (1964).

[2318] R. D. H. Murray, W. Parker, R. A. Raphael, and D. B. Jhaveri, *Tetrahedron*, **18**, 55 (1962).

[2319] R. D. Sands, *J. Org. Chem.*, **29**, 2488 (1964).

[2320] P. E. Papadakis, L. M. Hall, and R. L. Augustine, *J. Org. Chem.*, **23**, 123 (1958).

[2321] D. A. Tyner, U.S. pat. 3,090,792 [*C.A.*, **59**, 11606 (1963)].

[2322] K. Tanabe, R. Takasaki, and R. Hayashi, Japan. pat. 4877 (1963) [*C.A.*, **60**, 640 (1964)].

[2323] J. Mleziva, *Chem. Listy*, **51**, 2364 (1957) [*C.A.*, **52**, 6206 (1958)].

[2324] A. H. Richard and P. Langlais, *Bull. Soc. Chim. France*, [4] **7**, 454 (1910).

[2325] R. Y. Levina, N. P. Shusherina, E. Treshchova, and V. M. Tatevskiĭ, *Zh. Obshch. Khim.*, **22**, 199 (1952) [*C.A.*, **46**, 10095 (1952)].

[2326] R. V. Fedorova, O. D. Belova, and N. Y. Chestnykh, U.S.S.R. pat. 163,173 [*C.A.*, **61**, 13197 (1964)].

[2327] P. L. Southwick, L. A. Pursglove, and P. Numerof, *J. Am. Chem. Soc.*, **72**, 1604 (1950).

[2328] G. Sunagawa and T. Ichii, Japan. pat. 3847 (1964) [*C.A.*, **61**, 5663 (1964)].

[2329] M. De la Burde and R. De la Burde, *Chemotherapia*, **6**, 382 (1963) [*C.A.*, **64**, 9658 (1966)].

[2330] W. v. Miller and F. Kinkelin, *Ber.*, **19**, 525 (1893).

[2331] E. Erlenmeyer, Jr., *Ber.*, **23**, 74 (1890).

[2332] D. Vorländer, *Ber.*, **30**, 2261 (1897).

[2333] J. Sam, J. N. Plampin, and D. W. Alwani, *J. Org. Chem.*, **27**, 4543 (1962).

[2334] S. V. Tsukerman, V. M. Nikitchenko, and V. F. Lavrushin, *Zh. Org. Khim.*, **1**, 978 (1965) [*C.A.*, **63**, 6944 (1965).

[2335] V. F. Lavrushin, S. V. Tsukerman, and V. M. Nikitchenko, *Ukr. Khim. Zh.*, **27**, 379 (1961) [*C.A.*, **56**, 3437 (1962).

[2336] P. R. Hills and F. J. McQuillin, *J. Chem. Soc.*, 4060 (1953).

[2337] H. Wittmann and H. Uragg, *Monatsh. Chem.*, **97**, 891 (1966).

[2338] K. C. Joshi and A. K. Jauhar, *J. Indian Chem. Soc.*, **43**, 368 (1966).

[2339] D. R. Moore and A. Oroslan, *J. Org. Chem.*, **31**, 2620 (1966).

[2340] P. Leriverend and J.-M. Conia, *Bull. Soc. Chim. France*, 116, 121 (1966).

[2341] J. T. Braunholtz and F. G. Mann, *J. Chem. Soc.*, 398 (1955).

[2342] B. Puri and T. R. Seshadri, *J. Chem. Soc.*, 1589 (1955).

[2343] F. M. Dean and K. Manunapichu, *J. Chem. Soc.*, 3112 (1957).

[2344] Y. Poirier, L. Legrand, and N. Lozac'h, *Bull. Soc. Chim. France*, 1054 (1966).

[2345] Y. Poirier and N. Lozac'h, *Bull. Soc. Chim. France*, 1058, 1062 (1966).

[2346] T. Dumpis, I. Rodovia, and G. Vanags, *Latvijas PSR Zinatnu Akad. Vestis, Kim. Ser.*, (6) 733 (1965) [*C.A.*, **65**, 662 (1966)].

[2347] T. R. Govindachari, P. C. Parthasarathy, B. R. Pai, and P. S. Subramaniam, *Tetrahedron*, **21**, 2633 (1965).

[2348] S. Mitsui, Y. Senda, and H. Saito, *Bull. Chem. Soc. Japan*, **39**, 694 (1966).

[2349] W. J. Porter, Jr., J. A. Wingate, and J. A. Hanan, U.S. pat. 3,248,428 [*C.A.*, **65**, 2128 (1966)].

[2350] J. Majer and V. Rehak, *Sb. Ved. Praci, Vysoka Skola Chem. Technol., Pardubice*, (1), 123 (1965) [*C.A.*, **65**, 2254 (1966)].

[2351] S. V. Tsukerman, V. D. Orlov, V. P. Izvekov, V. F. Lavrushin, and Y. K. Yur'ev, *Khim. Geterotsikl. Soedin., Akad. Nauk Latv. SSR*, (1), 34 (1966) [*C.A.*, **65**, 674 (1966)].

[2352] G. P. Kugatova-Shemyakina and R. A. Poshkene, *Zh. Org. Khim.*, **2**, 447 (1965) [*C.A.* **65**, 7072 (1966)].

[2353] P. Rollin and R. Setton, *Compt. Rend.* (C), **263**, 1080 (1966).

[2354] K. C. Joshi and A. K. Jauhar, *Indian J. Chem.* **4**, 277 (1966).

[2355] S. V. Tsukerman, V. P. Izvekov, and V. F. Lavrushin, *Khim. Geterosikl. Soedin.*, **1**, 527 (1965) [*C.A.*, **64**, 676 (1966)].

[2356] S. V. Tsukerman, K. S. Chang, and V. F. Lavrushin, *Khim. Geterotsikl. Soedin.*, **1**, 537 (1965) [*C.A.*, **64**, 559 (1966)].

[2357] J. R. Merchant, J. B. Mehta, and V. B. Desai, *Indian J. Chem.*, **3**, 561 (1965).

[2358] G. L. Buchanan, A. McKillop, and R. A. Raphael, *J. Chem. Soc.*, 833 (1965).

[2359] V. S. Markevitch, S. M. Markevich, and M. P. Gerchuk, *Khim. Prom.*, **42**, 587 (1966) [*C.A.*, **65**, 16850 (1966)].

AUTHOR INDEX, VOLUMES 1–16

CHAPTER INDEX, VOLUMES 1–16

SUBJECT INDEX, VOLUME 16

Since the table of contents provides a quite complete index, only those items not readily found from the contents pages are listed here.

Numbers in **boldface** type refer to experimental procedures.

Middlesex Community College Library
Middletown, CT 06457
A2100 043421 5